U.S.NRC
United States Nuclear Regulatory Commission
Protecting People and the Environment

NUREG-1910
Supplement 5

Environmental Impact Statement for the Ross ISR Project in Crook County, Wyoming

Supplement to the Generic Environmental Impact Statement for *In-Situ* Leach Uranium Milling Facilities

Draft Report for Comment

Office of Federal and State Materials and
Environmental Management Programs

AVAILABILITY OF REFERENCE MATERIALS IN NRC PUBLICATIONS

NRC Reference Material

As of November 1999, you may electronically access NUREG-series publications and other NRC records at NRC's Public Electronic Reading Room at http://www.nrc.gov/reading-rm.html. Publicly released records include, to name a few, NUREG-series publications; *Federal Register* notices; applicant, licensee, and vendor documents and correspondence; NRC correspondence and internal memoranda; bulletins and information notices; inspection and investigative reports; licensee event reports; and Commission papers and their attachments.

NRC publications in the NUREG series, NRC regulations, and Title 10, "Energy," in the *Code of Federal Regulations* may also be purchased from one of these two sources.

1. The Superintendent of Documents
   U.S. Government Printing Office
   Mail Stop SSOP
   Washington, DC 20402–0001
   Internet: bookstore.gpo.gov
   Telephone: 202-512-1800
   Fax: 202-512-2250
2. The National Technical Information Service
   Springfield, VA 22161–0002
   www.ntis.gov
   1–800–553–6847 or, locally, 703–605–6000

A single copy of each NRC draft report for comment is available free, to the extent of supply, upon written request as follows:

Address: U.S. Nuclear Regulatory Commission
Office of Administration
Publications Branch
Washington, DC 20555-0001

E-mail: DISTRIBUTION.RESOURCE@NRC.GOV
Facsimile: 301–415–2289

Some publications in the NUREG series that are posted at NRC's Web site address http://www.nrc.gov/reading-rm/doc-collections/nuregs are updated periodically and may differ from the last printed version. Although references to material found on a Web site bear the date the material was accessed, the material available on the date cited may subsequently be removed from the site.

Non-NRC Reference Material

Documents available from public and special technical libraries include all open literature items, such as books, journal articles, transactions, *Federal Register* notices, Federal and State legislation, and congressional reports. Such documents as theses, dissertations, foreign reports and translations, and non-NRC conference proceedings may be purchased from their sponsoring organization.

Copies of industry codes and standards used in a substantive manner in the NRC regulatory process are maintained at—

The NRC Technical Library
Two White Flint North
11545 Rockville Pike
Rockville, MD 20852–2738

These standards are available in the library for reference use by the public. Codes and standards are usually copyrighted and may be purchased from the originating organization or, if they are American National Standards, from—

American National Standards Institute
11 West 42nd Street
New York, NY 10036–8002
www.ansi.org
212–642–4900

Legally binding regulatory requirements are stated only in laws; NRC regulations; licenses, including technical specifications; or orders, not in NUREG-series publications. The views expressed in contractor-prepared publications in this series are not necessarily those of the NRC.

The NUREG series comprises (1) technical and administrative reports and books prepared by the staff (NUREG–XXXX) or agency contractors (NUREG/CR–XXXX), (2) proceedings of conferences (NUREG/CP–XXXX), (3) reports resulting from international agreements (NUREG/IA–XXXX), (4) brochures (NUREG/BR–XXXX), and (5) compilations of legal decisions and orders of the Commission and Atomic and Safety Licensing Boards and of Directors' decisions under Section 2.206 of NRC's regulations (NUREG–0750).

U.S.NRC
United States Nuclear Regulatory Commission
Protecting People and the Environment

NUREG-1910
Supplement 5

Environmental Impact Statement for the Ross ISR Project in Crook County, Wyoming

Supplement to the Generic Environmental Impact Statement for *In-Situ* Leach Uranium Milling Facilities

Draft Report for Comment

Manuscript Completed: February 2013
Date Published: March 2013

Office of Federal and State Materials and
Environmental Management Programs

COMMENTS ON DRAFT REPORT

Any interested party may submit comments on this report for consideration by the NRC staff. Comments may be accompanied by additional relevant information or supporting data. Please specify the report number NUREG–1910, Supplement 5, in your comments, and send them by the end of the comment period specified in the Federal Register notice announcing the availability of this report to the following address:

Cindy Bladey, Chief
Rules, Announcements, and Directives Branch
Division of Administrative Services
Office of Administration
Mail Stop: TWB-05-B10M
U.S. Nuclear Regulatory Commission
Washington, DC 20555-0001

For any questions about the material in this report, please contact:

Johari Moore
Mail Stop T8F5
U.S. Nuclear Regulatory Commission
Washington, DC 20555-0001
Phone: 301-415-7694
E-mail: Johari.Moore@nrc.gov

Please be aware that any comments that you submit to the NRC will be considered a public record and entered into the Agencywide Documents Access and Management System (ADAMS). Do not provide information you would not want to be publicly available.

ABSTRACT

The U.S. Nuclear Regulatory Commission (NRC) issues licenses for the possession and use of source and byproduct materials provided that facilities meet NRC regulatory requirements and will be operated in a manner that is protective of public health and safety and the environment. Under the NRC environmental-protection regulations in the *Code of Federal Regulations* (CFR), Title 10, Part 51, which implement the *National Environmental Policy Act of 1969* (NEPA), issuance of a license to possess and use source and byproduct materials during uranium recovery and milling requires an environmental impact statement (EIS) or a supplement to an EIS (SEIS).

In May 2009, the NRC issued NUREG–1910, *Generic Environmental Impact Statement* (GEIS) *for In-Situ Leach Uranium Milling Facilities*. In the GEIS, the NRC assessed the potential environmental impacts from the construction, operation, aquifer restoration, and decommissioning of in situ recovery (ISR) facilities located in four specific geographic regions of the western U.S. As part of this assessment, the NRC determined which potential impacts would be essentially the same for all ISR facilities and which would result in varying levels of impacts for different facilities and would therefore require further site-specific information to determine potential impacts. The GEIS provides a starting point for the NRC's NEPA analyses for site-specific license applications for new ISR facilities as well as for applications to amend or to renew existing ISR licenses.

By a letter dated January 4, 2011, Strata Energy Inc. (referred to herein as Strata or the "Applicant") submitted a license application to the NRC for a new source and byproduct materials license for the proposed Ross Project. The Ross Project would be located in Crook County, Wyoming, which is in the Nebraska-South Dakota-Wyoming Uranium Milling Region identified in the GEIS. The NRC staff prepared this SEIS to evaluate the potential environmental impacts of the Applicant's proposal to construct, operate, conduct aquifer restoration, and decommission an ISR facility at the Ross Project. This SEIS describes the environment that could be affected by the proposed Ross Project activities, estimates the potential environmental impacts resulting from the Proposed Action and two Alternatives, discusses the corresponding proposed mitigation measures, and describes the Applicant's environmental-monitoring program. In conducting its analysis for this SEIS, the NRC staff evaluated site-specific data and information to determine whether the site characteristics and the Applicant's proposed activities were consistent with those evaluated in the GEIS. The NRC staff then determined relevant sections, findings, and conclusions in the GEIS that could be incorporated by reference, and identified the areas that needed additional analysis. Based on its environmental review, the preliminary NRC staff recommendation is that, unless safety issues mandate otherwise, the source and byproduct materials license be issued as requested.

Paperwork Reduction Act Statement

This NUREG contains and references information collection requirements that are subject to the Paperwork Reduction Act of 1995 (44 U.S.C. 3501 et seq.). These information collections were approved by the Office of Management and Budget (OMB), approval numbers 3150-0014, 3150-0020, 3150-0021, and 3150-0008.

Public Protection Notification

NRC may not conduct or sponsor, and a person is not required to respond to, a request for information or an information collection requirement unless the requesting document displays a current valid OMB control number.

References

10 CFR Part 51. Code of Federal Regulations, Title 10, *Energy*, Part 51. "*Environmental Protection Regulations for Domestic Licensing and Related Regulatory Functions.*" Washington, DC: U.S. Government Printing Office.

NRC. NUREG–1910, "Generic Environmental Impact Statement for *In-Situ* Leach Uranium Milling Facilities." Washington, DC: NRC. May 2009. Agencywide Documents Access and Management System (ADAMS) Accession Nos. ML091480244 and ML091480188.

TABLE OF CONTENTS

Section **Page**

TABLE OF CONTENTS
(Continued)

TABLE OF CONTENTS
(*Continued*)

Section **Page**

TABLE OF CONTENTS
(*Continued*)

Section **Page**

TABLE OF CONTENTS
(*Continued*)

Section **Page**

TABLE OF CONTENTS
(*Continued*)

Section **Page**

TABLE OF CONTENTS
(*Continued*)

Section **Page**

TABLE OF CONTENTS
(*Continued*)

Section **Page**

LIST OF FIGURES

LIST OF FIGURES
(*Continued*)

Figure **Page**

LIST OF TABLES

LIST OF TABLES
(*Continued*)

Table **Page**

EXECUTIVE SUMMARY

BACKGROUND

By a letter dated January 4, 2011, Strata Energy Inc. (Strata or the "Applicant") submitted an application to the U.S. Nuclear Regulatory Commission (NRC) for a new source and byproduct materials license for the proposed Ross Project, an in situ recovery (ISR) project to be located in Crook County, Wyoming. The proposed Ross Project includes a central processing plant (CPP) to produce yellowcake, corresponding injection and recovery wells, deep-disposal wells for liquid effluents, monitoring wells throughout the Ross Project area as well as other various infrastructure (e.g., pipelines, roads, and lighting).

The *Atomic Energy Act of 1954* (AEA), as amended by the Uranium Mill Tailings Radiation Control Act of 1978, authorizes the NRC to issue licenses for the possession and use of source material and byproduct material. The NRC must license facilities, including ISR operations, in accordance with NRC regulatory requirements. These requirements were developed to protect public health and safety from radiological hazards and to protect common defense and security. The NRC's environmental protection regulations are found at Title 10 of the *Code of Federal Regulations* (CFR), Part 51 (10 CFR Part 51); these regulations implement the National Environmental Policy Act of 1969 (NEPA). 10 CFR Part 51 requires that the NRC prepare an environmental impact statement (EIS) or supplement to another EIS (SEIS) or a generic EIS (GEIS) for its issuance of a license to possess and use source and/or byproduct materials for uranium milling (see 10 CFR Part 51.20[b][8]).

In May 2009, the NRC issued NUREG–1910, *Generic Environmental Impact Statement for In-Situ Leach Uranium Milling Facilities.* In this GEIS, the NRC assessed the potential environmental impacts of the construction, operation, aquifer restoration, and decommissioning of ISR facilities located in four specified geographic regions of the western U.S. The proposed Ross Project is located within the Nebraska-South Dakota-Wyoming Uranium Milling Region (NSDWUMR) identified in the GEIS. The GEIS provides a starting point for the NRC's NEPA analyses for site-specific license applications for new ISR facilities. This Draft SEIS incorporates by reference information from the GEIS. This document also uses information from the Applicant's license application and subsequent environmental report and its responses to the NRC's requests for additional information as well as other publicly available sources of information.

This Draft SEIS includes the NRC staff's analysis of the environmental impacts from the Proposed Action (i.e., for the NRC to license the Ross Project), the environmental impacts of two Alternatives to the Proposed Action (i.e., the "No-Action" Alternative and the "North Ross Project" Alternative), and the mitigation measures that are intended to either minimize or avoid adverse impacts. It also includes the NRC staff's preliminary recommendation regarding the Proposed Action.

PURPOSE AND NEED OF THE PROPOSED ACTION

The NRC regulates uranium milling, including the ISR process, under 10 CFR Part 40, Domestic Licensing of Source Material. The Applicant is seeking an NRC source and byproduct materials license to authorize commercial-scale in situ uranium recovery at the Ross Project area. The purpose and need for this Proposed Action is to provide an option that allows the Applicant to recover uranium and to produce yellowcake at the Ross Project area. Yellowcake is the uranium oxide product of the ISR uranium-milling process that is used to produce various products, including fuel for commercially operated nuclear power reactors.

This definition of purpose and need reflects the Commission's recognition that, unless there are findings in the safety review required by the AEA, as amended, or findings in the NEPA environmental analysis that would lead NRC to reject a license application, NRC has no role in a company's business decision to submit a license application to operate an ISR facility at a particular location.

THE PROJECT AREA AND FACILITY

Strata's Proposed Action, the Ross Project, would occupy 697 ha [1,721 ac] in the north half of the approximately 90-km^2 [56-mi^2] Lance District, where the Applicant is actively exploring for additional uranium reserves. Strata has also identified four other uranium-bearing areas that would extend the area of uranium recovery to the north with the Ross Amendment Area 1 and to the south of the Lance District with the Kendrick, Richards, and Barber satellite facilities. These areas are not a component of the Proposed Action in this SEIS.

The Lance District is located on the western edge in the northwest corner of the NSDWUMR. It is situated between the Black Hills uplift to the east and the Powder River Basin to the west. Both of these regional features are described in the GEIS. The environment of the Proposed Action is described in Section 3 of this SEIS.

The Proposed Action includes the ISR facility itself and its wellfields. The ISR facility consists of the following:

- A CPP that houses the uranium- and vanadium-processing equipment, drying and packaging equipment, and water-treatment equipment;
- A chemical storage area as well as other storage, warehouse, maintenance, and administration buildings; and
- Two double-lined surface impoundments, a sediment impoundment, and five Class I deep-injection wells.

The Proposed Action includes the option of the Applicant operating the Ross Project facility beyond the life of the Project's wellfields. The facility could be used to process uranium-loaded resins from satellite projects within the Lance District operated by the Applicant, or from other offsite uranium recovery projects not operated by the Applicant (i.e., "toll milling"), or from offsite water-treatment operations. With that option, the life of the facility would be extended to 14 years or more.

The Ross Project would also host 15 – 25 wellfield areas and would consist of a total of 1,400 – 2,000 recovery and injection wells. The wellfield areas would be surrounded by a perimeter ring of monitoring wells.

THE IN SITU URANIUM RECOVERY PROCESS

During the in situ uranium recovery process, an oxidant-charged solution, called a lixiviant, is injected into the ore-zone aquifer (or uranium "ore body") through injection wells. The ore zone is that portion of the aquifer that has been permanently exempted by the U.S. Environmental Protection Agency (EPA) from requirements as an underground source of drinking water under the *Safe Drinking Water Act*. Typically, a lixiviant uses native groundwater (from the ore-zone aquifer itself), carbon dioxide, and sodium carbonate/bicarbonate, with an oxygen or hydrogen peroxide oxidant. As it circulates though the ore zone, the lixiviant oxidizes and dissolves the mineralized uranium, which is present in a reduced chemical state. The resulting uranium-rich solution, the "pregnant" lixiviant, is drawn to recovery wells by pumping, and then transferred to the CPP via a network of pipes buried just below the ground surface. At the CPP, the uranium is extracted from the solution using an ion exchange process. The resulting "barren" (uranium-depleted) solution is then recharged with the oxidant and re-injected to recover more uranium from the wellfield.

During production, the uranium recovery solutions continually move through the aquifer from outlying injection wells to internal recovery wells. These wells can be arranged in a variety of geometric patterns depending on the ore-body's configuration, the aquifer's permeability, and the operator's selection based upon operational considerations. Wellfields are often designed in a five-spot or seven-spot pattern, with each recovery (i.e., production) well being located inside a ring of injection wells. Monitoring wells surround the wellfield pattern area, terminating in the ore-zone aquifer as well as in both the overlying and underlying aquifers. These monitoring wells are screened in appropriate stratigraphic horizons to detect lixiviant should it migrate out of the production, or ore, zone. The uranium that is recovered from the solution would be processed in the CPP to yellowcake. The yellowcake would be packaged into NRC-and U.S. Department of Transportation (USDOT)-approved 208-L [55-gal] steel drums, and trucked offsite to a licensed uranium-conversion facility.

Once uranium recovery is complete, the ore-zone's ground water is restored to NRC-approved ground-water protection standards, which are protective of the surrounding ground waters. The facility is decommissioned according to an NRC-approved decommissioning plan and in accordance with NRC-approved standards. Once decommissioning is approved by the NRC, the site may be released for public use.

THE ALTERNATIVES

The NRC environmental review regulations in 10 CFR Part 51, which implement NEPA, require the NRC to consider reasonable alternatives, including the No-Action alternative, to a proposed action. The NRC staff considered a range of alternatives that would fulfill the underlying purpose and need for the Proposed Action. From this analysis, a set of reasonable alternatives was developed, and the impacts of the Proposed Action were compared to the impacts that would result if a given alternative were implemented. This SEIS evaluates the potential environmental impacts of the Proposed Action and two

Alternatives, including the No-Action Alternative and the North Ross Project. Under the No-Action Alternative, the Applicant would neither construct nor operate a uranium recovery facility or wellfields at the proposed Ross Project. In Alternative 3, the proposed Ross Project's facility (i.e., the CPP, surface impoundments, and auxiliary structures) would be constructed at a site north of where it is proposed to be located in the Proposed Action, but the wellfields would remain in the same locations as in the Proposed Action. This alternative facility location would require additional, substantial earth-moving to construct the surface impoundments, but a containment barrier wall (CBW) (described later in this SEIS) would not be required. Alternatives considered and eliminated from detailed analysis include conventional mining and milling, conventional mining and heap leach processing, and alternate lixiviants. These alternatives were eliminated from detailed study because they either do not meet the purpose and need of the proposed Ross Project or would cause greater environmental impacts than the Proposed Action.

SUMMARY OF THE ENVIRONMENTAL IMPACTS

This Draft SEIS includes the NRC staff's analysis, which considers and weighs the environmental impacts resulting from the construction, operation, aquifer restoration, and decommissioning of an in situ uranium recovery facility at the proposed Ross Project area and the two Alternatives. This SEIS also describes mitigation measures for the reduction or avoidance of potential adverse impacts that either: 1) the Applicant has committed to in its NRC license application, 2) would be required under other State or Federal permits or processes, or 3) are additional measures that the NRC staff identified as having the potential to reduce environmental impacts, but the Applicant did not commit to in its license application. The SEIS uses the assessments and conclusions reached in the GEIS in combination with site-specific information to assess and categorize impacts.

As discussed in the GEIS and consistent with NUREG–1748 (NRC, 2003), the significance of potential environmental impacts is categorized as follows:

SMALL: The environmental effects are not detectable or are so minor that they will neither destabilize nor noticeably alter any important attribute of the resource considered.

MODERATE: The environmental effects are sufficient to alter noticeably, but not destabilize, important attributes of the resource considered.

LARGE: The environmental effects are clearly noticeable and are sufficient to destabilize important attributes of the resource considered.

Table ExS.1 provides a summary of the NRC's evaluation of the potential environmental impacts of the construction, operation, aquifer restoration, and decommissioning of the Ross Project, followed by a brief summary of impacts by environmental resource area and lifecycle phase. These potential impacts are more fully described in Section 4 of this SEIS, where the magnitude of impacts by phase of the Ross Project is provided for each resource area.

Table ExS.1
Summary of Environmental Impacts

Resource Area	Alternative 1: Proposed Action – Construction	Alternative 1: Proposed Action – Operation	Alternative 1: Proposed Action – Aquifer Restoration	Alternative 1: Proposed Action – Decommissioning	Alternative 2: No-Action	Alternative 3: North Ross Project
Land Use	SMALL	SMALL	SMALL	SMALL	SMALL	SMALL
Transportation	SMALL to MODERATE to LARGE With Mitigation: SMALL to MODERATE	SMALL to MODERATE to LARGE With Mitigation: SMALL to MODERATE	SMALL to MODERATE	SMALL to MODERATE With Mitigation: SMALL to MODERATE	SMALL	SMALL to MODERATE to LARGE with Mitigation: SMALL to MODERATE
Geology and Soils						
▪ Geology	SMALL	SMALL	SMALL	SMALL	SMALL	SMALL
▪ Soils	SMALL	SMALL	SMALL	SMALL	SMALL	SMALL
Water Resources						
▪ Surface Water						
Water Quantity	SMALL	SMALL	SMALL	SMALL	SMALL	SMALL
Water Quality	SMALL	SMALL	SMALL	SMALL	SMALL	SMALL
▪ Wetlands	SMALL	SMALL	SMALL	SMALL	SMALL	SMALL
▪ Ground Water	SMALL	SMALL to MODERATE (See OZ Aquifer Below)	SMALL to MODERATE (See OZ Aquifer Below)	SMALL	SMALL	SMALL to MODERATE (Excursions) (Short-Term Drawdowns)
Shallow Aquifers						
Water Quantity	SMALL	SMALL	SMALL	SMALL	SMALL	SMALL
Water Quality	SMALL	SMALL	SMALL	SMALL	SMALL	SMALL

Table ExS.1
Summary of Environmental Impacts

Resource Area	Alternative 1: Proposed Action				Alternative 2: No-Action	Alternative 3: North Ross Project
	Construction	Operation	Aquifer Restoration	Decommissioning		
▪ Ground Water (Continued)						
Ore-Zone Aquifers						
Water Quantity	SMALL	SMALL	SMALL to MODERATE (Short-Term Drawdown)	SMALL	SMALL	SMALL to MODERATE (Short-Term Drawdown)
Water Quality	SMALL	SMALL (Long-Term) SMALL to MODERATE (Excursion)	SMALL	SMALL	SMALL	SMALL (Long-Term) SMALL to MODERATE (Excursion)
Deep Aquifers						
Water Quantity	SMALL	SMALL	SMALL	SMALL	SMALL	SMALL
Water Quality	SMALL	SMALL	SMALL	SMALL		
Ecology					SMALL	SMALL
▪ Vegetation	SMALL	SMALL	SMALL	SMALL		
▪ Wildlife	SMALL	SMALL	SMALL	SMALL		
▪ Aquatic	SMALL	SMALL	SMALL	SMALL		
▪ Protected Species	SMALL	SMALL	SMALL	SMALL		
Air Quality	SMALL	SMALL	SMALL	SMALL	SMALL	SMALL
Noise	SMALL TO MODERATE (Short Term)	SMALL TO MODERATE (Short Term)	SMALL	SMALL TO MODERATE (Short Term)	SMALL	SMALL TO MODERATE
Historical, Cultural, and Paleontological Resources	SMALL to LARGE	SMALL	SMALL	SMALL	SMALL	SMALL to LARGE

Table ExS.1
Summary of Environmental Impacts

Resource Area	Alternative 1: Proposed Action				Alternative 2: No-Action	Alternative 3: North Ross Project
	Construction	Operation	Aquifer Restoration	Decommissioning		
Visual and Scenic Resources	SMALL (Long-Term) MODERATE (Short-Term) (First Year) (Nearest Residents)	SMALL	SMALL	SMALL	SMALL	SMALL (Long-Term) MODERATE (Short-Term) (First Year)
Socioeconomics	SMALL to MODERATE (Taxes Paid to Crook County)	SMALL to MODERATE (Taxes Paid to Crook County)	SMALL	SMALL	SMALL	SMALL to MODERATE (Taxes Paid to Crook County)
Environmental Justice	N/A No Minority or Low-Income Groups	N/A No Minority or Low-Income Groups	N/A No Minority or Low-Income Groups	N/A No Minority or Low-Income Groups	N/A No Minority or Low-Income Groups	N/A No Minority or Low-Income Groups
Public and Occupational Health and Safety	SMALL	SMALL	SMALL	SMALL	SMALL	SMALL
Waste Management						
▪ Liquid	SMALL	SMALL	SMALL	SMALL	SMALL	SMALL
▪ Solid	SMALL	SMALL	SMALL	SMALL	SMALL	SMALL

THE IMPACTS BY RESOURCE AREA AND PROJECT PHASE

Land Use

Construction: Impacts would be SMALL. The Ross Project area comprises a total of 697 ha [1,721 ac] in the north half of the approximately 90-km^2 [56-mi^2] Lance District. This area is currently used for livestock grazing, wildlife habitat, some agriculture, and some oil production. A total of 113 ha [280 ac] of land, which represents 16 percent of the Ross Project area, would be disturbed during the construction of a CPP, surface impoundments, and other auxiliary structures such as storage areas and parking lots. The wellfields would be sequentially developed over the Ross Project lifecycle. All disturbed areas would be fenced and, thus, somewhat limit grazing by livestock, access by wildlife, and recreational opportunities.

Operation: Impacts would be SMALL. Land-use impacts during the operations phase would be similar to, or less than, those during the construction phase because the buildings, surface impoundments, and infrastructure would be in place. Areas where Ross Project uranium-production activities would take place would remain fenced, somewhat limiting grazing and some crop production. No new facilities would be constructed that would result in additional land disturbance during operation, although well drilling would continue as the wellfields would be sequentially developed.

Aquifer Restoration: Impacts would be SMALL. Land-use impacts would be similar to, or less than, those during the construction and operation phases. Wellfield access would continue to be restricted from other uses such as livestock grazing and crop production, as described for the Ross Project's operation phase. No new facilities would be constructed that would result in additional land disturbance.

Decommissioning: Impacts would be SMALL. Land-use impacts during the Proposed Action's decommissioning as well as the site's reclamation would temporarily increase due to the additional equipment that would be used for dismantling and removing Ross Project components such as the CPP, surface impoundments, and wellfields. In addition, the reclamation of the site would involve significant earth moving, land disturbance, and access restrictions. However, these short-term impacts would not be greater than those experienced during the Ross Project's construction phase. At the end of the Ross Project's decommissioning and site reclamation, the preconstruction land uses would be restored.

Transportation

Construction: Impacts would be MODERATE TO LARGE on local and county roads, but would be SMALL on the Interstate-highway system of the U.S. With the identified mitigation measures, the transportation impacts on local and county roads would lessen and they would be MODERATE. The highest traffic volume resulting from the Ross Project would occur during its construction phase, because of the large workforce (200 workers) and frequent supply, building material, and equipment shipments. The increased traffic is expected to be 400 passenger cars and 24 trucks per day, which, when compared to 2010 volumes, represents a traffic increase of approximately 400 percent on the New Haven Road south of the Ross Project area. This significant increase in traffic could result in more traffic accidents as well as potentially significant wear and tear on the road surfaces.

Operation: Impacts would be SMALL to LARGE; however, with mitigation, the transportation impacts during the Ross Project's operation would be SMALL to MODERATE. Impacts such as the local road's deterioration would be less than during construction, because of a smaller workforce (i.e., approximately 60 workers); however, the traffic volume associated with facility and wellfield operation would still be double that of 2010. The effective mitigation measures taken during the construction phase would continue through the operation phase.

***Aquifer Restoration*:** Impacts would be MODERATE, and with the mitigation measures that would be implemented throughout the Ross Project's lifecycle, the transportation impacts of aquifer restoration would also be MODERATE. Transportation impacts during this phase would be similar to those during the operation phase, although the workforce would be smaller (40 workers), but similar volumes of truck traffic would occur as during operation, especially if the CPP is used for recovery of uranium-loaded ion-exchange (IX) resins from four potential satellite areas as well as for toll milling.

***Decommissioning*:** Impacts would be MODERATE, and with the continuing mitigation measures of the other lifecycle phases as well as the declining workforce, the impacts would be SMALL to MODERATE. The traffic volume during the decommissioning phase would be dominated by waste shipments for offsite disposal. Because of the reduced traffic volumes associated with this phase compared to the operations phase, there would be a reduced risk of transportation accidents. However, once the Ross Project has been fully decommissioned, all transportation impacts would be eliminated.

Geology and Soils

***Construction*:** Impacts to both geology and soils would be SMALL. Although the Ross Project's design for its CPP would include a CBW, the impacts of the wall's construction would be SMALL due to the relatively small and localized effects on the bedrock below it. The impacts on soils would occur largely during this phase of the proposed Ross Project, when most of the ground disturbance takes place. Potential soils impacts include soils loss (by wind and water erosion), soils compaction, increased salinity, soils-productivity loss, and soils contamination. Surface-disturbing activities would expose the soils and subsoils at the Ross Project area and would temporarily increase the potential for soil loss because of wind and water erosion. The Applicant, however, has proposed to remove vegetation only where necessary and would stockpile soils for reclamation during decommissioning. The Applicant has proposed to mitigate erosion by minimizing the required land disturbances, ensuring timely re-vegetation and reclamation of affected soils, and installing drainage controls. Finally, the Applicant has proposed to mitigate wind erosion by limiting traffic speeds, spraying unpaved roads, and implementing timely disturbed-area reclamation.

***Operation*:** Impacts to local geology and soils would be SMALL. The removal of uranium from the target sandstone (aquifer) during ISR operation would change the mineralogical composition of uranium-bearing rock formations. However, no significant matrix compression or ground subsidence would be expected during in situ uranium recovery. Because the proposed operation would result in small changes in the reservoir pressure, the operation would be unlikely to activate any geologic faults. The potential for spills during transfer of uranium-bearing lixiviant to and from the CPP would be mitigated by implementing onsite best management practices (BMPs), standard operating procedures, and compliance with NRC license and WDEQ permit requirements. The potential impacts from soil loss would be minimized by proper design and operation of surface-runoff features and implementation of BMPs.

Aquifer Restoration: Impacts would be SMALL. During aquifer restoration, the process of ground-water sweep, ground-water transfer, ground-water treatment, and recirculation would not remove rock matrix or structure. The formation pressure would be managed during restoration to ensure that the direction of ground-water flow is into the wellfields to reduce the potential for lateral migration of constituents. The change in pressure would not be significant enough to result in matrix compression, ground subsidence, or to reactivate the fault. The spill response and leak detection activities would be the same as described during the operation phase.

Decommissioning: Impacts would be SMALL. The potential impacts to the geology depend upon the density of plugged and abandoned drillholes and wells. At the end of decommissioning, the wellfields (whether recently operated or decommissioned some time ago) would contain approximately 3,000 drillholes and wells; these would include those drillholes from Strata's ore-zone delineation efforts and geotechnical investigations, ground-water monitoring wells used for site characterization, the injection and recovery wells from uranium-recovery activities, and Nubeth Joint Venture (Nubeth) drillholes and wells. This would represent an average density of approximately 4.3 wells/ha [1.7 wells/ac], which would be a low density with little geological impact. All areas of the Ross Project would be reclaimed and restored, so that the Project's impacts on the soils would be small as well.

Water Resources (Surface Water and Wetlands)

Construction: Impacts would be SMALL to both surface water quantity and quality as well as to wetlands. The Applicant intends to use surface water from either the Oshoto Reservoir or the Little Missouri River for dust control and construction. This equates to an annual use that is significantly less than the currently permitted annual appropriation for Oshoto Reservoir. Thus, the potential impacts of the Proposed Action's construction to surface-water quantity would be SMALL. Suspended-sediment concentrations in storm water at the Ross Project area could be increased due to vegetation removal and soil disturbance during construction of the Proposed Action. However, given the site-specific mitigation measures to be implemented by the Applicant, the potential impacts of the Ross Project's construction to surface-water quality would be SMALL. The potential impacts of the proposed Ross Project's construction to wetlands would also be SMALL.

Operation: Impacts would be SMALL. Release of process solutions from uranium-recovery wellheads, pipelines, module buildings, or process vessels; accidental discharge from surface impoundments; or release of yellowcake or IX resin during a transportation accident could result in surface-water contamination if the release(s) reached a surface-water body. Given mitigation measures that the Applicant would employ, however, the potential impacts to surface-water quality during the operation of the Ross Project would be SMALL. Surface-water monitoring and spill response would limit the impacts of potential surface spills to SMALL; however, impacts of spills to surface waters that are connected to shallow aquifers would be SMALL to MODERATE, depending upon the specifics of an incident. The Applicant's compliance with its permit conditions, use of BMPs, and implementation of other required mitigation measures, however, would reduce the impacts of the Ross Project's operation from MODERATE to SMALL, depending upon local conditions.

***Aquifer Restoration*:** Impacts would be SMALL. Potential risk of surface-water contamination associated with releases of process solutions and/or waste liquids as well as spills of other materials during aquifer restoration would be comparable to the operation phase of the Ross Project, but the uranium concentrations in such solutions would decline. Thus, the potential impacts of aquifer restoration to surface-water quantity and quality would be SMALL. The potential impacts during aquifer restoration to the wetlands on the Ross Project area would be the same as discussed under the Ross Project's construction and they would be SMALL.

***Decommissioning*:** Impacts would be SMALL. For the decommissioning of the Ross Project, the Applicant would use surface water from either the Oshoto Reservoir or the Little Missouri River for dust control during demolition activities. Potential surface-water contamination could occur from spilled or leaked fuel or lubricants from construction equipment and passenger vehicles that would be operated during decommissioning activities, although the equipment would generally be located away from surface-water bodies. The potential impacts from the Ross Project's decommissioning to surface-water quantity and quality would be SMALL. As during all of the earlier phases, the potential impacts to wetlands from the Ross Project's decommissioning would be SMALL.

Water Resources (Ground Water)

***Construction*:** Impacts would be SMALL. Potential impacts to the quantity of water in the shallow aquifers during construction of the Ross Project would be related to the quantity taken from the Oshoto Reservoir and the quantity involved in the installation of the CBW surrounding the facility. Any changes in ground-water levels due to water usage from Oshoto Reservoir would be small and restricted to the area around the Reservoir. Thus, the potential impacts during construction of the Ross Project to ground-water quantity in the shallow aquifers would be SMALL. Also, the potential impacts of the Proposed Action's construction to ground-water quality in the shallow aquifers would be SMALL. Based upon yields from regional baseline wells and other wells, ground-water modeling indicates that the ore-zone aquifer could support this level of withdrawal with little drawdown. Thus, the potential construction impacts on the ground-water quantity available from the confined aquifers (ore-zone, overlaying, and underlying aquifers) would be SMALL. Wells installed for further hydrologic studies, pre-licensing baseline site characterization (see SEIS Section 2.1.1.1), and production infrastructure would pass mechanical integrity testing (MIT) prior to use. Consequently, the potential impacts during construction on the ground-water quality in the confined aquifers would be SMALL. The potential impacts of construction on both the quantity and quality of ground water available from the deep aquifers would be SMALL.

***Operation*:** The impact would range from SMALL to MODERATE (depending upon whether excursions occur). Potential impacts from operation to ground-water quantity in the shallow aquifers would be similar to those as during the construction phase and would be SMALL. The Applicant would implement spill control, containment, and cleanup measures in the CPP and surface-impoundment areas (i.e., the facility). These measures would include secondary containment for process-solution vessels and chemical storage tanks, a geosynthetic liner beneath the CPP's foundation, dual liners with a leak-detection system for the surface impoundments, and a sediment impoundment to capture storm-water runoff. To reduce the risk of pipeline failure, the Applicant would hydrostatically test all pipelines prior to use and install leak-detection devices in manholes along the pipelines. The Applicant's implementation of BMPs during Ross Project operation would reduce the likelihood and magnitude of spills or leaks and facilitate expeditious cleanup. The potential impacts from the Ross Project's operation to ground-water quantity in the confined aquifers would be SMALL.

The potential impacts of ISR operation to ground-water quality in the confined aquifers above and below the ore zone would be SMALL. However, the short-term potential impacts of lixiviant excursions from uranium-recovery operation to the ore-zone aquifer outside the active ISR area would be SMALL to MODERATE. With respect to the deep aquifers where injection of liquid byproduct wastes would occur, regular monitoring of the water quality of the injected brine is required by the permit; thus, the potential impacts of the Ross Project's operation to ground-water quantity and quality in the deep aquifers would be SMALL.

***Aquifer Restoration*:** Impacts would be SMALL to MODERATE (due to potential significant drawdown in the ore-zone and confined aquifers, reducing ground-water quantity). The potential impacts to water quality would be reduced when compared to the Ross Project's operation because no lixiviant would be used in the injection stream and the concentration of chemicals in the recovered ground water would be significantly less than during ISR operations. The Applicant's implementation of BMPs during aquifer restoration would continue, and the other ground-water mitigation measures would be the same as those described for the operation of the Ross Project. Thus, the potential impacts of aquifer restoration to ground-water quantity and quality of the shallow aquifers would be SMALL. A conservative regional ground-water modeling analysis predicts a reduction in the available head in wells used for stock, domestic, and industrial use. These effects would be localized and short-lived. Consequently, the potential impacts of the Proposed Action's aquifer-restoration phase to ground-water quantity of the confined aquifers would be SMALL to MODERATE. In the deep aquifers, the volume of waste injected would be greater during the aquifer-restoration phase than during the Ross Project's operation phase, but the potential impacts would be similar. The impacts from aquifer restoration to ground-water quantity and quality of the deep aquifers would, therefore, be SMALL.

***Decommissioning*:** Impacts would be SMALL. After uranium-recovery operation is complete, unidentified, improperly abandoned wells (i.e., from previous subsurface explorations not associated with the Applicant or its activities could continue to impact aquifers above the ore-zone and adjacent aquifers by providing hydrologic connections between aquifers. Thus, the impacts to shallow aquifers during the Proposed Action's decommissioning would be SMALL. As decommissioning proceeds at the Ross Project area, and the concomitant land reclamation and restoration activities proceed, all monitoring, injection, and production wells would be plugged and abandoned as noted above. The wells would be filled with cement and/or bentonite and then cut off below plow depth to ensure ground water does not flow through the abandoned wells. Proper implementation of these procedures would isolate the wells from ground-water flow. Thus, the impacts to the ore-zone and adjacent confined aquifers would be SMALL. The Applicant estimates that very little brine and other liquid byproduct wastes would be disposed in the injection wells during the decommissioning (i.e., most wastes that would be generated during this phase would be solid). This small quantity would minimize potential impacts to ground-water quantity and quality during Ross Project's decommissioning and they would be SMALL to the deep aquifers.

Ecology

***Construction*:** Impacts would be SMALL. Potential environmental impacts to ecology of the Ross Project area, including both flora and fauna, could include removal of vegetation from the Ross Project area; reduction in wildlife habitat and forage productivity, and an increased risk of soil erosion and weed invasion; the modification of existing vegetative communities as a result of uranium-recovery activities; the loss of sensitive plants and habitats; and the potential spread of invasive species and noxious weed populations. Impacts to wildlife could include loss, alteration, or incremental fragmentation of habitat; displacement of and stresses on wildlife; and direct and/or indirect mortalities. Aquatic species could be affected by disturbance of stream channels, increases in suspended sediments, pollution from fuel spills, and habitat reduction. However, construction of the Ross Project would be phased over time, reducing the amount of surface area disturbed at any one time. Thus, the impacts to terrestrial vegetation and terrestrial wildlife would be SMALL. Because aquatic habitats would be avoided if at all possible during construction, impacts to reptiles, amphibians, and fish during the Ross Project's construction would also be SMALL.

***Operation*:** Impacts would be SMALL. Impacts would be similar to but less than those experienced during the construction phase because fewer earth-moving activities would occur and traffic would be less. Due to the Applicant's implementation of mitigation measures, such as wellfield perimeter and surface-impoundment fencing, leak-detection protocols, and wildlife protection and monitoring plans, the operation of the Ross Project would cause SMALL impacts to terrestrial vegetation and wildlife, including protected species, and to aquatic wildlife.

***Aquifer Restoration*:** Impacts would be SMALL. The potential impacts to ecological resources from aquifer-restoration activities would be similar to those experienced during the Ross Project's operation phase; therefore, the potential impact to vegetation and wildlife would be SMALL.

***Decommissioning*:** No loss of vegetative communities beyond that disturbed during the construction phase would occur. Pipeline removal would impact vegetation that could have re-established itself, although this, too, would be temporary as the disturbed areas are reseeded. Thus, the impacts of the Ross Project's decommissioning would not be expected to be greater than those experienced during its construction and would consequently be SMALL.

Air Quality

***Construction*:** Impacts would be SMALL. Combustion-engine emissions from diesel- and gas-powered equipment operation would occur during all phases of the Ross Project. The heaviest use of such equipment, however, would be the construction and decommissioning phases of the Ross Project. Fugitive dusts would also be generated by both construction, land-clearing activities as well as by commuters and delivery trucks. The largest workforce of the Ross Project's lifecycle would be employed on the Project's construction, and their respective commutes increase local traffic quite significantly. Combustion-engine emissions and fugitive dust would be generated by all of this traffic. However, the predominant winds (in terms of both speed and direction) in the region, the remote location of the Ross Project area, and the air-quality control systems and the BMPs that would be implemented by the Applicant would all minimize the air-quality impacts of the Ross Project's construction. In addition, the requirements of the Applicant's Air Quality Permit would require the Applicant to implement other specified mitigation measures as well, moderating the air emissions of the Ross Project. All anticipated gaseous-emission and fugitive-dust impacts would be limited in duration during the construction phase. Thus, the impacts of the Ross Project on air quality during construction would be SMALL and short-term.

***Operation*:** Impacts would be SMALL. Air-quality impacts during the Ross Project's operation phase would potentially include the same as those identified earlier for its construction phase (i.e., combustion-engine and fugitive-dust emissions). However, the quantity of the released air emissions would be reduced due to the reduced number of workers during ISR operation. Also, construction-equipment operation would decrease because most of the Ross Project area would have been cleared and graded during construction, so little earth movement would occur during operation; only the installation of wellfields would continue to generate fugitive dust. During uranium-recovery operation, several point sources of non-radioactive gaseous emissions would be located at the CPP. These would include process-pipelines, process-vessel, and storage-tank vents; emergency generators and space heaters; and other sources such as storage vessels and tanks containing acids and bases. However, these would all be very small point sources.

***Aquifer Restoration*:** Impacts would be SMALL. The emissions associated with the use of combustion-engine equipment would be limited in duration and result in small, short-term effects during the aquifer-restoration phase of the Ross Project. Vehicular traffic would be limited to delivery of supplies and commuting personnel; however, the workforce at the Ross Project would decrease to only 20 workers during aquifer restoration and, thus, the vehicular emissions of commuting traffic would substantially decrease. A significant decrease in the frequency of offsite yellowcake shipments would also occur as aquifer restoration proceeds.

***Decommissioning*:** Impacts would be SMALL. In the short term, emissions could increase somewhat, especially particulates because of decommissioning activities would generate particulate emissions such as fugitive dust. For example, the Applicant's dismantling and demolition of buildings, structures, surface impoundments, and process equipment; removing contaminated soils; moving construction equipment to the different areas where decommissioning activities would take place; and the grading and re-contouring during site reclamation and restoration could all generate air emissions, particularly fugitive dust. Combustion-engine emissions would also be produced by heavy equipment as well as vehicles transporting workers to and from the Ross Project, where the workforce would increase at the initiation of the decommissioning phase.

Noise

***Construction*:** Impacts would be SMALL to MODERATE. The nearest residents to the Ross Project area are substantially closer than those anticipated in the GEIS. Noise would be generated during construction activities as well as by vehicle traffic. Approximately 85 percent of the overall construction workforce (i.e., 200 workers) would commute to the Ross Project area. Heavy-equipment operation within the Ross Project area would peak during the Applicant's construction of the CPP, surface impoundments, wellfields, and associated infrastructure. In addition, the relocation of construction equipment to and from the Ross Project area and to and from different locations at the Ross Project area would generate noise. Impulse or impact noises from certain equipment, such as impact wrenches and pneumatic attachments on rock breakers, could be particularly loud as well. All of this noise could occasionally be annoying to the closest nearby residents. The overall noise impacts during the Proposed Action's construction would be SMALL to the general population, but the four closest residences to the Ross Project would experience MODERATE, but short-term, exposures to noise.

Operation: Impacts would be SMALL to MODERATE, with noise generated by construction activities greatly diminishing. The truck traffic associated with yellowcake, vanadium, and waste shipments would begin during the operation phase of the Ross Project; however, commuter-traffic noise would decrease due to the smaller workforce required during ISR operations (200 vs. 60 workers). However, because the county roads to and from the Ross Project area currently have very low average daily and annual traffic counts, there would be a continuing high relative increase in vehicular traffic and, thus, noise impacts to nearby residents would be MODERATE; the more distant local communities would experience only small, temporary impacts. The Applicant's compliance with the Occupational Safety and Health Administration's (OSHA's) noise regulations would minimize impacts to workers.

Aquifer Restoration: Impacts would be SMALL. During the Ross Project's aquifer-restoration phase, potential noise impacts would diminish to SMALL and would be only temporary for nearby residences. The workforce employed during aquifer restoration would be smaller (i.e., 20 worker) than during construction and operation phases of the Ross Project and, thus, there would be fewer workers, less traffic, and fewer noise-producing activities. The Applicant's continued compliance with OSHA's noise regulations would minimize impacts to workers.

Decommissioning: Impacts would be SMALL to MODERATE. Noise levels during the decommissioning phase of the Ross Project would be similar to those identified for the construction phase, for both onsite and offsite receptors. Most potential impacts to nearby residential receptors would occur as a result of the anticipated significantly increased commuter and truck traffic to and from the Ross Project area during decommissioning (i.e., 90 workers and additional waste shipments). At the Ross Project, despite the temporary nature of the decommissioning activities onsite, the short distance to the closest residences would make the noise impacts MODERATE.

Historical and Cultural Resources

Construction: Impacts would be SMALL to LARGE. Archaeological and historical sites may potentially be disturbed by construction. Within the area of potential effect at the proposed Ross Project, 25 sites are being treated as eligible for listing on the National Register of Historic Places (NRHP) for the purposes of this NEPA analysis. Avoidance of sites that are not within the proposed disturbance areas is recommended. For sites within the proposed disturbance areas, avoidance and mitigation, such as fencing and data recovery excavations are recommended.

Prior to an NRC license being granted, an agreement between the NRC, the Wyoming State Historic preservation Office (WY SHPO), BLM, interested Native American Tribes, the Applicant, and other interested parties will be established outlining the mitigation process for each affected resource. Additionally, prior to construction, the Applicant will develop an Unexpected Discovery Plan that will outline the steps required if unexpected historical and cultural resources are encountered.

Consultation efforts to identify properties of religious and cultural significance to Tribes have not been completed. Thus, the NRC cannot determine effects to these properties at this time. Section 106 consultation between NRC, WY SHPO, BLM, Tribal representatives, and the Applicant regarding potential impacts to these sites is ongoing.

Operation: Impacts would be SMALL. Minimal impacts will result during the operation phase because impacts to cultural resources will be mitigated before facility construction. If historical or cultural resources are encountered during operations, the Unexpected Discovery Plan will be implemented. Work would stop in the immediate area, and appropriate agencies would be notified.

Aquifer Restoration: Impacts would be SMALL. Impacts to historical and cultural resources during the aquifer restoration phase will be similar to operations. Minimal impacts will result because impacts to cultural resources will be mitigated before facility construction, and identified resources will be avoided. If historical or cultural resources are encountered during aquifer restoration, the Unexpected Discovery Plan will be implemented. Work would stop in the immediate area, and appropriate agencies would be notified.

Decommissioning: Impacts would be SMALL. Minimal impacts will result during the decommissioning phase because impacts to cultural resources will be mitigated prior to facility construction. If historical or cultural resources are encountered during decommissioning, the Unexpected Discovery Plan will be implemented. Work would stop in the immediate area, and appropriate agencies would be notified.

Visual and Scenic Resources

Construction: Impacts would be SMALL to MODERATE. The largest visible surface features of the Ross Project that would emerge during the construction phase would include wellhead covers and header houses; electrical and other utility distribution lines, which are mounted on 6-m [20-ft] wooden poles; more roads; the CPP; and the surface impoundments. There are protected visual resources near the Ross Project; the nearest such area is the Devils Tower National Monument, which is approximately 16 km [10 mi] east of the Ross Project. Although the Project itself would not be visible at the lower park portion of the Tower, climbers ascending to the top of the Tower may be able to see some of the Project's largest attributes as well as, in the night sky, the lights of the Project. These lights would also be visible at residences near the Ross Project. The short-term visual contrasts with the characteristic landscape of the Ross Project area would result from construction activities. However, the construction activities proposed for the Ross Project would be consistent with the U.S. Bureau of Land Management (BLM) visual classification of this area. The management objective of Visual Resource Management (VRM) Class III is to partially retain the existing character of the landscape so that the level of change to the characteristic landscape can be moderate. Also, prior to construction of the Ross Project, the Applicant would conduct baseline monitoring for potential light pollution and develop a light-pollution monitoring plan that would finalize the locations for both continuous and intermittent light sources. The short-term construction activities at the proposed Ross Project would result in SMALL to MODERATE visual impacts to the nearest four residences, each of which has a view of the Ross Project area. For the remaining 7 of the 11 nearby residences, the visual impacts would be SMALL.

Operation: Impacts would be SMALL. The overall visual impacts of an operating wellfield and the ISR facility itself would be small. In addition, the Ross Project would be located in gently rolling topography, where the visibility of aboveground infrastructure would vary and would be relative, depending upon the location and elevation of an observer as well as on nearby topography, total distance, and lighting characteristics. Lighting from the Ross Project would be visible from five of the residences to the east and from various locations directly to the west, north, and southeast. Mitigation measures for local light-pollution impacts would be the same as those described above for the construction phase of the Ross Project.

Aquifer Restoration: Impacts would be SMALL. Aquifer restoration activities would take place sequentially in the wellfields and last approximately two years per wellfield. There would be no modifications to either scenery or topography during aquifer restoration. Much of the same equipment and infrastructure used during operation would be employed during aquifer restoration, so that impacts to the visual landscape would be expected to be similar to or less than the impacts during the Proposed Action's operation phase. The mitigation measures presented above for both the Proposed Action's construction and operation phases would continue to be implemented during the aquifer-restoration phase, and these would continue to limit potential visual impacts.

Decommissioning: Impacts would be SMALL. The Ross Project would not result in significant impacts to the landscape that would persist after facility decommissioning and site restoration are completed. Most visual impacts during decommissioning would be temporary and diminish as structures, equipment, and other facility components are removed; the disturbed land surface is reclaimed and restored; and the vegetation is re-established.

Socioeconomics

Construction: Impacts would be SMALL to MODERATE. The Ross Project would employ approximately 200 people during construction, and this influx of workers would be expected to result in socioeconomic impacts, the greatest for communities with small populations. However, due to the short duration of construction, these workers would have only a limited effect on public services and community infrastructure. The Applicant is also committed to hiring locally—90 percent of the construction workforce would be local hires—so the overall socioeconomic impacts during the construction phase of the Ross Project would be SMALL. However the tax revenues paid to Crook County would be significant and, thus, that benefit would be a MODERATE impact of the Ross Project.

Operation: Impacts would be SMALL to MODERATE. If the majority of the operation workforce is local, the potential impacts to population and public services would continue to be SMALL. Because the Applicant is committed to hiring locally—80 percent of the operation workforce is expected to be local hires—the overall socioeconomic impacts during the Ross Project's operation phase would continue to be SMALL, with MODERATE impacts associated with the additional tax revenues that would accrue to Crook County.

Aquifer Restoration: Impacts would be SMALL. The Applicant indicates that there would be a smaller workforce of only approximately 20 workers during the aquifer-restoration phase, without concurrent operations. The need for regulatory, management, and health and safety personnel would continue throughout aquifer restoration, but this need would be met by personnel transitioning from operation-phase work to aquifer restoration and no new personnel would necessarily be required. Thus, the impacts of the Ross Project's aquifer-restoration phase would likely be at most the same, or, more likely, less than those noted above for the Ross Project's operation phase.

***Decommissioning*:** Impacts would be SMALL. Because the size of the workforce during the Ross Project's decommissioning phase would be initially be higher, but would subside as the decommissioning proceeds, there would be no significant socioeconomic impacts. In addition, socioeconomic impacts would no longer include tax revenues to Crook County during the decommissioning phase of the Ross Project and, thus, the earlier phases' moderate impacts would be eliminated.

Environmental Justice

***All Phases*:** No minority or low-income populations were identified in the vicinity of the proposed Ross Project. Therefore, there would be no disproportionately high and adverse impacts to minority and low-income populations from the construction, operation, aquifer restoration, and decommissioning of the Ross Project.

Public and Occupational Health and Safety

***Construction*:** Impacts would be SMALL. Construction activities, including the use of construction equipment and vehicles, would disturb the topsoil and create fugitive dust emissions. Fugitive dust generated from construction activities would be short term (1 to 2 years), and the levels of radioactivity in soils at the proposed project site are low; therefore direct exposure, inhalation, and ingestion of fugitive dust would not result in a significant radiological dose to workers or the public. Construction equipment would be diesel powered and would exhaust particulate diesel emissions. The potential impacts and potential human exposures from these emissions would be SMALL because of the short duration of the release and because the emissions would be readily dispersed into the atmosphere.

***Operation*:** The radiological impacts from normal operations would be SMALL. Public and occupational exposure rates at ISR facilities during normal operations have historically been well below regulatory limits. Dose assessments using the MILDOS computer code indicate that the 10 CFR Part 20 public dose limit of 1 mSv/yr [100 mrem/yr] would not be exceeded at any property boundary. The remote location of the proposed Ross Project site and the use of the proposed ISR technology coupled with the Applicant's proposed procedures to minimize exposure would cause the potential impact on public and occupational health and safety from facility operation to be SMALL. The radiological impacts from accidents would be SMALL for workers (if the Applicant's radiation safety and incident response procedures in an NRC-approved radiation protection plan are followed) and SMALL for the public because of the facility's remote location. The nonradiological public and occupational health and safety impacts from normal operations and accidents, due primarily to risk of chemical exposure, would be SMALL if handling and storage procedures are followed.

***Aquifer Restoration*:** Impacts would be SMALL. Impacts would be similar to, but less than, those during the operations phase. The reduction or elimination of some operational activities would further reduce the magnitude of potential worker and public health impacts and safety hazards.

Decommissioning: Impacts would be SMALL. Impacts would be similar to those experienced during construction. Soil and facility structures would be decontaminated, and lands would be restored to preoperational conditions.

Waste Management

Construction: Impacts would be SMALL. No significant liquid wastes would be generated during the construction of the Ross Project. Most of the solid wastes expected to be generated during the construction phase would be general construction debris including paper, wood, plastic, and scrap metal. These nonhazardous solid wastes would be disposed of at a permitted solid-waste facility. Hazardous wastes, such as organic solvents, paints, and paint thinners, would be disposed of in accordance with the requirements in the *Resource Conservation and Recovery Act* (RCRA). No radioactive (byproduct) wastes would be generated during this phase at the Ross Project, although technologically enhanced naturally occurring radioactive material (TENORM) wastes would be generated during well drilling and these wastes would be managed onsite.

Operation: Impacts would be SMALL. Wastes generated during the operation of the Ross Project would primarily be liquid waste streams consisting of process bleed, where, after reverse-osmosis treatment, some excess permeate early in the Project's operation and brine would be disposed of onsite at the five already permitted underground deep-injection wells. In addition, other liquid byproduct effluents would be generated as spent eluate, process-drains liquids, contaminated reagents, filter-backwash liquids, wash-down water, and decontamination shower water. State permitting actions, NRC license conditions, and NRC inspections would ensure that proper waste-management practices are implemented by the Applicant to comply with safety requirements to protect workers and the public. Nonhazardous solid waste such as facility trash, tires, piping, valves, and instrumentation, would be reused, recycled, or disposed of at a nearby landfill or other waste-disposal facility, each of which has available disposal capacity. Domestic wastes would be treated and disposed of in an onsite sewage-treatment system.

Aquifer Restoration: Impacts would be SMALL. Water from aquifer restoration would be treated through a combination of ion exchange and reverse osmosis (RO) and then would be re-injected into the ore-zone aquifer to limit the volume of water permanently withdrawn. Concentrated liquid effluents generated by these activities would be disposed of via deep well disposal. Ordinary trash would continue to be shipped offsite for disposal.

Decommissioning: Impacts would be SMALL. The goal of decommissioning is to reduce potential impacts by removing contaminants to allowable (regulatory) levels and restoring the land of the Ross Project area to pre-licensing baseline conditions. The Applicant proposes to decontaminate and recycle much of the process equipment or to reuse it at other uranium-recovery facilities. The Applicant would remove sludge from the storage ponds and liners and dispose of this material at a properly licensed radioactive-waste facility. Pre-operational agreements with a licensed radioactive-waste disposal facility to accept byproduct material would ensure the availability of sufficient disposal capacity for decommissioning activities. If hazardous waste is generated by decommissioning activities, it would be handled in accordance with applicable requirements.

SUMMARY OF THE CUMULATIVE IMPACTS

The cumulative impacts on the environment that would result from the incremental impact of the proposed Ross Project, when added to other past, present, and reasonably foreseeable future actions, was also considered. The NRC staff determined that the SMALL to LARGE incremental impacts of the Ross Project would not contribute perceptible increases to the SMALL to LARGE cumulative impacts, due primarily to the extensive exploration taking place in the area for uranium, oil, and gas, and from coal mining.

SUMMARY OF THE COSTS AND BENEFITS OF THE PROPOSED ACTION

The implementation of the Proposed Action would generate primarily regional and local costs and benefits. The regional benefits of building the proposed Ross Project would be increased employment, economic activity, and tax revenues to the region around the proposed Ross Project area (i.e., Crook County). Costs associated with the Ross Project are, for the most part, limited to the area immediately surrounding the Ross Project area and include small visual, air-quality, and noise impacts. The NRC staff determined that the benefit from constructing and operating the uranium-recovery facility would outweigh the environmental and social costs.

COMPARISON OF THE ALTERNATIVES

Under the No-Action Alternative, Alternative 2, the NRC would not approve the license application for the proposed Ross Project. The No-Action Alternative would result in the Applicant not constructing, operating, restoring the aquifer of, or decommissioning the proposed ISR project. However, even if the proposed Ross Project is not licensed, the Applicant has already accomplished certain preconstruction activities (those activities that do not require an NRC license) at the Ross Project area. These previously completed preconstruction activities are evaluated as part of Alternative 2: No Action.

Under Alternative 3, the NRC would issue the Applicant a license for the construction, operation, aquifer restoration, and decommissioning of the proposed ISR project at the Ross Project, except that the entire ISR facility, including all buildings, other auxiliary structures, and the surface impoundments would be located north of where it is to be situated for the Proposed Action. This alternate location for the ISR facility, referred as the "north site" by the Applicant (and referred to herein as the "North Ross Project"), was considered, but eliminated, by the Applicant in its license application. The north site is about 900 m [3,000 ft] northwest of where the facility would be located in the Proposed Action (referred to by the Applicant as the "south site"). An unnamed surface water drainage feature generally divides the north site. To avoid the floodplain of the drainage the Applicant would likely place the CPP and other buildings on one side of the drainage and the surface impoundments on the other side.

PRELIMINARY RECOMMENDATION

After weighing the impacts of the Proposed Action and comparing the Alternatives, the NRC staff, in accordance with 10 CFR Part 51.71(f), sets forth its preliminary NEPA recommendation regarding the Proposed Action. Unless safety issues mandate otherwise, the preliminary NRC staff recommendation to the Commission related to the environmental aspects of the Proposed Action is that a source and byproduct materials license for the Proposed Action be issued as requested. The NRC staff concludes that the applicable environmental monitoring program described in Chapter 6 and the proposed mitigation measures discussed in Chapter 4 will eliminate or substantially lessen the potential adverse environmental impacts associated with the Proposed Action.

The NRC staff has concluded that the overall benefits of the proposed action outweigh the environmental disadvantages and costs based on consideration of the following:

- Potential adverse impacts to all environmental resource areas are expected to be SMALL, with the exception of

  1. Transportation resources during all phases of the proposed action. Increases in traffic during construction and operation would have a MODERATE to LARGE impact. Impacts would be MODERATE with mitigation for construction, operation, aquifer restoration, and decommissioning (See SEIS Sections 4.3.1.1, 4.3.1.2, 4.3.1.3, and 4.3.1.4).

  2. Groundwater resources during operation and aquifer restoration. During operations there would be a MODERATE impact to ore-zone aquifer water quality due to excursions; however with measures in place to detect and resolve the excursions, the impacts would be reduced. During aquifer restoration there would be a MODERATE impact to ore-zone aquifer water quantity due to short-term drawdown (See SEIS Sections 4.5.1.2 and 4.5.1.3).

  3. Noise resources during construction, operations, and decommissioning. During these phases of the Ross Project there would be MODERATE impacts due to increased noise levels, however they would be intermittent and short term (See SEIS Sections 4.8.1.1, 4.8.1.2 and 4.8.1.4).

  4. Historical and cultural resources during construction. Section 106 consultation and efforts to identify and determine the eligibility of historical and cultural resources that could be adversely affected by the proposed Ross Project are currently ongoing. Therefore, to be conservative in this draft SEIS, the NRC staff considers that construction could have a MODERATE to LARGE impact on historic properties, sites currently listed or eligible for listing on the National Register of Historic Places (NRHP)—and other unevaluated historic, cultural, and religious properties in the project area (See SEIS Section 4.9.1.1). However, once identification efforts are complete, mitigation efforts, which could require an MOA, would be developed to reduce impacts. The final SEIS will include the outcome of Section 106 consultation and would discuss mitigation measures, including an MOA, if one is developed.

  5. Visual and scenic resources during construction. There would be MODERATE impacts to residents near the Ross Project for the first year, however over the long term, impacts would be reduced (See SEIS Section 4.10.1.1).

  6. Socioeconomic resources during construction and operations. There would be MODERATE impacts to Crook County during these phases of the Ross Project because taxes from the Project will be paid to the county (See Sections 4.11.1.1 and 4.11.1.2).

- Regarding groundwater, the portion of the aquifer(s) designated for uranium recovery must be exempted as underground sources of drinking water before ISR operations begin. Additionally, Strata would be required to monitor for excursions of lixiviant from the production zones and to take corrective actions in the event of an excursion. Prior to operations, the Applicant would be required to provide detailed hydrologic pumping test data packages and operational plans for each wellfield at the Ross Project. Strata would also be required to restore groundwater parameters affected by the ISR operations to levels that are protective of human health and safety.

- The costs associated with the Ross Project are, for the most part, limited to the area surrounding the site.

- The regional benefits of building the proposed Project would be: increased employment, economic activity, and tax revenues in the region around the proposed Project site.

LIST OF ABBREVIATIONS/ACRONYMS

AASHTO	American Association of State Highway and Transportation Officials
ACL	Alternate Concentration Limit
ADAMS	Agencywide Documents Access and Management System
AEA	Atomic Energy Act
ALARA	As Low As Reasonably Achievable
APE	Area of Potential Effect
APLIC	Avian Power Line Interaction Committee
AQD	Air Quality Division (Wyoming Department of Environmental Quality)
ARPA	Archaeological Resources Protection Act of 1979
ASTM	ASTM International (formerly American Society for Testing and Materials)
BACT	Best Available Control Technology
BGS	Below Ground Surface
BIA	Bureau of Indian Affairs
BLM	U.S. Bureau of Land Management (U.S. Department of the Interior)
BLMSS	BLM's Sensitive Species
BLS	Bureau of Labor Statistics (U.S. Department of Labor)
BMP	Best Management Practice
CAA	Clean Air Act
CBM	Coal-Bed Methane
CBW	Containment Barrier Wall
CCS	Center for Climate Strategies
CEQ	Council on Environmental Quality
CERCLA	Comprehensive Environmental Response, Compensation, and Liability Act
CESQG	Conditionally Exempt Small Quantity Generator
CFR	Code of Federal Regulations
CO	Carbon monoxide
CO_2	Carbon dioxide
CPP	Central Processing Plant
CR	County Road
CWA	Clean Water Act
dBA	A-Weighted Decibels
DM	Deep-Monitoring Zone or Unit
DOC	U.S. Department of Commerce
DOI	U.S. Department of the interior
EC	Electrical Conductivity
EIA	Energy Information Administration (U.S. Department of Energy)
EIS	Environmental Impact Statement
EMR	Emergency Medical Responder
EMT	Emergency Medical Technician
EO	Executive Order
EOR	Enhanced Oil Recovery
EPA	U.S. Environmental Protection Agency
ER	Environmental Report

LIST OF ABBREVIATIONS/ACRONYMS
(Continued)

ESA	Endangered Species Act
FHWA	Federal Highway Administration (U.S. Department of Transportation)
GCRP	U.S. Global Change Research Program
GEIS	Generic Environmental Impact Statement
HASP	Health and Safety Plan
HDPE	High-Density Polyethylene
HEC	Hydrologic Engineering Center
HMS	Hydrologic Modeling System
ISL	In situ Leach
ISR	In situ Recovery
IX	Ion Exchange
LOI	Letter of Intent
LQD	Land Quality Division (Wyoming Department of Environmental Quality)
LSA	Low Specific Activity
MARSSIM	Multi-Agency Radiation Survey & Site Investigation Manual
MCL	Maximum Contaminant Level
MIT	Mechanical Integrity Testing
MOA	Memorandum of Agreement
MOU	Memorandum of Understanding
MSDS	Material Safety Data Sheet
NAAQS	National Ambient Air Quality Standards
NEPA	National Environmental Policy Act
NHPA	National Historic Preservation Act
NPS	National Park Service
NRC	U.S. Nuclear Regulatory Commission
NRCS	Natural Resources Conservation Service (U.S. Department of Agriculture)
NRHP	National Register of Historic Places
NSDWUMR	Nebraska-South Dakota-Wyoming Uranium Milling Region
Nubeth	Nubeth Joint Venture
NWI	National Wetlands Inventory
NWP	Nationwide Permit (U.S. Army Corps of Engineers)
NWS	National Weather Service
OSHA	Occupational Safety & Health Administration (U.S. Department of Labor)
OSLI	(Wyoming) Office of State Lands and Investments
OZ	Ore Zone (Monitoring Interval or Aquifer)
Pb	Lead
PCB	Polychlorinated Biphenyl
PFYC	Potential Fossil Yield Classification System

LIST OF ABBREVIATIONS/ACRONYMS
(Continued)

pH	Hydrogen Ion Activity
PM_{10}	Particulate Matter 10 Microns or Less
$PM_{2.5}$	Particulate Matter 2.5 Microns or Less
PPE	Personal Protective Equipment
PRB	Powder River Basin
PSD	Prevention of Significant Deterioration
PSHA	Probabilistic Seismic Hazard Analysis
PVC	Polyvinyl Chloride
R	Range or Roentgens
RAI	Request for Additional Information
RCRA	Resource Conservation and Recovery Act
rem	Roentgen Equivalent Man
RFFA	Reasonably Foreseeable Future Actions
RMP	Resource Management Plan
RO	Reverse Osmosis
RPP	Radiation Protection Program or Plan
SA	Surficial Aquifer
SAR	Sodium Adsorption Ratio
SEIS	Supplemental Environmental Impact Statement
SER	Safety Evaluation Report
SHPO	State Historic Preservation Office
SM	Shallow-Monitoring Zone
SMC	USFWS's Migratory Bird Species of Management Concern in Wyoming
SOP	Standard Operating Procedure
SOW	Scope of Work
Strata	Strata Energy, Inc.
SWPPP	Storm Water Pollution Prevention Plan
TCP	Traditional Cultural Property
TDS	Total Dissolved Solids
TEDE	Total Effective Dose Equivalent
TENORM	Technologically Enhanced Naturally Occurring Radioactive Material
THPO	Tribal Historic Preservation Office
TLD	Thermo Luminescent Dosimeter
TR	Technical Report
TSCA	Toxic Substances Control Act
UCL	Upper Control Limit
UIC	Underground Injection Control
USACE	U.S. Army Corps of Engineers
USCB	U.S. Census Bureau (U.S. Department of Commerce)
USDA	U.S. Department of Agriculture
USDOT	U.S. Department of Transportation
USDW	Underground Source of Drinking Water

LIST OF ABBREVIATIONS/ACRONYMS
(*Continued*)

USFS	U.S. Forest Service
USFWS	U.S. Fish and Wildlife Service
USGS	U.S. Geological Survey
UW	University of Wyoming
VRI	Visual Resource Inventory
VRM	Visual Resource Management
WAAQS	Wyoming Ambient Air Quality Standards
WDAI	Wyoming Department of Administration and Information
WDEQ	Wyoming Department of Environmental Quality
WEUMR	Wyoming East Uranium Milling Region
WGFD	Wyoming Game and Fish Department
WOGCC	Wyoming Oil and Gas Conservation Commission
WQD	Water Quality Division (Wyoming Department of Environmental Quality)
WSEO	Wyoming State Engineer's Office
WSGS	Wyoming State Geological Survey
WSOC	Wyoming Species of Concern
WWC	WWC Engineering
WWDC	Wyoming Water Development Commission
WYCRO	Wyoming Cultural Records Office
WYDOT	Wyoming Department of Transportation
WYNDD	Wyoming Natural Diversity Database
WYPDES	Wyoming Pollutant Discharge Elimination System

SI* (MODERN METRIC) CONVERSION FACTORS

Approximate Conversions From SI Units				
Symbol	**When You Know**	**Multiply By**	**To Find**	**Symbol**
		Length		
cm	centimeters	0.39	inches	in
m	meters	3.28	feet	ft
m	meters	1.09	yards	yd
km	kilometers	0.621	miles	mi
		Areas		
mm^2	square millimeters	0.0016	square inches	in^2
m^2	square meters	10.764	square feet	ft^2
m^2	square meters	1.195	square yards	yd^2
Ha	hectares	2.47	acres	ac
km^2	square kilometers	0.386	square miles	mi^2
		Volume		
mL	milliliters	0.034	fluid ounces	fl oz
L	liters	0.264	gallons	gal
m^3	cubic meters	35.314	cubic feet	ft^3
m^3	cubic meters	1.307	cubic yards	yd^3
m^3	cubic meters	0.0008107	acre-feet	acre-feet
		Mass		
g	grams	0.035	ounces	oz
kg	kilograms	2.202	pounds	lb
Mg (or "t")	megagrams (or "metric ton")	1.103	short tons (2,000 lb)	T
		Temperature (Exact Degrees)		
^{0}	Celsius	1.8C + 35	Fahrenheit	^{0}

*SI is the symbol for the International System of Units. Appropriate rounding should be performed to comply with Section 4 of ASTM International's "Standard for Metric Practice Guide." West Conshohocken, Pennsylvania: ASTM International. Revised 2003.

1 INTRODUCTION

1.1 Background

The U.S. Nuclear Regulatory Commission (NRC) prepared this Supplemental Environmental Impact Statement (SEIS) in response to an application Strata Energy, Inc. (Strata) (referred to herein as the Applicant) submitted on January 4, 2011, to develop and operate the proposed Ross In Situ Uranium Recovery (ISR) Project (herein referred to as Ross Project), located in Crook County, Wyoming (Strata, 2011a; Strata, 2011b). The Applicant is a wholly owned subsidiary of Peninsula Minerals, Ltd. Figure 1.1 shows the geographic location of the proposed project. This site-specific SEIS supplements the Generic Environmental Impact Statement (GEIS) for In Situ Leach Uranium Milling Facilities (herein referred to as GEIS) and was prepared in accordance with the process described in GEIS Section 1.8 (NRC, 2009) and as detailed in Section 1.4.1 of this SEIS. The NRC's Office of Federal and State Materials and Environmental Management Programs prepared this SEIS as required by Title 10, Energy, of the *U.S. Code of Federal Regulations* (10 CFR), Part 51. These regulations implement the requirements of the *National Environmental Policy Act of 1969* (NEPA), as amended (Public Law 91-190), which requires the Federal government to assess the potential environmental impacts of major federal actions that may significantly affect the human environment.

The GEIS uses the terms "*in-situ* leach (ISL) process" and "11e.(2) byproduct material" to describe this uranium milling technology and the waste stream generated by this process. For the purposes of this SEIS, ISR is synonymous with ISL. The SEIS also uses the term "byproduct material" instead of "11e.(2) byproduct material" to describe the waste stream generated by this milling process to be consistent with the definition in 10 CFR Part 40.4.

1.2 Proposed Action

On January 4, 2011, Strata submitted an application for an NRC source and byproduct material license to construct and operate an ISR facility at the proposed Ross Project site and to conduct aquifer restoration, site decommissioning, and reclamation activities. Based on the application, the NRC's federal action is the decision to either grant or deny the license. The Applicant's proposal is described in detail in SEIS Section 2.1.1.

1.3 Purpose and Need of the Proposed Action

The NRC regulates uranium milling, including the ISR process, under 10 CFR Part 40, Domestic Licensing of Source Material. The Applicant is seeking an NRC source material license to authorize commercial-scale ISR at the proposed Ross Project site. The purpose and need for the proposed action is to provide an option that allows the Applicant to recover uranium and to produce yellowcake slurry at the Ross Project site. Yellowcake is the uranium oxide product of the ISR milling process that is used to produce various products, including fuel for commercially operated nuclear power reactors.

This definition of purpose and need reflects the Commission's recognition that, unless there are findings in the safety review required by the *Atomic Energy Act of 1954* (AEA), as amended, or findings in the NEPA environmental analysis that would lead NRC to reject a license application, NRC has no role in a company's business decision to submit a license application to operate an ISR facility at a particular location.

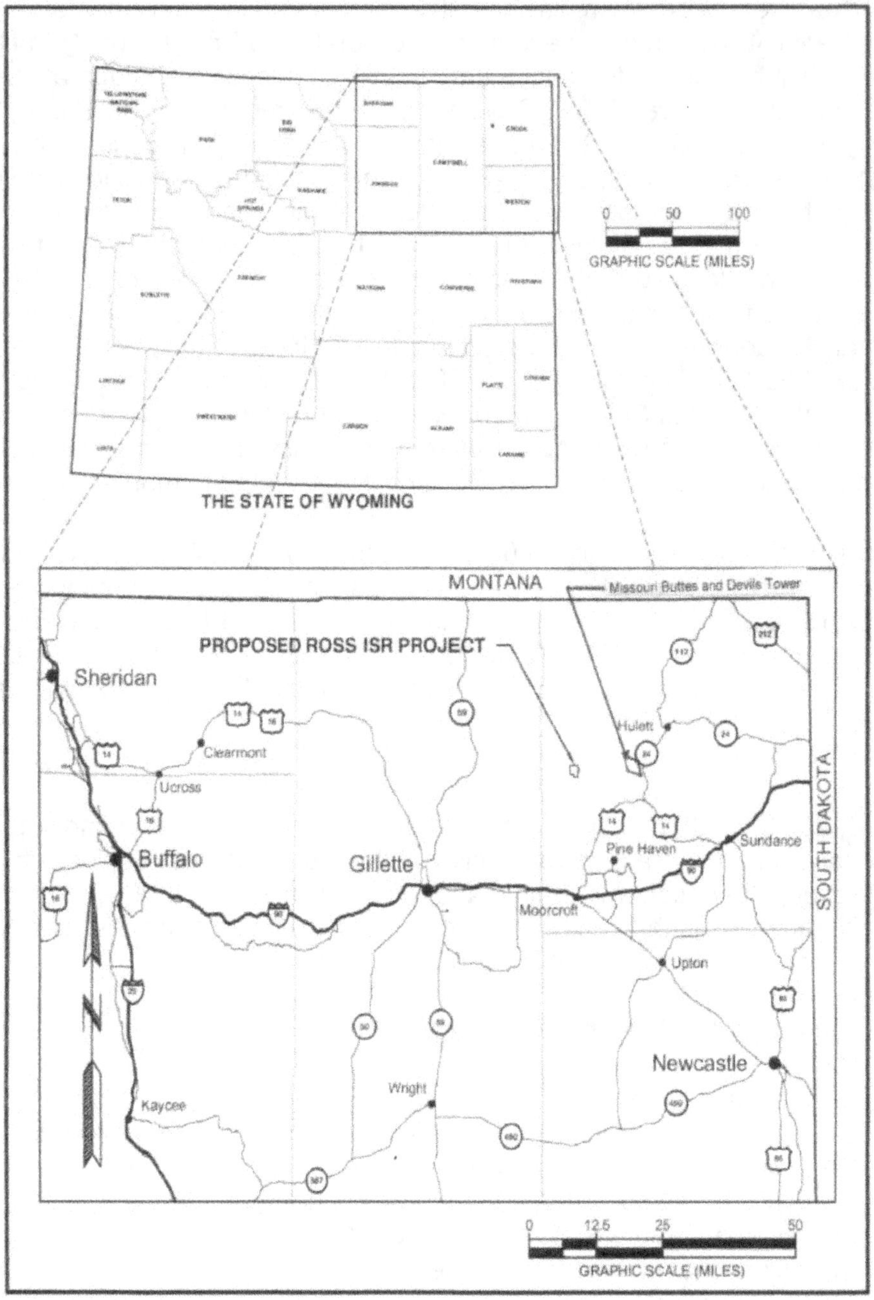

Source: Strata, 2011a.

Figure 1.1 Ross Project Location

1.3.1 BLM's Purpose and Need

The BLM purpose and need for the proposed action is to provide for orderly, efficient, and environmentally responsible mining of the uranium resource. The uranium resource is needed to fulfill market demands for this product for power generation and other needs. The proposed Ross Project area contains BLM-administered public lands open to mineral entry, and the Applicant has filed mining claims on them. The BLM federal decision is either to approve the Applicant's Plan of Operations subject to mitigation included in the license application and this draft SEIS, or deny approval of the Plan of Operations. BLM's responsibility to respond to the Applicant's Plan of Operations establishes the need for the action. The mining claimant (Strata) has the right to mine and to develop the mining claims as long as it can be done without causing unnecessary or undue degradation and is in accordance with pertinent laws and regulations under 43 CFR Part 3800.

1.4 Scope of the SEIS

The NRC staff prepared this SEIS to analyze the potential environmental impacts (i.e., direct, indirect, and cumulative impacts) of the proposed action and of reasonable alternatives to the proposed action. The scope of this SEIS considers both radiological and nonradiological (including chemical) impacts associated with the proposed action and its alternatives. This SEIS also considers unavoidable adverse environmental impacts, the relationship between short-term uses of the environment and long-term productivity, and the irreversible and irretrievable commitments of resources.

1.4.1 Relationship to the GEIS

As described in Section 1.1, this SEIS supplements the GEIS, which was published as a final report in May 2009 (NRC, 2009). The final GEIS assessed the potential environmental impacts associated with the construction, operation, aquifer restoration, and decommissioning of an ISR facility that could be located in four specific geographic regions of the western United States. The proposed Ross Project is located in the Nebraska/South Dakota/Wyoming Uranium Milling Region. Table 1.1 summarizes the expected environmental impacts by resource area in the Nebraska-South Dakota-Wyoming Uranium Milling Region based on the GEIS analyses.

The NRC conducted scoping activities for the purposes of defining the scope of GEIS and any future supplements to the GEIS. NRC staff accepted public comments on the scope of the GEIS from July 24, 2007, to November 30, 2007, and held three public scoping meetings, one of which was in the State of Wyoming. Additionally, NRC held eight public meetings to receive comments on the draft GEIS, published in July 2008. Three of these meetings were held in the State of Wyoming and one in nearby (Spearfish) South Dakota. Comments on the draft GEIS were accepted between July 28, 2008, and November 8, 2008. Comments received during scoping and on the draft GEIS were made available on the NRC website (http://www.nrc.gov/reading-rm/adams.html). Transcripts of the scoping meeting and draft GEIS comment meetings in Wyoming are available at http://www.nrc.gov/materials/uranium-recovery/geis/pub-involve-process.html.

Table 1.1

ISL GEIS Range of Expected Impacts in the Nebraska-South Dakota-Wyoming Uranium Milling Region

Resource Area	Construction	Operation	Aquifer Restoration	Decommissioning
Land Use	S	S	S	S to M
Transportation	S to M	S to M	S to M	S
Geology and Soils	S	S	S	S
Surface Water	S to M	S to M	S to M	S to M
Groundwater	S	S to L	S to M	S
Terrestrial Ecology	S to M	S	S	S
Aquatic Ecology	S	S	S	S
Threatened and Endangered Species	S to L	S	S	S
Air Quality	S	S	S	S
Noise	S to M	S to M	S to M	S
Historical and Cultural Resources	S to L	S	S	S
Visual and Scenic Resources	S	S	S	S
Socioeconomics	S to M	S to M	S	S to M
Public and Occupational Health and Safety	S	S to M	S	S
Waste Management	S	S	S	S

S: SMALL impact
M: MODERATE impact
L: LARGE impact
Source: NRC, 2009

A scoping summary report was provided as GEIS Appendix A and GEIS Appendix G and provides responses to public comments on the draft GEIS (NRC, 2009).

In addition to the scoping activities conducted by NRC during preparation of the GEIS, NRC published ads, soliciting scoping comments on the Ross Project SEIS, in four local newspapers (*Moorcroft Leader*, *Casper Star Tribune*, *Gillette News Record*, and *Sundance Times*). The newspaper ad ran on December 2, 2011 in the Casper Star Tribune and December 1, 2011 for the other three papers. Scoping comments were received until December 30, 2011. In total, 19 scoping comment letters were received containing a total of 53 individual comments.

This SEIS was prepared to fulfill the requirement at 10 CFR Part 51.20(b)(8) to prepare either an environmental impact statement (EIS) or supplement to an EIS (SEIS) for the issuance of a source material license for an ISR facility (NRC, 2009). The GEIS provides a starting point for the NRC's NEPA analyses for site-specific license applications for new ISR facilities, as well as for applications to amend or renew existing ISR licenses. As described in the GEIS, the GEIS provides criteria for each environmental resource area to assess the significance level of impacts (i.e., SMALL, MODERATE, or LARGE). The NRC staff applied these criteria to the site-specific conditions at the proposed Ross Project. This SEIS tiers from, and incorporates by reference, the GEIS relevant information, findings, and conclusions concerning environmental impacts. The extent to which NRC staff incorporates the GEIS impact conclusions depends on the consistency between: (i) the Applicant's proposed facilities and activities, and conditions at the Ross Project site; and (ii) the reference facility description, and activities, and information in the GEIS. NRC staff determinations regarding potential environmental impacts and the extent to which GEIS impact conclusions were incorporated by reference are described in Section 4 of this SEIS. GEIS Section 1.8.3 describes the relationship between the GEIS and a site-specific SEIS (NRC, 2009).

1.4.2 Public Participation Activities

As part of the preparation of this SEIS, NRC staff met with Federal, State, and local agencies and authorities, as well as public interest groups during a visit to the proposed Ross Project site and surrounding region in August 2011 (NRC, 2011a). The purpose of the meetings was to gather additional site-specific information to assist the NRC's environmental review.

The NRC staff published a Notice of Opportunity for Hearing on the proposed Ross Project license application in the *Federal Register* (FR) on July 13, 2011 (76 FR 41308). A hearing request from Petitioners Natural Resources Defense Council and Powder River Basin Resource Council was received on October 27, 2011. The NRC staff published a Notice of Intent (NOI) to prepare this SEIS on November 16, 2011 (76 FR 71082). In addition to the opportunities provided through the NEPA process, the NRC provided multiple opportunities for public involvement during the NRC staff's safety review. Specifically, the NRC staff held 10 public meetings or teleconferences with the Applicant from 2010 through 2012.

1.4.3 Issues Studied in Detail

To meet its NEPA obligations related to its review of the Ross Project license application, the NRC staff conducted an independent, detailed, comprehensive evaluation of the environmental impacts from construction, operation, aquifer restoration, and decommissioning of an ISR facility at the proposed Ross Project site and from reasonable alternatives. As described in GEIS Section 1.8.3, the GEIS: (i) evaluated the types of environmental impacts that may occur from ISR uranium milling facilities; (ii) identified and assessed generic impacts (i.e., the same or similar) at all ISR facilities (or those with specified facility or site characteristics); and (iii) determined the scope of environmental impacts that needed to be addressed in site-specific environmental reviews. Therefore, although all of the environmental resource areas identified in the GEIS would be addressed in site-specific reviews, certain resource areas would require a more detailed site-specific analysis, because the GEIS determined a range in the significance of impacts (e.g., SMALL to MODERATE, SMALL to LARGE) could result, depending upon site-specific conditions (see Table 1.1).

Based on the GEIS analyses, this SEIS provides a site-specific analysis of the following resource areas:

- Land Use
- Transportation
- Geology and Soils

- Transportation
- Surface Water
- Groundwater
- Ecology
- Threatened and Endangered Species
- Air Quality
- Noise
- Visual and Scenic Resources
- Historic and Cultural Resources
- Socioeconomics
- Environmental Justice
- Public Health and Safety
- Waste Management

Furthermore, certain site-specific analyses not conducted in the GEIS, such as assessment of cumulative impacts, were considered in this SEIS. Additionally, the NRC considers the potential effects from implementing the proposed action on global climate change by estimating the facility's greenhouse gas emissions, and also describes the potential effects of global climate change on the proposed action.

1.4.4 Issues Outside the Scope of the SEIS

Some issues and concerns raised during the scoping process on the GEIS (NRC, 2009, Appendix A) were determined to be outside the scope of the GEIS. These issues and concerns (e.g., general support or opposition for uranium milling, impacts associated with conventional uranium milling, comments regarding the alternative sources of uranium feed material, comments regarding energy sources, requests for compensation for past mining impacts, and comments regarding the credibility of NRC) are also outside the scope of this SEIS.

1.4.5 Related NEPA Reviews and Other Related Documents

A number of NEPA documents (environmental assessments [EAs] and environmental impact statements [EISs]) and other documents were reviewed and used in the development of this SEIS. The related documents are described below:

- **NUREG–1910, Generic Environmental Impact Statement for *In-Situ* Leach Uranium Milling Facilities, Final Report (NRC, 2009)**. As described previously, this GEIS was prepared to assess the potential environmental impacts from the construction, operation, aquifer restoration, and decommissioning of an ISR facility located in one of four different geographic regions of the western U.S. including the Nebraska/South Dakota/Wyoming Uranium Milling Region, where the proposed Ross Project would be located. The environmental analysis in this SEIS both tiers from the GEIS and incorporates it by reference.

- **NUREG–0706, Final Generic Environmental Impact Statement on Uranium Milling (NRC, 1980)**. This Generic EIS provides a detailed evaluation of the impacts and effects of anticipated conventional uranium milling operations in the United States through the year 2000, including an analysis of tailings disposal programs. NUREG–0706 concluded the environmental impacts from underground mining and conventional milling would be more severe than using ISR technology. As described in SEIS Section 2.2.1, conventional mining and milling were considered, but eliminated from detailed analysis.

- **NUREG–1508, Final Environmental Impact Statement To Construct and Operate the Crownpoint Uranium Solution Mining Project, Crownpoint, New Mexico (NRC, 1997)**. This EIS evaluates the use of ISR technology at the Church Rock and Crownpoint sites at Crownpoint, New Mexico. Alternative uranium mining methods were not evaluated because the uranium ore located at the proposed sites was too deep to be extracted economically and the Final EIS concluded underground mining would have more significant environmental impacts than ISR recovery.

- **NRC's Safety Evaluation Report**. The NRC is preparing a Safety Evaluation Report (SER) for the proposed Ross Project that evaluates the Applicant's proposed facility design, operational procedures, and radiation protection programs and whether the Applicant's proposed action can be accomplished in accordance with the applicable provisions in 10 CFR Part 20, 10 CFR Part 40, and 10 CFR Part 40, Appendix A. The SER also provides the NRC staff analysis of the Applicant's initial funding estimate to complete site decommissioning and reclamation.

- **Newcastle Resource Management Plan EIS (BLM, 2000).** This management plan addresses the Comprehensive Analysis of Alternatives for the Planning and Management of Public Land and Resources Administered by the U.S. Bureau of Land Management (BLM), Crook, Weston and Niobrara Counties, Wyoming. This EIS identifies activities occurring in the region surrounding the Ross Project site that could either affect or be affected by the proposed Ross Project.

1.5 Applicable Regulatory Requirements

NEPA establishes national environmental policy and goals to protect, maintain, and enhance the environment and provide a process for implementing these specific goals for those Federal agencies responsible for an action. This SEIS was prepared in accordance with NRC NEPA-implementing regulations in 10 CFR Part 51 and other applicable regulations that were in effect at the time of writing. GEIS Appendix B summarizes other Federal statutes, implementing regulations, and Executive Orders that are potentially applicable to environmental reviews for the construction, operation, aquifer restoration, and decommissioning of an ISR facility. GEIS Sections 1.6.3.1 and 1.7.5.1 summarize the State of Wyoming's statutory authority pursuant to the ISR process, relevant state agencies that are involved in the permitting of an ISR facility, and the range of state permits that would be required (NRC, 2009).

1.6 Licensing and Permitting

NRC has statutory authority through the AEA and the Uranium Mill Tailings Radiation Control Act of 1978 to regulate uranium ISR facilities. In addition to obtaining an NRC license, uranium ISR facilities must obtain the necessary permits from the appropriate Federal, State, local and Tribal governmental agencies. The NRC licensing process for ISR facilities is described in GEIS Section 1.7.1. GEIS Sections 1.7.2 through 1.7.5 describe the role of the other Federal, Tribal, and State agencies in the ISR permitting process (NRC, 2009). This section of the SEIS describes the NRC license application review process and summarizes the status of the NRC licensing process at the proposed Ross Project and the status of the Applicant's permitting with respect to other applicable Federal, Tribal, and State requirements.

1.6.1 NRC Licensing Process for the Ross Project

By letter dated January 4, 2011, the Applicant submitted a license application to NRC for the proposed Ross Project (Strata, 2011a; Strata, 2011b). As described in GEIS Section 1.7.1, NRC initially conducts an acceptance review of a license application to determine whether the application is complete enough to support a detailed technical review. The NRC staff accepted the Ross Project license application for detailed technical review by letter dated June 28, 2011 (NRC, 2011b).

The NRC's detailed technical review of the license application is composed of both a safety review and an environmental review. These two reviews are conducted in parallel (see GEIS Figure 1.7-1). The focus of the safety review is to assess compliance with the applicable regulatory requirements in 10 CFR Part 20 and 10 CFR Part 40, Appendix A. The environmental review is conducted in accordance with the regulations in 10 CFR Part 51. A Notice of Intent to prepare this SEIS was published in the Federal Register on November 16, 2011 (76 FR 71082).

The NRC hearing process (10 CFR Part 2) applies to licensing actions and offers stakeholders a separate opportunity to raise concerns associated with the proposed licensing actions. NRC published a Notice of Opportunity for Hearing related to the Ross Project license application on July 13, 2011 (76 FR 41308). NRC received a combined request for hearing from the Natural Resources Defense Council (NRDC) and Powder River Basin Resource Council (PRBRC) (collectively referred to as "Petitioners") on October 27, 2011 (NRDC and PRBRC, 2011).

Regulations in 10 CFR Part 2 specify that a petition for review and request for hearing must include a showing that the petitioner has standing and that the Atomic Safety and Licensing Board (ASLB) would rule on a petitioner's standing by considering (i) the nature of the petitioner's right under the AEA or NEPA to be made a party to the proceeding, (ii) the nature and extent of the petitioner's property, financial, or other interest in the proceeding, and (iii) the possible effect of any decision or order that may be issued in the proceeding on the petitioner's interest. Petitioners based their claim of standing on the possibility that the Ross Project would jeopardize the economic and environmental interests of at least one of their members (NRDC and PRBRC, 2011).

On February 10, 2012, the ASLB ruled that Natural Resources Defense Council (NRDC) and the Powder River Basin Resource Council (PRBRC) demonstrated standing to be parties to the licensing proceeding. The ASLB granted the petitioners' request for a hearing and admitted four contentions (ASLB, 2012).

1.6.2 Status of Permitting With Other Federal, Tribal, and State Agencies

In addition to obtaining a source material license from NRC prior to conducting ISR operations at the proposed Ross Project site, the Applicant is required to obtain necessary permits and approvals from other Federal and State agencies to address (i) the underground injection of solutions and liquid effluent from the ISR process, (ii) the exemption of all or a portion of the ore zone aquifer from regulation under the *Safe Drinking Water Act*, and (iii) the discharge of storm water during construction and operation of the ISR facility. Table 1.2 lists the status of the required permits and approvals.

1.7 Consultations

As a Federal agency, NRC is required to comply with consultation requirements in Section 7 of the *Endangered Species Act of 1973* (ESA), as amended, and Section 106 of the *National Historic Preservation Act of 1966* (NHPA), as amended. The GEIS took a programmatic look at the environmental impacts of ISR uranium milling within four distinct geographic regions and acknowledged that each site-specific review would include its own consultation process with relevant agencies. Section 7 (ESA) and Section 106 (NHPA) consultations conducted for the proposed Ross Project are summarized in Sections 1.7.1 and 1.7.2. A list of the consultation correspondence is provided in SEIS Appendix A. Section 1.7.3 describes NRC coordination with other Federal, Tribal, State, and local agencies conducted during the development of the SEIS.

1.7.1 Endangered Species Act of 1973 Consultation

The ESA was enacted to prevent the further decline of endangered and threatened species and to restore those species and their critical habitats. Section 7 of the ESA requires consultation with the U.S. Fish and Wildlife Service (USFWS) to ensure that actions it authorizes, permits, or otherwise carries out would not jeopardize the continued existence of any listed species or adversely modify designated critical habitats.

By letter dated August 12, 2011, NRC staff initiated consultation with USFWS requesting information on endangered or threatened species and critical habitat in the proposed Ross Project area. NRC received a response dated September 13, 2011, from the USFWS Ecological Services Cheyenne, Wyoming Field Office that: (i) listed the threatened and endangered species that may occur in the project area; (ii) provided recommendations for protective measures for threatened and endangered species; and (iii) provided recommendations concerning migratory birds (USFWS, 2011).

NRC staff also met with the Wyoming Game and Fish Department (WGFD) Sheridan Office on August 23, 2011, to discuss site-specific issues (NRC, 2011a). The Sheridan Office staff expressed concern about the potential impacts to water fowl, migratory birds, big game and small mammals, as well as sage grouse, a USFWS wait-list species for consideration as either threatened or endangered. WGFD staff also expressed concern about invasive species and impacts to wildlife due to power lines, evaporation ponds, and increased traffic. Impact mitigation measures were discussed. By letter dated, September 22, 2011, WGFD provided NRC with comments regarding the above concerns as follow up to the site visit (WGFD, 2011).

1.7.2 National Historic Preservation Act of 1966 Consultation

Section 106 of the NHPA requires that Federal agencies take into account the effects of their undertakings on historic properties and afford the Advisory Council on Historic Preservation (ACHP) an opportunity to comment on such undertakings. The Section 106 process seeks the views of consulting parties including the Federal agency, the State Historic Preservation Officer (SHPO), Indian tribes and Native Hawaiian organizations, Tribal Historic Preservation Officers (THPO), local government leaders, the Applicant, cooperating agencies, and the public.

Table 1.2
Environmental Approvals for the Proposed Ross Project

Issuing Agency	Description	Status
Wyoming Department of Environmental Quality	UIC Class III Permit (WDEQ, Title 35-11)	Received approval as part of Permit #802
	Underground Injection Control Class I (Deep Disposal Wells) (WDEQ, Title 35-11)	Application submitted June 2010 to UIC program in Cheyenne, Wyoming; TFN #WYS-011-00031, Approved April 2011, Permit #10-263
	Permit to Construct Domestic Wastewater System	To be prepared by Strata
	Storm Water Discharge Permit (industrial/mining)	To be prepared by Strata
	Storm Water Discharge Permit (construction)	Approved January 2013, Permit #WYR104738
	Storm Water Discharge Permit (discharge during well testing)	Approved April 2012, Permit #WYG720229, renewed December 2012
	Permit to Mine	Application submitted January 2011 to WDEQ District 3, Sheridan, Wyoming, TFN #5 6/110, Approved November 2012, Permit #802
	Mineral Exploration Permit (WDEQ, Title 35-11)	Approved #384DN
	Air Quality Permit	Approved CT-12198; September 2011
	Wastewater Pond Construction Permit (lined retention ponds and sediment pond)	To be prepared by Strata
	Public Water Supply System – Permit to construct	To be prepared by Strata
U.S. Bureau of Land Management	Plan of Operation	Submitted to BLM by Strata, January 2011; accepted for review July 2011, case file WYW170151
	Right of Way (roads)	To be prepared by Strata
	Notice of Intent to Explore	To be prepared by Strata
U.S. Nuclear Regulatory Commission	Source and Byproduct Materials License (10 CFR Part 40)	Application under review (submitted January 2011; accepted June 2011)

Table 1.2
Environmental Approvals for the Proposed Ross Project (Continued)

Issuing Agency	Description	Status
U.S. Environmental Protection Agency	Aquifer Exemption Permit for Class I Injection Wells (40 CFR 144, 146)	See WDEQ permits; Wyoming has primacy for the UIC program
	Aquifer Reclassification for Class III Injection Wells (WDEQ, Title 35-11)	
	Permit application to construct holding (storage) ponds (40 CFR 61.07)	
	Public Water Supply System	To be prepared by Strata
U.S. Army Corps of Engineers	Verification of Preliminary wetlands	Application submitted September 2010; Verification received December 2010
	Nationwide Permit Coverage authorization	Pre-construction notification submitted January 2013
Wyoming State Land & Farm Loan Office	Uranium Minerals Mining Lease	Approved #0-40979
Wyoming Department of Environmental Quality and State Engineer's Office	Permit to Appropriate Groundwater for ISR Wellfield	Under review, submitted December 2012
	Permit to Appropriate Groundwater for Mine Wells	Approved Permit #'s 191679-191702; 192703-192705 (regional baseline monitor wells) To be prepared for ISR monitor wells
	Permits to Appropriate Surface Water and/ or Lined Retention Ponds and Sediment Pond	To be prepared by Strata
Crook County	County Development Permits (access road approach and emergency services agreement)	Memorandum of Understand between Crook County and Strata executed April 2011.
Source: WWC Engineering, 2013		

The goal of consultation is to identify historic properties potentially affected by the undertaking, assess the effects of the undertaking on these properties, and seek ways to avoid, minimize, or mitigate any adverse effects on historic properties. As detailed in 36 CFR Part 800.2(c)(1)(i), the role of the Wyoming SHPO in the Section 106 process is to advise and assist Federal agencies in carrying out their Section 106 responsibilities.

NRC initiated consultation with the Wyoming SHPO by letter dated August 19, 2011, requesting information from the SHPO to facilitate the identification of historic and cultural resources that could be affected by the proposed project (NRC, 2011c). The NRC staff continues to consult with the Wyoming SHPO to evaluate the effects of the proposed project on historic and cultural resources.

NRC is also consulting with potentially affected Native American Tribes as part of the Section 106 consultation process per 36 CFR Part 800.2(c). These interactions are detailed in Section 1.7.3.3 of the SEIS.

1.7.3 Coordination with Other Federal, Tribal, State, and Local Agencies

The NRC staff interacted with Federal, Tribal, State, and local agencies and/or entities during preparation of this SEIS to gather information on potential issues, concerns, and environmental impacts related to the proposed ISR facility at the Ross Project site. The consultation and coordination process included discussions with BLM, National Park Service (NPS), Tribal governments, the Wyoming Department of Environmental Quality (WDEQ), WGFD, the Wyoming State Engineer's Office (SEO), and local organizations (PRBRC, City of Moorcroft First Responders, and Crook County).

1.7.3.1 Coordination with the Bureau of Land Management

In its letter dated January 27, 2011, U.S. Bureau of Land Management (BLM) indicated its intent to serve as a cooperating agency in the NEPA assessment and licensing process for the proposed Ross Project, with the NRC serving as the lead agency. The proposed Ross Project site contains approximately 16 ha [40 ac] of BLM-administered surface lands. Additionally, BLM has jurisdiction over locatable mineral rights within the proposed project area. As discussed in Section 1.3, BLM's responsibility for the proposed action is to fulfill its statutory responsibilities to regulate mining on federal lands as described in 43 CFR Part 3809. A Memorandum of Understanding between NRC and BLM (75 FR 1088), signed by BLM on October 16, 2009 and by NRC on November 30, 2009, provides the framework for the cooperating agency relationship.

BLM is responsible for administering the National System of Public Lands and the federal minerals underlying these lands. BLM is also responsible for managing split estate situations where federal minerals underlie a surface that is privately held or owned by state or local government. In these situations, operators on mining claims, including ISR facilities, must submit a Plan of Operations and obtain BLM approval before beginning operations beyond those for casual use {for surface disturbance of more than 2 ha [5 ac]}.

The NRC has coordinated with BLM during preparation of this SEIS. Regular conference calls and meetings have been held. The NRC staff met with the staff of BLM Newcastle, Wyoming field office on August 24, 2011 to discuss the Applicant's Plan of Operations for the proposed Ross Project. BLM familiarized the NRC staff with the Plan of Operations review process and shared some of the comments and the concerns BLM had received from individuals commenting on the Plan of Operations.

1.7.3.2 Interactions with Tribal Governments

Pursuant to Section 106 of the NHPA, the NRC staff initiated discussions with potentially affected Native American Tribes that possess heritage and cultural interest to the proposed Ross Project area. On November 19, 2010, NRC sent a letter to 14 Tribes, notifying them of Strata's intent to submit an application for a license for the Ross Project and soliciting input from the Tribes (NRC, 2010). NRC sent letters, dated February 9, 2011, to the following 24 Tribes, inviting the Tribes to participate in formal consultations for the proposed Ross Project (NRC, 2011d):

- Apache Tribe of Oklahoma
- Blackfeet
- Cheyenne and Arapaho Tribes of Oklahoma
- Cheyenne River Lakota
- Crow
- Crow Creek Sioux
- Eastern Shoshone
- Flandreau Santee Lakota
- Fort Belknap Community
- Fort Peck Assiniboine/Sioux
- Kiowa Tribe of Oklahoma
- Lower Brule Lakota
- Northern Arapaho
- Northern Cheyenne
- Oglala Lakota (Sioux)
- Rosebud Sioux
- Salish, Pend d'Oreille and Kootenai Tribes
- Santee Sioux Nation
- Sisseton-Wahpeton Lakota
- Spirit Lake
- Standing Rock Sioux
- Three Affiliated Tribes (Mandan, Hidatsa, and Arikara Nation)
- Turtle Mountain Band of Chippewa Indians
- Yankton Lakota

The NRC staff continued its efforts to engage in consultation with Tribes that might be affected by the proposed action with follow-up telephone calls and by sending emails.

On April 15, 2011, the Rosebud Sioux Tribe notified the NRC via email that it was interested in consultation and had concerns about the proposed project (Rosebud Sioux Tribe, 2011). On April 29, 2011, the Standing Rock Sioux Tribe notified NRC via email of its desire to consult (Standing Rock Sioux Tribe, 2011). On May 5, 2011 the Northern Cheyenne Tribe notified NRC via email of its interest to consult (Northern Cheyenne Tribe, 2011). On May 17, 2011, the Cheyenne River Sioux Tribe notified NRC via email of its interest to consult on the proposed project (Cheyenne River Sioux Tribe, 2011).

By letter dated April 14, 2011 the Tribal Historic Preservation Officer (THPO) for the Turtle Mountain Band of Chippewa Indians, informed NRC that it does not likely have any traditional cultural properties that would be of National Register significance at the Ross Project site (Turtle Mountain Band of Chippewa Indians, 2011). NRC was notified by email on August 19, 2011 that the Apache Tribe of Oklahoma, was not interested in consultation on the Ross Project (Apache Tribe of Oklahoma, 2011). The Salish, Pend d'Oreille and Kootenai Tribes notified NRC by email on December 29, 2011 that it would defer to nearer Tribes for consultation on the Ross Project (Salish, Pend d'Oreille and Kootenai Tribes, 2011).

The NRC staff, along with BLM staff, and the Applicant, conducted a site visit with representatives from the Northern Arapaho, the Northern Cheyenne, and the Fort Peck Assiniboine Sioux Tribes on September 13, 2011. The NRC staff and the BLM staff participated in a consultation meeting with the Northern Arapaho and the Northern Cheyenne Arapaho Tribes on September 14, 2011. On November 2, 2011, the NRC staff along with BLM staff, NPS staff for Devils Tower National Monument, and the Applicant conducted a second site visit with representatives from the Chippewa Cree, Crow Creek Sioux, Santee Sioux Nation, and the Fort Peck Assiniboine Sioux Tribes. On November 3, 2011, the NRC staff, BLM staff, and NPS staff participated in a consultation meeting with representatives from the Crow Creek Sioux, Santee Sioux Nation, and the Fort Peck Assiniboine Sioux Tribes. The Chippewa Cree Tribe expressed interest in consulting during planning for the second consultation meeting.

During the September 2011 and November 2011 consultation meetings, the Tribes requested that a survey for properties of religious and cultural significance [or a Traditional Cultural Property (TCP) survey] of the Ross Project area be conducted. During the November 2011 site visit, Strata indicated that it would be willing to support such a survey. On December 6, 2011, the NRC sent a letter to Strata requesting a written proposal to acquire TCP information. Strata responded with a letter, dated January 12, 2012, in which it stated that in lieu of submitting a proposal for a TCP assessment of the Ross Project area, Strata would like to issue a Request for Proposals from consultants to prepare the TCP assessment. During conversations with several THPOs, the NRC staff was informed that the Tribes did not want to work with a third-party consultant hired by the Applicant. Therefore, the NRC staff enlisted support from its own third-party consultant to work with the Tribes to obtain information on TCPs.

At this time, the NRC staff was also working with many of the same Tribes to obtain TCP information for other ISR projects under NRC review. The Tribes consulting on the Ross Project suggested using a Scope of Work (SOW) that was being prepared for one of the other ISR projects under NRC review and revising it to be applicable for the Ross Project. The Tribes requested background information on the Ross Project area to assist them in developing a draft SOW for the Ross Project. This information was provided to the Tribes via email on July 25, 2012. In August 2012, the NRC's third-party consultant began reaching out to Tribes via phone and email to invite them to meet in Bismarck, North Dakota in early September to discuss the SOW as many of the Tribes were planning to be Bismarck at that time for a meeting with another agency. Strata provided a draft SOW to the NRC to be shared with the Tribes during the meeting. Sixteen Tribal representatives indicated that they would attend the meeting. On September 4, 2012, the NRC's third-party consultant met with representatives from the Standing Rock Sioux Tribe and the Three Affiliated Tribes in Bismarck, North Dakota. The Standing Rock Sioux Tribe representative indicated during this meeting that the Tribes did not want to use the SOW developed by Strata and would develop a draft SOW for the Ross Project. The Tribal representatives also indicated that a separate cost proposal would need to be developed for the TCP survey. In October and November 2012, the NRC staff worked with the representative from the Standing Rock Sioux Tribe to revise the SOW provided to the NRC by the Tribes for another ISR project under NRC review to be applicable for the Ross Project. Also, on October 23, 2012, Strata hosted three representatives from the Makoche Wowapi company at the Ross Project site to facilitate the company's preparation of a cost proposal for the TCP survey. The Makoche Wowapi company had submitted a cost proposal for a TCP survey for another ISR project under NRC review and many of the THPOs were discussing naming the company as the preferred consultant to conduct the TCP survey at the Ross Project site.

On November 13, 2012 and November 14, 2012, the NRC staff provided the draft SOW for the TCP survey to the THPOs and Strata, respectively, via email for review and comment. The THPOs held a teleconference to discuss the draft SOW on November 14, 2012 and invited the NRC staff to participate to answer questions. During the November 14, 2012 teleconference several THPOs indicated that the draft Scope of Work was acceptable and recommended that the Makoche Wowapi company was their preferred consultant to conduct the survey.

The NRC staff shared the final SOW with the consulting THPOs via email on November 30, 2012. After no comments were received, the NRC staff also shared the final SOW with the Makoche Wowapi company on December 4, 2012. On December 12, 2012, the Makoche Wowapi company submitted a cost proposal for the survey to the NRC. Strata notified the NRC staff, by email dated February 15, 2013, that its negotiations with Makoche Wowapi had come to an end and an agreement had not been reached. The NRC staff iscurrently consulting with the Tribes and Strata on an alternative approach to conduct a TCP survey. The survey is expected to be conducted during spring 2013.

The Section 106 consultation process is ongoing. Results of the consultation will be presented in the final SEIS.

1.7.3.3 Coordination with National Park Service

NRC staff met with NPS staff at Devils Tower on August 25, 2011 (NRC, 2011a). NPS staff discussed the use of the monument by Tribes for cultural activities and prayers. NPS staff shared concerns about the night-sky viewshed and noise as well as potential impacts to groundwater quality. NPS is a "commenting agency" for this SEIS.

1.7.3.4 Coordination with the Wyoming Department of Environmental Quality

NRC staff met with WDEQ in Sheridan, Wyoming, on August 23, 2011, to discuss the WDEQ role in the NRC environmental review process for ISR facilities (NRC, 2011a). WDEQ staff participating in the meeting included representatives from the Land Quality Division (LQD), Water Quality Division (WQD), and the Air Quality Division (AQD). Topics discussed during the meeting included the WDEQ air quality review and permitting as well as other required permits. The WDEQ expressed concern regarding the proposed location of the Central Processing Plant (CPP) and the evaporation ponds along with fugitive dust and emissions.

NRC staff also met with personnel from the WDEQ in Casper, Wyoming on August 24, 2011 (NRC, 2011a). WDEQ staff participating in the meeting included representatives from the WQD as well as the Solid and Hazardous Waste Division. The WDEQ explained the permitting process for land application of waste water and discussed solid waste management.

1.7.3.5 Coordination with the Wyoming Game and Fish Department

WGFD is responsible for controlling, propagating, managing, protecting, and regulating all game and nongame fish and wildlife in Wyoming under Wyoming Statute (W.S.) 23-1-301-303 and 23-1-401. Regulatory authority given to WGFD allows for the establishment of hunting, fishing, and trapping seasons, as well as the enforcement of rules protecting nongame and state-listed species.

NRC staff met with a representative of the Sheridan Regional WGFD office on August 23, 2011 (NRC, 2011a). As discussed in Section 1.7.1, WGFD staff expressed concerns about big game animals, raptors, migratory birds, and small mammals that may be affected by the proposed Ross Project and suggested mitigation strategies to minimize or eliminate impacts.

1.7.3.6 Coordination with the City of Moorcroft First Responders

NRC staff met with the City of Moorcroft First Responders on August 25, 2011 (NRC, 2011a). The City of Moorcroft First Responders briefed the NRC on the availability of local emergency equipment, personnel, and medical facilities. The emergency personnel discussed their need for additional training. The availability of land use plans and socioeconomic data was also discussed.

1.7.3.7 Coordination with the Powder River Basin Resource Council

NRC staff met with PRBRC on August 23, 2011 (NRC, 2011a). PRBRC shared several concerns regarding the proposed Ross Project including concerns about the Applicant's experience, potential direct and cumulative impacts to water quality, air quality, and ecology from operations, the potential for accidents and long-term effects, and restoration and excursion monitoring.

1.7.3.8 Coordination with Localities

NRC staff met with Crook County officials and staff on August 25, 2011, including representatives from the Crook County Sheriff's Office, Crook County Attorneys, Crook County Road & Bridge, Crook County Natural Resource District, Crook County Weed & Pest, Crook County Commissioner, Crook County Growth & Development, and Crook County Emergency Management (NRC, 2011a). The Crook County officials and staff shared several concerns and asked many questions about the proposed Ross Project. Topics discussed included the chemical and radiological hazards associated with the project, the management of boreholes and the potential for drinking water contamination, water use, financial assurance, solid waste management, invasive species, decommissioning, and cumulative impacts.

1.8 Structure of the SEIS

As noted in Section 1.4.1 of this document, the GEIS (NRC, 2009) evaluated the broad impacts of ISR projects in a four-state region where such projects are anticipated, but did not reach site-specific decisions for new ISR projects. The NRC staff evaluated the extent to which information and conclusions in the GEIS could be incorporated by reference into this SEIS. The NRC staff also determined whether any new and significant information existed that would change the expected environmental impact beyond what was evaluated in the GEIS.

SEIS Section 2 describes the proposed action and reasonable alternatives considered for the proposed Ross Project, Section 3 describes the affected environment, and Section 4 evaluates the environmental impacts from implementing the proposed action and alternatives. Cumulative impacts are discussed in Section 5, while Section 6 describes the environmental measurement and monitoring programs proposed for the Ross Project. A cost-benefit analysis is provided in Section 7, and the environmental consequences from the proposed action and alternatives are summarized in Section 8.

1.9 References

10 CFR Part 2. Code of Federal Regulations, Title 10, *Energy,* Part 2, "Rules of Practice for Domestic Licensing Proceedings and Issuance of Orders." Washington, DC: U.S. Government Printing Office.

10 CFR Part 20. Code of Federal Regulations, Title 10, *Energy*, Part 20, "Standards for Protection Against Radiation." Washington, DC: U.S. Government Printing Office.

10 CFR Part 40. Code of Federal Regulations, Title 10, Energy, Part 40, Domestic Licensing of Source Material, Washington, DC: U.S. Government Printing Office. 2010.

10 CFR Part 40. Appendix A. Code of Federal Regulations, Title 10, *Energy*, Part 40, Appendix A, "Criteria Relating to the Operation of Uranium Mills and to the Disposition of Tailings or Wastes Produced by the Extraction or Concentration of Source Material from Ores Processed Primarily from their Source Material Content." Washington, DC: U.S. Government Printing Office.

10 CFR Part 51. Code of Federal Regulations, Title 10, *Energy*, Part 51, "Environmental Protection Regulations for Domestic Licensing and Related Regulatory Functions." Washington, DC: U.S. Government Printing Office.

75 FR 1088. *Federal Register*, Vol. 75, No. 1497, p. 1088, "Notice of Availability of a Memorandum of Understanding between the Nuclear Regulatory Commission and the Bureau of Land Management." January 8, 2010.

76 FR 41308. *Federal Register*, Vol. 76, No. 134, p. 41308-41312. "Strata Energy, Inc. Ross In Situ Recovery Uranium Project, Crook County, WY; Notice of Materials License Application, Opportunity to Request a Hearing and To Petition for Leave To Intervene, and Commission Order Imposing Procedures for Document Access to Sensitive Unclassified Non-Safeguards Information for Contention Preparation." July 13, 2011. Agencywide Documents Access and Management System (ADAMS) Accession No. ML111940012.

76 FR 71082. *Federal Register*, Vol. 76, No. 221, p. 71082-71083. "Strata Energy, Inc., Ross Uranium Recovery Project; New Source Material License Application; Notice of Intent to Prepare a Supplemental Environmental Impact Statement." November 16, 2011.

Apache Tribe of Oklahoma. "RE: Ross Consultation." E-mail (August 19) from L. Guy to A. Bjornsen, Project Manager, Office of Federal and State Materials and Environmental Management Programs, U.S. Nuclear Regulatory Commission. Anadarko, OK: Apache Tribe of Oklahoma. 2011. Accession No. ML11336A224.

ASLB (Atomic Safety and Licensing Board). "Memorandum and Order (Ruling on Standing and Contention Admissibility)." In the Matter of Strata Energy, Inc. (Ross In Situ Recovery Uranium Project). Docket No. 40-9091-MLA. ASLBP No. 12–915–01–MLA–BD01. February 10, 2012. Accession No. ML12041A295.

BLM (Bureau of Land Management). "Newcastle Resource Management Plan." Newcastle, WY: BLM. 2000. ADAMS Accession No. ML12209A101.

Cheyenne River Sioux Tribe. E-mail (May 17) from S. Vance to A. Bjornsen, Project Manager, Office of Federal and State Materials and Environmental Management Programs, U.S. Nuclear Regulatory Commission. Eagle Butte, SD: Cheyenne River Sioux Tribe. 2011. ADAMS Accession No. ML11336A442.

(US)FWS (U.S. Fish and Wildlife Service). Cheyenne, Wyoming: USFWS. September 13, 2011. ADAMS Accession No. ML112770035.

Northern Cheyenne Tribe. "Re: Ross Letter Inviting Section 106 Consultation." E-mail (May 5) from C. Fisher to A. Bjornsen, Project Manager, Office of Federal and State Materials and Environmental Management Programs, U.S. Nuclear Regulatory Commission. Lame Deer, MT: Northern Cheyenne Tribe. 2011. ADAMS Accession No. ML11337A064.

(US)NRC (U.S. Nuclear Regulatory Commission). "Site Visit and Informal Information Gathering Meetings Summary Report for the Proposed Ross In Situ Recovery Project (Docket No. 040-09091)." Memorandum (November 28) to K. Hsueh, Branch Chief from A. Bjornsen, Project Manager, Office of Federal and State Materials and Environmental Management Programs. Washington, DC: NRC. 2011a. ADAMS Accession No. ML112980194.

(US)NRC. "Acceptance for Review of Materials License Application, Strata Energy, Inc., Ross In Situ Recovery Uranium Project, Crook County, Wyoming (TAC J00640)." Washington, DC: NRC. 2011b. ADAMS Accession No. ML111721948.

(US)NRC. "Initiation Of Section 106 Consultation For Strata Energy, Inc's Proposed Ross Uranium Recovery Project – License Request (Docket 040-09091)." Washington, DC: NRC. 2011c. ADAMS Accession No. ML112150393.

(US)NRC. "Invitation For Formal Section 106 Consultation Pursuant To The National Historic Preservation Act Regarding the Strata Energy, Inc. License Application for the Proposed Uranium In-Situ Recovery Facility, in Oshoto, Crook County, Wyoming to Cheyenne River Lakota Tribe (ML110040131), Blackfeet Tribe (ML110040076), Yankton Lakota (ML110040312), Spirit Lake Tribe (ML110040484), Apache Tribe (ML110310152) Oglala Sioux Tribe (ML110400125),Three Affiliated Tribes (ML110400293), Fort Peck Assiniboine and Sioux Tribe (ML110400344), Standing Rock Sioux Tribe (ML110400258), Eastern Shoshone Tribe (ML110400199), Cheyenne and Arapaho of OK Tribe (ML110400090), Flandreau Santee Lakota Tribe (ML110400285), Crow Creek Sioux Tribe (ML110400225), Northern Cheyenne (ML110400529), Santee Sioux Nation (ML110400181), Fort Belknap Community Tribe (ML110400311), Sisseton-Wahpeton Oyate (ML110400222), Northern Arapaho Tribe (ML110400508), Rosebud Sioux Tribe (ML110400154), Confederated Salish and Kootenai Tribe (ML110400176), Turtle Mountain Band of Chippewa Tribe (ML110400279), Crow Tribe (ML110400256), Kiowa (ML110400461), and Lower Brule Tribe (ML110400489)." Washington, DC: NRC. 2011d.

(US)NRC. "Notification of Proposed Facility – Strata Energy, Inc., Ross Uranium In Situ Recovery Facility, Oshoto, Crook County, Wyoming." Washington, DC: NRC. 2010. ADAMS Accession No. ML103160580.

(US)NRC. NUREG–1910. "Generic Environmental Impact Statement for In Situ Leach Uranium Milling Facilities—Final Report." Washington, DC: NRC. May 2009. ADAMS Accession Nos. ML091480244 and ML091480188.

(US)NRC. NUREG–1508. "Final Environmental Impact Statement to Construct and Operate the Crownpoint Uranium Solution Mine Project, Crownpoint, New Mexico." Washington, DC: NRC. February 1997. ADAMS Accession No. ML082170248.

(US)NRC. NUREG–0706. "Final Generic Environmental Impact Statement on Uranium Milling Project M-25." Washington, DC: NRC. September 1980. ADAMS Accession Nos. ML032751663, ML0732751667, and ML032751669.

NRDC and PRBDC. "Petition to Intervene and Request for Hearing by the Natural Resource Defense Council and Powder River Basin Resource Defense Council." Washington, DC: NRDC. 2011. ADAMS Accession No. ML11300A188.

Rosebud Sioux Tribe. "RE: Ross Uranium ISR Project." E-mail (April 15) from R. Bordeaux to A. Bjornsen, Project Manager, Office of Federal and State Materials and Environmental Management Programs, U.S. Nuclear Regulatory Commission. Rosebud Sioux Tribe. 2011. ADAMS Accession No. ML111220299.

Salish, Pend d'Oreille and Kootenai Tribes. "December 6th Letter." E-mail (December 29) from F. Auld to A. Bjornsen, Project Manager, Office of Federal and State Materials and Environmental Management Programs, U.S. Nuclear Regulatory Commission. Salish, Pend d'Oreille and Kootenai Tribes. 2011. ADAMS Accession No. ML120050552.

Standing Rock Sioux Tribe. E-mail (April 29) from W. Young to A. Bjornsen, Project Manager, Office of Federal and State Materials and Environmental Management Programs, U.S. Nuclear Regulatory Commission. Fort Yates, ND: Standing Rock Sioux Tribe. 2011. ADAMS Accession No. ML11337A071.

Strata "Ross ISR Project USNRC License Application, Crook County, Wyoming, Environmental Report, Volumes 1, 2 and 3 with Appendices." Docket No. 40-09091. Gillette, WY: Strata Energy, Inc. 2011a. ADAMS Accession Nos. ML110130342, ML110130344, and ML110130348.

Strata. "Ross ISR Project USNRC License Application, Crook County, Wyoming, Technical Report, Volumes 1 through 6 with Appendices." Docket No. 40-09091. Gillette, WY: Strata. 2011b. ADAMS Accession Nos. ML110130333, ML110130335, ML110130314, ML110130316, ML110130320, and ML110130327.

Turtle Mountain Band of Chippewa Indians. E-mail (April 14) from K. Ferris to A. Bjornsen, Project Manager, Office of Federal and State Materials and Environmental Management Programs, U.S. Nuclear Regulatory Commission. Belcourt, ND: Turtle Mountain Band of Chippewa Indians. 2011. ADAMS Accession No. ML111080059.

WWC Engineering. "Re: Request Update of ER Table 1.6-a." E-mail (February 1) from B. Schiffer to J. Moore, Project Manager, Office of Federal and State Materials and Environmental Management Programs, U.S. Nuclear Regulatory Commission. Sheridan, Wyoming: WWC Engineering. 2013. ADAMS Accession No. ML13035A012.

Wyoming Game and Fish Department (WGFD). Cheyenne, Wyoming: WGFD. 2011. ADAMS Accession No. ML112660130.

2 IN SITU URANIUM RECOVERY AND ALTERNATIVES

This section describes the Proposed Action, which is to issue a U.S. Nuclear Regulatory Commission (NRC) source and byproduct material license to Strata for the proposed Ross Project in northeastern Wyoming. Strata would use its NRC license in connection with the construction, operation, aquifer restoration, and decommissioning of the proposed Ross Project. This section also discusses alternatives to the proposed action, including the No-Action alternative as required under the National Environmental Policy Act of 1969 (NEPA).

> **What is source material?**
>
> "Source material" means either the element thorium or the element uranium, provided that the uranium has not been enriched with the radioisotope uranium-235.
>
> **What is byproduct material?**
>
> "Byproduct materials" are tailings or wastes generated by extraction or concentration of uranium or thorium processed ores, as defined under Section 11e.(2) of the Atomic Energy Act (AEA).

Figure 2.1 indicates the proposed location of the Ross Project. Section 2.1 of this Supplemental Environmental Impact Statement (SEIS) describes the Alternatives that are included for detailed analysis, including the Proposed Action; Section 2.2 describes those alternatives that were considered but eliminated from detailed analysis; Section 2.3 summarizes the potential environmental impacts of the Proposed Action and the two Alternatives; and Section 2.4 discusses the NRC staff's preliminary recommendation that the NRC issue a source and byproduct materials license for the Proposed Action unless safety issues mandate otherwise.

2.1 Alternatives Considered for Detailed Analysis

In addition to the Proposed Action, two alternatives to the Ross Project are also considered in this SEIS. All alternatives are evaluated with regard to the four phases of an uranium-recovery operation: construction, operation, aquifer restoration, and decommissioning. The range of alternatives has been established based on the purpose and need statement as described in Section 1.3 of this SEIS. In addition, this SEIS adopts many of the conclusions reached in the GEIS that was prepared for in situ recovery (ISR) projects (NRC, 2009).

Alternatives examined in this SEIS are:

- Alternative 1 is the Proposed Action, as described in the Applicant's license application. The Proposed Action is described in SEIS Section 2.1.1.
- Alternative 2 is the No-Action Alternative, as required by the *National Environmental Policy Act* (NEPA), where the Applicant would not construct, operate, restore the aquifer, or decommission the Ross Project. Alternative 2 is described in SEIS Section 2.1.2.
- Alternative 3 is the same as the Proposed Action, except that the Ross Project facility (i.e., the central processing plant [CPP], auxiliary and support buildings and structures, and the surface impoundments) would be situated at a different location to the north of the Proposed Action (i.e., at the "north site"). Alternative 3 is identified in this SEIS as the "North Ross Project" and is described in SEIS Section 2.1.3.

The sources of information used in the development of this SEIS include the following: the Applicant's license application, including its *Environmental Report* (ER) (Strata, 2011a) and its *Technical Report* (TR) (Strata, 2011b) as well as its Responses to Requests for Additional

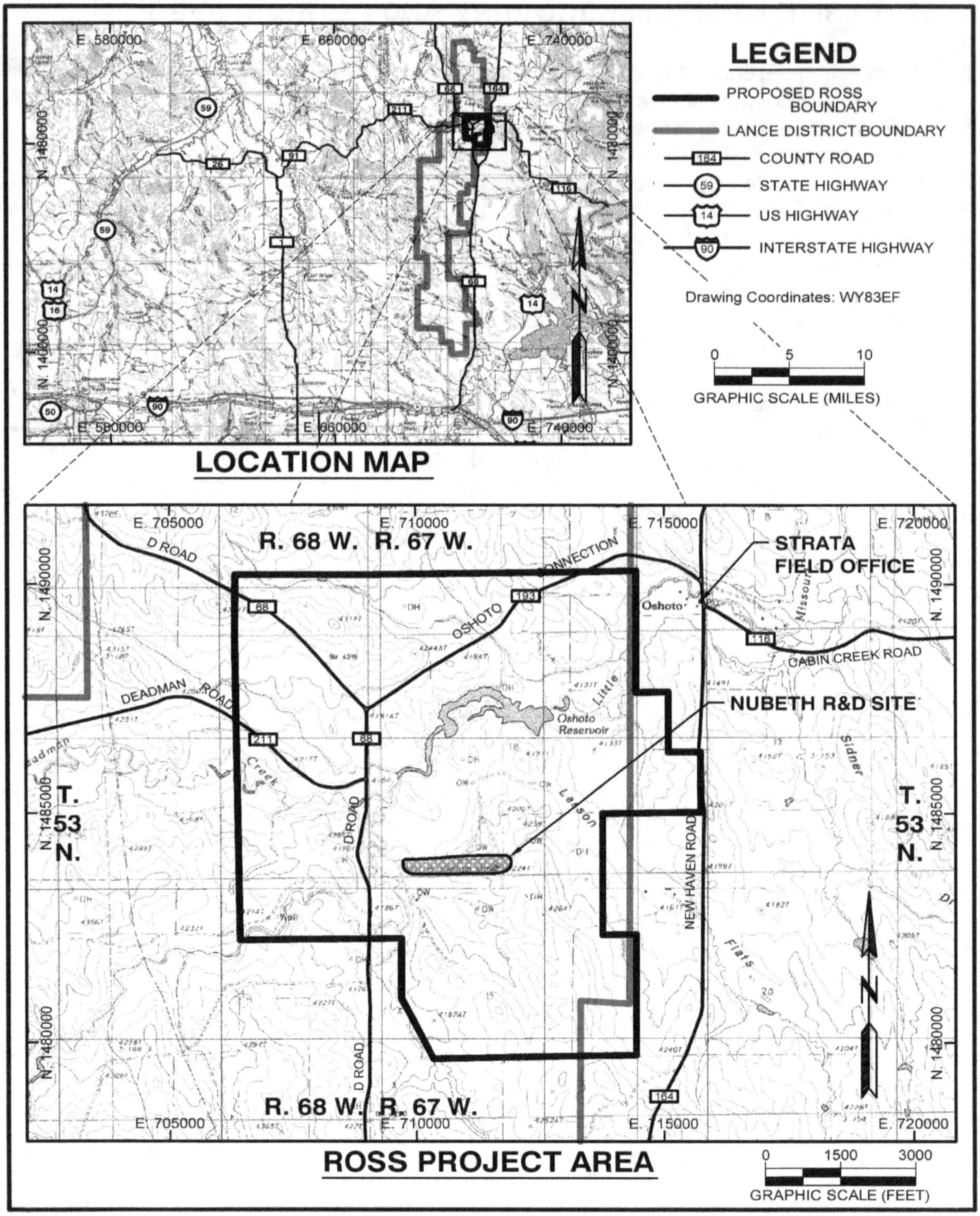

Source: Strata, 2012a.

Figure 2.1

Ross Project Within the Lance District

Information (RAIs) (Strata, 2012a; Strata, 2012b); the information and scoping comments gathered during the NRC staff's and NRC consultants' site visit in August 2011 (NRC, 2011); information independently researched by the NRC staff from publicly available sources; multidisciplinary discussions held among NRC staff and various stakeholders; and the Generic Environmental Impact Statement (GEIS) itself (NRC, 2009).

2.1.1 Alternative 1: Proposed Action

Under the Proposed Action, the NRC would issue the applicant a source material license. The Applicant would use its NRC license in connection with the construction, operation, aquifer restoration, and decommissioning of the ISR facility at the Ross Project area as described in its license application (Strata, 2011a; Strata, 2011b). Also, under the proposed action, the U.S. Bureau of Land Management (BLM) would approve the Applicant's Plan of Operations (POO). The Ross Project would occupy 697 ha [1,721 ac] in the north half of the approximately 90-km^2 [56-mi^2] Lance District, an area where the Applicant is actively exploring to determine whether there are additional uranium deposits. As Figure 2.2 shows, Strata has also identified four other uranium-bearing areas that would extend the area of uranium recovery to the north with the Ross Amendment Area 1 and to the south of the Lance District with the Kendrick, Richards, and Barber satellite facilities (Strata, 2012a).

The Lance District is located on the western edge in the northwest corner of the Nebraska-North Dakota-Wyoming Uranium Milling Region (NSDWUMR) (see Figure 2.3). It is situated between the Black Hills uplift to the east and the Powder River Basin to the west (Strata, 2011a). Both of these regional features are described in the GEIS (NRC, 2009). However, the Powder River Basin has been described as part of the Wyoming East Uranium Milling Region (WEUMR) and the Black Hills uplift as part of the NSDWUMR. The uranium ore zone at the Ross Project is situated in the upper Cretaceous Fox Hills and Lance Formations. Although these stratigraphic units are not specifically described in the GEIS, they share key attributes that are important for ISR with the uranium-hosting Wasatch Formation in the Powder River Basin described for the WEUMR and the Inyan Kara Group described for the NSDWUMR (NRC, 2009). These key attributes include alternating layers of sandstone, which allow hydraulic circulation, and shale, which prevent hydraulic circulation. The environment of the Proposed Action is described in Section 3 of this SEIS.

The Proposed Action includes the ISR facility itself and its wellfields (see Figures 2.4 and 2.5). The ISR facility consists of the following:

- A CPP that houses the uranium- and vanadium-processing equipment, drying and packaging equipment, and water-treatment equipment.
- A chemical storage area as well as other storage, warehouse, maintenance, and administration buildings.
- Two double-lined surface impoundments, a sediment impoundment, and five Class I deep-injection wells.

The schedule for the Proposed Action is shown in Figure 2.6. The Proposed Action includes the option of the Applicant's operating the Ross Project facility beyond the life of the Project's wellfields.

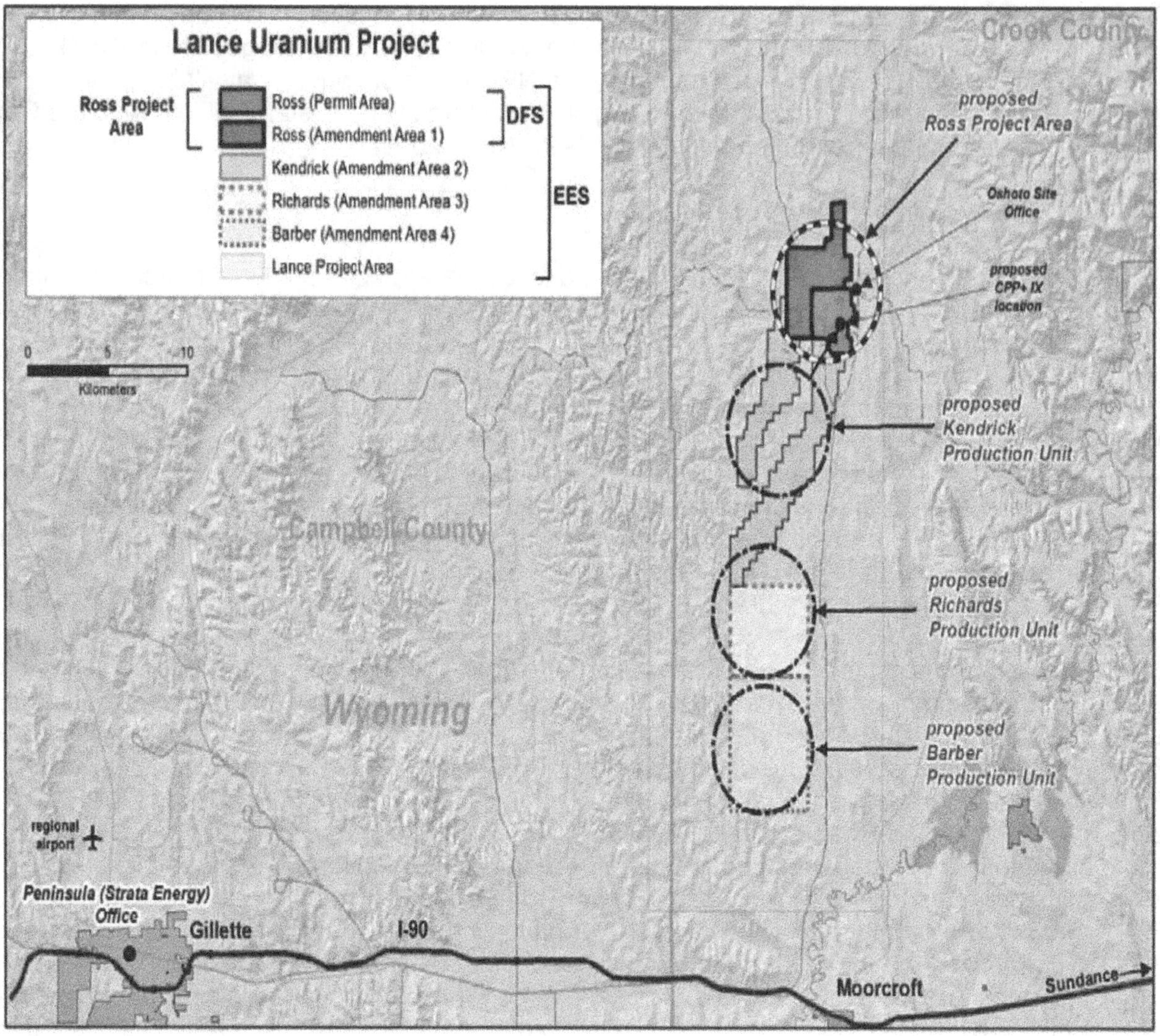

Source: Strata, 2012a.

Figure 2.2

Potential Satellite Areas in the Lance District

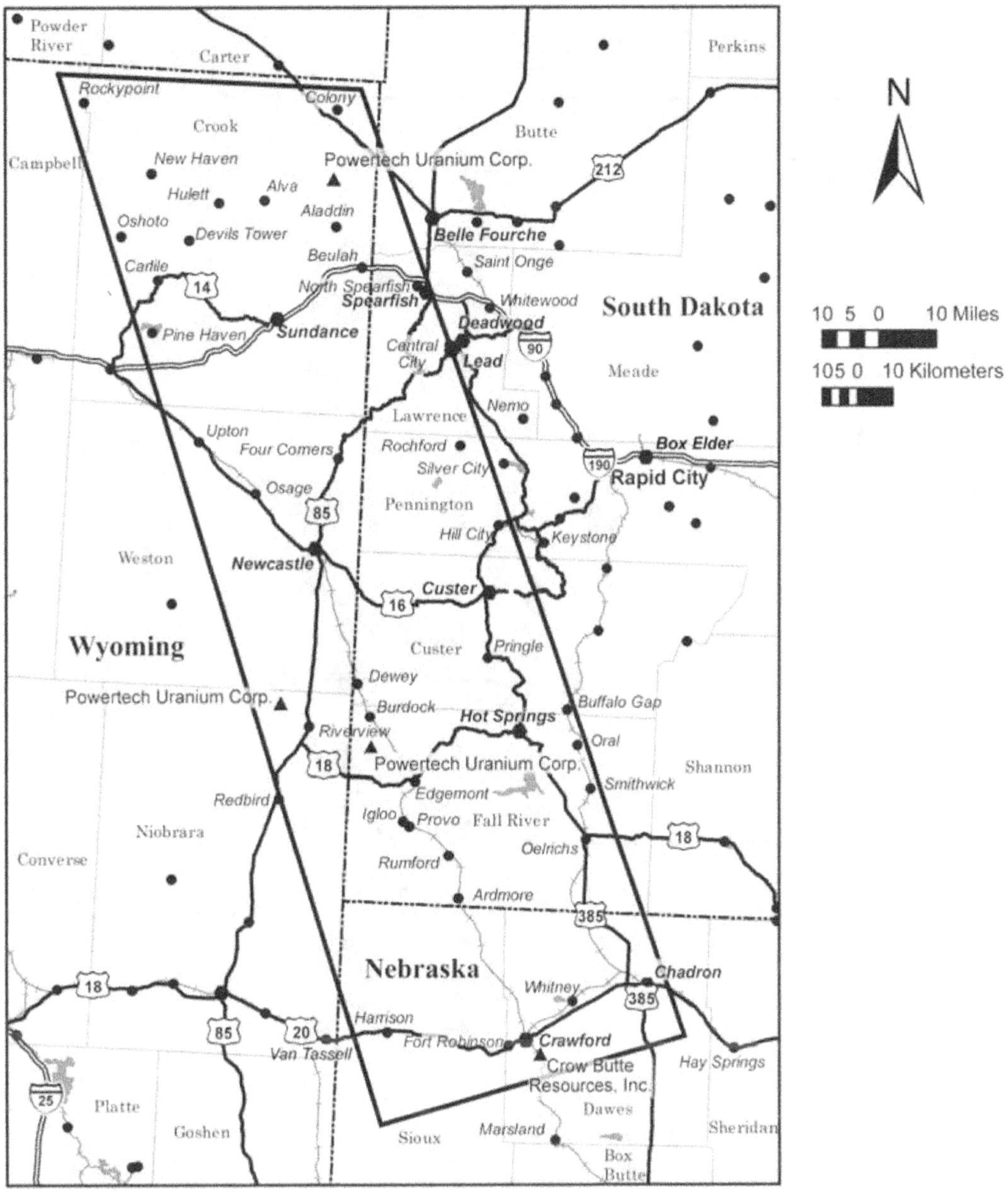

Source: NRC, 2009

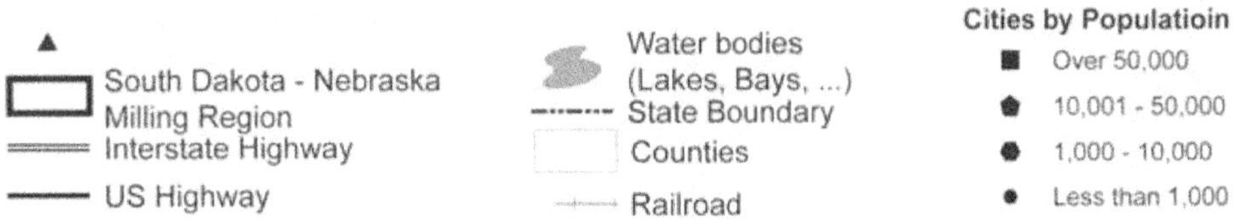

Figure 2.3

Nebraska-South Dakota-Wyoming Uranium Milling Region

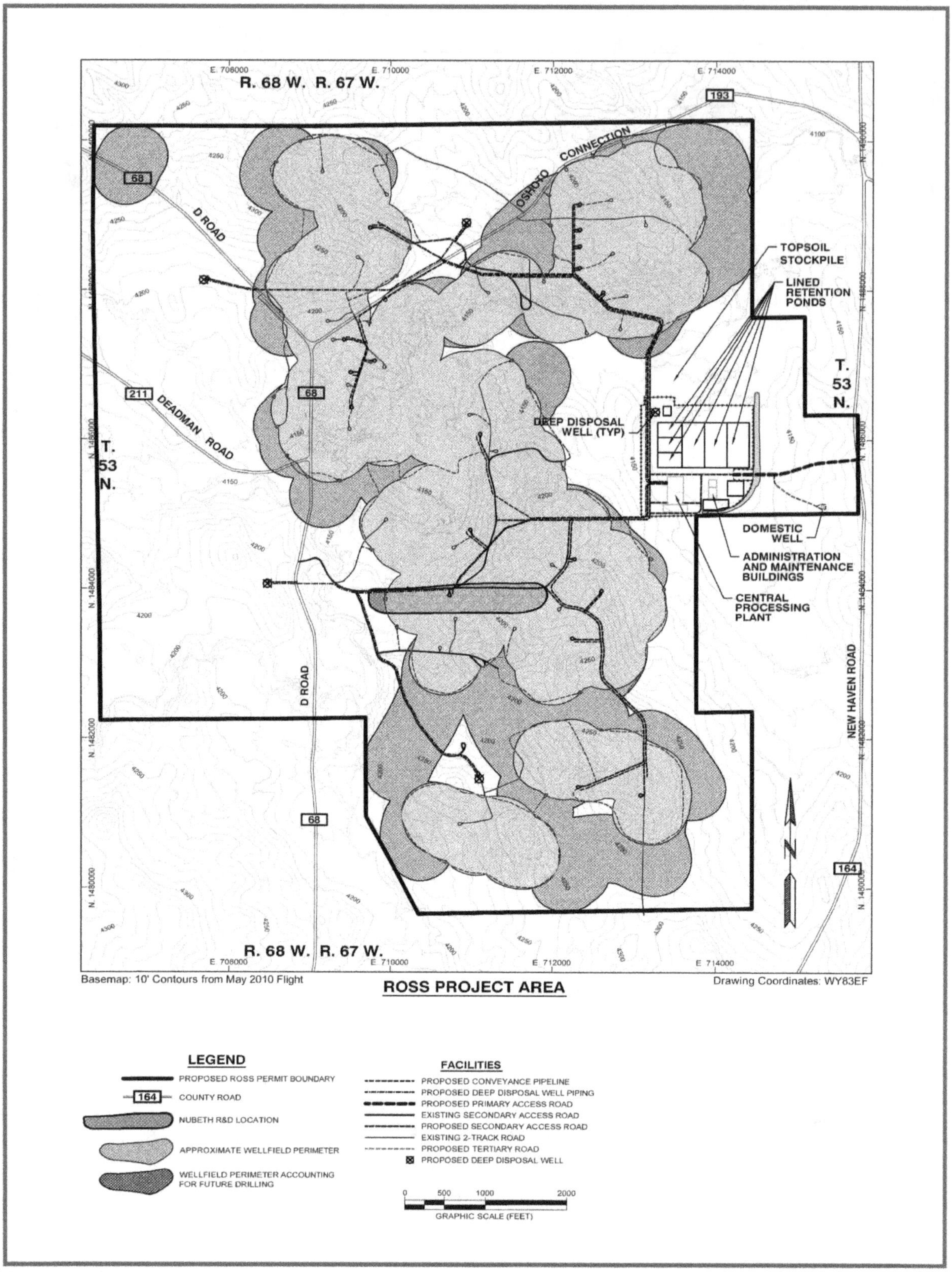

Source: Strata, 2011b.

Figure 2.4

Proposed Ross Project Facility and Wellfields

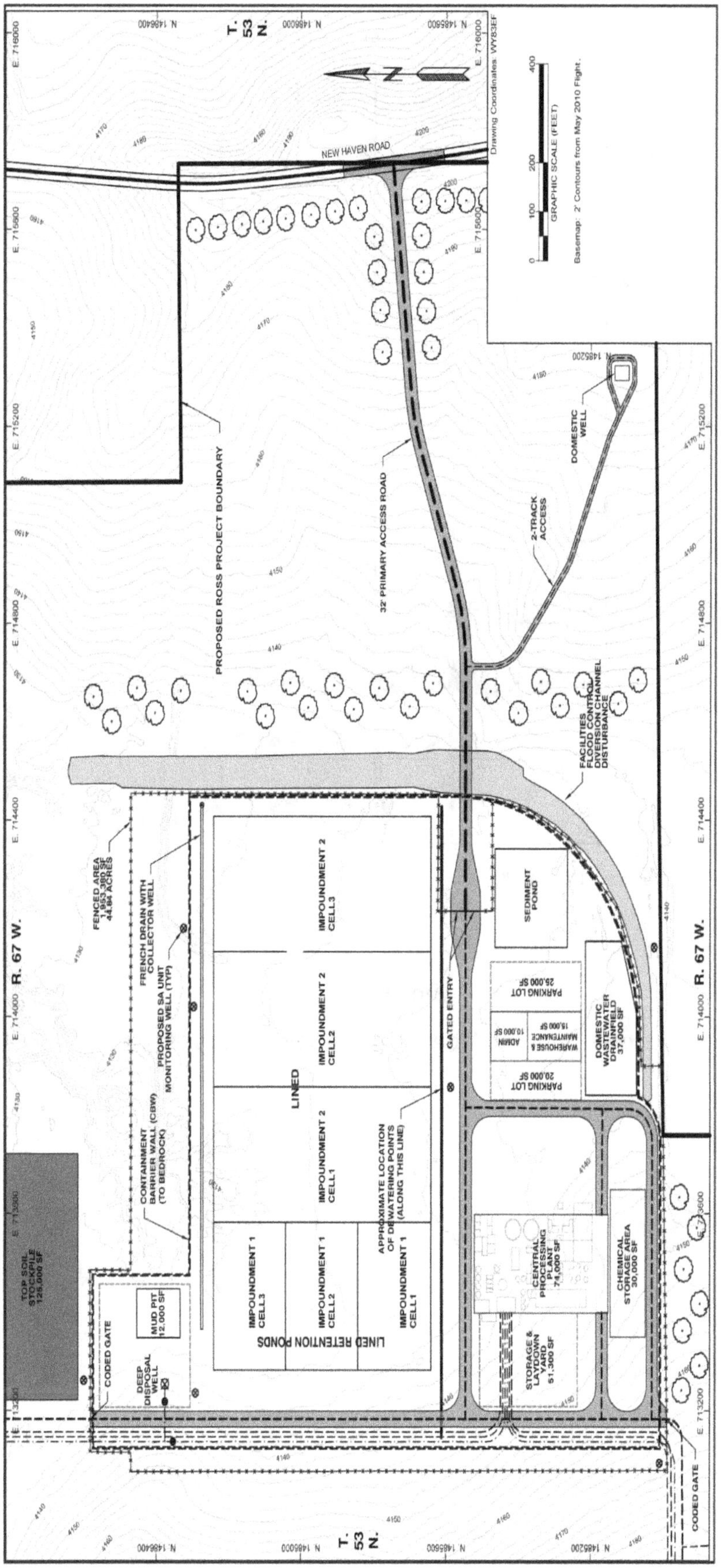

Source: Strata, 2012a.

Figure 2.5
General Layout of Proposed Ross Project Facility

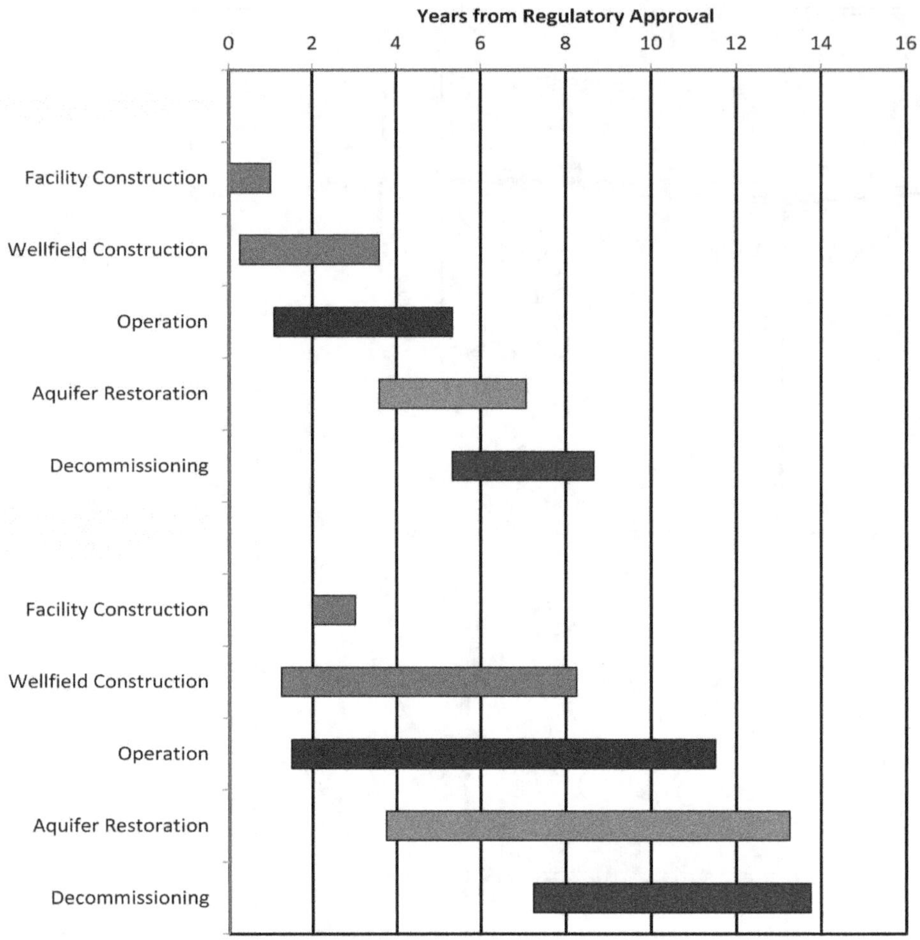

Source: Strata, 2012a.

Figure 2.6

Schedule for Potential Lance District Development

The facility could be used to process uranium-loaded resins from satellite projects within the Lance District operated by the Applicant, or from other offsite uranium-recovery projects not operated by the Applicant, or from offsite water-treatment operations. In this case, the life of the facility would be extended to 14 years or more (Strata, 2012a).

The Ross Project would host 15 – 25 wellfield areas and would consist of a total of 1,400 – 2,000 recovery and injection wells (Strata, 2011a). Groups of wells ("well modules") within a wellfield would be connected with piping to a central collection facility called a "module building," or a "header house." The wellfields would also be surrounded by a perimeter ring of monitoring wells.

This type of uranium extraction, in situ uranium recovery, consists of water to which chemicals have been added, referred to as "lixiviant," that is injected into the aquifer bearing the uranium ore (the "ore zone" or "ore body") (see Section 2.1.1.2). The chemicals in the lixiviant dissolve the uranium from the rock within the aquifer. Ground water containing dissolved uranium is then pumped from the ore-zone aquifer, processed through ion-exchange (IX) columns to remove the uranium from the lixiviant, and then the uranium is precipitated into a solid material called "yellowcake" (U_3O_8). Most of the water is then reused for uranium recovery.

> **What is lixiviant?**
> A solution composed of native ground water and chemicals added during the ISR operations. Lixiviant is then pumped underground to mobilize (dissolve) uranium from a uranium-bearing ore zone, or the ore body.

ISR is not hydraulic fracturing or "hydrofracking." Hydrofracking is a technique that is used by oil companies to increase the production of petroleum and natural gas by creating cracks in tight rocks containing oil and gas. A hydraulic fracture is formed by a fracturing fluid that is pumped into a well at a rate sufficient to increase pressure in the well, so that it exceeds the in situ pressure of the rock. The fracturing fluid is a slurry of water, chemicals to aid in cracking, and a proppant, a material such as sand grains or ceramic particulates that keep the fractures open when the injection is stopped and oil recovery occurs. In contrast, ISR operates at much lower pressure in the injection well. In situ pressures in ISR injection wells are only slightly above the in situ aquifer pressure. In addition, ISR is only used in aquifers with sufficient porosity and permeability to allow water flow from an injection well with a slightly positive pressure to the recovery well with a slightly negative pressure. This difference in pressure causes the ground water to move toward the recovery well. Finally, the chemicals in the water injected in ISR are for the purpose of dissolving the uranium, not to affect the porosity or permeability of the rock as are those during hydrofracking.

The Ross Project would be located in Crook County, Wyoming, 35 km [22 mi] north of the town of Moorcroft and Interstate-90 (see Figure 2.1). Other nearby towns and approximate direct distances to the Ross Project area include Pine Haven (27 km [17 mi] southeast), Gillette (53 km [33 mi] southwest), and Sundance (48 km [30 mi] southeast). The Ross Project area is adjacent to the unincorporated ranching community of Oshoto. The Oshoto community includes 11 residences within 3.2 km [2 mi] of the Proposed Action's boundary. Access to the Ross Project area is by either County Road (CR) 68 (D Road) or CR 164 (New Haven Road), both of which proceed north.

The Ross Project encompasses approximately 697 ha [1,721 ac] in portions of Sections 7, 17, 18, and 19, Township 53N, Range 67 West, and portions of Sections 12, 13, and 24, Township 53N, Range 68 West.

Table 2.1 Surface Ownership at Ross Project Area			
Surface Ownership	**Total Acres within Ross Project Area**	**Acres Disturbed During Year Preceding Operation**	**Acres Disturbed Over Life of Proposed Action**
U.S. Bureau of Land Management	40.0	1.3	1.3
State of Wyoming	314.1	40	80
Private	1,367.2	69	199
TOTAL	**1,721.3**	**110.3**	**280.3**

Source: Table 1.2-1 in Strata, 2011a.

Surface ownership within the Ross Project area is primarily private, with small tracts of land owned by the State of Wyoming and the BLM (Strata, 2011a). Approximately 16 ha [40 ac] are BLM land. The Wyoming Office of State Lands and Investments (WOSLI) administers 127 ha [314 ac]. In addition to the surface ownership, the BLM manages the subsurface mineral rights under 65 ha [160 ac] of privately owned land. Table 2.1 indicates the respective landowners of the Ross Project area. Current land uses are discussed in Section 3.2.

The Ross Project area is located in the upper reaches of the Little Missouri River, which flows northeasterly into southeastern Montana, through northwest South Dakota, and into North Dakota where it empties into the Missouri River at Lake Sakakawea. The area is characteristic of northwestern Wyoming: It is sparsely populated rangeland used primarily for grazing and some dry-land agricultural production. Oil development from the Minnelusa Formation in western Crook County began in the 1970s. There are three oil-recovery wells within the Ross Project area; oil production from these wells peaked in 1985 – 1986, but production has generally declined since then (Strata, 2011a).

As noted earlier, uranium targeted for production within the Ross Project is located in permeable sandstones of the Upper Cretaceous Lance and Fox Hills Formations. The uranium in the Oshoto area resides in roll-front deposits typical of those across the Powder River Basin as described in the WEUMR (NRC, 2009). Roll fronts are formed in sandstone formations when uranium-bearing ground water, moving down-gradient, encounters changing conditions. As the aquifer changes from oxygenated to oxygen-deficient, uranium precipitates as a coating on sand grains. The precise geometry of the uranium-ore deposits is controlled by the site-specific characteristics of the host sandstones. At the Ross Project area, the ore zones are generally thicker and more massive in the deeper Fox Hills compared to the deposits in the Lance Formation (Strata, 2011a).

Exploration of uranium deposits in the Lance Formation began in late 1970 (Strata, 2011a). The Nubeth Joint Venture (Nubeth), a joint venture between Nuclear Dynamics (later named ND Resources, Inc.) and Bethlehem Steel, received a License to Explore (No. 19) from the Wyoming Department of Environmental Quality's (WDEQ's) Land Quality Division (LQD) in

August 1976, with subsequent modifications to accommodate research and development activities in 1978 (Strata, 2011a). ND Resources, Inc. filed for an NRC source materials license in November 1977, and the license was approved in April 1978. Nubeth constructed a research and development operation in Section 18 of Township 53 North, Range 67 West, which is located within the Ross Project area (see Figure 2.1).

The research and development operation consisted of a single five-spot well pattern, with four injection wells and one recovery well, and a small facility with an IX, elution, and precipitation circuit capable of producing yellowcake slurry. The research and development facility could process 340 L/min [90 gal/min] of uranium-bearing lixiviant. Hydraulic control during the operation was accomplished with "buffer" wells, which were meant to form a hydraulic barrier to keep the lixiviant within the well pattern. Nubeth operated from August 1978 through April 1979 and recovered small amounts of uranium. No precipitation of a uranium product took place, and all of the recovered uranium was stored as a solution. After uranium-recovery tests were completed, the single five-spot used in the test was restored. Restoration was completed in February 1983 and Nubeth was notified by the WDEQ on April 25, 1983 that the restoration was satisfactory. Final approval for the research and development project's final operation decommissioning was granted by the NRC and WDEQ/LQD during the time period from 1983 through 1986 (Strata, 2011b).

Undesirable plugging of the aquifer, which was attributed to the build-up of fine particles, restricted injection rates and eventually led to the Nubeth operation's premature shutdown. A summary report on production feasibility estimated that uranium production could average about 360 kg/d [800 lb/d] in a facility sized to process 11,000 – 15,000 L/min [3,000 – 4,000 gal/min] (Strata, 2011a). However, due to the declining price of uranium at the time, commercial-scale licensing, construction, and operation did not occur. Two of Nubeth's wells (Well Nos. 789V and 19XX) have been used by oil companies since 1980 (Strata, 2011b); currently, the Merit Oil Company (Merit) is operating these two wells in addition to one more on the Ross Project area.

The Applicant notes that information obtained from the Nubeth research and development project was used in its decision to develop the Ross Project at the location described in this SEIS (Strata, 2011a). Nubeth's operation contributed the following information:

- Demonstration of the probability of an aquifer exemption of the mineralized zone
- Determination of strong geologic confinement above and below the identified ore body(ies)
- Confirmation of fundamental hydrogeologic hypotheses regarding ground-water flow and behavior
- Validation of information on potential regulatory and operational technical issues
- Determination of site geology, hydrology, soils, ecology, climate, and background radiological conditions
- Decrease of disturbance to both the surface and subsurface based on data collected in the past
- Demonstration of successful ground-water restoration and site reclamation

Peninsula Energy Ltd. (formerly Peninsula Minerals Ltd.) initiated acquisition of mineral rights in the Lance District in 2007 and 2008 (Peninsula, 2011). Exploration drilling programs, which were conducted in 2008 and 2009, confirmed significant uranium resources in the Ross Project area. Strata was incorporated in 2009; in 2010, Strata submitted applications for an NRC combined source and byproduct materials license, a Permit to Mine to WDEQ/LQD, and a POO to BLM. WDEQ/LQD approved Strata's Permit to Mine application in November 2012. The BLM is currently reviewing Strata's application, as is the NRC through the development of this SEIS and its SER. BLM is participating as a "cooperating agency" to the NRC under a Memorandum of Understanding (MOU) for the Ross Project.

In Section 2 of the GEIS, the four stages in the life of an ISR facility are described: 1) construction, 2) operation, 3) aquifer restoration, and 4) decommissioning (NRC, 2009). The decommissioning phase would include facility decontamination, dismantling, demolition, and disposal as well as site reclamation and restoration. Although NRC recognizes that these four phases could be performed concurrently, and in practice early wellfields would undergo aquifer restoration while other wellfields are being installed, the GEIS determined that describing the ISR process in terms of these stages aids in the discussion of the ISR process and in the evaluation of potential environmental impacts from an ISR facility.

2.1.1.1 Ross Project Construction

Construction of the Ross Project would be consistent with the general construction activities described in Section 2.3 of the GEIS (NRC, 2009). The Applicant discusses certain preconstruction activities that could be performed prior to its receiving a license from the NRC (Strata, 2011a); however, for the purposes of this evaluation of environmental and other impacts, this SEIS assumes that these preconstruction activities would occur at the same time as the Proposed Action such that the impacts of the preconstruction activities are considered as part of Alternative 1: Proposed Action. These preconstruction activities could include site excavation and preparation, such as clearing, grading, and constructing design components intended to control drainage and erosion as well as other mitigation measures; erection of fences and other access control measures that are not related to the safe use of, or security of, radiological materials; support-building construction; infrastructure construction, such as paved roads and parking lots, exterior utility and lighting systems, domestic-sewage facilities, and transmission lines; and other activities which have no measurable relationship to radiological health and safety nor common defense and security. In addition, the Applicant has indicated its intent to construct one Class I deep-injection well to better characterize the hydrologic and geochemical properties of the targeted geologic formation (i.e., ore zone) (Strata, 2011a). No radioactive materials would be present at the Ross Project during preconstruction activities.

After some or all of these activities, actual construction of the Proposed Action would begin and include: 1) the ISR facility that would consist of the CPP as well as administration, warehouse, and maintenance buildings, including storage and other structures, and lined surface impoundments; 2) wellfields including piping and module buildings; and 3) deep-disposal wells (see Figure.2.5) (Strata, 2011b; Strata, 2012b).

The Applicant anticipates construction of the facility and initial wells within one year of receiving an NRC license (see Figure 2.6). Main access roads would be constructed at the same time as the facility (Strata, 2011a). Secondary wellfield access roads would be constructed as necessary, as each wellfield is developed. It is estimated that the facility would encompass 21 ha [51 ac] (Strata, 2011b). A total of 44 ha [110 ac] would be disturbed by construction activities during the year preceding ISR facility operation and 113 ha [280 ac] over the life of the Proposed Action (see Table 2.1) (Strata, 2011a).

The Ross Project would employ approximately 200 people during construction. The Applicant anticipates that most employees would be from Crook and Campbell Counties (Strata, 2011a).

Further information on employment and other socioeconomic issues are described in Section 3.11.

Ross Project Facility

The Applicant proposes to construct and operate a single facility to serve the Ross Project as well as other potential ISR satellites (i.e., wellfields) within the Lance District. It could also process uranium-loaded resins from other ISR and water-treatment operations, which would be trucked into the facility (Strata, 2011a). The facility would include an administration building of 900 m^2 [10,000 ft^2], 1,400 m^2 [15,000 ft^2] of warehouse and maintenance space, 1,800 m^2 [20,000 ft^2] of parking, and a 3,400 m^2 [37,000 ft^2] for a domestic waste-water drainfield as well as the CPP mentioned earlier.

The proposed CPP would be a large, 6,900 m^2 [74,000 ft^2] pre-engineered metal building. The size of the CPP is about twice the size of a typical processing facility described in the GEIS (NRC, 2009). Adjoining the CPP would be 2,800 m^2 [30,000 ft^2] of chemical storage space and 4,800 m^2 [51,300 ft^2] of storage and work space (see Figure 2.5). The CPP would contain a control room housing the master-control system to allow remote monitoring and control of ISR process operations, wellfield operations, and deep-well disposal (Strata, 2011b). Operators in the CPP control room, who would be present 24 hours a day, would use a computer-based station to command the control system.

Proposed operations in the CPP would be generally consistent with typical processing involving three primary stages as described in the GEIS (NRC, 2009; Strata, 2011b):

- Uranium would be mobilized by the distribution of "barren" (containing no uranium) lixiviant from the CPP to injection wells and return of "pregnant" (containing dissolved uranium) lixiviant from the recovery wells to the CPP for processing.
- Dissolved uranium would be processed to yellowcake through a multi-step process involving IX resins, elution, precipitation, washing, drying, and packaging which would produce waste water.
- Waste water would be treated as necessary and then recirculated as lixiviant.

What is yellowcake?

Yellowcake is the product of the uranium-recovery and milling process; early production methods resulted in a bright yellow compound, hence the name "yellowcake." The material is a mixture of uranium oxides that can vary in proportion and in color from yellow to orange to dark green (blackish) depending on the temperature at which the material was dried (level of hydration and impurities). Higher drying temperatures produce a darker, less soluble material. Yellowcake is commonly referred to as U_3O_8 and is assayed as pounds U_3O_8 equivalent. This fine powder is packaged in 208-L [55-gal] drums and sent to a conversion plant that uses yellowcake to produce uranium hexafluoride (UF_6) as the next step in the manufacture of nuclear fuel.

This uranium-recovery process would be continued in a particular wellfield until the uranium concentration in the recovered solution becomes uneconomical.

The IX circuit proposed by the Applicant would be designed for a maximum of 28,400 L/min [7,500 gal/min] of pregnant lixiviant from Ross Project wells (Strata, 2011a). The elution, precipitation, and drying and packaging circuits would be designed to process approximately 1.4 million kg/yr [3 million lb/yr] of yellowcake (Strata, 2011b), which is about four times the capacity necessary to recovery uranium from the Ross Project. The excess capacity in the yellowcake production circuit would allow processing of loaded IX resins brought to the Ross Project from other ISR or water-treatment facilities. Except for the Smith Ranch-Highland operation that has a yellowcake capacity of 2.5 million kg/yr [5.5 million lb/yr], the capacity of the Ross Project exceeds the capacity of other facilities in Wyoming, which range from 0.2 million kg/yr [0.5 million lb/yr] to 0.9 million kg/yr [2 million lb/yr] (EIA, 2012).

The Applicant also proposes a vanadium-recovery circuit within the CPP to recover vanadium from uranium-depleted solutions (Strata, 2011b). The GEIS did not include vanadium recovery in its discussion of a typical uranium-recovery operation (vanadium recovery is discussed in Section 2.1.12 of this SEIS).

In addition to the uranium- and vanadium-recovery circuits, the CPP would house the water-treatment circuit for ground-water restoration. Water treatment would utilize an IX column to remove the uranium, followed by two reverse-osmosis (RO) units in series. The circuit would be designed for a maximum flow rate of 4,200 L/min [1,100 gal/min]. Operation of the first RO stage is expected to return approximately 70 percent of the flow as "permeate" (relatively clean water) and 30 percent of the flow as "brine" (water containing high concentrations of salts, which were mostly introduced to water to form the lixiviant, and contaminants, which were picked up during the lixiviant's residence time in the aquifer). When the remaining brine is run through the second RO stage, it would generate 50 percent permeate and 50 percent brine. Only 15 percent of waste water would be brine after the two-stage RO processing.

The ISR process requires chemical storage and feeding systems to introduce chemicals at various stages in the lixiviant extraction and processing as well as during the waste-treatment processes. Space for chemical storage would be built adjacent to the CPP (see Figure 2.5) (Strata, 2011b). The chemical-storage area would be constructed with secondary containment, which will consist of a concrete berm as part of the floor area that would be able to contain at least 110% of the volume of the largest tank (Strata, 2011b). The space would be divided into two areas, one inside the CPP and one outside. Chemicals stored outside would include oxygen, ammonia, and carbon dioxide. Chemicals stored inside would include some or all of the following: sulfuric acid, hydrochloric acid, sodium hydroxide, hydrogen peroxide, sodium chloride, sodium carbonate, and barium chloride.

The proposed location for the facility is currently on a relatively flat, currently used, dry-land hayfield. To route surface storm-water runoff around the facility, a diversion structure consisting of a berm, concrete-box culvert, and drainage channel would be constructed east of the proposed ISR facility. This system would be designed to manage runoff from a 100-year, 24-hour runoff event (Strata, 2011b; Strata, 2012b).

The Applicant's design calls for paving the areas adjacent to the CPP. Paved areas would be sloped to direct runoff water to slot drains. From the slot drains, storm water would be conveyed through pipes to a smaller, sediment-settling surface impoundment also designed to contain the runoff from a 100-year, 24-hour runoff event. The sediment impoundment would be constructed with the same double-liner and leak-detection configurations as the larger surface impoundments that would be used to store permeate and brine. After a significant storm event, water in the sediment impoundment would be immediately routed to the deep-disposal well (Strata, 2011b).

The facility is proposed to be located in an area of shallow ground water (Strata, 2012b). Shallow ground water directly beneath the facility could present construction and operational issues and create a higher risk of ground-water contamination in the event of a spill. To mitigate these concerns, the Applicant's proposed facility design would include a containment barrier wall (CBW). The CBW and associated dewatering system would be designed to prevent contaminated liquids from entering and contaminating shallow ground water outside of the facility, in the event of a process solution spill, hazardous-chemical spill, or a disposal-system failure. The CBW would restrict the flow of ground water from traveling beneath the facility and any water that seeps or flows into the area would be drained away. The design calls for the CBW to be constructed around approximately two-thirds of the facility's boundary along the north, east, and south. The CBW would be 0.7 m [2 ft] wide and extend from the ground surface to a minimum of 0.7 m [2 ft] into bedrock. It would be constructed of a soil-bentonite mixture. The configuration of the CBW is shown in Figure 2.5 and is described in Addendum 3.1-A of the TR (Strata, 2012b). Three French drains (i.e., trenches filled with very porous material, such as gravel) would be installed to drain the area within the CBW, when needed (Strata, 2011b; Strata, 2012b). The Applicant proposes approximately eight wells to monitor water levels and water quality inside and outside the CBW (Strata, 2012b). Any seepage and/or spillage collected on the facility side of the CBW would be discharged to the surface impoundments for storage or disposal with excess permeate and brine (Strata, 2011b). Construction of a CBW to mitigate impacts to shallow ground water beneath impoundments is not included in the GEIS's description of a typical ISR facility design (NRC, 2009).

The Proposed Action would also include the construction of two double-lined surface impoundments (retention ponds) over a 6.5 ha [16 ac] area; these impoundments would be used for process-solution and waste-water management (Strata, 2011b). Each surface impoundment would include three cells, built with common containment berms. At full capacity the impoundments' surface area would be about 5.3 ha [13.2 ac]. Interconnected pipes between the cells would allow the controlled transfer of solutions or water between cells. The impoundments would have double geomembrane liners and a leak-detection system. The design for the impoundment, including the liners, leak-detection systems, freeboard requirements, and reserve capacity are in accordance with the GEIS, but the size of the impoundments is about twice the upper range of typical surface impoundment sizes described in the GEIS (NRC, 2009).

The surface impoundments would be designed to meet the requirements of NRC Regulatory Guide 3.11 (NRC, 1980a), all conditions established by the NRC in the Applicant's license, and all requirements found in *Wyoming Water Quality Rules and Regulations*, Chapter 11, for lined waste-water surface impoundments (Strata, 2011b; Strata, 2012b; WDEQ/WQD, 1984).

The Applicant's surface-impoundment design calls for rectangular cells with maximum internal slopes of 3 horizontal to 1 vertical (Strata, 2011b; Strata, 2012b). The impoundments would be 4.6 m [15 ft] deep with 1 m [3 ft] of freeboard and a maximum hydraulic depth of 3.6 m [12 ft]. The primary liner would be impermeable high-density polyethylene (HDPE) or polypropylene, with a minimum thickness of 36 mils (0.9 mm [0.036 in]). The secondary liner would be a geosynthetic material with a minimum thickness of 36 mils (0.9 mm [0.036 in]) or native clay. The leak-detection system would be installed between the primary and secondary liners. The system would consist of a permeable drainage layer such as sand and perforated collection pipes.

The primary purpose of the surface impoundments would be to manage liquid, byproduct material (i.e., the permeate and brine described above) to optimize disposal techniques, and to provide capacity for liquid-waste storage in the event of "upset," or accident, conditions. In addition, the impoundments would provide some evaporation of stored brine. Under normal operating conditions, the water levels in the surface-impoundment cells would be maintained such that the volume of liquid in any one cell can be transferred to one of the other two cells to facilitate leak repair.

Ross Project Wellfields

Wellfields are the areas over the ore zone(s) where the injection and recovery wells for uranium recovery would be located. The proposed wellfields of the Ross Project are expected to encompass approximately 36.4 ha [90 ac] in portions of Sections 7, 17, 18, and 19, in Township 53N, Range 67W and in portions of Sections 12 and 13 in Township 53N, Range 68W. The Applicant notes that the final areal extent of the constructed wellfields is expected to be greater as additional ore-zone delineation occurs (Strata, 2011b).

The proposed wellfields would be divided into two units (Strata, 2011b). Each unit would be further divided into 15 to 20 modules with approximately 40 recovery wells per wellfield module (Strata, 2011b). The flow capacity of each wellfield module would range from 2,300 L/min [600 gal/min] to 3,800 L/min [1,000 gal/min]. The wellfields would be fenced to exclude livestock, wildlife, and other intruders.

Wells would be constructed to recover uranium from ore deposits found in permeable sand zones in stacked roll fronts and tabular ore zones described as "stratabound" deposits in the GEIS (NRC, 2009). The geology of the ore zone at the Ross Project area is described in SEIS Section 3.4.1. The average depth to the top of the ore zone ranges from less than 91 m [300 ft] to more than 213 m [700 ft] with an average depth of 149 m [490 ft] (Strata, 2011b). The ore-zone thickness averages 2.7 m [8.9 ft]. The sand units hosting uranium are saturated with ground water and are confined aquifers (Strata, 2011b). The hydrogeology of this area is described in SEIS Section 3.5.3.

The features and design of the wellfields proposed by the Applicant are generally consistent with the wellfields described in the GEIS (NRC, 2009). The primary components of a wellfield module are illustrated in Figure 2.7; these are:

- Injection wells to introduce lixiviant into the ore zone.
- Production (or recovery) wells to recover the uranium-enriched (or pregnant) lixiviant for subsequent processing at the CPP.
- Module buildings (or header houses) to manage the pipes (or "flow lines") that route the lixiviant between the injection and recovery wells within a module and the "feeder lines" that carry fluids between the module building to a manhole containing a valve.
- Valve manholes to manage the pipes to the module buildings, to the CPP, and to other value manholes (or "trunk lines").
- Perimeter monitoring wells to detect excursions of lixiviant outside the exempted portion of the aquifer from which uranium is recovered, should they occur.

The Applicant proposes three well-construction methods that would each comply with WDEQ/LQD requirements (see Figures 2.8, 2.9, and 2.10) (Strata, 2011b).

1

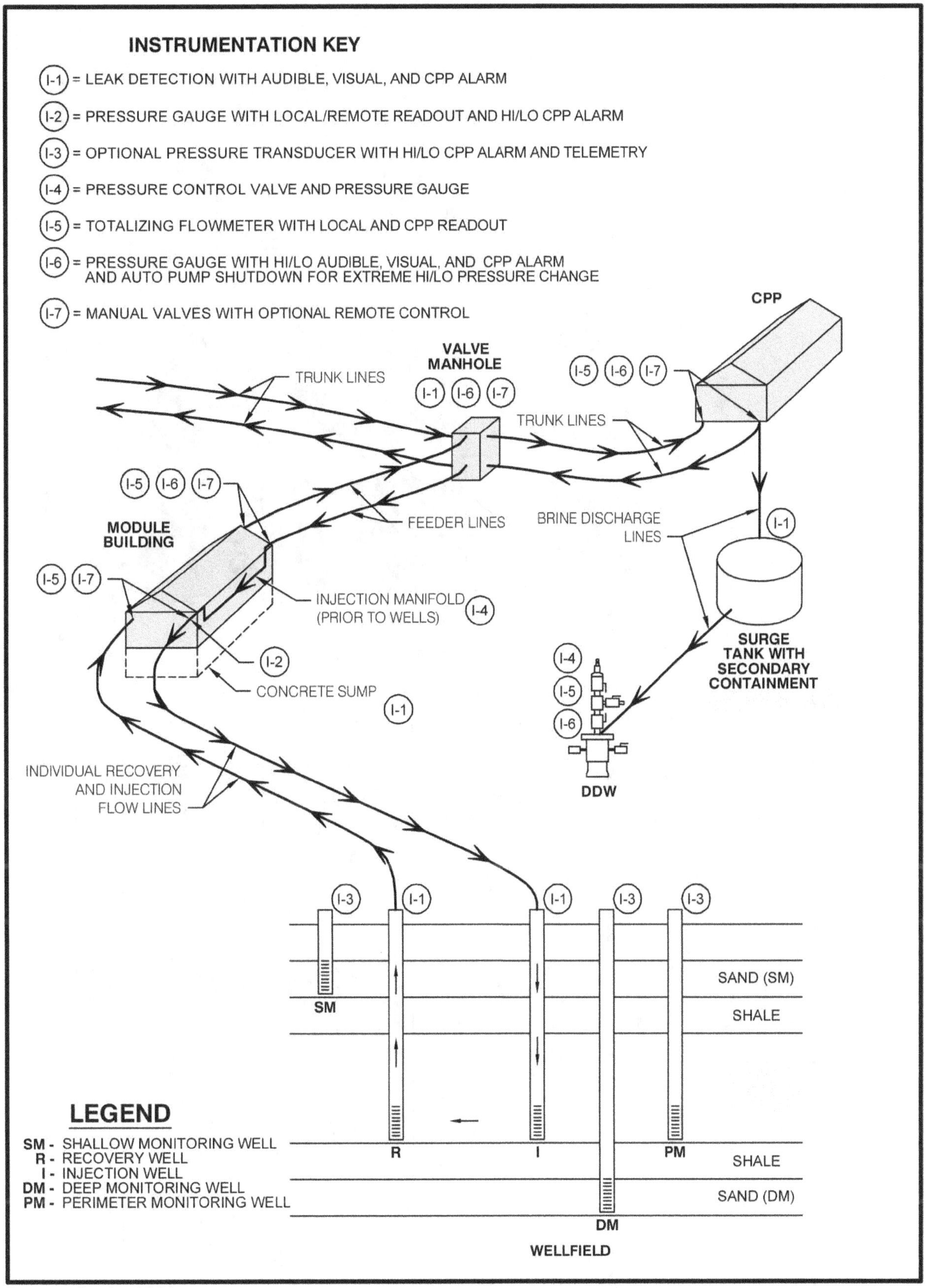

Source: Strata, 2011a.

Figure 2.7

Primary Components of a Ross Project Wellfield Module

Step	Description of Activity
1	A pilot hole 5 to 6.5 inches in diameter is drilled through the projected mineralization zone. Geophysical logs consisting of gamma, resistivity, spontaneous potential, and deviation are then completed. From the geophysical logs, the grade of each mineralized intercept is calculated.
2	If, after geophysical logging, it is determined that the mineralization is not of sufficient quality or that the ore continuity is inadequate to warrant completion, the hole is sealed from the bottom to the top with neat cement slurry. An Abandonment Record is then completed for each sealed hole.
3	Assuming the decision is reached to complete the well, the hole is reamed to a diameter of 8 to 10 inches (a minimum of 3 inches larger than the casing OD) to a depth approximately 15 feet below the bottom of the mineralization. Alternatively, in areas where the geologist is more confident in intercepting mineralization, the initial hole may be drilled at the final diameter of 8 to 10 inches in one pass followed by the geophysical logging. Fiberglass or PVC casing (minimum rating of SDR 17) with an outside diameter (OD) of 5 to 6.5 inches is placed in the reamed hole to a depth approximately 10 feet below the mineralization. PVC centralizers are placed on the casing string at a maximum spacing of one per 40 feet.
4	A calculated amount of neat cement slurry mixed to the required specifications (approximate unit weight of 15 lbs/gallon) is placed inside the casing through a cementing or pump-down head. A calculated volume of displacement water is then pumped into the casing forcing the cement slurry out the bottom of the casing and up the annulus between the casing and the reamed hole until cement reaches the surface. After displacement, the valve on the cementing head is closed which holds the cement in place while hardening occurs.
5	After a minimum of four days, the well is underreamed through the mineralized zones to a diameter of 10 to 14 inches. The well annulus will be topped off with cement to the surface prior to reentry by the drilling rig. The underreaming is completed by a specialized tool utilizing retractable blades. The blades are closed for the trip down the well and are opened by pressure from the rig mud pump. The blades are held open by the weight of the drill string. After underreaming the designated zone through the casing and cement, the blades are again retracted for the trip out of the well. The well may be caliper logged as necessary to verify the correct interval has been opened. If deemed necessary, to support sand zones that are not competent, PVC screen is telescoped into the casing using a J-collar hooked to the drill pipe. The uppermost screen openings will be placed below the top of the underreamed interval and below the bottom of the annular seal. A PVC riser pipe is extended from the top of the screen approximately 10 feet. One or more k-packer(s) will provide a seal between the riser pipe and the casing. Filter sand may be placed between the screen and the underreamed hole.
6	The well is developed to remove contaminants and fines from the drilling and completion process and maximize the flow rate. A Well Installation Record is completed which contains all the details on drilling, geophysical logging, completion materials, casing depth, completion interval, and the cement mix.
7	After drying, the drill cuttings contained in the pits are covered with subsoil and the stockpiled topsoil. The ground surface is then recontoured and reseeded.
8	The well is integrity tested as discussed in Section 3.1.2.3 below.

Source: Strata, 2012a.

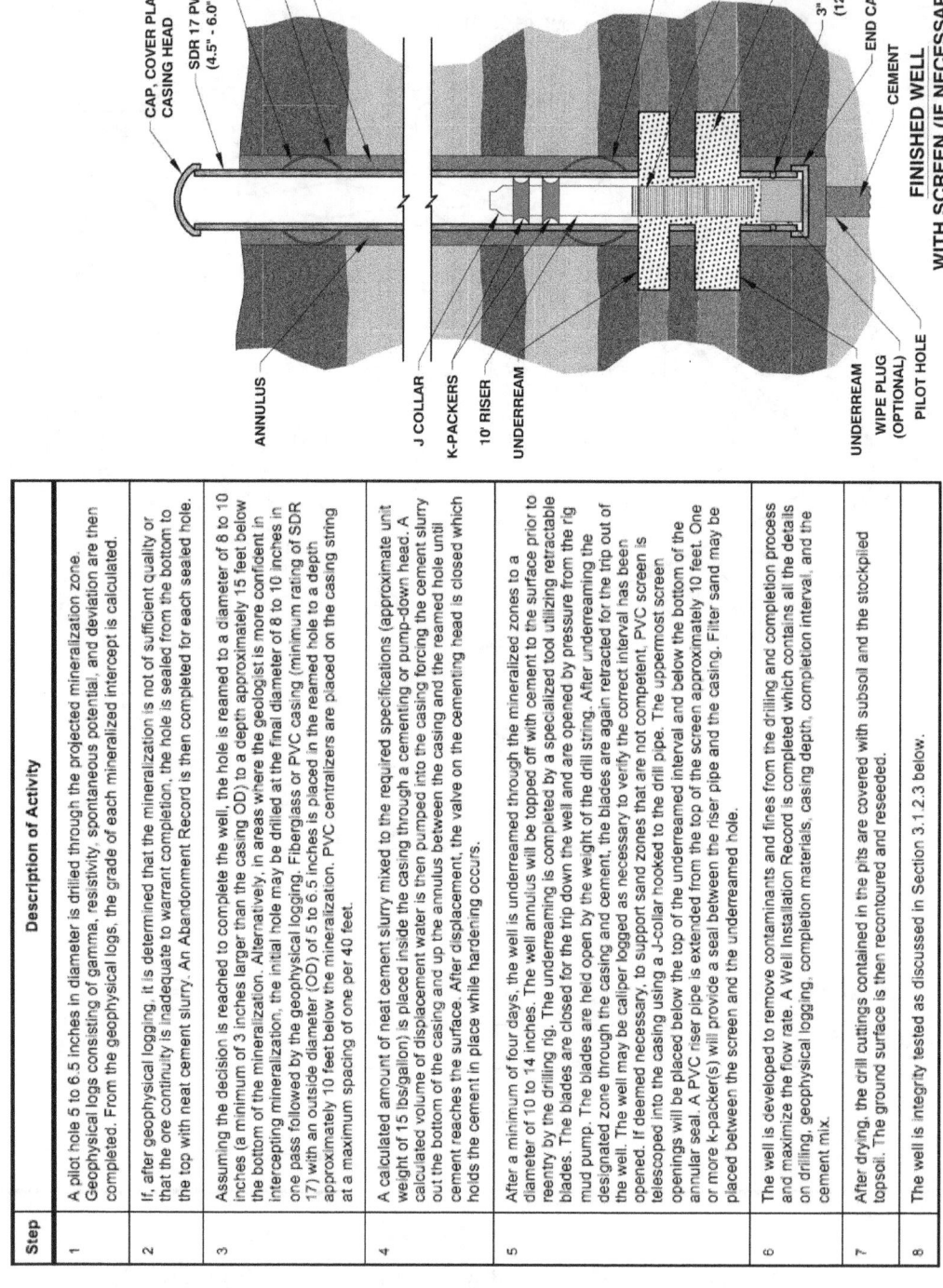

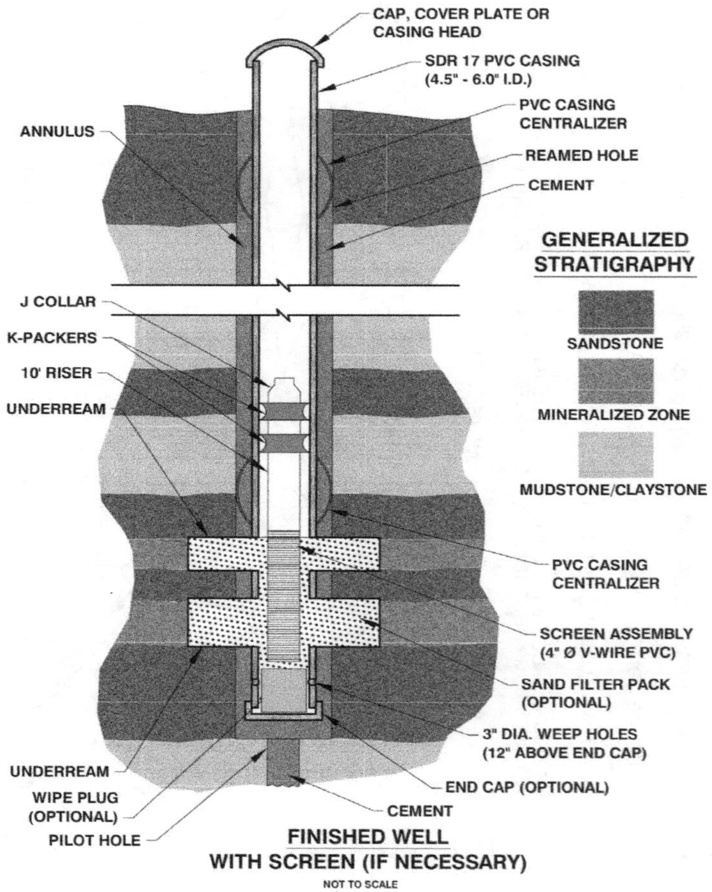

Figure 2.8

Proposed Well-Installation Method 1 for Ross Project Injection and Recovery Wells

Step	Description of Activity
1	A pilot hole 5 to 6.5 inches in diameter is drilled to the top of the projected completion interval. Geophysical logs consisting of a minimum of gamma, resistivity, and self potential are then completed.
2	The hole is reamed to a diameter of 8 to 10 inches (a minimum of 3 inches larger than the casing OD). An option for this method is to drill to the final hole diameter of 8 to 10 inches in one pass followed by the geophysical logging.
3	Fiberglass or PVC casing (minimum rating of SDR 17) with an OD of 5 to 6.5 inches is placed in the reamed hole. PVC centralizers are placed on the casing string at a maximum spacing of one per 40 feet.
4	A calculated amount of neat cement slurry mixed to the required specifications (approximate unit weight of 15 lbs/gallon) is placed inside the casing through a cementing head. A calculated volume of displacement water is then pumped into the casing forcing the cement slurry out the bottom of the casing and up the annulus between the casing and the reamed hole until cement reaches the surface. After displacement, the valve on the cementing head is closed which holds the cement in place while hardening occurs.
5	After a cement-hardening period of at least two days, the designated completion interval is drilled below the casing with a bit that is smaller than the casing inside diameter (ID). The well annulus will be topped off with cement to the surface prior to reentry by the drilling rig. Geophysical logs consisting of gamma, resistivity, spontaneous potential, and deviation are then completed in the newly drilled hole. If the sand zone is competent, the completed interval may be left open and unsupported. If PVC screen is necessary, the completion interval may be underreamed to a larger diameter prior to the installation of the screen. The uppermost screen openings will be placed below the bottom of the casing and the annular seal. A PVC riser pipe is extended from the top of the screen approximately 10 feet. A seal between the riser pipe and the casing is provided by one or more k-packer(s). Filter sand may be placed between the screen and the underreamed hole.
6	The well is developed to remove contaminants and fines from the drilling and completion process and maximize the flow rate. A Well Installation Record is completed which contains all the details on drilling, geophysical logging, completion materials, casing depth, completion interval, and the cement mix.
7	After drying, the drill cuttings contained in the pits are covered with subsoil and the stockpiled topsoil. The ground surface is then recontoured and reseeded.
8	The well is integrity tested as discussed in Section 3.1.2.3 below.

Source: Strata, 2012a.

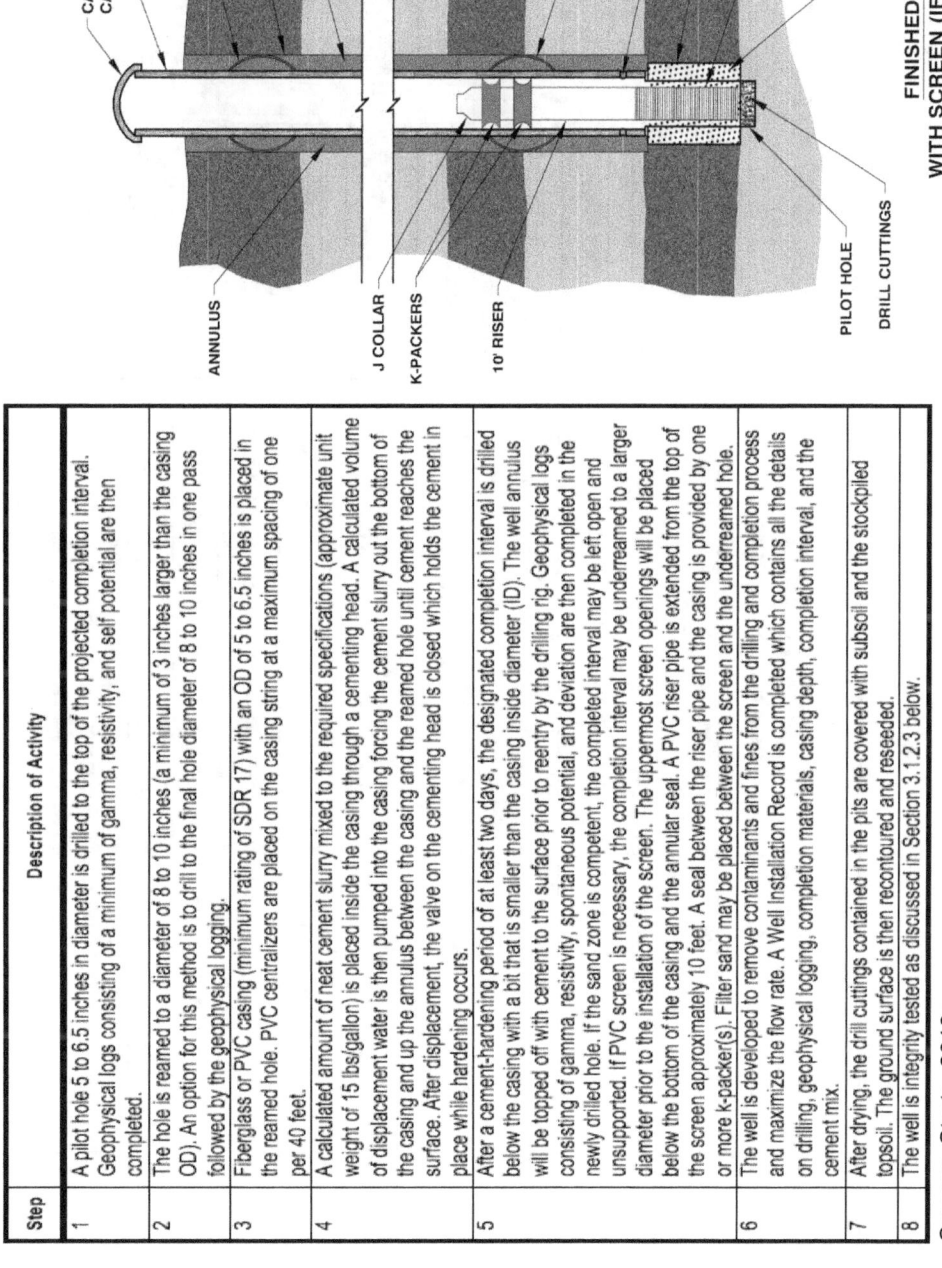

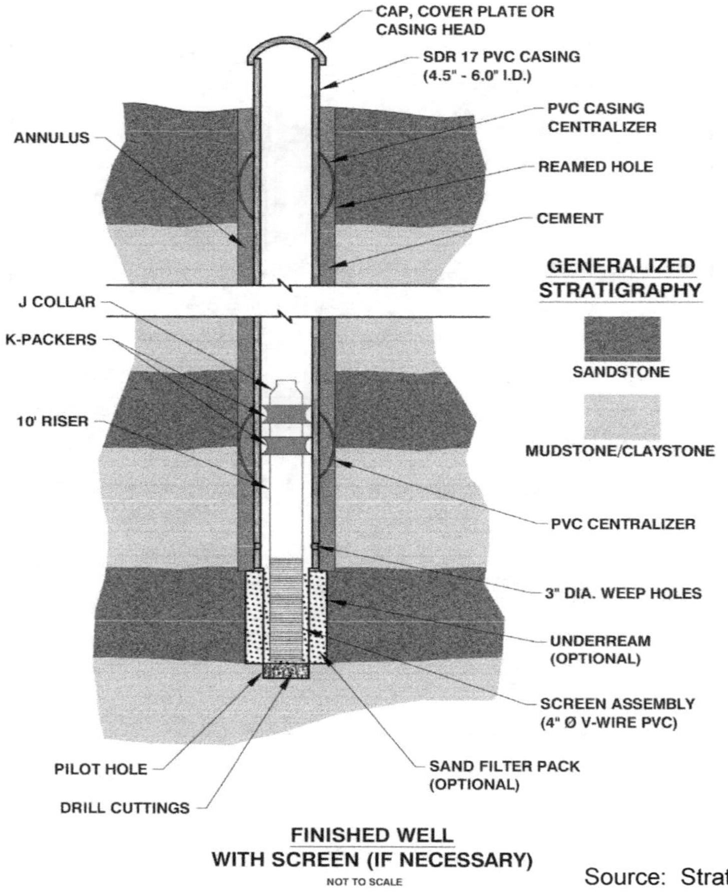

Source: Strata, 2011b

Figure 2.9

Proposed Well-Installation Method 2 for Ross Project Monitoring Wells

Step	Description of Activity
1	A pilot hole 5 to 6.5 inches in diameter is drilled to the top of the projected completion interval. Geophysical logs consisting of a minimum of gamma, resistivity, and self potential are then completed.
2	The hole is reamed to a diameter of 8 to 10 inches (a minimum of 3 inches larger than the casing OD). An option for this method is to drill to the final hole diameter of 8 to 10 inches in one pass followed by the geophysical logging.
3	Fiberglass or PVC casing (minimum rating of SDR 17) with an OD of 5 to 6.5 inches is placed in the reamed hole. PVC centralizers are placed on the casing string at a maximum spacing of one per 40 feet.
4	A calculated amount of neat cement slurry mixed to the required specifications (approximate unit weight of 15 lbs/gallon) is placed inside the casing through a cementing head. A calculated volume of displacement water is then pumped into the casing forcing the cement slurry out the bottom of the casing and up the annulus between the casing and the reamed hole until cement reaches the surface. After displacement, the valve on the cementing head is closed which holds the cement in place while hardening occurs.
5	After a cement-hardening period of at least two days, the designated completion interval is drilled below the casing with a bit that is smaller than the casing inside diameter (ID). The well annulus will be topped off with cement to the surface prior to reentry by the drilling rig. Geophysical logs consisting of gamma, resistivity, spontaneous potential, and deviation are then completed in the newly drilled hole. If the sand zone is competent, the completed interval may be left open and unsupported. If PVC screen is necessary, the completion interval may be underreamed to a larger diameter prior to the installation of the screen. The uppermost screen openings will be placed below the bottom of the casing and the annular seal. A PVC riser pipe is extended from the top of the screen approximately 10 feet. A seal between the riser pipe and the casing is provided by one or more k-packer(s). Filter sand may be placed between the screen and the underreamed hole.
6	The well is developed to remove contaminants and fines from the drilling and completion process and maximize the flow rate. A Well Installation Record is completed which contains all the details on drilling, geophysical logging, completion materials, casing depth, completion interval, and the cement mix.
7	After drying, the drill cuttings contained in the pits are covered with subsoil and the stockpiled topsoil. The ground surface is then recontoured and reseeded.
8	The well is integrity tested as discussed in Section 3.1.2.3 below.

Source: Strata, 2012a.

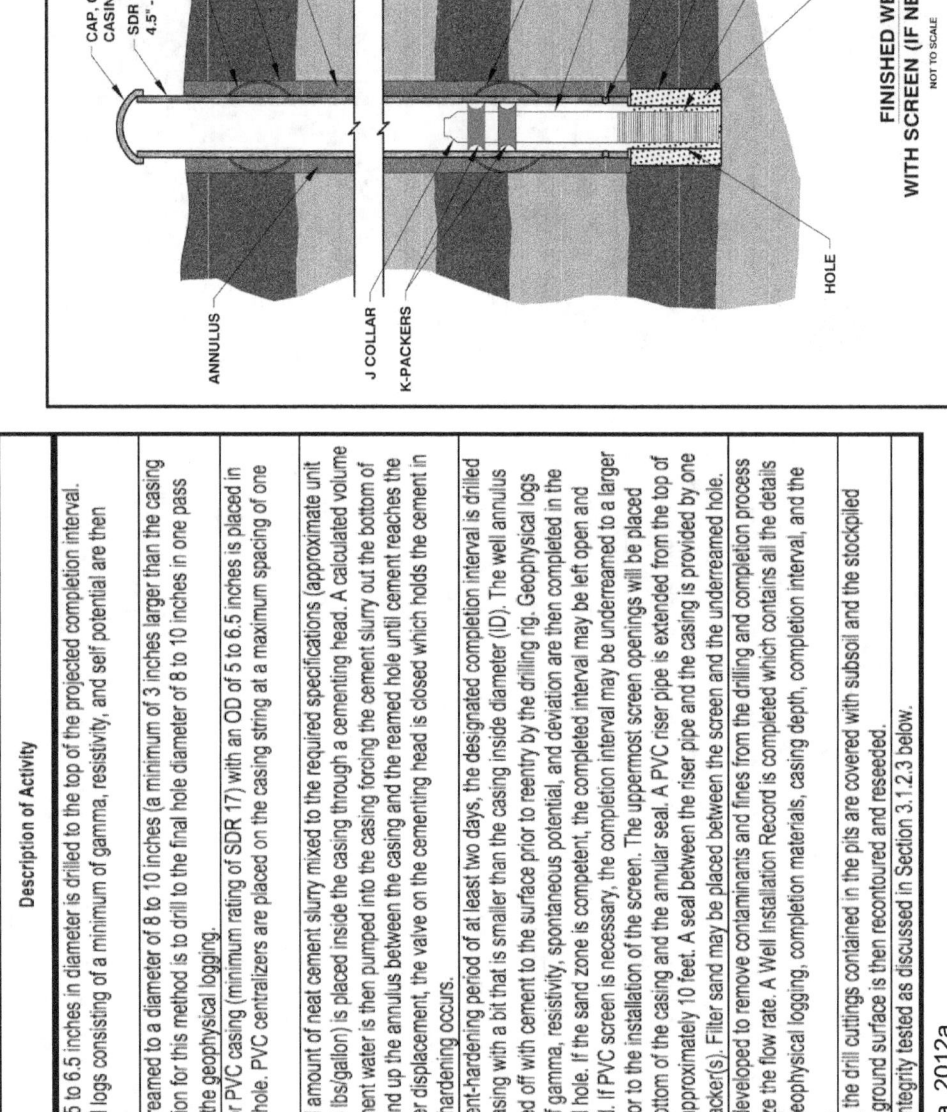

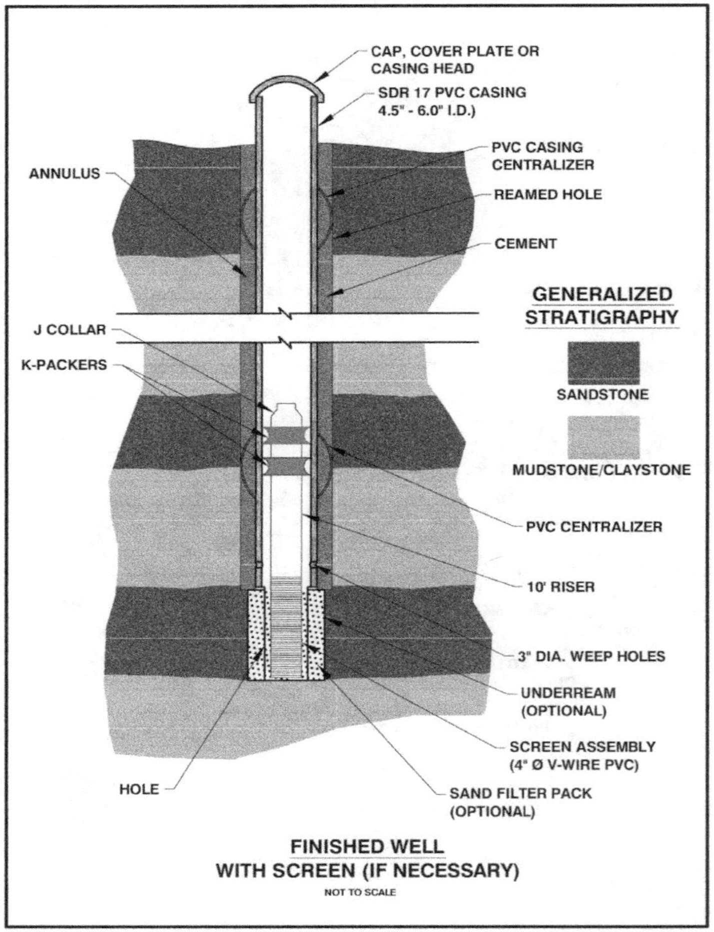

Source: Strata, 2011b

Figure 2.10

Proposed Well-Installation Method 3 for Ross Project Monitoring Wells

1

These methods all conform to the typical well-completion standards described in the GEIS (NRC, 2009). Wells would be constructed of polyvinyl chloride (PVC) or fiberglass with a sufficient pressure rating to withstand the maximum anticipated injection pressure, the maximum external collapsing pressure, and the maximum pressure of cementing; they would be constructed in accordance with WDEQ rules (WDEQ/LQD, 2005). The casings would be joined using an O-ring and spline modified to fit the ore zone, and well spacing would range from 15 – 46 m [50 – 150 ft]. The Applicant proposes that wells configured in a line-drive pattern would likely require increased aquifer restoration efforts; therefore, the Applicant would make limited use of line-drive patterns. Where it is not possible to avoid the use of line-drive patterns, the Applicant would perform additional computer modeling to determine the most efficient well spacing so as to facilitate aquifer restoration.

The Underground Injection Control (UIC) program administered by WDEQ/LQD regulates the design, construction, testing, and operation of all injection and recovery wells (WDEQ/LQD, 2005). WDEQ has primary regulatory authority for such actions as delegated by the U.S. Environmental Protection Agency (EPA). Wells for uranium extraction are classified under the UIC program as Class III wells; the Proposed Action would therefore require a UIC permit from WDEQ to use Class III injection wells. Before ISR operations could begin at any wellfield, the Applicant would be required by a license condition to provide the NRC with documents clearly delineating the approved aquifer exemption areas. (Portions of the aquifers designated for uranium recovery must be exempted as an underground source of drinking water [USDW] by EPA and reclassified by WDEQ/Water Quality Division (WQD) in accordance with the *Safe Drinking Water Act* [SDWA].)

Consistent with the typical design described in the GEIS (NRC, 2009), the Applicant proposes that each wellhead would be covered by an insulated fiberglass box in order to provide freeze protection and spill containment (Strata, 2011b). The protective box would include a solid base with access tunnels for well casing, electrical, and water-flow lines as well as a leak-detection system. Each recovery well would contain a submersible pump properly sized to carry solutions from the well to the module building. Injection wells would be equipped with air-release valves to permit relief of any excess pressure that could occur in the wells.

In the event that recovery, injection, and/or monitoring wells must be located within a floodplain, engineered controls and instrumentation would act to prevent leakage to the environment or contamination to the wells from a flood event (Strata, 2011b). The well seals would prevent inflow of flood waters down the well casing, while the fiberglass structure and bottom containment feature would limit exposure of the well to the environment. Erosion-control measures, such as rip-rap, grading, contouring, and water bars, would be utilized where appropriate in order to reduce sediment mobilization and runoff velocities.

Following installation, the well would be “developed” by pumping, air lifting, jetting, and/or swabbing to clean it and improve its hydraulic efficiency. The goal of these activities would be to remove drilling fluids and any small, fine particles from the well-completion zone, to provide good hydraulic communication, and to maintain the natural geochemical conditions. The Applicant expects that the water produced during well development would meet Wyoming’s temporary Wyoming Pollution Discharge Elimination System (WYPDES) discharge standards, which would allow this water to be discharged directly to the ground surface (WDEQ/WQD, 2007).

What is mechanical integrity testing (MIT)?

After each well is completed, and before the well is brought into service, all injection and recovery wells are tested for mechanical integrity. A "packer" is set above the well screen, and the well casing is filled with water. At the surface, the well is pressurized with either air or water to 125 percent of the maximum operating pressure, which is calculated based upon the strength of the casing material and depth. The well pressure is monitored to ensure significant pressure drops do not occur through drillhole leaks. A pressure drop of no more than 10 percent in a period of 10 to 20 minutes indicates that the casing and grout are sound (i.e., do not leak) and that the well is fit for service. Well integrity tests are also performed if a well has been damaged by nearby surface or subsurface activities or has been serviced with equipment or procedures that could damage the well casing, such as insertion of a drill bit or cutting tool. Additionally, each well is retested periodically (once each 5 years or less) to ensure its continued integrity. If a well casing fails an MIT, the well is taken out of service, repaired, and retested. If an acceptable test cannot be obtained after repairs, the well is plugged and properly abandoned.

Prior to operation, the integrity of each well would be verified by a pressure-based mechanical-integrity testing (MIT) that conforms to the procedure described in the GEIS and required by WDEQ (NRC, 2009; Strata, 2011b; WDEQ/LQD, 2005). After initial testing by the Applicant, the well would be retested at five-year intervals. In addition, the MIT would be repeated if the well is entered by a drilling bit or an under-reaming tool, or if well damage is suspected for any reason. The well-integrity test results would be documented and filed onsite and provided to WDEQ/LQD on a quarterly basis.

The Applicant proposes that MIT be conducted by placing inflatable packers or a comparable device near the top of the casing and above the screened interval (Strata, 2011b). The packers are inflated, and the interval between the packers is pressurized with water to the designated test pressure (maximum allowable injection pressure plus a safety factor of 25 percent). This pressure must be maintained within 10 percent for 10 minutes in order for the well to pass the MIT. A well-integrity record would be completed for each tested well. If a well demonstrates an unacceptable pressure drop during the MIT, the packers would be reset, the equipment checked for leaks, and the test repeated. If in subsequent tests the well passes the integrity requirements, the well would be deemed acceptable for use as an injection, recovery, or monitoring well. If a well continues to fail the MIT, it would be plugged and properly abandoned (i.e., sealed with cement slurry). Any well excluded due to MIT failure, or any that have arrived at the end of their useful life, would be properly abandoned. A well-abandonment record would be completed and retained onsite until the termination the Applicant's license, as would be required in NRC's license.

The Applicant's proposed design for pipes and module buildings is consistent with the industry standard described in the GEIS (NRC, 2009). Module buildings (referred to as pump and header houses in the GEIS) would be located throughout the wellfield and would be approximately 4.6 m x 12.2 m [15 ft x 40 ft] in size (see Figure 2.7) (Strata, 2011b). Piping from the module building to the CPP is referred to as feeder lines and trunk lines. Flow to injection wells and from recovery wells would be conveyed through 2.5 – 5 cm [1 – 2 in] HDPE pipelines (flow lines) that are connected through a manifold in the module building. Pipes inside the module buildings would be HDPE, PVC, or stainless steel rated for an operating pressure greater than the proposed maximum injection pressure. Feeder-line and trunk-line junctions would be contained in valve manholes located along the trunk lines. Each module building would have the capability of being isolated from the trunk lines by manually operated butterfly valves contained in the valve manholes. Piping would be buried below the frost line.

Each well flow line would have a meter to record the total flow passing through each flow line, pressure transmitter, and manual valve to control the flow rate. A small sample-collection valve for each well would be included on the recovery flow lines. The recovery-well flow lines would enter a manifold on one side of the module building, and the injection well lines would enter a manifold on the other side. A manifold building would house: 1) electrical equipment required to control the recovery pumps; 2) a pressure-limiting valve, a pressure transmitter, and equipment to add the oxidant to lixiviant on the injection manifold; and 3) flow meters that would indicate rate and totalizer readings on the trunk lines (Strata, 2011b). Each module building would have a manhole to access flow lines and feeder lines (see Figure 2.7). The manholes would also contain leak-detection systems.

The Applicant would test for leaks with fresh water on the pipelines prior to their burial, in order to ensure the pipelines' mechanical integrity (Strata, 2011b). The tests would be conducted in accordance with the manufacturer's recommendations or industry standards prior to final burial. In the event of leakage from pipelines or fittings, the defective component would be replaced. Prior to backfilling the trench dug to install a pipeline, the Applicant would perform a final inspection of all pipes and valves, the quality of the pipe embedment material, and the suitability of the backfill. Pipeline installation and trench backfilling would follow standard procedures that would be designed to ensure the quality of the installation and backfilling (Strata, 2011b). These procedures include the Applicant:

- Laying of pipe at required grades and lines
- Minimizing accumulation of water during laying or backfilling
- Limiting lateral displacement with use of embedment material
- Preventing contamination of the trench with foreign, unsuitable material
- Covering pipe with at least 0.6 – 2 m [2 – 6 ft] of material
- Using insulated tracer wire and warning tape
- Using properly sized and placed bedding material
- Using proper backfill material, which would not impose undue shock or unbalance to the pipe (i.e., frozen soils, mud, or snow)
- Using trench plugs at the appropriate spacing, particularly at or near areas of elevated ground water

> **What are pre-licensing baseline water-quality concentrations?**
>
> Prior to the submittal by an Applicant of its license application to the NRC, an Applicant performs site-characterization environmental-monitoring efforts for at least a year at the site at which it wishes to conduct uranium recovery prior to major Project construction. 10 CFR Part 40, Appendix A, Criterion 7 requires this monitoring (10 CFR Part 40). In addition, other regulations, such as those promulgated by the U.S. Environmental Protection Agency (e.g., 40 CFR Part 192, 40 CFR Part 141, and 40 CFR Part 143) and/or pertinent authorized State regulations, such as Wyoming Department of Environmental Quality's Hydrology Guidelines for Permitting Mines, Appendix 1, *Pre-mining Water Quality Sampling in the Guideline No. 8* may also inform an Applicant's environmental-monitoring strategies (WDEQ/LQD, 2005). Finally, NRC's guidance, Regulatory Guide 4.14, also makes recommendations regarding environmental monitoring efforts.
>
> As part of site-characterization efforts, ground-water monitoring wells are installed and ground-water samples are obtained. These samples are analyzed for certain water-quality constituents, or parameters, that are important to the characterization of existing conditions at a particular site. These concentrations are known as the "pre-licensing baseline" values of the respective water-quality constituents.
>
> These values are also sometimes known as "background" values. However, in the case of the Ross Project, because an earlier uranium-recovery operation was conducted within the Ross Project area, this operation could potentially have impacted "background values." Thus, the values measured by Strata prior to its submitting its license application are called "pre-licensing baseline" values in this SEIS.

As NRC license conditions would require, the Applicant would install a monitoring-well ring around the perimeter of each wellfield that would be used to detect horizontal and vertical excursions of uranium-recovery solutions during ISR operations (see SEIS Section 2.1.1.2) (Strata, 2011b). Prior to commencing ISR operations, these wells would allow sampling and analysis of ground water and, in this SEIS, this type of monitoring is called "post-licensing, pre-operational." The resulting post-licensing, pre-operational data would be used to determine concentration-based levels that would permit identification of any excursions from the respective wellfields; these would be called the Ross Project's upper control limits (UCLs). These post-licensing, pre-operational baseline values would be established for each separate wellfield (and they would be codified in the Applicant's NRC license). During uranium-recovery wellfield operation, the Applicant would then sample ground water from the wells and compare the analytical values to the NRC-specified baseline constituent concentrations to determine whether an excursion of any solution (such as lixiviant) into the surrounding aquifers has occurred. The Applicant would use Methods 2 or 3 (shown in Figures 2.9 and 2.10) to install these ground-water monitoring wells.

The Applicant's site-characterization efforts, which were conducted prior to its license-application submittal to the NRC, established "pre-licensing baseline" values of certain ground-water constituents; these values represent the baseline constituent concentrations currently present in the ground water under the Ross Project area (Strata, 2011a; Strata, 2011b). (See the text box above.) Later, prior to actual uranium-recovery wellfield operation, but after the initial NRC license is issued for wellfield construction, the ground water in each wellfield would be analyzed for the post-licensing, pre-operational baseline concentrations of constituents specified by the NRC (NRC, 2003a).

Within each wellfield, the well spacing that the Applicant proposes is in accordance with the minimum requirement described in the GEIS as necessary to detect excursions (NRC, 2009). Typical well spacing for a five-spot or seven-spot pattern is between 12 and 50 m [40 and 150 ft] apart. Wells completed in the aquifer underlying the ore body and wells completed in the aquifer overlying the ore body would be installed at an interval of one well per 0.8 ha [2 ac] of wellfield to detect vertical migration (Strata, 2011b). The Applicant also proposes a spacing of the perimeter monitoring wells of 122 m [400 ft] apart and at a distance of approximate 122 [400 ft] from the edge of the wellfield, to detect potential horizontal excursions. Simulations by the Applicant demonstrate that the proposed spacing successfully detects hydraulic anomalies in the form of water-level increases well before lixiviant has moved beyond the active uranium-recovery areas.

To reduce the possibility of lixiviant excursions, all previously drilled exploration and/or delineation drillholes that can be located on the Ross Project area and that are within a monitoring-well ring would be re-entered to each drillhole's total depth and sealed with cement slurry, per standard well-abandonment protocols (Strata, 2011b). These historic exploration and/or delineation drillholes would be located through the use of a hand-held metal detector that would locate the brass cap associated with each drillhole with its identification number. After a drillhole is located, a small drilling rig would be set up over the hole to ream them out to their total depth. The drillholes would then be cemented from the bottom to the ground surface. Details of each drillhole's abandonment would be documented in a record (examples in Strata, 2011b, Addendum 2.7-F), which would be filed at Strata's Oshoto field office in the appropriate drillhole file and provided with the respective wellfield data package, as appropriate.

What are underground injection control permits?

The EPA has delegated authority to the State of Wyoming, to administer its own Underground Injection Control (UIC) Permits. Classes I and III are most applicable to ISR operations.

- **Aquifer Exemption:** UIC criteria for the exemption of an aquifer that might otherwise be defined as an underground source of drinking water are found at 40 CFR Part 146.4. These criteria include whether the aquifer is currently a source of drinking water and whether the water quality is such that it would be economically or technologically impractical to use the water to supply a public water system.
- **Industrial and Municipal Waste Disposal Wells (UIC Class I):** Wells in this Class are used for the deep disposal of industrial, commercial, or municipal waste below the deepest usable aquifer. This type of well uses injection and requires applied pressure. This Class includes all wells that dispose of waste on a commercial basis. For ISR operations, this type of UIC Permit is necessary to use deep-well injection for waste disposal.
- **Mining Wells (UIC Class III):** This type of UIC Permit governs injection wells used to recover minerals. They include experimental technology wells; underground coal gasification wells; and wells for the in situ recovery of materials such as copper, uranium, and trona. For ISR operations, this type of UIC Permit covers wells that inject lixiviant into the uranium-bearing aquifer.

<u>Deep-Injection Wells</u>

The Applicant plans to dispose of liquid effluent generated during uranium-recovery operations via Class I UIC disposal wells. The Applicant has received a ten-year permit (UIC Permit No.10-263), dated April 4, 2011, for up to five Class I deep-disposal wells from WDEQ (WDEQ/WQD, 2011b). This Permit authorizes the injection of liquids into the Flathead and Deadwood Formations within specified intervals at depths of about 2,488 – 2,669 m [8,163 – 8,755 ft] below the ground surface; these formations are at least 500 ft below the lowermost potential USDW (the Madison Formation).

Under the terms of the UIC Class I Permit, the Applicant is allowed to inject into the Class I deep-disposal wells the following: operation bleed streams, yellowcake wash water, sand-filter and ion-exchange wash water onsite laboratory waste water, RO brine, aquifer-restoration ground water, facility wash-down water, wash waters used in cleaning or servicing waste-disposal-system equipment, and storm water—all generated during uranium-recovery activities—as well as fluids produced during the drilling, completion, testing, or stimulation of wells or test drillholes related to uranium-recovery operations, or during the work-over or abandonment of any such well, and drilling-equipment wash water. Under the terms of the UIC Permit, the Applicant is also prohibited from injecting certain materials into these wells. For example, hazardous wastes as defined by EPA or WDEQ cannot be injected into these wells (WDEQ/WQD, 2011b). Well construction, operation, MIT inspection, and well abandonment plugging and requirements are defined in this Permit as well. The Applicant would need to obtain written acceptance of financial-assurance methods from WDEQ prior to construction of each of the proposed wells.

The Applicant proposes that each well location would consist of a 76 m x 76 m [250 ft x 250 ft] pad with a storage tank (Strata, 2011b; Strata, 2012b). Surface equipment for the deep-disposal wells would include storage tanks, pumps, filtration systems, instrumentation and control systems, and equipment for injection of process chemicals (Strata, 2011b). Pads would either be asphalt pavement or gravel and would be retained through the life of the disposal well in order to conduct maintenance. Access roads to well sites with widths up to 4.3 m [14 ft] would be constructed on existing roads where possible. The supply pipelines to the wells would be 15 – 25 cm [6 – 10 in] HDPE plastic.

Pressures and flow rates for the pipes and disposal wells would be constantly monitored at the CPP. Instrumentation details for the deep-disposal wells are provided in Addendum 4.2-A of the TR (Strata, 2011b). System instrumentation would provide the necessary measures to ensure safe operation of the disposal system. At a minimum, instrumentation would include a flow totalizer, flow meter, pressure regulator, pressure indicator, pressure switch, annular tank level indicator, and injection pressure chart recorder. Water quality, fluid quantity, and injection rates would be reported to the WDEQ/LQD UIC program as required by the UIC Permit.

Injection rates up to the maximum are controlled by surface-injection pressures that are limited to the fracture pressure. Exceeding the limiting surface pressure set forth in the permit or creating or propagating fractures within the receiving zone would be a permit violation. The permit requires the installation of a kill switch on the injection tubing to preclude violation of the pressure limits.

2.1.1.2 Ross Project Operation

As shown by the proposed schedule in Figure 2.6, uranium recovery during the proposed Ross Project would follow a "phased" approach, where one group of well modules could be in operation, while preceding well modules are being engaged in aquifer restoration (Strata, 2011b). During the operation phase, three major phases would occur involving the wellfields: an operation-only phase, a concurrent operation- and aquifer-restoration phase, and an aquifer-restoration-only phase.

Uranium Mobilization

The Applicant proposes the use of an alkaline lixiviant to dissolve the uranium as described in Section 2.4 of the GEIS (NRC, 2009; Strata, 2011b). Gaseous oxygen (O_2) or hydrogen peroxide (H_2O_2) is used as the oxidant and sodium bicarbonate ($NaHCO_3$) or carbon dioxide (CO_2) is added to aid in keeping uranium in its dissolved state. Native ground water would be fortified with sodium bicarbonate at the CPP and then pumped to the module buildings where the oxidant and, potentially, CO_2 would be added at the injection manifolds located inside the module buildings (see Figure 2.7).

What are the basic steps of uranium mobilization?

- **Ground-Water Injection**

  Uranium mobilization is accomplished by the injection of a non-uranium-bearing ("barren") solution, or "lixiviant," through "injection" wells into the uranium-bearing ore zone. The lixiviant moves through pores in the ore-zone aquifer, dissolving uranium and other metals.

- **Ground-Water Extraction**

  Recovery, or "production," wells extract the now "pregnant" lixiviant, which contains uranium and other dissolved metals, and the solution is then pumped to a central processing plant (CPP) for further uranium recovery and purification.

The Applicant proposes the carbonate/bicarbonate lixiviant because of its compatibility with minerals within the ore zone. In addition, carbonate/bicarbonate lixiviants are generally considered more amenable to aquifer restoration than other acidic lixiviants (NRC, 2009). Preliminary leach testing performed by the Applicant in 2010 demonstrated that this type of lixiviant successfully mobilized uranium into solution. Comparison of the Applicant's expected concentration ranges of chemical constituents in the pregnant lixiviant with the typical lixiviant chemistry presented in Table 2.4-1 of the GEIS shows consistency between the Ross Project and the GEIS, except for higher concentrations of uranium and vanadium that could be present in the pregnant lixiviant at the Ross Project (Strata, 2011b; NRC, 2009).

As described in Section 2.4.3 of the GEIS, the recovery wells extract slightly more water than is injected into the ore-containing aquifer, which creates a "cone of depression" within the respective wellfield and, thus, maintains an inward flow of ground water. This inflow prevents migration of lixiviant toward the perimeter monitoring wells. The excess water, referred to as "production bleed," is a radioactive byproduct material that must be properly managed and disposed (NRC, 2009). For the Ross Project, the Applicant proposes a production-bleed range from 0.5 percent to 2 percent, and averaging 1.25 percent of the injection volume (Strata, 2011b). At the maximum flow rate, approximately 360 L/min [94 gal/min] of production bleed would be generated.

The Applicant proposes to use actual wellfield data and reservoir-engineering software to predict a sufficient bleed rate to minimize water consumption while the potential for hydraulic anomalies outside of the uranium-recovery area is minimized (Strata, 2011b). The wellfield flows would be balanced to produce appropriate bleed based upon the module-injection and recovery feeder-line meters. The individual well-flow targets would be determined on a per-pattern basis to ensure that local wellfields are balanced on at least a weekly basis.

The Applicant proposes a maximum injection pressure of 970 kPa [140 lb^2/in] measured at the injection manifold. This pressure is less than the formation-fracture pressure, which is approximately 2,240 kPa [325 lb^2/in] at the Ross Project and less than the pressure rating for operation of the pipes and other equipment (Strata, 2011b). Although injection pressures are initially expected to be relatively low, pressure requirements within a specific wellfield generally tend to increase with time. The Applicant suggests that, in order to maintain flow rates and wellfield balance, some wells would require flexibility in their allowable injection pressure. To specifically avoid the injection-restriction problems that plagued the Nubeth operation, the Applicant has proposed several improvements to well design, well development, and filtration (Strata, 2011a; Strata, 2011b).

Flows and pressures for the injection and recovery pipeline network would be monitored continuously at the module building, valve manhole, and CPP; the pressures would also be displayed in the CPP's control room (Strata, 2011b). Changes in flow or pressure that are outside of normal operating ranges would result in the activation of visual and audible alarms in the CPP, and eventually automatic sequential shutdown of pumps and control valves, if the condition is not corrected promptly.

In addition, the leak-detection sensors that would be located in the module-building sumps and the valve manholes would trigger audible and visual alarms at that location and in the CPP if fluid is detected (Strata, 2011b). The Applicant could also utilize dual leak detection in these areas, which would consist of two sensors at high and low levels within a module building. If fluid is detected by the low-level sensor, an audible and visual alarm would be triggered at that location and in the CPP. If fluid is detected by the high-level sensor, automatic pump shutdown would occur to prevent the fluid from overflowing the containment system and contaminating the surrounding environment.

Pipe and fitting leaks at the wellheads would be detected by sensors located in the wellhead sumps. In addition, a system would be instituted in the facility's operating plan for personnel to inspect the interior of each well module on a weekly basis. Minor leaks or other problems would be detected in this manner and then promptly repaired to reduce the likelihood of major releases.

As noted in SEIS Section 2.1.1, NRC regulations at 10 CFR Part 40, Appendix A, as well as the individual NRC license that would be issued to the Applicant, would require licensees to have an operational monitoring-well system to detect excursions. NRC guidance defines an excursion as occurring when two or more excursion indicators or parameters are present in a monitoring well or if one excursion parameter exceeds the respective UCLs by 20 percent (NRC, 2009). GEIS Section 2.4.1.4 described how ISR operations can potentially affect the ground-water quality near a site, when, during an excursion, lixiviant escapes the production zone, where uranium recovery is underway, and is not recovered by the intended recovery wells (NRC, 2009). This would result in either a vertical or horizontal excursion. Excursions can be caused by an improper water balance between injection and recovery wells, undetected high-permeability strata or geological faults, improperly plugged and abandoned exploration drillholes, discontinuity within the confining layers, poor well integrity, or unintended fracturing in the well zone or surrounding units (NRC, 2009). The monitoring of water levels that would be performed would serve to avert a potential excursion. Water-quality indicators in the ground water from monitoring wells that would be established after wellfield installation (i.e., post-licensing, pre-operational baseline concentrations defined as excursion indicators) would also be used to detect whether an excursion has occurred.

What are excursion indicators and upper control limits?

Prior to the commencement of injection of lixiviant into a wellfield and actual uranium recovery, an Applicant must propose excursion indicators (which are water-quality parameter concentrations, such as chloride, that are measured to describe the quality of the ground water) as well as upper control limits (UCLs) per 10 CFR Part 40, Appendix A, and as per the license the NRC would issue (10 CFR Part 40). These indicator chemical constituents, or "excursion indicators," would be based upon post-licensing, pre-operational baseline ground-water-quality parameters (i.e., chemical constituents occurring in the ground water) and lixiviant chemistry.

Only after a wellfield and its monitoring-well ring are installed would several ground-water samples would be obtained and analyzed by the Applicant. The results of these analyses provide post-licensing, but pre-operational, baseline values for the respective ground-water-quality parameters that would be used to indicate contemporary ground-water quality. If, during ISR operations, two indicator constituents' are exceeded, or if one is exceeded by 20 percent, (with respect to the corresponding UCLs), then an excursion of lixiviant would be defined as occurring.

UCLs are set on a wellfield-by-wellfield basis and are stated in constituent concentrations for selected excursion indicators so as to provide early warning if uranium-bearing solutions (lixiviant) are moving away from a particular wellfield. The UCLs are subject to the NRC's staff review and approval and their establishment would be required in the NRC license. As described by the NRC (2003a), the best excursion indicators are easily measurable parameters that are found in higher concentrations during uranium recovery than in the natural ground water.

At most in situ uranium-recovery operations, for example, chloride is often selected because it does not interact strongly with the minerals in the ore zone; it is easily measured; and chloride concentrations are significantly increased during ISR operations. Conductivity, which is correlated to total dissolved solids (TDS), is also considered a good excursion indicator because of the high concentrations of dissolved constituents in the lixiviant as compared to the surrounding aquifers (Staub et al., 1986, and Deutsch et al., 1985, as cited in NRC, 2009b). Total alkalinity (carbonate plus bicarbonate plus hydroxide) is used as an indicator in wellfields where sodium bicarbonate or carbon dioxide is used in the lixiviant.

At least three excursion indicators are selected to be monitored in each wellfield, and the UCLs are determined using statistical analyses of the post-licensing, pre-operational baseline water quality in the respective wellfield. The NRC staff has identified several statistical methods that can be used to establish UCLs. For example, in areas with good water quality (TDS less than 500 mg/L), the UCL could be set at a value of 5 standard deviations above the mean of the measured concentrations. Conversely, if the chemistry or a particular excursion indicator is very consistent, a specific concentration could be specified as the UCL. If post-licensing, pre-operational baseline data indicate that the ground water is homogeneous across the wellfield, the same UCLs could be used for all monitoring wells. Alternatively, if the water chemistry in the wellfield is highly variable, unique UCLs could be set for individual wells.

An excursion is defined to occur when two or more excursion indicators in a monitoring well exceed their UCLs (NRC, 2003a). Alternate excursion detection procedures (e.g., one excursion indicator exceeded in a monitoring well by a specified percentage) could also be used, if approved by the NRC.

The NRC would require in its license that the Applicant conduct sampling of its monitoring wells twice each month and to analyze those samples for the excursion indicators (i.e., select baseline water-quality constituent concentrations) specified in its license, so it can be determined whether an excursion has occurred. The Applicant has proposed such an operational ground-water monitoring program (Strata, 2011b). Water levels would be routinely measured during the sampling of the perimeter, overlying, and underlying monitoring wells in order to provide an early warning for impending wellfield problems. An increasing water level in a perimeter monitoring well has been shown to be an indication of a local flow imbalance within the wellfield, which could result in an excursion (Strata, 2011b). An increasing water level in an overlying or underlying monitoring well could be caused by the migration of fluid from the ore zone or by an injection well-casing failure. As stated above, samples would also be collected from the appropriate monitoring wells once every two weeks and would be analyzed for the license-established excursion parameters. In addition, the Applicant expects that dedicated pressure transducers and/or in situ water-quality instruments could be used in the perimeter monitoring wells to provide the earliest detection of potential excursions or hydraulic anomalies. The Applicant anticipates that this monitoring effort would allow corrective action to be immediately taken to balance locally the injection and recovery flows or to shut down individual injection well(s) or the entire wellfield, as necessary (Strata, 2011b).

Per conditions that the NRC would include in the Ross Project's license, the Applicant would be required to notify the NRC within 24 hours if an excursion were confirmed in the Project's ground-water monitoring wells. If a vertical excursion occurs, then the Applicant's injection of lixiviant would cease and, for any excursion, corrective action would be initiated (the GEIS documented that vertical excursions tend to be more difficult to recover than horizontal excursions) (NRC, 2009). The NRC would require in the Applicant's license that verification and progress ground-water samples are collected by the Applicant weekly until the excursion indicators are at or below their respective UCLs (i.e., the excursion is "recovered") as indicated by three consecutive weekly samples.

The Applicant would also be required to provide a report to NRC within 60 days, including a confirmation of an excursion, a description of the excursion, a discussion of the corrective actions taken, and the results of those corrective actions. If an excursion cannot be recovered within 60 days of confirmation (measured by a concentration of more than 20 percent of any excursion indicator), the Applicant would be required either to terminate lixiviant injection within the wellfield until aquifer cleanup is complete (for horizontal excursions) or to increase the surety for the ISR project by an amount sufficient to cover the full third-party cost of correcting and remediating the excursion. As the GEIS described in Section 2.11.4, licensees typically retrieve horizontal excursions back into the production zone by repairing and reconditioning wells and adjusting pumping rates in the wellfield.

Uranium and Vanadium Processing

Uranium and vanadium in pregnant lixiviant would be extracted from solution by IX resin, stripped from the loaded IX resin ("eluted"), precipitated into a slurry, thickened, de-watered, dried, and packaged as yellowcake (Strata, 2011b). Prior to introduction to the IX columns, pregnant lixiviant could be passed through a de-sanding filtration system (Strata, 2011b). Carbon dioxide could also be added to the pregnant lixiviant to optimize the IX resin-loading capacity. The filtered, pregnant lixiviant would then be passed through two-stage, pressurized, down-flow IX columns, where the uranium and the vanadium dissolved in the lixiviant would be selectively adsorbed onto the IX resin beads. In exchange of uranium and vanadium, the resin releases chloride, bicarbonate, or sulfate ions into the lixiviant. The barren lixiviant exiting the second IX column would be monitored and would normally contain less than 2 mg/kg ("parts per million" or "ppm") of uranium. When the resin beads in the IX column become saturated with uranium and vanadium, the columns would be taken offline for resin elution.

Prior to elution ("elution" is the process whereby the resin beads are "washed" with water to remove uranium and vanadium), the loaded uranium-bearing resin would be transferred to vibrating screens to wash away sand, silt, broken resin, scale, and other process contaminants. The solid material recovered during this step would be collected, stored, and disposed of as a byproduct waste. The elution process would then consist of four stages. The first three sequential stages are where a single batch of resin is contacted with a volume of eluant (water containing approximately 10 percent sodium chloride and 2 percent sodium carbonate) three times the volume of the batch of loaded resin. The fourth stage is a final rinse where the batch of resin is contacted with four bed volumes, or pore volumes, of fresh water (i.e., four bed volumes is equal to four times the amount of pore space [i.e., empty space] in the resin) (Strata, 2011b). In addition to processing resin from the Ross Project wellfields, the elution circuit would have the capacity to process loaded resin from other uranium-recovery operations owned either by the Applicant or another company as well as from water-treatment facilities that use IX resin to filter or condition water (Strata, 2011b).

The precipitation circuit produces a slurry of uranium solids from the eluant. The Applicant proposes a design consisting of multiple precipitation tanks plumbed in series, with mechanical agitation. The sequential addition of chemicals to bring about precipitation would be as follows: 1) sulfuric acid, 2) sodium hydroxide (caustic soda), 3) hydrogen peroxide, and 4) sodium hydroxide. The slurry containing the uranium precipitate would then be pumped to a yellowcake thickener, which separates the solids particles from the liquid. The "underflow" from this thickener (i.e., the still-wet separated solids) would then undergo a second stage of dissolution and precipitation to remove any impurities entrained in the first precipitate (the underflow). The "overflow" (i.e., the liquid with few solid particles remaining after precipitation) from both thickener stages would then go to the vanadium-recovery circuit.

After precipitation, the yellowcake slurry would be washed in a filter press to remove excess chloride and other soluble contaminants. After multiple washings, the filter cake would be transferred to a radiologically controlled area for drying and packaging (Strata, 2011b). Drying would be accomplished in completely enclosed low-temperature vacuum dryers. The GEIS describes the type of dryer proposed by the Applicant as the standard for newer ISR facilities (NRC, 2009). The off-gases generated during the drying cycle would be filtered and scrubbed to remove entrained particulates. The GEIS noted that the drying, filtration, and scrubber process proposed by the Applicant is designed to capture virtually all escaping particles (NRC, 2009).

The dryers would be batch type, and drying would typically take 16 hours per batch. Batch dryers create the potential for the escape of yellowcake during loading and unloading of the dryer. The Applicant proposes to reduce this potential by the design of the equipment. A water-sealed vacuum pump would provide ventilation during loading of the yellowcake slurry into the dryer and transferring the dried product into 208-L [55-gal] drums by facility personnel (Strata, 2011b). Transfer equipment would be located directly below the dryer and would include a discharge chute, rotary airlock valve, ventilated drum hood, and a drum conveyor. A drum would be placed beneath the dryer discharge chute; the ventilation hood would be secured over the drum opening to prevent escape of yellowcake into the surrounding environment. After a drum is in place and securely covered, the rotary airlock valve would be activated to start the loading process. A viewport in the hood would allow personnel to determine when the drum is full. The loaded drum would be weighed and labeled, and then moved to the side to cool and off-gas before it is sealed and stored for offsite shipment.

The uranium-depleted solutions from the uranium thickeners would be pumped to a vanadium precipitation tank (Strata, 2011b). Steam, facility air, ammonia, and ammonium sulfate would be added to cause precipitation of crystals containing vanadium. The precipitate slurry would be pumped to a horizontal belt filter, where the solution is removed from the crystals. The filter cake would be washed and transferred to a batch vacuum rotary dryer similar to the dryer that would be used to dry uranium yellowcake. Off-gas from the precipitation tanks and dryer would be filtered to remove particulates and directed to a wet scrubber to capture ammonia for reuse. The dried product would then be packaged for offsite shipment. The Applicant estimates that 0.1 – 2 kg [0.2 – 4.4 lb] of V_2O_5 would be produced for every 1 kg [2.2 lb] of U_3O_8.

The waste water would be treated by reverse osmosis (RO) (Strata, 2011b). The water quality of permeate that is anticipated by the Applicant is provided in Table 2.2. Most of the permeate from the RO system would be recycled back to the wellfield as lixiviant. The lined surface impoundments within the facility would be used to store and manage excess permeate and brine. Permeate and brine would be managed as radioactive byproduct materials. Brine would be disposed in the deep-injection wells.

Table 2.2
Permeate Water Quality

Parameter	Unit	Typical Value	Minimum Value	Maximum Value
EC	µS/cm	300	180	400
TDS	mg/L	200	100	250
pH	s.u.	8	6	6.5
Alkalinity as $CaCO_3$	mg/L	100	50	200
Sulfate	mg/L	15	10	20
Bicarbonate	mg/L	150	50	200
Chloride	mg/L	15	5	25
Calcium	mg/L	0	0	1
Sodium	mg/L	50	20	100
Manganese	mg/L	0	0	0.1
Selenium	mg/L	0	0	0.1
Arsenic	mg/L	0	0	0.1
Uranium	mg/L	0	0	0.1
Radium	pCi/L	30	5	100

Source: Table 4.2-2 in Strata, 2011b.

2.1.1.3 Ross Project Aquifer Restoration

After uranium recovery has ended, each wellfield that is to undergo aquifer restoration would contain ground-water constituents that would have been mobilized by the lixiviant. The purpose of aquifer restoration is to restore the respective aquifer to its baseline conditions, as defined by post-licensing, pre-operational constituent concentrations (see Section 2.1.1.2), so as to ensure public health and safety. The Applicant would be required to provide a financial-surety instrument that would cover planned and delayed aquifer-restoration costs in compliance with 10 CFR Part 40, Appendix A, Criterion 9 to cover the ISR facility's decontamination and decommissioning. NRC would review the adequacy of this financial-surety annually (see SEIS Section 2.1.1.7) (10 CFR Part 40).

Under the Federal UIC program, the exempted production aquifer would no longer be used as a USDW under the SDWA (40 CFR Part 145). In accordance with the requirements for a Class I-V well under 40 CFR Part 146.4, the exempted aquifer does not currently serve as a source of drinking water and cannot now and would not in the future serve as a source of drinking water (40 CFR Part 146). Hence, ground water in exempted aquifers cannot be considered as a source of drinking water after restoration.

The aquifer-restoration activities proposed for the Ross Project are the same as those methods described in Section 2.5 of the GEIS: 1) ground-water transfer, 2) ground-water sweep, 3) RO with permeate injection, 4) ground-water recirculation, and 5) stabilization monitoring (Strata, 2011a; NRC, 2009). The Applicant proposes that concurrent ISR operations and aquifer restoration would occur when several of the first well modules have been depleted and are ready for restoration activities (Strata, 2011b). As aquifer restoration occurs in depleted well modules, ISR operations would be ongoing in subsequent well modules.

The Applicant has proposed a ground-water restoration schedule that is benchmarked to production schedules and waste-water disposal capacity, but it estimates that aquifer restoration for each wellfield would take approximately eight months (Strata, 2011b). The Applicant's proposed restoration methodology would include ground-water sweep, permeate injection, and ground-water recirculation.

During ground-water sweep, water is pumped from injection and recovery wells to the facility without reinjection, as the GEIS described in Section 2.5.2. In response to this pumping, water from outside the wellfield flows into the ore zone, flushing contaminants from areas that have been affected by the horizontally spreading lixiviant in the respective aquifer during uranium recovery (NRC, 2009). Ground water produced during the sweep phase would contain uranium and other contaminants mobilized during uranium recovery as well as residual lixiviant. The initial concentrations of these constituents would be similar to those during uranium recovery, but the concentrations would decline gradually with time. The water removed from the aquifer during the sweep first would be passed through the IX system to recover the uranium and then be disposed of as excess permeate. The pumping rates used would depend on the hydrologic conditions at the Ross Project, and the duration of the aquifer sweep and the volume of water removed would depend on the volume of the aquifer affected by the ISR process.

Aquifer volume typically is described in terms of "pore volumes," a term used by the ISR industry to represent the volume of water that fills the void space in a given volume of rock or sediment. The Applicant's aquifer-restoration plan calls for removing up to 0.5 pore volumes of water during ground-water sweep (Strata, 2011b). Additional pumping would occur in select areas that would be identified during facility operation. The pumping rate is estimated at 284 L/min [75 gal/min] from well modules in the ground-water sweep stage. The Applicant proposes to use ground-water sweep selectively (for example, around the perimeter of the wellfield) rather than throughout the entire well module to minimize the consumptive use of ground water (Strata, 2011a).

The Applicant proposes to use ground-water treatment and permeate injection would be used after the ground-water sweep process, as described in Section 2.5.3 of the GEIS (Strata, 2011b). This phase would return total dissolved solids (TDS) (a water-quality parameter), trace-metal concentrations, and aquifer pH to the pre-operational baseline values that would have been determined during the Applicant's post-licensing, pre-operational sampling and analysis program; these concentrations would be required by the NRC license (NRC, 2009). Ground water recovered from a depleted portion of the ore zone would be treated with sulfuric acid or other chemicals to prevent scaling on the RO circuit (Addendum 6.1-A in Strata, 2011b). Low concentrations of uranium in the ground water would be removed by passing the water through the IX circuit, as during operations. Following the IX circuit, other chemical constituents are removed by passing the ground water through the two-phase RO system consisting of pressurized, semi-permeable membranes. The RO process yields two fluids: permeate (approximately 85 percent), which would be re-injected into the aquifer, and brine (approximately 15 percent), which would be managed as liquid waste.

The pumping and injection rates during this process would be similar to those during the sweep phase, but depending upon site hydrology, many pore volumes (often more than 10) could be circulated to achieve aquifer restoration goals (NRC, 2009). For the Ross Project, the Applicant estimates that aquifer restoration would average 3,880 L/min [1,025 gal/min] from well modules in the RO and permeate-injection process of aquifer restoration (Strata, 2011b). During aquifer restoration (except during ground-water sweep), all permeate would be used as lixiviant or injected into the aquifer for restoration.

The ground-water recirculation process would begin after completion of the permeate-injection process. In this phase, ground water from the production zone would be pumped from recovery wells and re-circulated into injection wells in the same well module. This process homogenizes the ground water within the aquifer to minimize the risk of "hot-spots," areas of the aquifer with unusually high concentrations of dissolved metal concentrations. The Applicant proposes that the only water treatments that would occur during recirculation are filtration and removal of uranium and vanadium (Strata, 2011a).

The purpose of stabilization during aquifer restoration is to establish a chemical environment that would reduce the solubility of dissolved constituents such as uranium, arsenic, and selenium, as described in GEIS Section 2.5.4. An important component of aquifer stabilization during the aquifer-restoration phase is to convert metals to their insoluble forms (NRC, 2009). If the oxidized (i.e., the more soluble) state is allowed to persist after uranium recovery is complete, metals and other constituents such as arsenic, selenium, molybdenum, uranium, and vanadium could continue to leach and remain at elevated levels. To stabilize these constituents' concentrations, the pre-operational oxidation state in the ore zone must be reestablished as much as is possible. This stabilization often requires adding an oxygen scavenger or a reducing agent, such as hydrogen sulfide (H_2S) or a biodegradable organic compound such as ethanol, into the production zone during the later stages of recirculation (NRC, 2009).

The need for aquifer stabilization would be determined on a case-by-case basis and would depend upon how effectively the sweep and recirculation processes restore the affected aquifer to the license-required standards. Following aquifer restoration, the Applicant would monitor the ground water by quarterly sampling to demonstrate that the approved standard for each constituent has been met and that any adjacent nonexempt aquifers are unaffected. The Applicant would reinitiate the entire aquifer restoration phase if stabilization monitoring determines it is necessary. Both WDEQ and the NRC must review and approve all monitoring results before aquifer restoration would be considered to be complete.

All injection, recovery, and monitoring wells and drillholes would be plugged and abandoned in place according to applicable regulations after ground-water restoration is approved by the NRC and WDEQ (WDEQ/LQD, 2005). To comply with these regulations, the Applicant proposes standard operating procedures (SOPs) of well abandonment that includes plugging all wells with cement containing 2 percent bentonite clay (Strata, 2011b).

2.1.1.4 Ross Project Decommissioning

Prior to the Ross Project's facility decontamination, dismantling, and decommissioning; the wellfields' aquifer restoration; and the Project site's reclamation and restoration; appropriate cleanup criteria for surfaces would need to be established in concert with NRC requirements, and a Ross Project-specific decommissioning plan (DP) would need to be accepted by the NRC (NRC, 2003b). The Applicant has committed to satisfying these NRC requirements for decontamination and decommissioning (Strata, 2011b).

To begin the Ross Project's decommissioning phase, the Applicant would conduct a series of radiation surveys to identify those areas at the Ross Project that would need decontamination to meet applicable cleanup criteria or those that cannot economically meet the criteria (Strata, 2011b). These surveys would include building, structural, and equipment surfaces as well as potentially contaminated environmental media such as soil and water (NRC, 1999; NRC, 2003a). The onsite excavated pits, or "mud pits," used for the disposal of drilling fluids and muds (or "cuttings") during the installation of wells, would be included in the survey to ensure no long-term radiological impacts (Strata, 2011a). In addition, records of radiation surveys and the entire cycle of decontamination, dismantling, decommissioning, and disposal activities would be maintained in accordance with the Applicant's license.

Based upon the results of the radiation surveys, decontamination and dismantling of buildings, structures, and equipment would be conducted in accordance with the DP. Contaminated surfaces, including processing and water-treatment equipment such as tanks, filters, IX columns, pipes, and pumps, would be decontaminated (Strata, 2011b). High-pressure washing would be used to remove loose contamination from the surfaces. If required, secondary decontamination would consist of washing with dilute acid or equivalent compatible solution (Strata, 2011b). All successfully decontaminated buildings and equipment could be released for unrestricted use (NRC, 2003b).

The buildings, structures, and equipment that are not or no longer contaminated would be moved to a new location within the Ross Project for further use or storage, removed to another facility for either reuse or salvage, or taken to a properly permitted, permanent solid-waste disposal facility. Concrete flooring, foundations, and foundation materials, if uncontaminated, would be broken up and disposed of at an appropriately permitted solid-waste facility. All radioactively contaminated buildings and structural materials that cannot be successfully decontaminated would be dismantled and then disposed of at a properly licensed radioactive waste disposal facility (i.e., a facility licensed by the NRC or an Agreement State). Contaminated soils would also be disposed of at the same or similar licensed facility. A final-status radiation survey would then be performed to ensure that any residual contamination on the surfaces is below the cleanup criteria. All disturbed lands would be reclaimed (NRC, 1999). Section 2.6 of the GEIS describes the general process for decontamination, dismantling, and decommissioning of an ISR facility and the restoration and reclamation of the land itself (NRC, 2009).

During decommissioning of the facility, all UIC Class III injection and recovery wells, monitoring wells, and the UIC Class I injection wells would be abandoned according to the DP. The total number of wells would number between 750 and 1,000 based upon the Applicant's estimate of 40 recovery wells per each of 15 – 20 wellfield modules plus monitoring wells (Strata, 2012a). Decontamination, decommissioning, and restoration of a wellfield would begin approximately five years after its construction (refer to Figure 2.6) (Strata, 2011a). However, at the Ross Project, complete decontamination, dismantling, and decommissioning of the ISR facility itself, and restoration and reclamation of the Ross Project area, could occur years after the wellfields begin to be decommissioned and the aquifer begins to be restored, in order to accommodate the Applicant's continuing recovery of uranium and production of yellowcake from its future satellite projects and/or from other uranium-recovery or waste-water-treatment operations (Strata, 2011a).

During the decommissioning phase, the Applicant proposes that all primary, secondary, and tertiary roads and other temporary access routes to and within the Ross Project would be removed and the land reclaimed, unless a request by the respective landowners or lessees to not do so is received by the Applicant. In this case, then, the landowners or lessees would assume responsibility for the long-term maintenance and ultimate reclamation of the roads and routes, after the NRC license has been withdrawn (Strata, 2011b).

All contaminated soil or gravel that is determined to be a byproduct radioactive waste would be disposed at a radioactive waste disposal facility licensed by the NRC or an Agreement State, as necessary, while petroleum-contaminated soil would be disposed at a WDEQ-permitted facility. Removal of roads would be accomplished by the Applicant removing excess road surfacing material, and then ripping the road and the underlying shallow subsoil to loosen the base. Culverts would be removed and preconstruction drainages would be re-established. The vicinity would be graded to a contour consistent with the surrounding landscape. Finally, topsoil would be applied in a uniform manner and the area seeded to achieve WDEQ/LQD reclamation standards.

The Class I deep-disposal wells would be plugged and abandoned in accordance with the requirements of the Applicant's UIC Class I Permit (Strata, 2011b). All wastes and the equipment associated with the surface impoundments, such as accumulated sludge, impoundment liners, and leak-detection pipes and lines, would be surveyed for radioactive contamination and then disposed of appropriately or released for unrestricted use (Strata, 2011b). The soil beneath the surface impoundments would be analyzed for radioactive contamination, and any areas that exceed the cleanup criteria for unrestricted release would be excavated and disposed of at a licensed radioactive waste disposal facility.

The natural flow of shallow ground water beneath the facility and in the immediate vicinity outside of the CBW would also be re-established during decommissioning (Strata, 2011b). Flow through the CBW would be accomplished by the Applicant's creating a series of breaches, also known as finger drains, along the up-gradient and down-gradient reaches of the CBW. Each finger drain would

consist of a 0.5 m [1.5 ft] wide by 7.6 m [25 ft] long trench that is cut through the CBW at a right angle and to a depth that is 0.6 m [2 ft] below the lowest historical ground-water level. Gravel would be placed in the trench from the bottom to a point 0.6 m [2 ft] above the highest recorded ground-water level such that a highly permeable flow path is created through the CBW. The remaining trench would be backfilled with native topsoil and seeded. Selected monitoring wells that would have been used by the Applicant to characterize the shallow aquifer in the area, before its installation of the CBW, would be retained. Water levels would be monitored following CBW reclamation to verify that the natural flow of shallow ground water through the CBW and beneath the facility has been restored.

> **What types of wastes would be generated at the proposed Ross Project?**
>
> **Liquid Wastes**
>
> ***Liquid Byproduct Waste*** is all liquid-phase wastes generated by the proposed Ross Project, except for sanitary waste water and well development and testing waste water. This waste is contaminated with byproduct material.
>
> ***Liquid Hazardous Waste*** is regulated under the Resource Conservation and Recovery Act or is a State-defined hazardous waste that is a non-byproduct waste. This waste includes universal hazardous wastes and used oil.
>
> ***Sanitary Waste Water*** is ordinary sanitary septic-system waste water; this waste water is non-hazardous, non-byproduct waste water.
>
> ***Well Development and Testing Waste Water*** is waste water generated during well development and during pumping tests; this waste water is non-hazardous, non-byproduct waste water. Such waste water does not require treatment before disposal.
>
> **Solid Wastes**
>
> ***Solid Byproduct Waste*** is all solid-phase wastes generated by the Ross Project that exceed NRC limits at 10 CFR Part 20 for unrestricted release. This waste is contaminated with byproduct material.
>
> ***Hazardous Waste*** is regulated under the Resource Conservation and Recovery Act or is a State-defined hazardous waste that is non-byproduct waste. This waste includes universal hazardous wastes.
>
> ***Nonhazardous Solid Waste*** is domestic, office, and municipal waste (i.e., trash), construction and demolition debris, septic solids, and materials such as equipment and soils that have been determined to meet NRC criteria in 10 CFR Part 20 for unrestricted (i.e., unregulated) release.

The Applicant proposes to re-contour, as necessary, the disturbed areas within the Ross Project area to blend in with the natural terrain and to be consistent with the preconstruction topography (Strata, 2011b). Revegetation would be accomplished in accordance with the WDEQ/LQD Permit to Mine requirements and would be required by the NRC license. Topsoil that was salvaged prior to construction activities and stored in a stockpile would be used for reclamation to the extent possible (Strata, 2011b); the topsoil would be spread over the area to be reclaimed and would be seeded with a native seed mix. During ISR facility operation the topsoil stockpiles and as much as is practical of the disturbed wellfield, would be seeded to establish vegetative cover to minimize wind and water erosion. At the completion of decommissioning, the Applicant commits to reclaiming the entire area to equal or better conditions than existed prior to ISR (Strata, 2011b, Addendum 6.1-A). Reclaimed land would be capable of supporting livestock grazing, dry-land farming, and wildlife habitat. The respective landowners and WDEQ would be consulted as the Applicant selects the seed mix. Seeding would be conducted by drill or broadcast methods depending upon the type of seed being used. Mulch could also be used to cover the seed (Strata, 2011b).

2.1.1.5 ISR Effluents and Waste Management

Section 2.7 of the GEIS describes the airborne effluents as well as the liquid and solid wastes that are typically generated at ISR facilities and corresponding waste-management practices

(NRC, 2009). The effluents and wastes expected from the proposed ISR project and the waste-management practices the Applicant proposes are consistent with the industry standards reported in the GEIS. The types of liquid and solid wastes, the quantities of these wastes anticipated by the Applicant, and the Applicant's proposed management systems are provided in Strata (2012a). (See also Table 4.9 in SEIS Section 4.14.) Impacts from liquid and solid waste management are described in SEIS Section 4.14.

Airborne Emissions

There would be both radioactive and non-radioactive airborne particulates and gases emitted during all phases of the Proposed Action (Strata, 2011b). As discussed below, the design features proposed by the Applicant to control all airborne effluents are consistent with the industry standards presented in the GEIS (NRC, 2009).

Non-Radioactive Emissions

Emissions from internal combustion engines would be the primary source of non-radioactive gaseous effluents (i.e., emissions). Releases would be anticipated from drilling rigs, drilling support equipment (e.g., backhoes, water trucks, pipe trucks, and cement units), utility trucks employed for wellfield service, light vehicles used for personal transport through the wellfields, in addition to vehicles used by ISR facility personnel to and from the Ross Project area (Strata, 2011b). The emissions from these types of vehicles would include carbon monoxide (CO), CO_2, sulfur dioxide (SO_2), nitrogen species (NO_x), and total hydrocarbon (THC) as well as particles less than 10 μm in diameter (PM_{10}) (Strata, 2011a). These emissions are consistent with those from a generic ISR project described in the GEIS (NRC, 2009).

Smaller sources of airborne non-radioactive gaseous and particulate emissions during operation would also include fugitive dust from cementing operations; welding fumes; particulates from grinding steel during construction and during operation; salt and soda ash during process-chemical delivery; and fumes from chemicals used in the laboratory, in addition to the carbon dioxide, oxygen, and water vapor that would be vented from the Ross Project. Vanadium precipitation, drying, and packaging would also present a potential for non-radioactive particulate emissions.

Fugitive dust would also be generated during all phases of the Proposed Action due to the mechanical disturbance of soil by heavy equipment, from transport vehicles traveling on access roads, and from wind blowing over disturbed areas and stockpiles. The Applicant has proposed to mitigate fugitive-dust emissions with its use of speed limits, strategic placement of water-loading facilities near access roads, suppression of dust with chemicals such as magnesium chloride, selection of road-surface materials that would minimize dust, and prompt revegetation of disturbed areas (Strata, 2011a).

Radioactive Emissions

Radon gas would be the primary radioactive gaseous effluent from the Ross Project. Radon is a radioactive, colorless, and odorless gas that occurs naturally as the decay product of radium, which is found where there is uranium as radium itself is a radioactive decay product of uranium. Radon would be found in the lixiviant solution that is extracted from the wellfields and piped to the CPP for processing. Radon gas could potentially be released in the CPP as a result of uranium-recovery fluid spills, filter changes, IX resin-transfer operations, and maintenance activities. Routine monitoring of radon progeny (i.e., the products of radon's own radioactive decay) within the CPP would identify exposure levels and would allow timely corrective actions to be initiated, if necessary (Strata, 2011b). The sources of radon described by the Applicant and the design features proposed by the Applicant to limit radon concentrations (e.g., the use of proper ventilation systems and radon detectors) are consistent with the industry standard described in the GEIS (NRC, 2009).

All exhaust points in the CPP would be ducted through a common system to a wet scrubber and discharged to the atmosphere (Strata, 2011b). The Applicant has committed that these discharges would meet all local, State, and Federal requirements related to air quality as well as occupational health and safety (Strata, 2012b). A performance-monitoring station would be located at the CPP's exhaust fan's point of discharge at the roof. The ambient air within the facility would be gravity ventilated up through a ridge vent. The CPP and other buildings would also be passively ventilated by the opening and closing of doors during periods of time when radon could be released.

Radon gas could also be released outside of the CPP from wellheads, other auxiliary buildings such as well modules, and the surface impoundments (Strata, 2011b). At the wellheads and the surface impoundments, radon would be released directly to the atmosphere, where it would rapidly disperse and decrease in concentration. Wellhead enclosures, such as the module buildings, would be vented to reduce radon buildup that could otherwise expose wellfield personnel to radon during inspection and maintenance activities. The Applicant proposes that, if vents are not installed on wellhead enclosures, SOPs would be developed for accessing wellheads to ensure radon exposures are below the regulatory limits of the EPA and the NRC. Such buildings would have ventilation systems consisting of a roof- or wall-mounted fan as well as a separate radon ventilation system with an intake located in the building's sump and an exhaust point on the building's roof.

Potential radioactive particulate emissions would consist primarily of airborne yellowcake in the uranium drying and packaging process (Strata, 2011b). This potential would be mitigated by design features to prevent releases into the atmosphere as described earlier in this section of this SEIS.

Liquid Effluents

The GEIS, Section 2.7.2, describes the liquid effluents generated during all phases of uranium recovery: construction, operation, aquifer restoration, and decommissioning. During most of these phases, liquid wastes could contain elevated concentrations of radioactive and chemical constituents. The composition and quantities of liquid waste from Ross Project processes related to uranium recovery are similar to those ranges provided in Table 2.7-3 of the GEIS (NRC, 2009); however, representative water quality parameter(s) for permeate are not included in the GEIS for comparison. The methods that the Applicant proposes for treatment of liquid wastes, such as RO as well as its disposal and management practices, are similarly noted as industry standards in the GEIS (NRC, 2009).

The Proposed Action would generate liquid effluents classified as byproduct wastes as well as other liquid effluents that are not (Strata, 2011b; Strata, 2012a). Liquid wastes would be categorized as follows:

- Brine and permeate from the RO treatment of lixiviant bleed and ground water from aquifer restoration. Most of the permeate would be reused as lixiviant in the wellfields and as process make-up water.
- Other liquids such as spent eluate, collected fluids from drains in the processing areas at the CPP, contaminated reagents, IX resin wash water, filter back wash, facility wash-down water, decontamination water (e.g., employee showers), and fluids generated from work-over and enhancement operations on injection and recovery wells.
- Non-byproduct liquid wastes would include drilling fluids and ground water collected during construction and development of injection, recovery, and monitoring wells as well as during environmental sampling and aquifer testing; storm-water runoff; toxic and hazardous wastes such as petroleum products and spent chemicals; and domestic sewage.

The Applicant proposes the use of surface impoundments for the collection and management of byproduct waste liquids (Strata, 2011a). Production of liquid byproduct wastes would vary over the three phases of operations and ground-water restoration: 1) operation only; 2) concurrent operations and aquifer restoration; and 3) aquifer restoration (Strata, 2011b).

GEIS Section 2.7.2 described four disposal options for use at ISR facilities: evaporation, land application, deep-well injection, and surface-water discharge (NRC, 2009). Of these disposal options, the Applicant proposes to rely on deep-well injection, with supplemental disposal by evaporation of brine and disposal of excess permeate from the surface impoundments (Strata, 2011b; Strata, 2012a). Land application is not currently proposed as a method for permeate disposal by the Applicant (Strata, 2012b). The surface impoundments would primarily provide transient storage of liquids with little evaporation actually occurring during the liquids' residence time.

Excess permeate could be produced during two relatively brief periods of operations (Strata, 2011b): the first two and one-half years of uranium production without reinjection of permeate into the aquifer for wellfield restoration and the two months when ground-water sweep is occurring in the first wellfield modules to undergo aquifer restoration. The Applicant proposes that excess permeate during the periods of uranium-recovery would be disposed of by deep-well injection (WWC Engineering, 2013). As noted earlier, the Applicant would utilize Class I deep-well injection for disposal of brine and other liquid wastes (Strata, 2011b). WDEQ has approved a UIC Class I Permit for up to five wells to be installed in the Deadwood and Flathead Formations (Permit No. 10-263) (WDEQ/WQD, 2011b). The Applicant expects the capacity of each of the five Class I wells to range between 132.5 – 302.8 L/min [35 – 80 gal/min]. The Applicant proposes a storage tank that, along with the lined impoundments, would provide surge capacity for management of the brine (Strata, 2012b).

Net annual evaporation of brine in the surface impoundments would be approximately 5.3 L/min-ac [1.4 gal/min-ac] which would reduce the volume of brine injected in the disposal wells (Strata, 2011b). The Applicant estimates typical flow rates of brine mixed with other byproduct liquid waste to the deep-disposal wells of 235 L/min [62 gal/min] during the operation-only phase; 859 L/min [227 gal/min] during the phase where the ISR facility is operating concurrently with aquifer restoration; and 719 L/min [190 gal/min] during the aquifer-restoration-only phase (Strata, 2011a). Brine produced during decontamination and decommissioning would be less than 38 L/min [10 gal/min] (Strata, 2011a). The Applicant's estimated flow rate of brine, permeate, and other liquid wastes for disposal would be less than noted in the GEIS (Table 2.7-3) (NRC, 2009).

The following non-byproduct (non-radioactive) liquid wastes would be generated at the Ross Project:

- Storm water from the paved areas of the proposed Ross Project facility
- Domestic sewage from the proposed facility
- Drilling fluids from construction of the proposed wellfields

Storm-water management would be controlled under a WYPDES Permit from WDEQ. As part of this permit, best management practices (BMPs) would be developed to restrict contaminants from the surface water and storm drains. Runoff from the facility would be diverted by the storm-drain system to a sediment surface impoundment near the CPP (Strata, 2011b).

The Applicant estimates that the volume of domestic sewage would range between 1,100 L/d [300 gal/d] and 4,500 L/d [2,600 gal/d] depending upon the number of workers during each project phase (Strata, 2012a). Domestic waste water would be collected in a gravity-sewer collection system serving the administration building, CPP, maintenance building, and any other buildings or structures with restrooms. This system would be designed according to WDEQ/WQD standards and would include one or more septic tanks for primary treatment. Septic-tank effluent would be disposed in a drainfield or in an enhanced treatment system (Strata, 2011b).

Drilling fluids of ground water and drilling muds would be produced only during the construction phase from the drilling and development of injection, recovery, and monitoring wells. The Applicant estimates that a volume of 22,000 L [6,000 gal] of water and 12 m^3 [15 yd^3] of drilling muds would be produced per well. The fluid would be stored onsite in mud pits constructed adjacent to the respective drilling pad(s) and evaporated. The Applicant expects the production of ground water during operation and decommissioning from wells completed outside of the aquifer exempted for uranium recovery (Strata, 2011a). This ground water would be discharged under a temporary WYPDES Permit. The Applicant was authorized to discharge these fluids under a temporary WYPDES Permit (No. WYG720229) issued during installation and sampling of monitoring wells (WDEQ/WQD, 2011a). This Permit was renewed in December 2012.

Solid Effluents

The GEIS describes the solid-phase wastes that would be generated during all phases of uranium-recovery operations. These solid wastes would be hazardous, radioactive, or typical solid waste. The projections of solid-waste generation and management methods proposed by the Applicant for the Proposed Action are within the industry standards described in Section 2.7 of the GEIS (Strata, 2011b; Strata, 2012b; NRC, 2009). The Applicant provides a list of anticipated waste disposal facilities with adequate capacity that could be used for waste generated at the Ross Project (Strata, 2012a).

The Applicant estimates the production of 19 L/mo [5 gal/mo] of used oil and less than 9 kg/mo [20 lb/mo] of used oil filters and oily rags. These wastes would be stored in a designated used-oil storage area and would be shipped to a commercial recycling facility for disposal, such as Tri-State Recycling Services, Newcastle, Wyoming (Strata, 2012a). Petroleum-contaminated soil, estimated as less than 1 m^3/wk [1 yd^3/wk], would be transported by a waste-disposal contractor to a permitted land farm in northeast Wyoming such as the Campbell County Landfill (Strata, 2012a).

Less than 100 kg/mo [220 lb/mo] of waste designated as hazardous by the EPA and WDEQ, such as used batteries, expired laboratory reagents, burnt-out fluorescent light bulbs, spent solvents, certain cleaners, and used degreasers, would also be generated (Strata, 2012a). The hazardous waste would be stored at the Ross Project in secure, specially designed containers inside the maintenance shop. The Applicant expects the Ross Project to be classified as a conditionally exempt small quantity generator (known as a CESQG) of hazardous waste (Strata, 2011b). Hazardous waste would be transported by a hazardous waste contractor to an appropriately permitted commercial recycling facility outside Wyoming (Strata, 2012a). The Applicant proposes onsite disposal contaminated laboratory reagents in the lined retention impoundments and deep-well injection (Strata, 2012a).

Radioactive byproduct solid waste that would be generated at the Ross Project include filtrate and spent filter media from production and restoration circuits; general sludge, scale, etc. from maintenance operations; affected soil collected from any spill or leak areas; spent/damaged ion exchange resin; well solids from injection/recovery well work-over operations; contaminated PPE; wellfield decommissioning waste such as pipelines, pumps, and impacted soil; affected concrete floors, sumps and berms in the CPP; equipment and piping in the CPP; pond sludge, pond liners, and leak detection systems; and disposal well piping and equipment (Strata, 2012a). Byproduct solid wastes would be generated during all Proposed Action phases, except construction. During facility operation and aquifer restoration, the Applicant estimates the production of 80 m^3/yr [100 yd^3/yr] of solid byproduct waste. The largest volumes of byproduct waste, including contaminated soil requiring licensed disposal, would be generated during facility decommissioning, which is estimated to be 4,000 m^3 [5,000 yd^3] (Strata, 2012a). The Applicant has identified four facilities with sufficient capability located in Wyoming, Utah, and Texas that are permitted to accept byproduct waste from ISR facilities (Strata, 2012a).

During all phases of the Proposed Action, when any byproduct wastes are generated, they would be stored inside a locked and posted room within the CPP (i.e., this area would be a restricted area). The wastes would be placed inside 208-L [55-gal], lined drums, sealed and placed inside a 15-m^3 [20-yd^3] roll-off container. The sealed roll-off containers containing the waste would be transported by a licensed transporter to a licensed radioactive waste facility for disposal. The Applicant anticipates about five annual shipments of byproduct wastes during the facility-operation and aquifer-restoration phases. During decommissioning, which is expected to last 12 to 18 months, up to 200 shipments per year would be expected (Strata, 2011b).

Non byproduct solid wastes generated at the Ross Project include ordinary trash, petroleum-contaminated soil, construction debris, and decontaminated material and equipment. The Applicant estimates that 12 m^3/wk [15 yd^3/wk] of ordinary municipal solid waste such as office trash along with 4 m^3/wk [5 yd^3/wk] of recyclable wastes (plastic, glass, paper, aluminum, and cardboard) would be generated throughout the life of the Ross Project (Strata, 2012b). Small amounts (less than 0.8 m^3/wk [5 yd^3/wk]) of petroleum-contaminated soil would also be generated. The generation of solid waste consisting of construction debris and decontaminated materials and equipment would be less than 4 m^3/wk [5 yd^3/wk] during facility construction and operation, and aquifer restoration. During the decommissioning phase, the Applicant estimates up to 1,500 m^3 [2,000 yd^3] of such solid waste (Strata, 2012a).

During facility operation and aquifer restoration, non-hazardous solid wastes would be collected daily from work areas and disposed in trash receptacles located within the facility, but near a primary access road for convenient access for a waste-disposal contractor. Non-hazardous solid waste would be disposed offsite in the Moorcroft landfill or the Campbell County landfill in Gillette, Wyoming (Strata, 2011a). Solid waste of construction and demolition debris would be disposed in the municipal or country landfills in the three towns nearest the Ross Project: Moorcroft, Sundance, and Gillette.

2.1.1.6 Transportation

Primary transportation activities would involve truck shipping and personnel commuting. A variety of truck shipments are planned to support proposed activities during all phases of the Proposed Action. Light-duty trucks and automobiles would transport construction contractors and the operations workforce. Baseline transportation conditions and impact of the Ross Project are discussed in SEIS Sections 3.3 and 4.3, respectively.

Transportation routes within 80 km [50 mi] of the Proposed Action include interstate highways, other U.S. highways, Wyoming highways, county roads, and local roads (Strata, 2011a). The major transportation corridors that could be used to access the Ross Project area include Interstate-90, approximately 32 km [20 mi] south; U.S. Highway 14, approximately 16 km [10 mi] southeast; State Highway 59, approximately 32 km [20 mi] west; and U.S. Highway 212, approximately 64 km [40 mi] northeast. Regional and local transportation routes are shown on Figure 2.1.

The primary access to the Ross Project area is from D Road [CR 68] from the New Haven Road (CR 164). The primary access road to the ISR facility would be constructed to flow from New Haven Road (CR 164). The design of the road includes a 9 m [30 ft] top width with 5 horizontal to 1 vertical side slopes. According to American Association of State Highway and Transportation Officials (AASHTO), a 5:1 slope is traversable and recoverable; therefore, no guardrails would be used on the access road (AASHTO, 2002; Strata, 2011b).

2.1.1.7 Financial Surety

Prior to commencement of operations, the Applicant would be required to provide assurance that sufficient funds will be available to cover decontamination, dismantling, and decommissioning as well as to cover aquifer restoration of the Ross Project, including all costs of site reclamation and decommissioning waste disposal (10 CFR Part 40, Appendix A, Criterion [9]). A decommissioning funding plan (DFP) would be required from the Applicant as an NRC license condition; the DFP would contain a decommissioning cost estimate, the amount of which the Applicant would be required to maintain in a financial-surety arrangement. The initial decommissioning cost estimate would be based upon the first year of operation, which includes the construction of the CPP, and would be fully described in the DFP. NRC license conditions and the WDEQ/LQD Permit to Mine would also require, on a forward-looking basis, annual revisions to the decommissioning cost estimate and the related financial surety. When NRC,

WDEQ, and the Applicant have agreed to the initial cost estimate and DFP, the Applicant would submit a surety instrument acceptable to both NRC and WDEQ. Details of NRC's requirement for financial surety would be part of the Safety Evaluation Report (SER) for the Ross Project and the surety would be required by the Applicant's NRC license. The Applicant would be required to maintain these surety arrangements until the NRC determined that the Applicant had complied with its reclamation plan. For additional information on decommissioning funding plans and financial-surety requirements, see 10 CFR Part 40, Appendix A; NUREG–1757, *Consolidated NMSS Decommissioning Guidance*; and the GEIS in Section 2.10 (NRC, 2003b; NRC, 2009).

2.1.2 Alternative 2: No Action

Under the No-Action Alternative, the NRC would not issue a license for the proposed ISR project and BLM would not approve the Applicant's Plan of Operations (POO). The No-Action Alternative would result in the Applicant's not constructing, operating, restoring the aquifer of, or decommissioning the proposed ISR project. However, even if the proposed Ross Project is not licensed, the Applicant has already accomplished certain preconstruction activities that do not require an NRC license or BLM POO at the Ross Project area. At no time would radioactive materials be present at the Ross Project during any preconstruction activities. These previously completed preconstruction activities are evaluated as part of Alternative 2: No Action.

Preconstruction activities that have already been accomplished include the Applicant's locating and properly abandoning the former Nubeth's exploration drillholes. As of October 2010, the Applicant has located 759 of the 1682 holes thought to exist from Nubeth exploration activities and has plugged 55 of them (Strata, 2011b). In addition, Strata has drilled and then properly abandoned 512 holes used to delineate the ore zone. The Applicant has also drilled and completed 51 wells for ground-water monitoring and testing (Strata, 2011a) as well as installed 3 surface-water monitoring stations and a meteorology station. Data collection activities from the ground-water wells, surface-water stations, and the meteorological station are continuing. In August 2011, an additional 74 drillholes and 4 ground-water monitoring wells were installed to support a geotechnical investigation of the area proposed for the Ross Project (Strata, 2012b). These drillholes have also been properly plugged and abandoned, and the four ground-water monitoring wells are being used for ongoing ground-water monitoring. Finally, a ranch house that was present on the property has been remodeled to serve as the Applicant's Field Office at the Ross Project area.

In the No-Action Alternative, no uranium would be allowed to be recovered from the subsurface ore zone, and no injection, production, or monitoring wells would be installed. No lixiviant would be introduced to the subsurface, and no recovered uranium would be extracted and no facilities would be constructed to process extracted uranium or store chemicals. The No-Action Alternative is included to provide a benchmark for the NRC to compare and evaluate the potential impacts of the other alternatives, including the Proposed Action.

2.1.3 Alternative 3: North Ross Project

Under Alternative 3, the NRC would issue the Applicant a license for the construction, operation, aquifer restoration, and decommissioning of the proposed ISR project, except that the entire ISR facility itself, which includes all buildings, other auxiliary structures, and the surface impoundments would be located north of where it is to be situated during the Proposed Action, but the locations of the wellfields would not change. This alternate location for the ISR facility, referred as the "north site" by the Applicant (and referred to herein as the "North Ross Project"), was considered, but eliminated, by the Applicant in its license application (Strata, 2011a). The north site is located about 240 m [800 ft] north of the Oshoto Reservoir in S½SW¼ Section 7, T53N, R67W (see Figure 2.11). It is about 900 m [3,000 ft] northwest of where the facility would be located in the Proposed Action (referred to by the Applicant as the "south site"). An unnamed surface water drainage feature generally divides the north site. To avoid the floodplain of the drainage an actual design of the facility at this site would likely place the CPP and other buildings on one side of the drainage and the surface impoundments on the other side.

The Applicant documents its decision to select the south site over the north site with the following comparisons (Strata, 2011a):

- The south site is situated on relatively flat topography, which would minimize the amount of earthwork and surface disturbance required to prepare the site for construction of the CPP, auxiliary buildings, surface impoundments, and parking areas.
- The south site's surface is entirely privately owned and onsite instrumentation is currently adequate for all required pre-operational baseline environmental studies (see 10 CFR Part 40, Appendix A).
- The south site has little uranium mineralization beneath it, and what there is would be accessible without major modification of the wellfield- and monitoring-well layout.
- The preliminary geotechnical studies at the south site indicate that subsoil materials are relatively impermeable and have adequate strength for the proposed buildings and structures.
- The preliminary estimates of the radionuclide release rates from the entire project, including the south site, indicates that the average annual radiation dose to the nearest receptor would be less than 5 percent of the NRC's 1 mSv/yr [100 mrem/yr] annual limit.
- The owner of the south site is also the owner of the Oshoto Reservoir, so a surface-use agreement, lease, or purchase of this area would afford Strata control over the Reservoir as well.

The North Ross Project is included as an Alternative in this SEIS because of the expected differences in the depth to ground water between the north and south sites. Based upon the water levels measured in a nearby well cluster, Well No. 12-18, and the surface topography, shallow ground water of the north site is likely to be greater than 15 m [50 ft] below the ground surface (Strata, 2011a). In contrast, shallow ground water beneath the south site ranges from 2 – 4 m [8 – 12 ft] below the ground surface and necessitates the construction of the CBW (Strata, 2011b).

Certain factors related to the north site as a location for the proposed Ross Project facility are considered in this SEIS's impact analyses. These factors include:

- The north site's deeper ground-water levels, which could eliminate the need for a CBW and dewatering in order to protect ground water.

- The north site's more pronounced topography, which could require more earthwork and surface disturbance for construction of the facility and surface impoundments.
- The north site's greater distance to the Little Missouri River, which could mitigate potential impacts on surface-water resources.
- The north site's natural screen provided by the ridges to the west, north, and east, which could decrease impacts on visual and scenic resources.
- The north site's increased uranium mineralization beneath it, which could potentially require a reconfiguration of the facility to allow uranium recovery.

2.2 Alternatives Eliminated from Detailed Analysis

This section describes alternatives to the Proposed Action that were considered for this SEIS, but were not carried forward for detailed analysis. Section 2.2.1 describes the recovery of uranium by conventional mining and milling; Section 2.2.2 discusses the use of a lixiviant with different chemistry; and Section 2.2.3 compares alternative methods of waste management.

2.2.1 Conventional Mining and Milling

The GEIS includes an evaluation of conventional mining and milling as an alternative to ISR (NRC, 2009). Although the characteristics of the uranium deposits of the proposed Ross Project are amenable to ISR extraction, evaluating the Proposed Action against the conventional mining and milling allows comparison of impacts of the two uranium-recovery methods. Conventional mining practices (open-pit and underground) to recover uranium ore in addition to conventional milling were considered and eliminated as an alternative to ISR operations at the proposed Ross Project, as they were in the GEIS (NRC, 2009; Strata, 2011a).

Conventional mining refers to the physical removal of uranium ore by either underground mining methods or from an open pit. Uranium is extracted and converted to yellowcake in a processing facility; this process is referred to as uranium "milling." Open-pit mining is suitable for shallow ore deposits, generally deposits less than 170 m [550 ft] below ground surface (bgs), such as those found at the Ross Project area.

Underground mining could be used for deeper deposits; however, the cost of underground mining and milling requires a higher grade of ore to be economically feasible compared to open pit-mining and ISR (EPA, 2008). Uranium-ore grade in the Lance District is low-grade (Strata, 2011a; Peninsula, 2011). The ore zone at the Ross Project is approximately 30 – 60 m [90 – 180 ft] thick (Strata, 2011b). The base of the ore is generally at depths of 150 – 200 m [500 – 700 ft], which is nearly the maximum depth for surface mining to practically recover uranium from an open pit.

In addition to the depths involved with open-pit mining, water consumption of open-pit mining likely would be greater than at an ISR facility because of the required dewatering down to the depth of the pit's floor. At the Ross Project, dewatering of several aquifers above the ore zone and the ore zone itself would be required for open-pit mining and large amounts of water would be produced (Strata, 2011a).

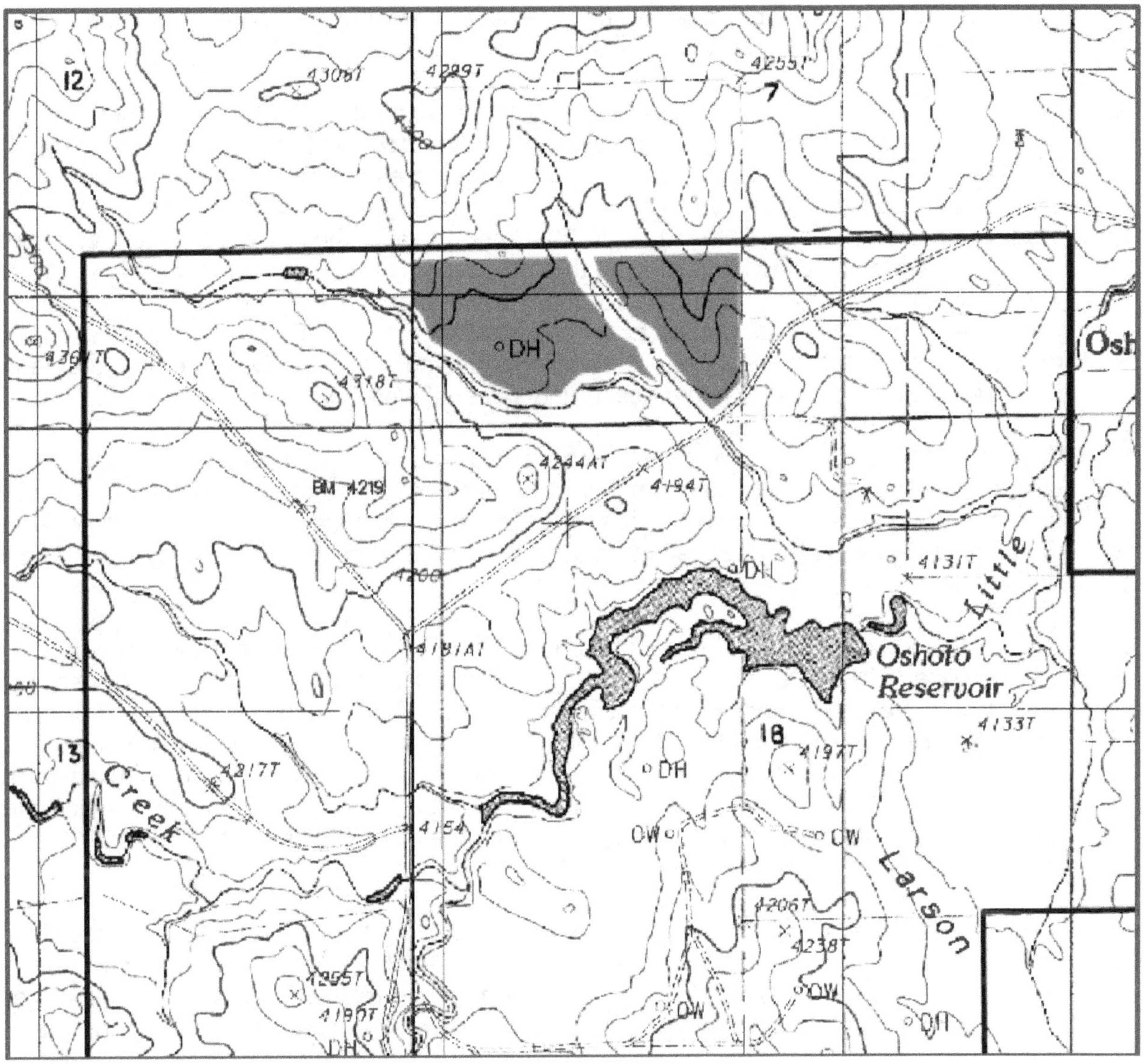

Source: Strata, 2012a

Figure 2.11

Alternative 3: North Ross Project (CPP on Right and Surface Impoundments on Left)

Far greater areas of land disturbance would occur from an open-pit mine compared with the Ross Project and the required restoration of the open pit would be far more extensive. Even though overburden could be backfilled into the pit, the pit would permanently impact the surface's appearance and its land use.

Conventional uranium milling requires construction of a facility that would be larger than the proposed facility at the Ross Project. As described in Appendix C of the GEIS (NRC, 2009), ore processing at a conventional uranium mill involves a series of steps (handling and preparation, concentration, and product recovery). Uranium ore is crushed, ground, and classified to produce uniform-sized particles (EPA, 2008). After grinding, the ore is added to a series of tanks for leaching by a lixiviant similar to that proposed by the Applicant for the Ross Project. The precipitation of uranium from the pregnant lixiviant, drying the product, and packaging the yellowcake follow the same processes as proposed for the Ross Project. Emissions containing radiological constituents generated by handling, grinding, and classifying the ores creates the potential for greater impacts to the health and safety of workers.

Wastes generated by milling include the spent ore, which are referred to as "tailings." The volume of tailings is roughly 95 percent of the volume of the ore brought to the mill. Wastes from conventional uranium milling, such as well waste water, spent resins, and filtrate, would be the same as the wastes generated by Applicant's proposed processing of pregnant lixiviant from ISR wellfields.

Wet tailings are disposed in surface impoundments constructed with liners and covers to prevent escape to the environment. Although the chemical character of tailings depends upon the uranium ore and lixiviant, tailings generally contain soluble metals, radium, and high levels of dissolved solids. Reclamation of a tailings pile generally involves evaporation of any liquid in the tailings, settlement of the tailings over time, and protection of the pile with a thick radon barrier and earthen material or rocks for erosion control. An area surrounding the reclaimed tailings piles would be fenced off in perpetuity, and the site transferred to either a State or Federal agency for long-term care (EIA, 1995).

As an alternative to conventional milling, uranium from low-grade ore that is recovered by open-pit mining can be recovered by heap leaching. Heap leaching occurs at or very near the mine site itself. The low-grade ore is crushed to a fine size and mounded above grade on a prepared pad. A sprinkler or drip system distributes lixiviant over the mound. The lixiviant trickles through the ore and mobilizes uranium into solution. The solution is collected at the base of the mound and processed to produce yellowcake. The processing to yellowcake of the pregnant lixiviant would be the same as for the Ross Project.

Given the uranium ore grade and depth to the ore, open-pit mining and conventional milling would be possible at the Ross Project; however, the costs, environmental impacts, and potential health and safety impacts to workers are more substantial than impacts from the ISR process (see SEIS Section 4).

As noted in the GEIS on uranium milling (NRC, 1980b), besides cost considerations, the environmental impacts of open-pit mining, and tailings impoundment would be greater than from an ISR project. Greater impacts such as those listed below would affect land use and soils as well as ecological, water, and air resources. Some of these impacts are:

- A larger area of surface disturbance for an open-pit mine and uranium mill, which could increase environmental impacts.
- A permanent tailings pile, which would require long-term care and maintenance to prevent impacts to air and water.
- A permanent mine pit if an open-pit mining were to be used, into which groundwater would flow creating a lake of poor water quality.
- A greater consumptive water use, which would result from the ground water's intruding into the mine and its needing to be pumped (i.e., dewatered) with the excess water then discharged to the environment.
- A greater surface discharge of water, which would result from the pumping and treatment of excess water from the mine pit.

The mine workers' excavating the uranium ore during the mining operation, through the uranium milling process itself, and the disposal of the tailings also increase the potential impacts to workers' health and safety.

Based upon these greater impacts, the alternatives of conventional uranium mining and milling have been eliminated from further analysis in this SEIS.

2.2.2 Alternate Lixiviant Chemistry

The lixiviant proposed for the Ross Project is consistent with the assumption in the GEIS that the ISR process would employ alkaline lixiviants (NRC, 2009). Alkaline solutions are typically used to dissolve uranium in the ore zone when the lime content of the host rock in the ore zone is above 12 percent, which is the case for the Ross Project site (Strata, 2011b). Other lixiviants can be made with sulfuric acid or ammonia, and these have been shown to dissolve uranium (NRC, 2009). However, the lixiviant that is selected for a specific ISR project must be able to dissolve uranium from the host rock while it maintains the permeability of the aquifer. In addition, the lixiviant and its reaction products must be amenable to ground-water restoration.

How do you select a proper lixiviant?

The geology and ground-water chemistry determine the proper ISR techniques and chemical reagents used for uranium recovery. For example, if the ore-bearing aquifer is rich in calcium (e.g., limestone or gypsum), alkaline (carbonate), lixiviant might be used (Hunkin, 1977, as cited in NRC, 2009). Otherwise, an acid (sulfate) lixiviant might be preferable. The lixiviant chemistry chosen for ISR operations could affect the type of potential contamination and the vulnerability of aquifers during and after ISR operations.

Typical ISR operations in the U.S. use an alkaline sodium bicarbonate system to remove the uranium from ore-bearing aquifers. In addition, aquifers where an alkaline-based lixiviant was used were considered to be easier to restore than those where acid lixiviants were used (Tweeton and Peterson, 1981, and Mudd, 1998, as cited in NRC, 2009).

Acidic lixiviant has been used most broadly in conventional milling. These acid-based fluids have generally achieved high yield and efficient, rapid uranium recovery, but they also dissolved other metals associated with the uranium in the host rock, and this dissolution can contribute to adverse environmental impacts. In Wyoming, acid lixiviants have been used for small-scale research and development operations, but they have not been used in commercial operations (NRC, 2009). Tests with acid lixiviants have identified two major problems: 1) gypsum (a calcium mineral) precipitates on well screens and within the aquifer during uranium recovery, plugging wells and reducing the aquifer's permeability, which is critical for economic operation; and 2) the precipitated gypsum gradually dissolves after aquifer restoration, increasing the salinity and sulfate levels in the ground water. Because of the potential impacts of soluble metals and increased salinity in the aquifer as well as the potential for plugging of the aquifer by their use, acid-based lixiviants have been eliminated from further analysis in this SEIS.

Ammonia-based lixiviants have been used at some ISR operations in Wyoming. However, operational experience has shown that ammonia tends to adsorb onto clay minerals in the ore zone and then slowly dissolves from the clay during aquifer restoration, therefore requiring that a much larger volume of ground water be removed and processed during the aquifer restoration phase (NRC, 2009). Traces of the ammonia from the lixiviant have remained in affected aquifers even after extensive aquifer restoration. Because of the greater consumption of ground water to meet aquifer-restoration requirements, the use of an ammonia-based lixiviant has been eliminated from further analysis in this SEIS.

2.2.3 Alternate Waste Management Methodologies

Liquid-effluent disposal practices that the NRC has previously approved for use at specific ISR sites include waste evaporation from surface impoundments, application of waste on land, injection of waste into deep wells, and discharge of waste to surface water (NRC, 2009).

The Proposed Action would employ injection into a UIC-permitted Class I well as the primary method of disposal of the brine and other process waste waters excluding permeate from the RO process. The Proposed Action would include surface impoundments located near the CPP to store and manage the brine and to allow reuse of permeate as lixiviant or process water. Of the approximately 6.5 ha [16 ac] of impoundment surface area in the Proposed Action, 2.5 ha [6.3 ac] would be available for evaporation (Strata, 2011b). The Applicant predicts that the evaporation of brine during the time it is stored in the surface impoundments would reduce the volume for deep disposal by 20 percent during the operation-only phase and about 5 percent during the concurrent operation- and aquifer-restoration phases. Excess permeate while stored in the surface impoundments would evaporate at an average annual rate of 1.5 gpm per surface acre (Strata, 2012b).

Reliance on evaporation to dispose of all the brine and other liquid byproduct wastes generated at the CPP, and thus eliminating the need for deep-well injection, would require a larger surface area of the impoundments. The maximum production of brine and other process waste occurs during the concurrent facility operation and aquifer-restoration phases. During this time, 859 L/min [227 gal/min] of byproduct liquid would be generated (Strata, 2011a). The remaining surface-impoundment volume in the Proposed Action would be used for permeate management and reserve capacity in the event of upset conditions.

The Applicant has estimated that the 2.5 ha [6.3 ac] available for evaporation in the Proposed Action would provide 33.3 L/min [8.8 gal/min] of average annual evaporation. Linear extrapolation suggests that 65 ha [160 ac] is the minimum surface area required for evaporation of all brine and other byproduct waste generated at the CPP. Considering the requirement to maintain reserve capacity to manage upset conditions and the natural fluctuations, the necessary surface impoundments would exceed 80 ha [200 ac]. Impoundments of sufficient size to eliminate the need for deep-well injection would nearly double the disturbed area. In the Proposed Action, approximately 113 ha [280 ac] would be disturbed during the entire Ross Project. The disturbed area required for only evaporation would be present throughout the entire construction, operation, aquifer restoration and decommissioning phases. It is likely that the CBW would need to be constructed around these large surface impoundments. Because the CPP and the surface impoundments would be expected to remain operational after the life of the proposed wellfields of the Ross Project, the surface impoundments would likely be in place for more than 10 years.

These large-scale surface impoundments could potentially impact land use and soils as well as ecological, water, air, and visual resources. These impacts and related occupational health impacts could require mitigation. In contrast, the GEIS concluded that the permit process required for a Class I injection well provides confidence that the impacts from deep-well disposal would be SMALL. For these reasons, the alternative of the elimination of waste disposal in Class I deep-injection wells in favor of surface impoundments over more than 12 times the area of impoundments in the Proposed Action has not been carried forward for impact analysis in this SEIS.

2.3 Comparison of Predicted Environmental Impacts

The GEIS categorized the significance of potential environmental impacts as described in the adjacent text box (NRC, 2009). The large table, presented in the "Executive Summary" as Table ExS.1, summarizes the potential environmental impacts to each resource area for all four of the Ross Project's phases: construction, operation, aquifer restoration, and decommissioning. The levels of significance—SMALL, MODERATE, and LARGE—are noted for each resource area.

How is the significance of identified impacts classified?

- ***Small Impact*:** The environmental effects are not detectable or are so minor that they will neither destabilize nor noticeably alter any important attribute of the resource considered.
- ***Moderate Impact*:** The environmental effects are sufficient to alter noticeably, but not destabilize, important attributes of the resource considered.
- ***Large Impact*:** The environmental effects are clearly noticeable and are sufficient to destabilize important attributes of the resource considered.

The respective resource areas, as they currently exist at the Ross Project area, which is called the "affected environment," are described in Section 3 of this SEIS. The potential environmental impacts of the Ross Project are evaluated in Section 4 of this SEIS. The measures intended to mitigate any impacts are also discussed in SEIS Section 4 of this SEIS.

2.4 Preliminary Recommendation

After weighing the impacts of the Proposed Action and comparing the Alternatives, the NRC staff, in accordance with 10 CFR Part 51.71(f), sets forth its preliminary NEPA recommendation regarding the Proposed Action. Unless safety issues mandate otherwise, the preliminary NRC staff recommendation to the Commission related to the environmental aspects of the Proposed Action is that a source and byproduct materials license for the Proposed Action be issued as requested. The NRC staff concludes that the applicable environmental monitoring program described in Chapter 6 and the proposed mitigation measures discussed in Chapter 4 will eliminate or substantially lessen the potential adverse environmental impacts associated with the Proposed Action.

The NRC staff has concluded that the overall benefits of the proposed action outweigh the environmental disadvantages and costs based on consideration of the following:

- Potential adverse impacts to all environmental resource areas are expected to be SMALL, with the exception of

    1. Transportation resources during all phases of the proposed action. Increases in traffic during construction and operation would have a MODERATE to LARGE impact. Impacts would be MODERATE with mitigation for construction, operation, aquifer restoration, and decommissioning (See SEIS Sections 4.3.1.1, 4.3.1.2, 4.3.1.3, and 4.3.1.4).

    2. Groundwater resources during operation and aquifer restoration. During operations there would be a MODERATE impact to ore-zone aquifer water quality due to excursions; however with measures in place to detect and resolve the excursions, the impacts would be reduced. During aquifer restoration there would be a MODERATE impact to ore-zone aquifer water quantity due to short-term drawdown (See SEIS Sections 4.5.1.2 and 4.5.1.3).

    3. Noise resources during construction, operations, and decommissioning. During these phases of the Ross Project there would be MODERATE impacts due to increased noise levels, however they would be intermittent and short term (See SEIS Sections 4.8.1.1, 4.8.1.2 and 4.8.1.4).

    4. Historical and cultural resources during construction. Section 106 consultation and efforts to identify and determine the eligibility of historical and cultural resources that could be adversely affected by the proposed Ross Project are currently ongoing. Therefore, to be conservative in this draft SEIS, the NRC staff considers that construction could have a MODERATE to LARGE impact on historic properties, sites currently listed or eligible for listing on the National Register of Historic Places (NRHP)—and other unevaluated historic, cultural, and religious properties in the project area (See SEIS Section 4.9.1.1). However, once identification efforts are complete, mitigation efforts, which could require an MOA, would be developed to reduce impacts. The final SEIS will include the outcome of Section 106 consultation and would discuss mitigation measures, including an MOA, if one is developed.

    5. Visual and scenic resources during construction. There would be MODERATE impacts to residents near the Ross Project for the first year, however over the long term, impacts would be reduced (See SEIS Section 4.10.1.1).

    6. Socioeconomic resources during construction and operations. There would be MODERATE impacts to Crook County during these phases of the Ross Project because taxes from the Project will be paid to the county (See Sections 4.11.1.1 and 4.11.1.2).

- Regarding groundwater, the portion of the aquifer(s) designated for uranium recovery must be exempted as underground sources of drinking water before ISR operations begin. Additionally, Strata would be required to monitor for excursions of lixiviant from the production zones and to take corrective actions in the event of an excursion. Prior to operations, the Applicant would be required to provide detailed hydrologic pumping test data packages and operational plans for each wellfield at the Ross Project. Strata would also be required to restore groundwater parameters affected by the ISR operations to levels that are protective of human health and safety.

- The costs associated with the Ross Project are, for the most part, limited to the area surrounding the site.

- The regional benefits of building the proposed Project would be: increased employment, economic activity, and tax revenues in the region around the proposed Project site.

2.5 References

10 CFR Part 20. Title 10, "Energy," *Code of Federal Regulations*, Part 20, "Standards for Protection Against Radiation," Subpart K, "Waste Disposal." Washington, DC: Government Printing Office. 1991, as amended.

10 CFR Part 40. Title 10, "Energy," *Code of Federal Regulations*, Part 40, "Domestic Licensing of Source Material," Appendix A, "Criteria Relating to the Operation of Uranium Mills and the Disposition of Tailings or Wastes Produced by the Extraction or Concentration of Source Material from Ores Processed Primarily for their Source Material Content." Washington, DC: Government Printing Office. 1985, as amended.

10 CFR Part 40. Title 10, "Energy," *Code of Federal Regulations*, Part 40, "Domestic Licensing of Source Material." Washington, DC: Government Printing Office. 1961, as amended.

40 CFR Part 145. Title 40, "Protection of the Environment," *Code of Federal Regulations*, Part 145, "State UIC Program Requirements." Washington, DC: Government Printing Office. 1983, as amended.

40 CFR Part 146. Title 40, "Protection of the Environment," *Code of Federal Regulations*, Part 146, "Underground Injection Control Program: Criteria and Standards." Washington, DC: Government Printing Office. 1980, as amended.

(US)EIA (U. S. Energy Information Administration). *Decommissioning of U.S. Uranium Production Facilities*. DOE/EIA-0592. Washington, DC: Office of Coal, Nuclear, Electric, and Alternate Fuels, EIA. February 1995. Agencywide Documents Access and Management System (ADAMS) Accession No. ML13011A269.

(US)EIA. *2011 Domestic Uranium Production Report*. Washington, DC: Office of Electricity, Renewables, and Uranium Statistics, EIA. May 2012. ADAMS Accession No. ML13011A271.

(US)EPA (U.S. Environmental Protection Agency). *Technical Report on Technologically Enhanced Naturally Occurring Radioactive Materials from Uranium Mining: Mining and*

Reclamation Background. Volume 1. EPA–402–R–08–005. Washington, DC: Office of Radiation and Indoor Air/Radiation Protection Division, USEPA. 2008. ADAMS Accession No. ML13015A579.

(US)NRC (U.S. Nuclear Regulatory Commission). *Operational Inspection and Surveillance of Embankment Retention Systems for Uranium Mill Tailings.* Regulatory Guide 3.11.1, Revision 1. Washington, DC: USNRC. October 1980a.

(US)NRC. NUREG–0706. "Final Generic Environmental Impact Statement on Uranium Milling Project M-25." Washington, DC: NRC. September 1980b. ADAMS Accession Nos. ML032751663, ML0732751667, and ML032751669.

(US)NRC. *Residual Radioactive Contamination From Decommissioning.* Volumes 1, 2, 3, and 4. NUREG–CR-5512. Washington, DC: USNRC. October 1999.

(US)NRC. *Standard Review Plan for In Situ Leach Uranium Extraction License Applications, Final Report.* NUREG–1569. Washington, DC: USNRC. June 2003a. ADAMS Accession No. ML032250177.

(US)NRC. *Consolidated NMSS Decommissioning Guidance/Decommissioning Process for Materials Licensees.* Volumes 1 Rev. 2, 2, and 3 Rev. 1. NUREG–1757. Washington, DC: USNRC. September 2003b. ADAMS Accession Nos. ML063000243, ML053260027, and ML12048A683.

(US)NRC. *Generic Environmental Impact Statement for In-Situ Leach Uranium Milling Facilities.* Volumes 1 and 2. NUREG–1910. Washington, DC: USNRC. May 2009. ADAMS Accession Nos. ML091480244 and ML091480188.

(US)NRC. "Site Visit and Informal Information Gathering Meetings Summary Report for the Proposed Ross In-Situ Recovery Project (Docket No. 040-09091)." Memorandum to K. Hsueh, Branch Chief, from A. Bjornsen, Project Manager, Office of Federal and State Materials and Environmental Management Programs. November 28, 2011. Washington, DC: USNRC. 2011. ADAMS Accession Nos. ML112980194.

Peninsula Energy, Ltd. "Lance Project, Wyoming." 2011. At www.pel.net.au/projects/lance_project_wyoming_usa.phtml (as of June 25, 2012).

Strata (Strata Energy, Inc.) *Ross ISR Project USNRC License Application, Crook County, Wyoming, Environmental Report, Volumes 1, 2 and 3 with Appendices.* Docket No. 40-09091. Gillette, WY: Strata Energy, Inc. 2011a. ADAMS Accession Nos. ML110130342, ML110130344, and ML110130348.

Strata. *Ross ISR Project USNRC License Application, Crook County, Wyoming, Technical Report, Volumes 1 through 6 with Appendices.* Docket No. 40-09091. Gillette, WY: Strata. 2011b. ADAMS Accession Nos. ML110130333, ML110130335, ML110130314, ML110130316, ML110130320, and ML110130327.

Strata. *Ross ISR Project USNRC License Application, Crook County, Wyoming, RAI Question and Answer Responses, Environmental Report, Volume 1 with Appendices.* Docket No. 40-09091. Gillette, WY: Strata. 2012a. ADAMS Accession No. ML121030465.

Strata. *Ross ISR Project USNRC License Application, Crook County, Wyoming, RAI Question and Answer Responses, Technical Report, Volumes 1 and 2 with Appendices.* Docket No. 40-09091. Gillette, WY: Strata. 2012b. ADAMS Accession Nos. ML121020357 and ML121020361.

WDEQ/LQD (Wyoming Department of Environmental Quality/Land Quality Division). "Noncoal In Situ Mining," *Rules and Regulations*, Chapter 11. Cheyenne, WY: WDEQ/WQD. 2005.

WDEQ/WQD (Wyoming Department of Environmental Quality/Water Quality Division). "Design and Construction Standards for Sewage Systems, Treatment Works, Disposal Systems or other Facilities Capable of Causing or Contributing to Pollution." *Rules and Regulations*, Chapter 11. Cheyenne, WY: WDEQ/WQD. 1984.

WDEQ/WQD. "Wyoming Surface Water Quality Standards," *Rules and Regulations*, Chapter 1. Cheyenne, WY. WDEQ/WQD. 2007.

WDEQ/WQD. *Authorization to Discharge Wastewater Associated with Pump Testing of Water Wells Under the Wyoming Pollutant Discharge Elimination System.* Authorization #WYG720229. Cheyenne, WY: WDEQ/WQD. March 2011a. ADAMS Accession No. ML13015A695.

WDEQ/WQD. *Strata Energy, Inc. – Ross Disposal Injection Wellfield, Final Permit 10-263, Class I Non-hazardous, Crook County, Wyoming.* Cheyenne, WY: WDEQ/WQD. April 2011b.

WWC Engineering. "Re: Request Update of ER Table 1.6-a." E-mail (February 1) from B. Schiffer to J. Moore, Project Manager, Office of Federal and State Materials and Environmental Management Programs, U.S. Nuclear Regulatory Commission. Sheridan, Wyoming: WWC Engineering. 2013. ADAMS Accession No. ML13035A012.

3 DESCRIPTION OF AFFECTED ENVIRONMENT

3.1 Introduction

The Ross Project would be located in northeastern Wyoming, in a rural area of western Crook County, approximately 35 km [22 mi] north of the town of Moorcroft, Wyoming (see Figure 2.1 in SEIS Section 2). This section describes the existing conditions at the Ross Project area, the 697-ha [1,721-ac] area that is addressed in this Supplemental Environmental Impact Statement (SEIS), and its vicinity. The resource areas described in this section include land use; transportation; geology and soils; water, both surface water and ground water; ecology; noise; meteorology, climatology, and air quality; historical and cultural resources; visual and scenic resources; socioeconomics; public and occupational health and safety; and waste management. This description of the affected environment is based upon information provided in the Applicant's license application and its Responses to the U.S. Nuclear Regulatory Commission's (NRC's) Requests for Additional Information (RAIs) and supplemented by additional information identified by NRC and others in the public domain (Strata, 2011a; Strata, 2011b; Strata, 2012a; Strata, 2012b). The information in this section forms the basis for the evaluation discussed in Section 4, Environmental Impacts and Mitigation Measures, which discusses the potential impacts of the Proposed Action and of each of the Alternatives in each resource area, as defined in SEIS Section 2.1.

3.1.1 Relationship between the Proposed Project and the GEIS

As shown on Figure 2.3 in SEIS Section 2.1.1, the Ross Project area is located in the northern end and on the western edge of the Nebraska-South Dakota-Wyoming Uranium Milling Region (NSDWUMR), as defined in the GEIS (NRC, 2009b). However, in defining the NSDWUMR, the Generic Environmental Impact Statement (GEIS) focused on potential in situ recovery (ISR) sites located in the Black Hills area of South Dakota, which is east of the Ross Project area. As a result, some of the affected environment discussion in the GEIS for the NSDWUMR does not reflect actual site conditions at the Ross Project area (in particular, the subsurface geology and water resources information). However, the GEIS's discussion of the Wyoming East Uranium Milling Region (WEUMR), located west of the Ross Site, does provide germane information with respect to the Ross Project area's subsurface geology and water resources. These differences are described in the subsequent sections below.

3.2 Land Use

The Ross Project area encompasses approximately 697 ha [1,721 ac], as described in SEIS Section 2.1.1. Nearby towns include Pine Haven, 27 km [17 mi] southeast; Moorcroft, 35 km [22 mi] south; Sundance, 48 km [30 mi] southeast; and Gillette, 53 km [33 mi] southwest. The Ross Project area is adjacent to the unincorporated ranching community of Oshoto. There are 11 residences within 3 km [2 mi] of the Ross Project, but no residences within the Project area. The closest residence is approximately 210 m [690 ft] north-northeast of the Ross Project boundary (see Figure 3.1). Existing land uses include livestock grazing, oil production, crop agriculture, communication and power transmission infrastructure, transportation infrastructure,

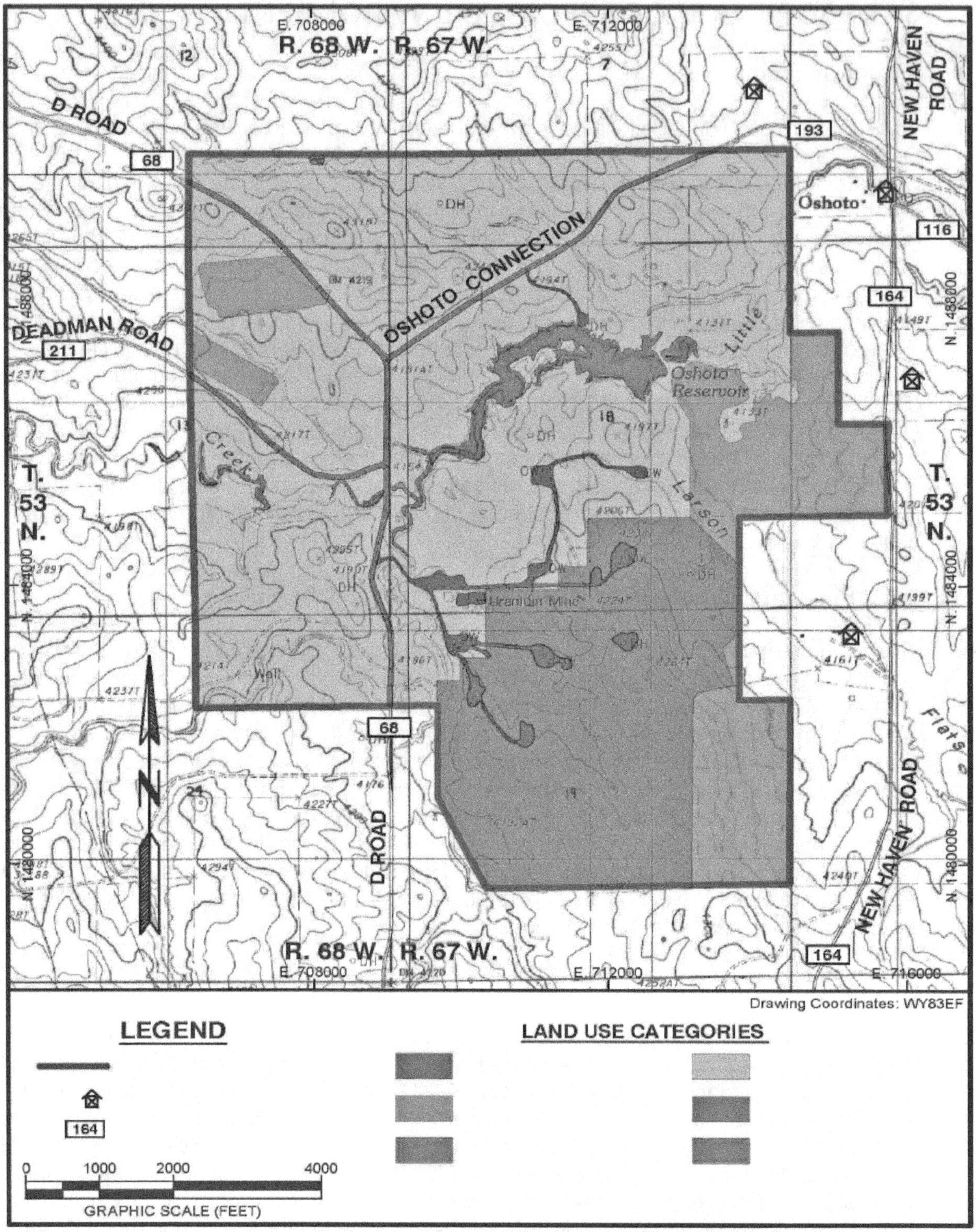

Source: Strata, 2012a.

Figure 3.1

Current Land Use of Ross Project Area

limited recreational opportunities, stock and other reservoirs, and wildlife habitat (see Figure 3.2). The actual land ownership of the Ross Project area's surface differs from general land ownership in the region, in that 97.6 percent is owned by private landowners or the State of Wyoming, and 2.3 percent is owned by the Federal Government (as described in Section 3.3.1 of the GEIS, 53.3 percent of Wyoming land is public land). The proposed Ross Project facility would be located on private property, and the wellfields would be located on private, State, and Federal lands.

The State of Wyoming owns all of the mineral rights below State-owned land, and the Federal Government controls all of the mineral rights below U.S. Bureau of Land Management (BLM)-owned land. There are private lands where the Federal Government (through the BLM) controls the mineral rights below the Ross Project area, a situation known as a "split estate." Between land ownership and split estate, the Federal Government through the BLM therefore controls 11.7 percent of the total mineral rights under the Ross Project area (see Table 3.1), as opposed to 2.3 percent of the surface. All of the Federal rights are managed by the BLM.

Table 3.1
Distribution of Surface Ownership and Subsurface Mineral Ownership

Ownership	Surface Ownership		Subsurface Mineral Ownership	
	Ha / Ac	Percent	Ha / Ac	Percent
Private	553.3 / 1367.2	79.4	488.2 / 1206.4	70.1
State	127.1 / 314.1	18.2	127.1 / 314.1	18.2
Federal	16.2 / 40.0	2.3	81.3 / 200.9	11.7
TOTAL	696.6 / 1721.3	--	696.6 / 1721.3	--

Source: Strata, 2011a.

3.2.1 Pasture-, Range-, and Croplands

Approximately 95 percent of the Ross Project area is used for rangeland, cropland, or pastureland. The largest portion, over 80 percent, is rangeland, while 14 percent is used for agriculture. In Crook County, rangeland is primarily used for cattle, with some grazing of sheep. Crops grown in the vicinity include hay, oats, and wheat.

3.2.2 Hunting and Recreation

There are many hunting and recreational opportunities within Crook County. However, there are limited opportunities for hunting and recreation within the Ross Project area because the majority of the land is privately owned. The State-owned land within the Ross Project area is accessible from County Road (CR) 193, but the Federal BLM land is not served by public roads so the public cannot access the BLM land to hunt. Large-game hunting in the area includes antelope (North Black Hills herd), mule deer (Powder River and Black Hills herds), and white-tailed deer (Black Hills herd). Other hunting opportunities in the vicinity include sage-grouse, wild turkeys, and small game such as cottontail rabbits and snowshoe hares as well as red, gray, and fox squirrels. There are hunting seasons specific to each type of game; however, because of the predominantly private ownership of the land, hunting within the Ross Project area is limited.

Recreational areas in the Ross Project vicinity include Devils Tower National Monument (Devils Tower), Black Hills National Forest, and Keyhole State Park. These areas offer access to hiking, camping, boating, biking, horseback riding, fishing, and hunting. The nearest of these is Devils Tower, approximately 16 km [10 mi] east of the Ross Project.

Although native fish have been observed in the Oshoto Reservoir, there are no fisheries in the Ross Project area because of the ephemeral or intermittent nature of the streams. The Oshoto Reservoir is partially located on State land; however, the Wyoming Game and Fish Department (WGFD) does not stock the Reservoir and it is not managed by any private agencies. However, fishing has been reported downstream of the Little Missouri River, outside of the Ross Project area (Strata, 2011a).

3.2.3 Minerals and Energy

There are three operating oil wells within the Ross Project area, producing from depths between 1,800 – 2,000 m [5,900 – 6,500 ft] below ground surface (bgs) (see Figure 3.2). Oil production is currently the only mineral extraction activity within the Ross Project area, although Crook County has other mineral resources which include coal, gas, bentonite (mine located 8 km [5 mi] to the northeast), sand, gravel, gypsum, and limestone in addition to uranium and vanadium.

There are currently no licensed or operating uranium-recovery facilities within 80 km [50 mi] of the proposed Ross Project, although four potential projects are under preliminary consideration and are in the very early planning stages (Strata, 2011a). These include the Bayswater Uranium Corporation's (Bayswater's) Elkhorn, Wyoming, project approximately 27 km [17 mi] to the northeast of the Ross Project area; Bayswater's Alzada, Montana, project at 58 km [36 mi] to the north-northeast; the UR-Energy/Bayswater's Hauber, Wyoming, project at 21 km [13 mi] to the north-northeast; and Powertech Uranium Corporation's (Powertech) Aladdin project at 64 km [40 mi] to the east-northeast (see Figure 3.3).

3.3 Transportation

The Proposed Action would rely on existing roads for supply and material transport, workforce commuting, and yellowcake and waste shipments to and from the Ross Project. The existing transportation network is discussed in this section; Figure 3.4 depicts this network. The primary access road to the Ross Project area is from Exit 153 on I-90. From that point the Ross Project is reached by a vehicle's travelling south on US 14/16, west on WY 51, north on Bertha Road, north on CR 68 (also known as D Road), and north on CR 164 (also known as New Haven Road). The distance from the I-90 exit to D Road is 2.6 km [1.6 mi]. D Road is a two-lane asphalt and gravel road approximately 9 – 11 m [30 – 35 ft] wide with posted speed limits of 89 km/hr [55 mi/hr] for cars and 72 km/hr [45 mi/hr] for trucks. The asphalt pavement extends to 4.8 km [3 mi] north of Bertha Road, where it changes to a reclaimed-asphalt pavement, which has been rotomilled and blended with crushed base and subgrade. This surface continues for 11.7 km [7.3 mi] after which D Road has only a gravel surface. New Haven Road is a two-lane, crushed-shale road approximately 7.6 – 9.1 m [25 – 30 ft] wide, with a posted speed limit of 72 km/hr [45 mi/hr]. CR 193, also known as the Oshoto Connection, is a two-lane, crushed-shale

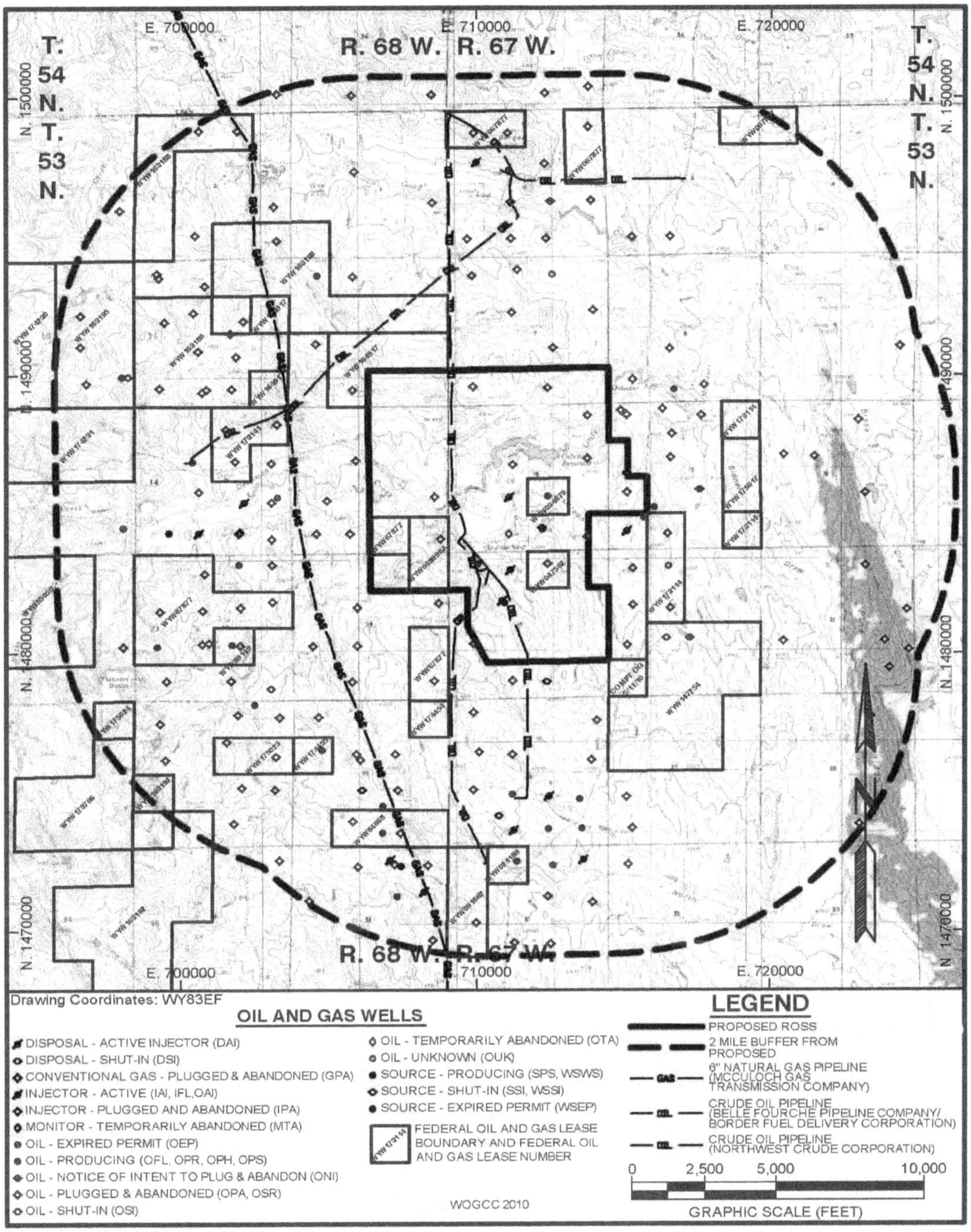

Source: WOGCC, 2010, as shown in Strata, 2012a.

Figure 3.2

Oil and Gas Wells within Two Miles of Ross Project Area

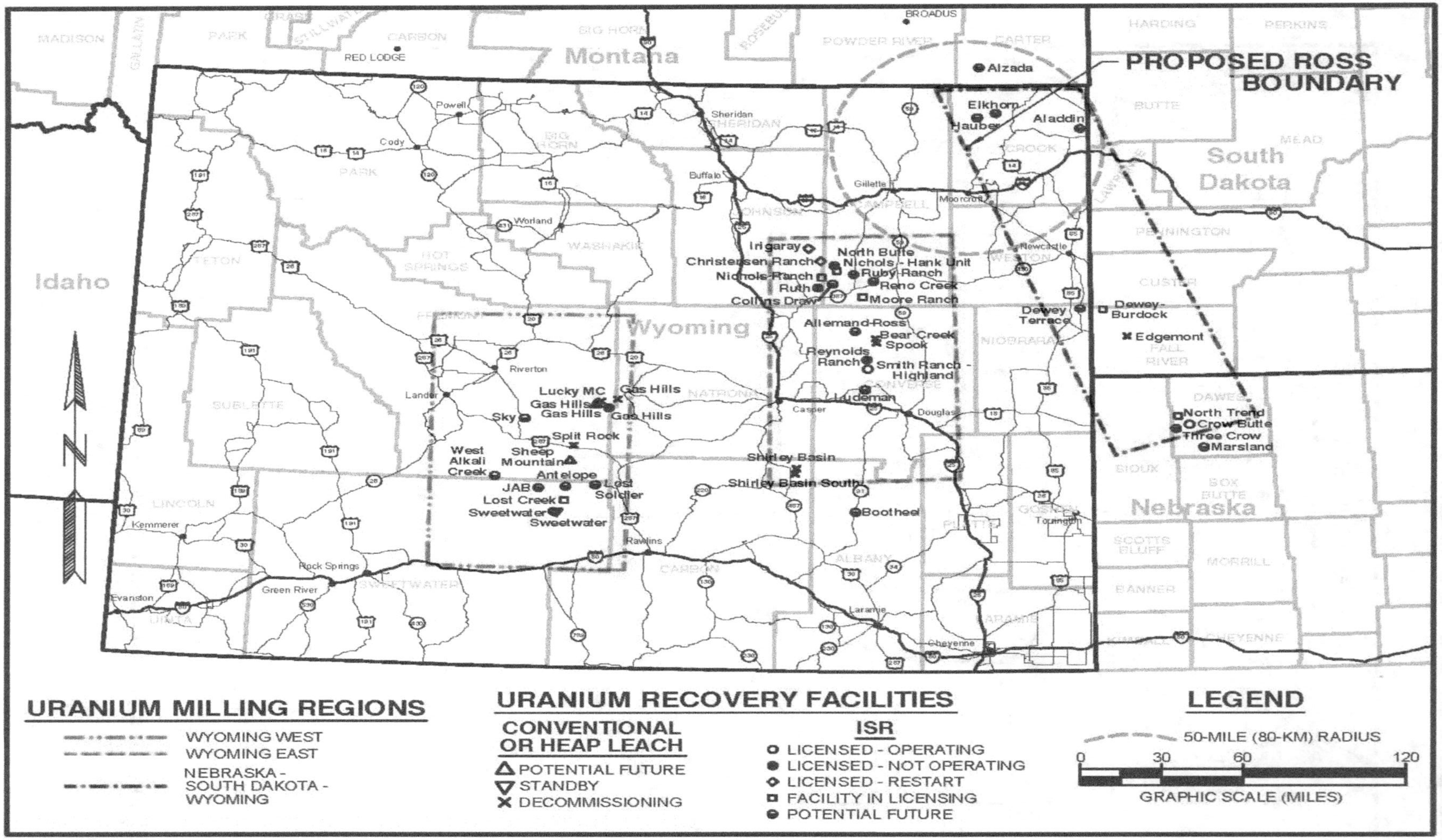

Sources: Bayswater, 2010a; NRC, 2009b; NRC, 2010a; NRC, 2010b; Powertech, 2010; and UR-Energy, 2010 as shown in Strata, 2012a.

Figure 3.3

Existing and Planned Uranium-Recovery Facilities

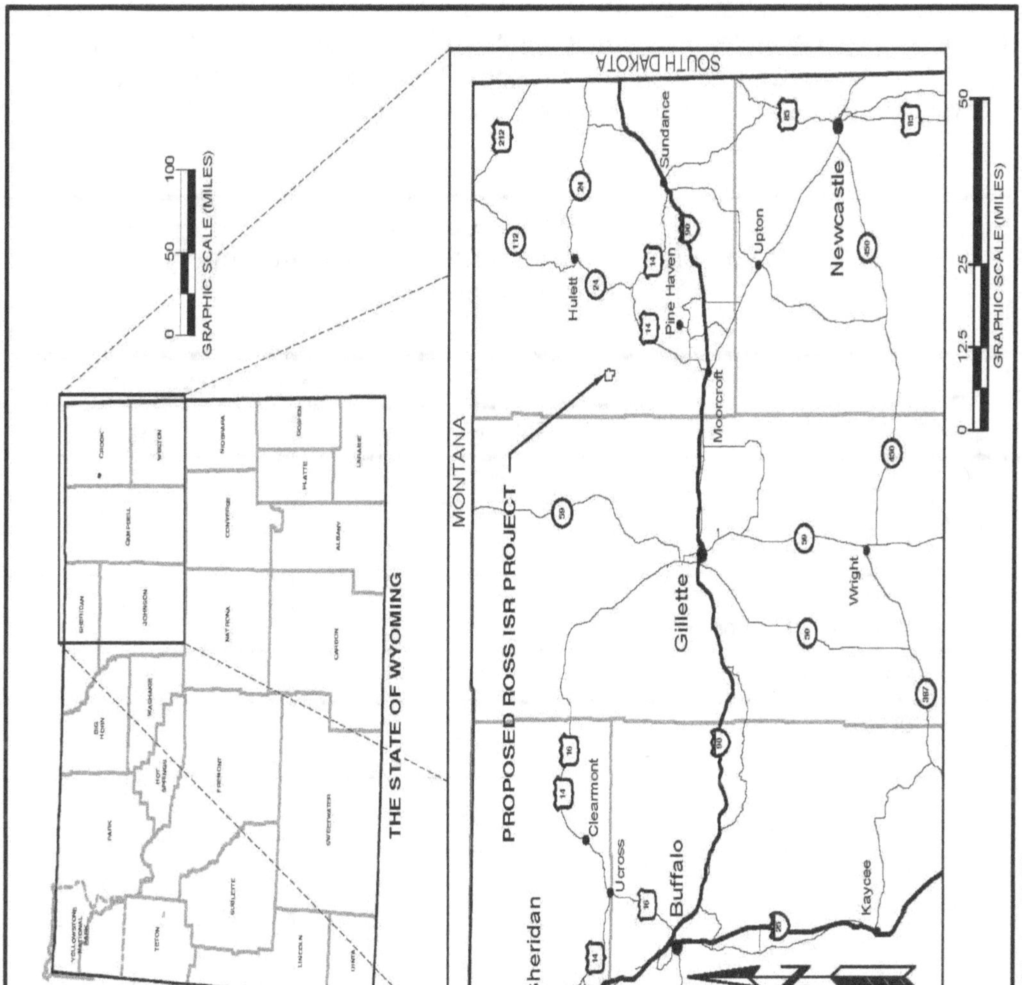

Source: Strata, 2012a.

Figure 3.4
Existing Transportation Network in Northeast Wyoming

road that connects New Haven Road to D Road along the northern portion of the Ross Project area. Other county roads in the local vicinity that can be used to access the Ross Project area include CR 26 (Cow Creek Road), CR 91 (Spring Creek Road), and CR 211 (Deadman Road). Figure 2.1 shows the relative locations of these roads. Crook County conducts year-round routine maintenance of all CRs, including snow and debris removal, blading and grading, and miscellaneous repair.

The Applicant has completed traffic studies on the county roads near the Ross Project area (Strata, 2011a), as has the State of Wyoming for its highways (see Table 3.2). Much of the existing truck traffic on the CRs adjacent to the Ross Project is due to local oil- and gas-recovery activities as well as to a bentonite mine approximately 8 km [5 mi] northeast of the Project.

Table 3.2
Traffic Volumes on Roads and Highways in Vicinity of Ross Project Area (2010)

Road/Highway	Vehicles per Day	
	All Vehicles	Trucks
I-90 at Moorcroft	4,744	906
New Haven Road South of Ross Project Area	108	10.8
New Haven Road South of Oshoto Connection	138	11
On-Site Measurements		
D Road South of Deadman Road	25	1.5
D Road North of Deadman Road	49	2.3
D Road North of Oshoto Connection	62	6.2
Oshoto Connection between D Road and New Haven Road	87	11.3

Sources: Strata, 2011a, and Wyoming Department of Transportation (WYDOT), 2011.

3.4 Geology and Soils

The Lance District, which includes the Ross Project area (refer to Figure 2.1), is structurally situated between two major tectonic features: the Black Hills uplift to the east and the Powder River Basin to the west (Strata, 2011a). Both of these regional features are described in the GEIS (NRC, 2009b). The Black Hills uplift is generally allocated to the NSDWUMR, and the Powder River Basin to the WEUMR. The Project area's structural geology, stratigraphy, uranium mineralization, and seismology as well as the types and characteristics of the soils present at the Project area are described in this section.

3.4.1 Ross Project Geology

The uranium-bearing units targeted for recovery within the Ross Project area are located in permeable sandstones of the Late Cretaceous Lance and Fox Hills Formations. The uranium roll fronts deposited in the Oshoto area demonstrate patterns similar to those across the Powder River Basin. The Ross Project area's roll fronts were created by precipitation of uranium from ground water as a coating on sand grains primarily due to changes in aquifer conditions and ground-water flow (Buswell, 1982). The roll-front geometry at the Project area can vary as a result of differences of the host sandstones. The deeper Fox Hills roll fronts are generally thicker and more massive due to the near-shore environment into which the sediments were deposited. The lower Lance Formation sandstones were deposited in a fluvial environment (i.e., deposited by rivers or streams), resulting in narrower, often stacked channel systems containing uranium mineralization. Because of the variability of the depositional environment, the roll fronts near or at the Ross Project area are complex, and new exploration activities consistently yield increasing total uranium estimates. At this time, estimates of recoverable uranium within the Ross Project area exceed 2,495 t [5.5 million lb] of uranium and, based on current projections, these estimates are likely to increase as more exploration and characterization results become available.

3.4.1.1 Structural Geology

The Black Hills uplift is a broad north-trending dome-like structure approximately 290 km [180 mi] long (north to south) and 121 km [75 mi] wide (west to east) whose core is composed of Precambrian basement rocks (NRC, 2009b). The western flank of the uplift is characterized by a monoclinal (a one-limbed or step-like flexure) break near the Ross Project area (Lisenbee, 1988). The eastern edge of the Ross Project area lies along the hinge of the Black Hills monocline. Because of the Black Hills monocline, the regional stratigraphic dip goes from essentially horizontal within the Powder River Basin, to steeply dipping along the eastern edge of the Ross Project area (see Figure 3.5). As indicated in the bedrock geologic map, Figure 3.6, the entire Ross Project area lies within the outcrop of the Lance Formation. The Cretaceous Formations below the Lance Formation all outcrop within roughly 3 km [2 mi] east of the Ross Project area.

Devils Tower, which is discussed later in the visual and scenic resources section of this section (Section 3.10), is located approximately 16 km [10 mi] east of the Ross Project area. Devils Tower and the Missouri Buttes (15 km [9.5 mi] northeast of the Ross Project) are geologic features formed by the intrusion of igneous material (i.e., magma) through the earth's crust during the Tertiary Period (i.e., subsequent to the deposition of the upper Cretaceous formations hosting the Lance District's uranium deposits) (Robinson, 1964).

With the exception of the Black Hills monocline, there are no significant structural features within the Ross Project area. No faults of major displacement are known to exist within the Ross Project area; however, minor localized slumps, folds, and differential compaction features that formed shortly after deposition are common (Strata, 2011a).

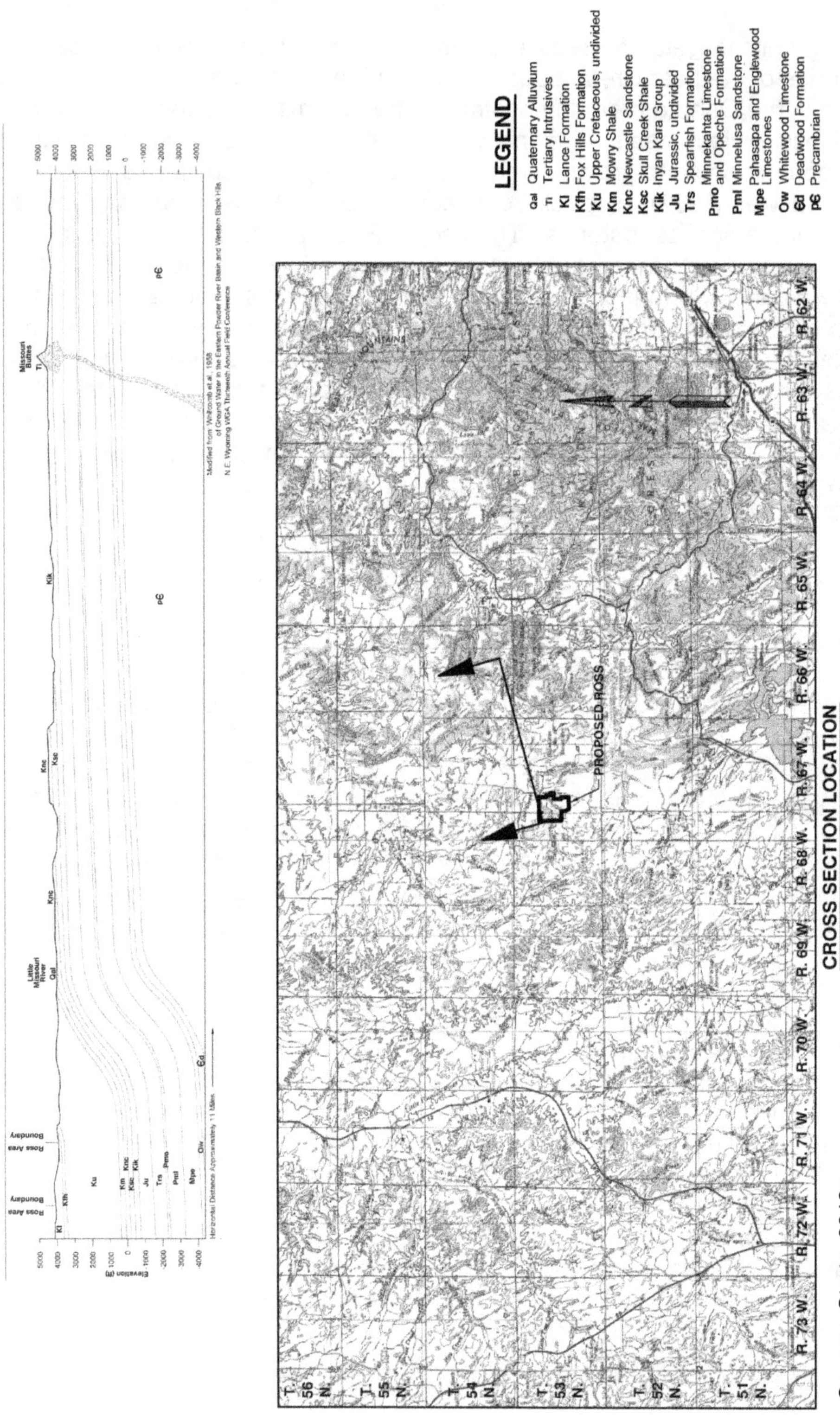

Source: Strata, 2012a.

Figure 3.5

Generalized Cross Section of Black Hills Monocline in the Oshoto Area

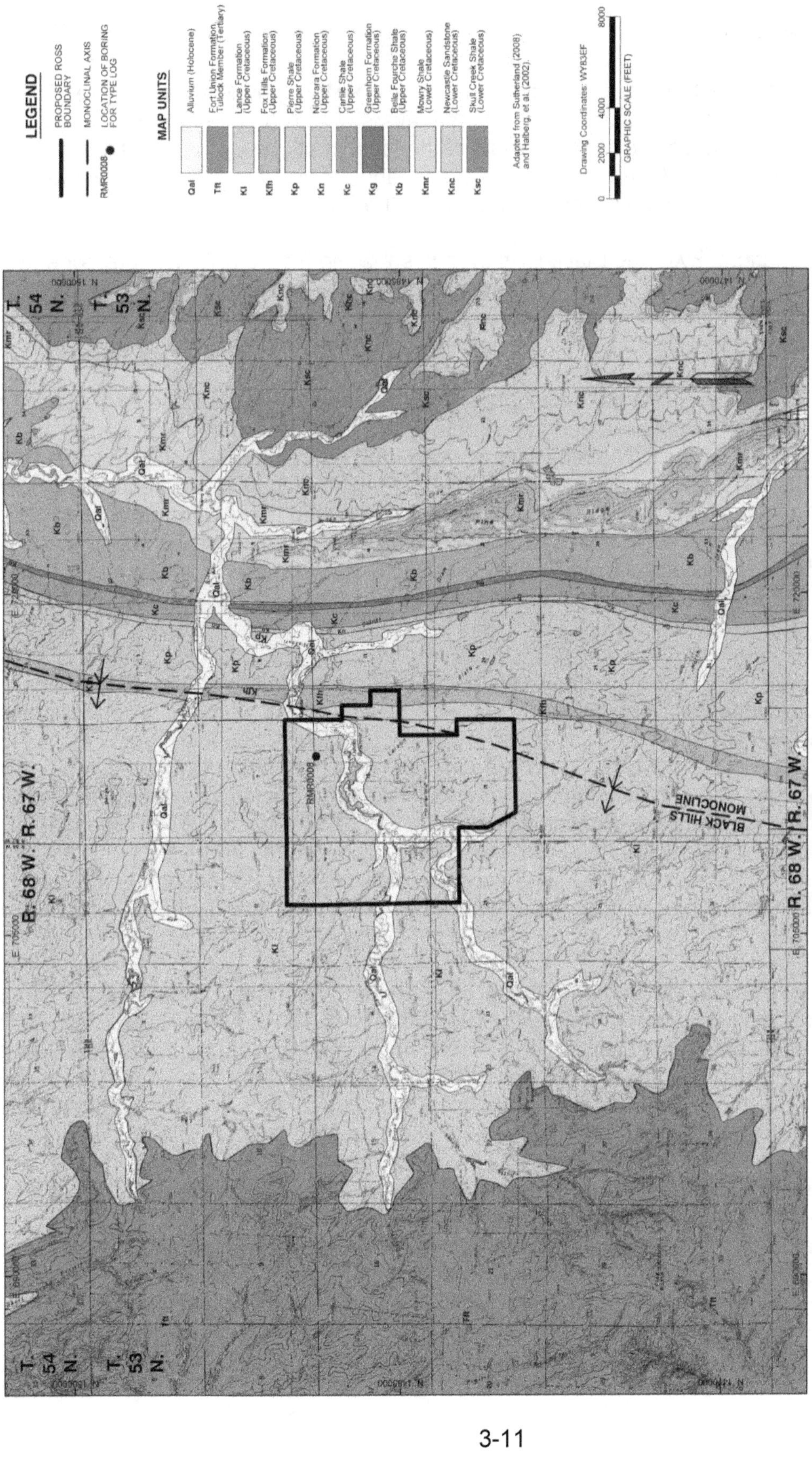

Source: Strata, 2012a.

ROSS PROJECT AREA
SCALE: 1" = 4,000'

Figure 3.6
Surface Geology of Ross Project Area

3.4.1.2 Stratigraphy

Stratigraphy describes the layers of rocks and soils below the ground's surface (i.e., the subsurface) that host the ore zone as well as the layers of rock that separate the ore zone from the aquifers above and below it. An analysis of the local stratigraphy is used in assessments of whether the ore zone is adequately confined above and below by rock layers of low permeability that would prevent vertical movement of water from the ore zone.

The regional stratigraphy of the Black Hills area is shown in Figure 3.7. The ore zone, which would be the "production zone" (i.e., the deposits from which uranium would be recovered) at the Ross Project, is within the upper Cretaceous stratigraphic units, including the lower Lance (Hell Creek) and upper Fox Hills Formations.

Detailed analysis of the subsurface stratigraphy and mineralogy of the Ross Project area began with the first uranium exploration and development efforts in the Oshoto area during the 1970s by the Nubeth Joint Venture (Nubeth) as described in SEIS Section 2.1.1 (Strata, 2011a). In 2008 and 2009, the Applicant began confirmation and exploration drilling at the Ross Project (Strata, 2011a). As of October 2010, the Applicant possessed information from the 1,682 holes drilled by Nubeth as well as its own 540 recent exploration drillholes, which are all located within a 0.8-km [0.5-mi] radius of the Ross Project area. The logs of these drillholes were used by the Applicant to characterize the site-specific stratigraphy of the Ross Project area (Strata, 2011a; Strata, 2011b).

The Pierre Shale in this area is a massively bedded, relatively uniform, thick marine shale that is considered a regional confining layer (or "unit" or "interval") (NRC, 2009b). This unit outcrops approximately 0.4 km [0.3 mi] east of the Ross Project's eastern boundary (see Figure 3.6). Based upon the width of the outcrop and geophysical logs from oil wells located in the general area, the Applicant has estimated the thickness of the Pierre Shale to be approximately 670 m [2,200 ft] thick under the Ross Project area (Strata, 2011a; Robinson, 1964). Because of its thickness and low permeability, the Pierre Shale is considered the lower ground-water-confining unit within the Ross Project vicinity, separating the older, deeper Formations below the Pierre Shale from the Ross Project's target ore zones which are in the overlying Fox Hills and Lance Formations.

Below the Pierre Shale, the Cambrian-age Deadwood and Flathead Formations are encountered at depths of approximately 2,490 – 2,600 m [8,160 – 8,560 ft] bgs (WDEQ/WQD, 2011). The Applicant proposes that these Formations are the optimum target interval for the Underground Injection Control (UIC) Class I deep-injection wells that would be used for waste-water disposal at the Ross Project. The Applicant has already received its UIC Class I Permit for this type of disposal (Strata, 2011a).

The Fox Hills Formation, which lies between the Pierre Shale and the Lance Formation, outcrops along the proposed eastern boundary of the Ross Project (refer to Figure 3.6). The Fox Hills Formation is a sequence of marginal marine to estuarine sand deposits that were deposited during the eastward regression of the upper Cretaceous interior seaway (Dunlap, 1958; Merewether, 1996). In the vicinity of Oshoto, the Fox Hills Formation is divided into lower and upper units, which are based on differences in color, bedding, trace fossil concentrations, lithology, and texture (Dodge and Spencer, 1977).

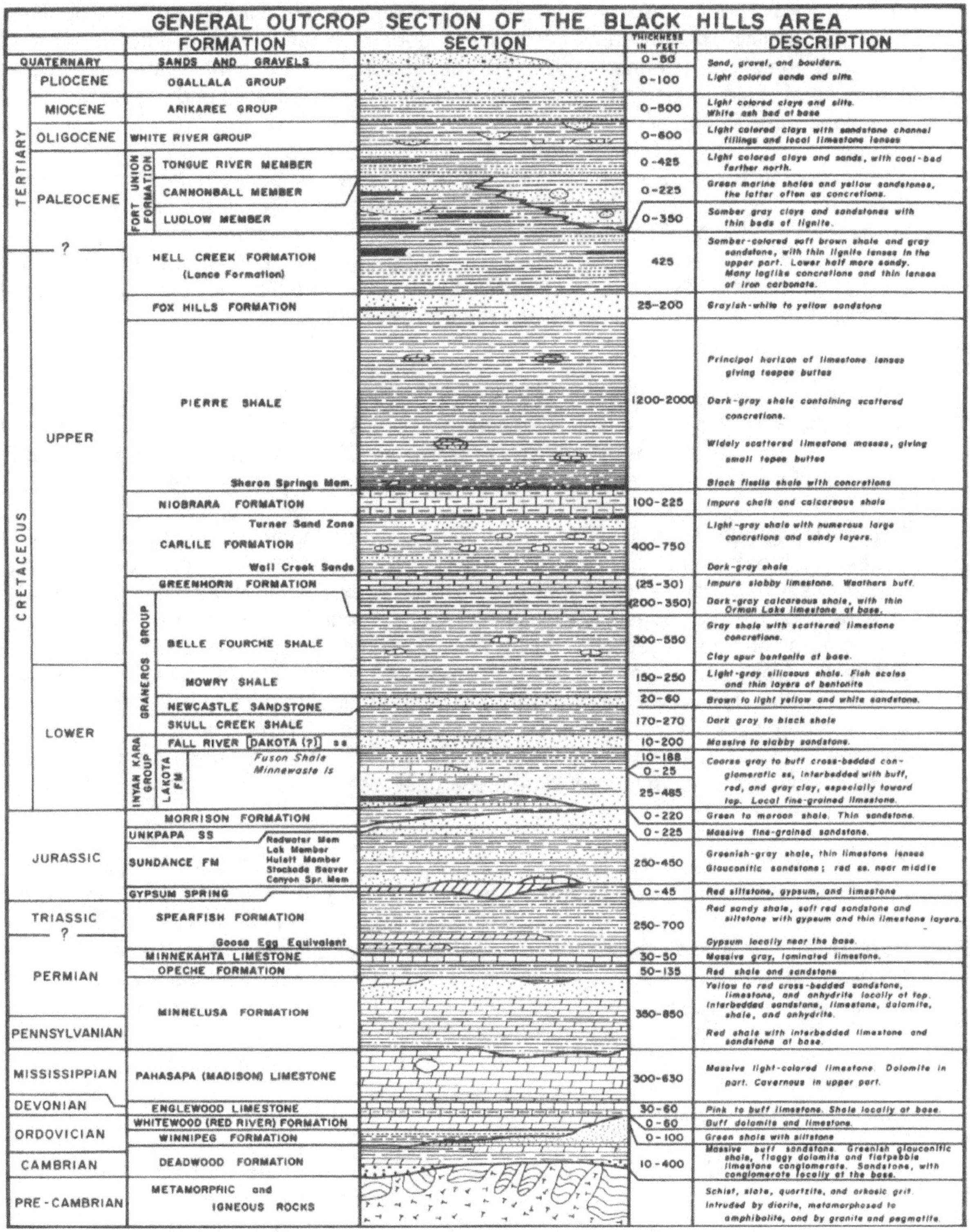
GENERAL OUTCROP SECTION OF THE BLACK HILLS AREA
FORMATION
SECTION
THICKNESS IN FEET
DESCRIPTION
QUATERNARY
SANDS AND GRAVELS
0-50
Sand, gravel, and boulders.
TERTIARY
PLIOCENE
OGALLALA GROUP
0-100
Light colored sands and silts.
MIOCENE
ARIKAREE GROUP
0-500
Light colored clays and silts. White ash bed at base
OLIGOCENE
WHITE RIVER GROUP
0-600
Light colored clays with sandstone channel fillings and local limestone lenses
PALEOCENE
FORT UNION FORMATION
TONGUE RIVER MEMBER
0-425
Light colored clays and sands, with coal-bed farther north.
CANNONBALL MEMBER
0-225
Green marine shales and yellow sandstones, the latter often as concretions.
LUDLOW MEMBER
0-350
Somber gray clays and sandstones with thin beds of lignite.
?
CRETACEOUS
HELL CREEK FORMATION (Lance Formation)
425
Somber-colored soft brown shale and gray sandstone, with thin lignite lenses in the upper part. Lower half more sandy. Many loglike concretions and thin lenses of iron carbonate.
FOX HILLS FORMATION
25-200
Grayish-white to yellow sandstone
UPPER
PIERRE SHALE
Sharon Springs Mem.
1200-2000
Principal horizon of limestone lenses giving teepee buttes
Dark-gray shale containing scattered concretions.
Widely scattered limestone masses, giving small tepee buttes
Black fissile shale with concretions
NIOBRARA FORMATION
100-225
Impure chalk and calcareous shale
CARLILE FORMATION
Turner Sand Zone
Wall Creek Sands
400-750
Light-gray shale with numerous large concretions and sandy layers.
Dark-gray shale
GREENHORN FORMATION
(25-30)
Impure slabby limestone. Weathers buff.
(200-350)
Dark-gray calcareous shale, with thin Orman Lake limestone at base.
GRANEROS GROUP
BELLE FOURCHE SHALE
300-550
Gray shale with scattered limestone concretions.
Clay spur bentonite at base.
MOWRY SHALE
150-250
Light-gray siliceous shale. Fish scales and thin layers of bentonite
NEWCASTLE SANDSTONE
20-60
Brown to light yellow and white sandstone.
LOWER
SKULL CREEK SHALE
170-270
Dark gray to black shale
INYAN KARA GROUP
FALL RIVER [DAKOTA (?)] ss
10-200
Massive to slabby sandstone.
LAKOTA FM
Fuson Shale
Minnewaste ls
10-188
0-25
25-485
Coarse gray to buff cross-bedded conglomeratic ss, interbedded with buff, red, and gray clay, especially toward top. Local fine-grained limestone.
JURASSIC
MORRISON FORMATION
0-220
Green to maroon shale. Thin sandstone.
UNKPAPA SS
0-225
Massive fine-grained sandstone.
SUNDANCE FM
Redwater Mem
Lak Member
Hulett Member
Stockade Beaver
Canyon Spr. Mem
250-450
Greenish-gray shale, thin limestone lenses
Glauconitic sandstone; red ss. near middle
GYPSUM SPRING
0-45
Red siltstone, gypsum, and limestone
TRIASSIC
?
SPEARFISH FORMATION
Goose Egg Equivalent
250-700
Red sandy shale, soft red sandstone and siltstone with gypsum and thin limestone layers.
Gypsum locally near the base.
PERMIAN
MINNEKAHTA LIMESTONE
30-50
Massive gray, laminated limestone.
OPECHE FORMATION
50-135
Red shale and sandstone
PENNSYLVANIAN
MINNELUSA FORMATION
350-850
Yellow to red cross-bedded sandstone, limestone, and anhydrite locally at top.
Interbedded sandstone, limestone, dolomite, shale, and anhydrite.
Red shale with interbedded limestone and sandstone at base.
MISSISSIPPIAN
PAHASAPA (MADISON) LIMESTONE
300-630
Massive light-colored limestone. Dolomite in part. Cavernous in upper part.
DEVONIAN
ENGLEWOOD LIMESTONE
30-60
Pink to buff limestone. Shale locally at base.
ORDOVICIAN
WHITEWOOD (RED RIVER) FORMATION
0-60
Buff dolomite and limestone.
WINNIPEG FORMATION
0-100
Green shale with siltstone
CAMBRIAN
DEADWOOD FORMATION
10-400
Massive buff sandstone. Greenish glauconitic shale, flaggy dolomite and flatpebble limestone conglomerate. Sandstone, with conglomerate locally at the base.
PRE-CAMBRIAN
METAMORPHIC and IGNEOUS ROCKS
Schist, slate, quartzite, and arkosic grit. Intruded by diorite, metamorphosed to amphibolite, and by granite and pegmatite.

Source: South Dakota School of Mines, 1963. **Figure 3.7**

Regional Stratigraphic Column of Area Containing the Lance District

Above the Fox Hills Formation, the Lance Formation has been interpreted as being fluvio-deltaic in origin, consisting of a mixture of non-marine-deposited sandstones and floodplain mudstones with thin beds of coal (Connor, 1992). This depositional environment created a stratigraphic sequence of shale, mudstones, and sandstones that is complicated and vertically heterogeneous (Dodge and Powell, 1975).

The horizontal continuity of the various stratigraphic intervals beneath the Ross Project is clearly depicted on the geologic cross-sections and fence diagrams provided by the Applicant (Strata, 2011a; Strata, 2012b). The upper Fox Hills and lower Lance Formations are stratigraphically continuous and hydraulically isolated from the overlying upper Lance Formation by continuous and impermeable mudstones and claystones as well as from the underlying units by the basal Fox Hills siltstone-claystone interval and the Pierre Shale.

3.4.2 Soils

Soils at the Ross Project are typical for semi-arid grass- and shrublands in the western U.S. (Strata, 2011a). Most of these soils are classified as Aridic Argiustolls, Ustic Haplargids, or Ustic Torrifluvents that were derived from the Lance Formation over time.

General topography of the Ross Project area ranges from nearly level uplands to steep hills, ridges, and breaks. The soils occurring on hills, ridges, and breaks at the Ross Project are generally sandy or coarse texture with clayey or fine-textured soils occurring on nearly level uplands and near drainages. The Ross Project area contains moderate and deep soils on level upland areas and drainages with shallow soils located on hills, ridges, and breaks. Figure 3.8 depicts the types of pre-licensing baseline soils located on the Ross Project area (Strata, 2011a; Strata, 2012b). The area of the Ross Project is about equally divided between sandy loam soils and clay loam soils (Strata, 2011a; Table 2.6-9 in Strata, 2012b). The soil characteristics of both the Proposed Action's south site (Alternative 1) and the north site (Alternative 3) are of particular interest since these would be the largest areas of soils disturbance during the Ross Project (see Table 3.3).

Approximate topsoil salvage depths range from 0.13 – 1.5 m [0.42 – 5 ft] with an average of 0.5 m [1.7 ft]. Factors that affect the suitability of a soil as a vegetation-growth medium are: texture, soil-adsorption ratio (SAR), electrical conductivity (EC), and pH as well as selenium and calcium carbonate concentrations. Based upon a comparison of laboratory analysis results and field observations with the respective Wyoming Department of Environmental Quality (WDEQ)/Land Quality Division (LQD) standards, suitable and marginally suitable material was found in 19 of the 26 samples within the Ross Project area (Strata, 2011a; WDEQ/LQD, 1994); unsuitable material was found in 7 of the 26 samples. The parameters that exceeded topsoil suitability criteria in those seven samples were high clay texture, high SAR, alkaline pH, and high concentration of selenium.

The hazard for wind and water erosion at the Ross Project varies from negligible to severe, based upon the soil-mapping descriptions. The potential for wind and water erosion is primarily dependent on the surface characteristics of the soils, including texture and organic-matter content. Given the slightly coarser texture of the surface horizons at the majority of the Ross Project, the soils are slightly more susceptible to erosion from wind than water.

Table 3.3 Soil Coverage and Characteristics for Ross Project Area					
Soil Name	**Soil Map Symbol**	**Alternative 1 (South) Site (ha [ac])**	**Alternative 3 (North) Site (ha [ac])**	**Water Erosion Hazard**	**Wind Erosion Hazard**
Absted very fine sandy loam	AB	3.7 [9.1]	N/A	Moderate	Moderate
Bidman loam	BI	9.3 [23.1]	2.2 [5.4]	Moderate	Moderate
Cushman very fine sandy loam	CU	N/A	2.0 [5.0]	Moderate	Slight
Forkwood loam	FO	7.1 [17.5]	3.4 [8.4]	Moderate	Slight
Nunn clay loam	NU	N/A	2.4 [5.9]	Slight	Slight
Shingle clay loam	SH	N/A	2.3 [5.7]	Moderate	Moderate
Tassel fine sandy loam	TA	N/A	2.7 [6.7]	Slight	Moderate

Source: Strata, 2011a.

Notes:

N/A = The type of soil is not present at the south or north site as indicated.
"Water Erosion Hazard" describes the susceptibility of the soil type to erosion by water, and "Wind Erosion Hazard" describes the susceptibility of the soil type to erosion by wind.

Although laboratory analyses for non-radioactive, chemical constituents in the soils at the Ross Project are not required by WDEQ/LQD to establish pre-operational baseline values, radioactive constituents in some soils were measured in order to establish such a pre-licensing baseline for radioactive species concentrations. These concentrations of specific radioactive elements are presented in Table 3.21 (see Section 3.12.1).

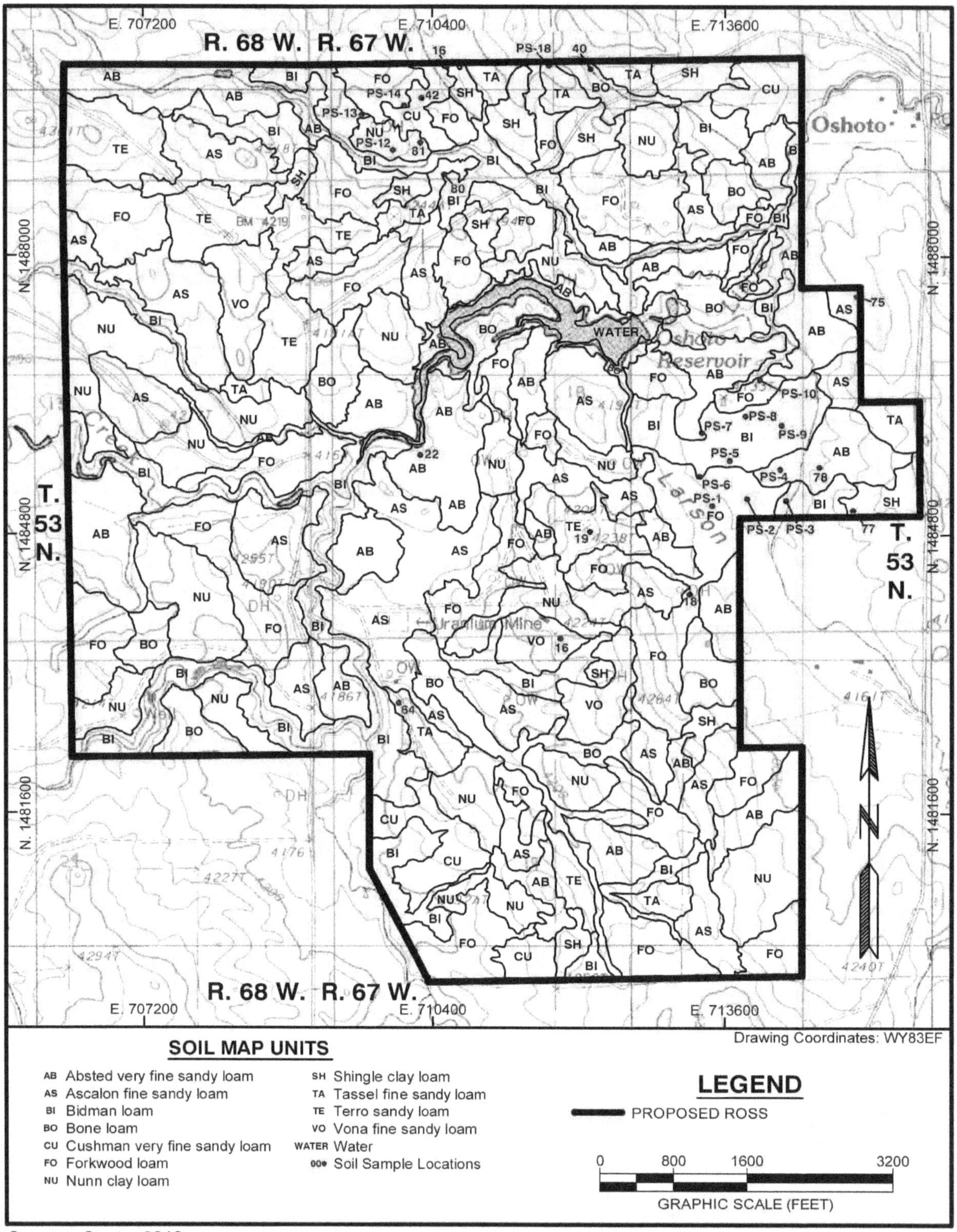

Source: Strata, 2012a.

Figure 3.8

Soil Mapped Units at Ross Project Area

3.4.3 Uranium Mineralization

> **What are the characteristics of uranium deposits that make them amenable to in situ uranium recovery?**
>
> Certain geologic and hydrological features make a uranium deposit in an ore zone suitable for in situ uranium recovery (based on Holen and Hatchell, 1986, as cited in NRC, 2009b):
>
> - **Deposit geometry:** For ISR operations, the wellfield boundaries are defined based upon the geometry of the specific uranium mineralization. The deposit should generally be horizontal and have sufficient size and lateral continuity to enable economic uranium extraction.
> - **Permeable host rock:** The host rock of the ore-zone aquifer must be permeable enough to allow the solutions (the lixiviant) to access and interact with the uranium mineralization. Preferred flow pathways, such as fractures in the rock, may short circuit portions of the mineralization and reduce the recovery efficiency. The most common host rocks are sandstones.
> - **Confining layers:** Hydrogeologic (formation) geometry must prevent lixiviant from vertically migrating. Typically, low permeability layers such as shales or clays "confine" the uranium-bearing sandstone(s) both above and below. This confinement isolates the uranium-producing zone from overlying and underlying aquifers.
> - **Saturated conditions:** For ISR uranium-recovery techniques to work, the uranium mineralization should be located in a hydrologically saturated zone (in an aquifer).

The process of uranium mineralization in the Lance District in general and specifically at the Ross Project is consistent with the characteristics of the uranium deposits that are identified in the GEIS as amenable to in situ uranium recovery. This mineralization includes fluvial sandstones (NRC, 2009b).

The lithological variability within the upper Fox Hills and Lance Formations would allow the geometric definition of ore deposits (i.e., areas of uranium mineralization) with sufficient size and continuity to make economic recovery viable. The saturated sandstone lithology of the ore zone would provide adequate permeability to allow uranium-recovery solutions access and interaction with uranium in the ore zone. In addition, the presence of impermeable intervals above and below the ore zone would prevent vertical migration of lixiviant or other fluids. Thus, the geology of the deposits would provide the characteristics required for an effective uranium-extraction project.

The mineralogy and petrography determined by the Applicant indicated that the ore zone is suitable for ISR (Strata, 2011a). The sandstone in the ore zone consists of 60 percent quartz, 35 percent feldspar, 5 percent montmorillonite clay, approximately 1 percent organic material, and less than 1 percent of pyrite and carbonate minerals (Strata, 2011a). The presence of pyrite confirms the geochemical conditions necessary for formation of the roll front. Petrographic analyses show that the ore zone has sufficient porosity (or reservoir quality) for movement of lixiviant from injection to recovery wells (Strata, 2011a). The ore zone is composed of fine grained, moderately well sorted, argillaceous sandstone with subangular to subrounded grains that are lightly to moderately compacted.

Consistent with the GEIS and typical of roll-front deposits (NRC, 2009b), analysis of the samples from the ore zone at the Ross Project shows that the principal uranium minerals are uraninite, an uranium oxide (UO_2), and coffinite, an uranium silicate ($U[SiO_4][OH])_4$) (Strata, 2011a). Vanadium in the form of vanadinite (a lead chlorovanadate [$Pb_5\{VO_4\}_3Cl$]) and carnotite [a hydrated potassium uranyl vanadate ($K_2[UO_2]_2[VO_4]_2\ 3H_2O$)] is also found in association with the uranium at an average ratio of 0.6 (vanadium) to 1.0 (uranium).

3.4.4 Seismology

There are no active faults with surface expression mapped within or near the Ross Project, according to the U.S. Geological Survey (USGS) (USGS, 2011). The closest capable faults to the Project area are located in central Wyoming, 270 km [170 mi] to the west-southwest. Six east-west trending structural faults through the Ross Project area were mapped by Buswell (1982). These faults are due to heterogeneity of the lithology among the shale and sandstone intervals within the upper Cretaceous Formations. However, these were based upon limited observations and information from one core sample and one aquifer test. The Applicant's examination of multiple geological cross-sections developed from stratigraphic information obtained from exploration drillholes do not appear to support this interpretation of the Ross Project area's faults (see SEIS Section 3.4.1.2) (Strata, 2011a).

Two earthquakes with magnitudes greater than 2.5 (on the Richter Magnitude Scale) have been recorded in Crook County and nine in Campbell County (Strata, 2011a). Of those with magnitudes greater than 2.5, 3 had magnitudes 3.0 and greater (Case, Toner, and Kirkwood, 2002). The first reported earthquake in Crook County with a magnitude of greater than 3 occurred near Sundance on February 3, 1897, severely shaking the Shober School on Little Houston Creek southwest of Sundance. On November 2004, an earthquake of magnitude of 3.7 was recorded near Moorcroft in Crook County. On February 18, 1972, a magnitude 4.3 earthquake occurred approximately 30 km [18 mi] east of Gillette near the Crook-Campbell County line (Case, Toner, and Kirkwood, 2002). No damage was reported. The occurrence of few, low-magnitude events is consistent with the predicted low probability of seismic-induced or earthquake-caused ground motion in northeastern Wyoming (Algermissen et al., 1982).

Earthquakes generally do not result in ground-surface rupture unless the magnitude of the event is greater than 6.5 (Case and Green, 2000). Because of this, areas of Wyoming that do not have active faults exposed at the surface, such as the Ross Project area, are generally thought not to be capable of having earthquakes with magnitudes over 6.5. As shown on Figure 3.9, the probability of an earthquake with magnitude greater than or equal to 6.5 in the vicinity of the Ross Project is less than 0.001. This figure was prepared using the USGS Probabilistic Seismic Hazard Analysis (PSHA) model (USGS, 2010). Earthquakes with magnitudes less than 6.5 would cause little damage in specially built structures, but they could cause considerable damage to ordinary buildings and even severe damage to poorly built structures. Some walls could collapse, but underground pipes would generally not be broken, and ground cracking would not occur or would be minor (USGS, 2010).

3.5 Water Resources

Water resources in the vicinity of Ross Project include both surface water and ground water. Both the quantity and the quality of both surface and ground waters are described in this section.

Pre-licensing baseline water-quality data have been collected and analyzed by the Applicant in accordance with the following guidelines:

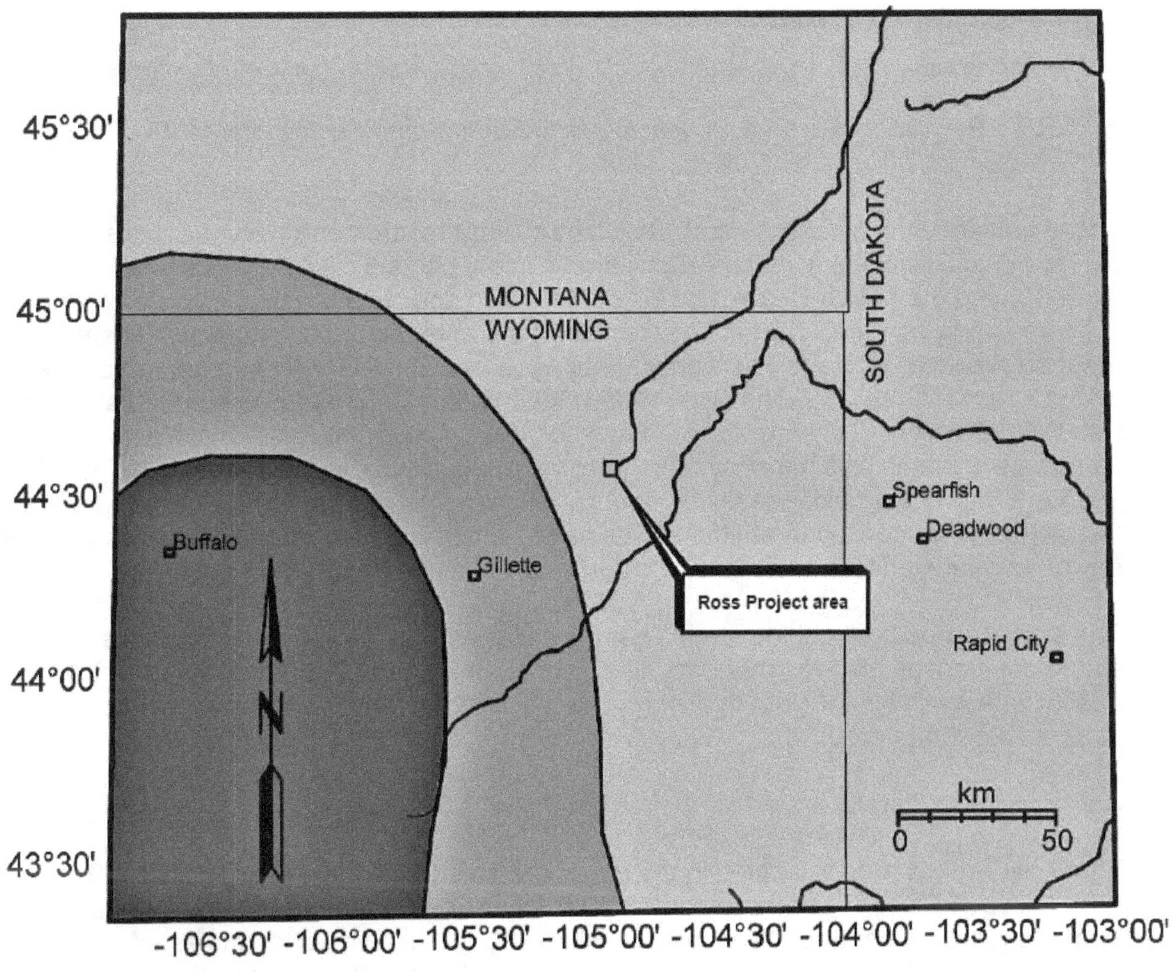

Source: Strata, 2011a.

Note: Darkest shaded area indicates probability between 0.003 and 0.002; lighter shaded area indicates probability between 0.002 and 0.001; lightest shaded area indicates probability between 0.001 and 0.000.

Figure 3.9

Probability of Earthquake with Magnitude of Greater Than or Equal to 6.5 in 50 Years

- American Society for Testing and Materials (ASTM) International's Standard D449-85a, *Standard Guide for Sampling Groundwater Monitoring Wells*, as recommended in the NRC's guidance document, NUREG–1569, *Standard Review Plan for In Situ Leach Uranium Extraction License Applications* (NRC, 2003). This ASTM Standard was replaced by ASTM Standard D4448-01 in 2007.
- WDEQ's "Hydrology, Coal and Non Coal," Guideline No. 8 (WDEQ/LQD, 2005b).
- NRC's Regulatory Guide 4.14, *Radiological Effluent and Environmental Monitoring at Uranium Mills*, Revision 1 (NRC, 1980).

These guidance documents by both NRC and WDEQ recommend water samples be filtered before the analysis of any metals each sample might contain. ASTM D449-85a (now ASTM 4448-01) and the NRC's Regulatory Guide 4.14 also specify analysis of radiological parameters in filtered samples (NRC, 1980). The results of the analysis of constituents in filtered samples are then reported as "dissolved" concentrations (versus "unfiltered" samples, which are reported as "total" concentrations). The filtering of water samples before analysis for metals is consistent with WDEQ/WDQ's *Groundwater Sampling for Metals: Summary*, which explains that filtering samples eliminates bias that may arise from variable turbidity in the samples (WDEQ/WQD, 2005a). The NRC's guidance on filtering samples applies to both pre-licensing baseline site-characterization monitoring efforts as well as post-licensing, pre-operational and operational environmental monitoring efforts during ISR operation and aquifer restoration.

The standardized protocol for filtering samples that will be analyzed for metals also allows a sound comparison among other data sets. For example, pre- and post-ISR operation water-quality data available for Nubeth also reported dissolved metal concentrations (i.e., filtered samples were analyzed).

3.5.1 Surface Water

The Ross Project area is located in the upper reaches of the Little Missouri River Basin. The Little Missouri River originates in northeastern Wyoming, flows through southeastern Montana, through northwestern South Dakota, and into North Dakota where it empties into the Missouri River at Lake Sakakawea. The total river length is 652 km (405 mi), and the total drainage area (i.e., the area where all surface waters flow toward the Little Missouri River) is approximately 24,500 km^2 [9,470 mi^2]. Figure 3.10 depicts the Little Missouri River Basin. The drainage area of the Little Missouri River at the downstream boundary of the Ross Project area is approximately 47 km^2 [18.2 mi^2].

A surface-water monitoring system has been employed by the Applicant to characterize surface-water quantity and quality at the Ross Project area. This system includes three monitoring stations and was designed to monitor the major surface-water drainages to the Little Missouri River and to establish pre-licensing baseline, site-characterization surface-water quality.

Surface-Water Features

The surface-water features located within the Ross Project are depicted in Figure 3.11 and consist of several reservoirs and minor stream channels. Oshoto Reservoir, located in the channel of the Little Missouri River, is the main hydrologic feature of the Project area (Water

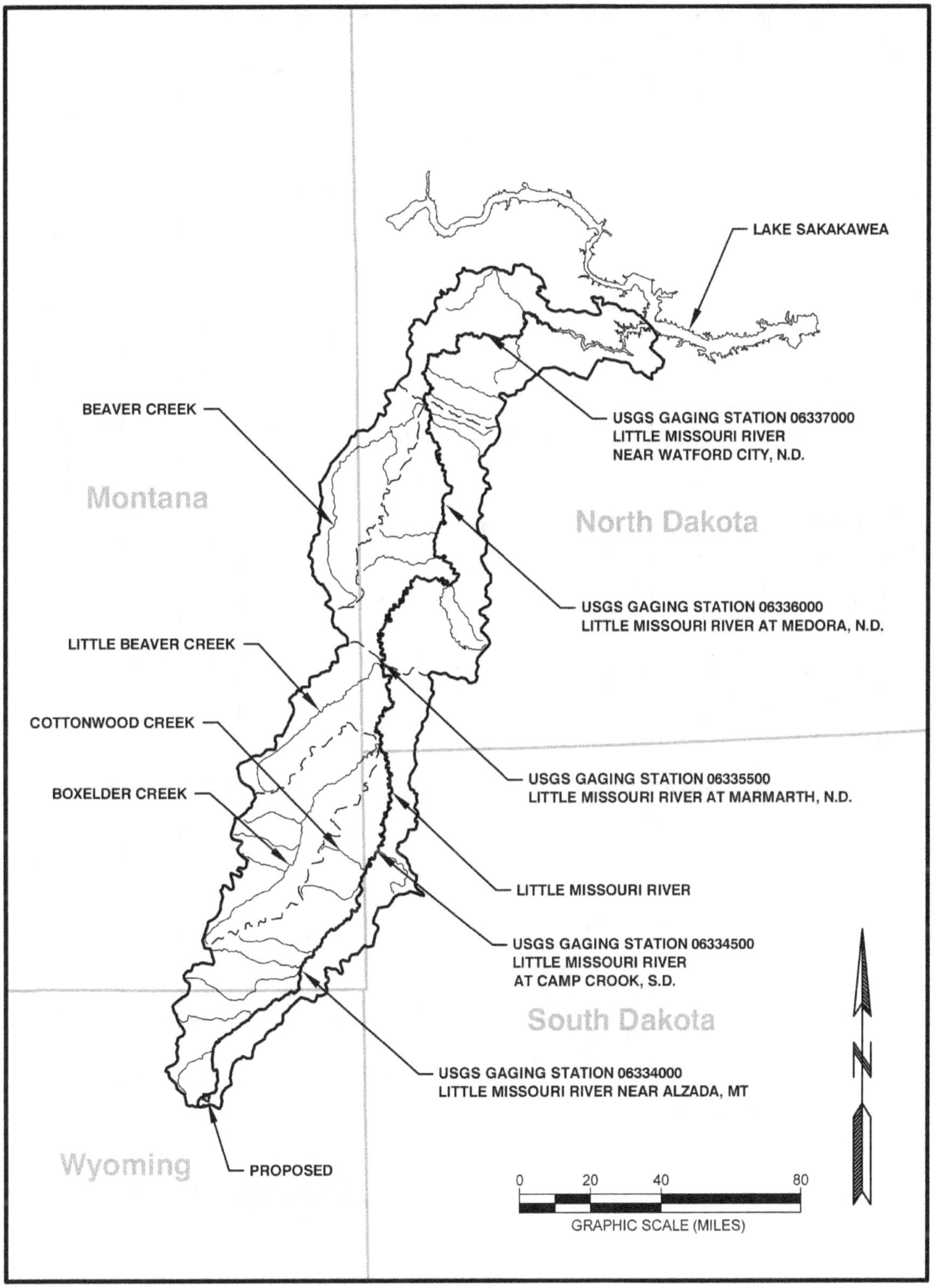

Source: Strata, 2012a.

Figure 3.10

Little Missouri River Basin and Surface-Water Gaging Stations

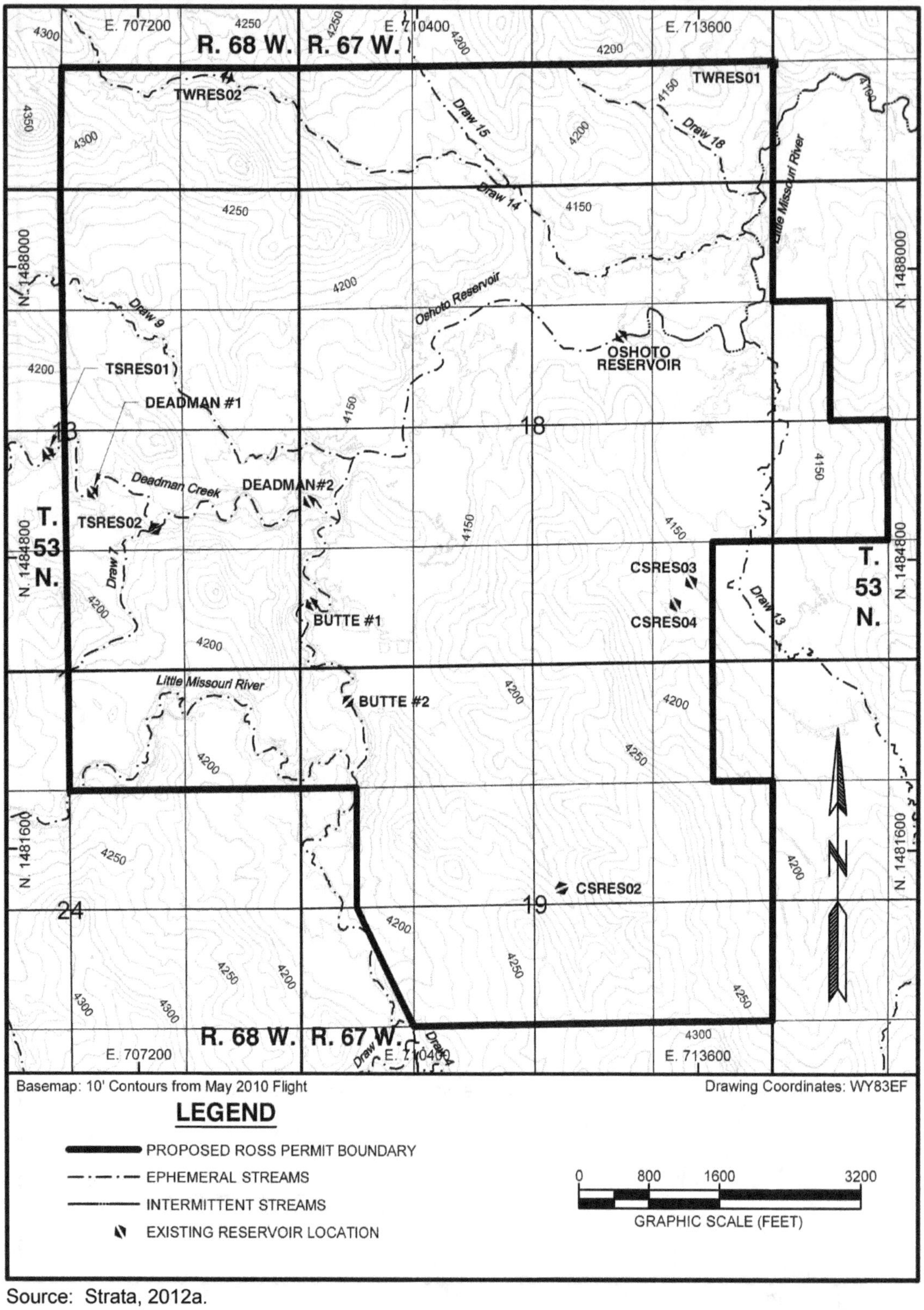

Source: Strata, 2012a.

Figure 3.11

Surface-Water Features of Ross Project Area

Right Permit No. P6046R) (WSEO, 2006). The only potential springs identified within the Ross Project area are associated with nearby wetlands (see Section 3.5.2 of this SEIS) or with the Little Missouri River in the vicinity of the Oshoto Reservoir.

The Applicant has identified 12 existing reservoirs within or just outside the Ross Project area using aerial photography, Wyoming State Engineer's Office (WSEO) permits, and landowner interviews (see Figure 3.11). Other than the Oshoto Reservoir, which has a maximum capacity of 21 ha-m [173 ac-ft] and an area of 11.3 ha [28 ac], all the identified reservoirs have a capacity of less than 1.2 ha-m [10 ac-ft] and a surface area of less than 1 ha [2.5 ac] (Strata, 2011a). The Oshoto Reservoir has the potential to affect stream flow and appears to influence water-table elevations in its proximity (Strata, 2011a).

There are three Wyoming Pollution Discharge Elimination System (WYPDES)-permitted outfalls associated with the oil-production operations within the watershed that includes the Ross Project area: two upstream from the Ross Project (Permit Nos. WY0044296 and WY0033065) and one downstream (Permit No. WY0034592) (Strata, 2011a). Discharge rates from the outfalls are relatively low, approximately 0 – 150 m^3/d [0 – 5,300 ft^3/d].

Surface-Water Flow

As shown in Figure 3.10, five USGS gaging stations are located on the Little Missouri River downstream of the Ross Project (USGS, 2012a). The mean annual discharges range from 2 m^3/s [77 ft^3/s] at the most upstream gaging station (near Alzada, Montana) to 15.1 m^3/s [533 ft^3/s] at the most downstream gaging station (near Watford City, North Dakota). The discharges are typically lowest from November through January and highest during the months of March through June (Strata, 2011a). The peak flow for the Alzada, Montana, gaging station occurred in April 1944 when an estimated discharge of 170 m^3/s [6,000 ft^3/s] occurred. The peak flow at the Camp Crook, South Dakota, gaging station took place in March 1978 with a flow of 267 m^3/s [9,420 ft^3/s]. The timing of these events indicates that snow melt and spring runoff typically result in the highest flows for this portion of the Little Missouri River.

The Applicant has established three surface-water monitoring stations and installed continuous stage recorders and pump samplers at each station within the Ross Project area in 2010 (see Figure 3.12) (Strata, 2011a). The stations were located at two sites on the Little Missouri River and one site on Deadman Creek, a tributary to Little Missouri River. The stage recorders are designed to continuously measure discharge and are integrated with the pump samplers that collect water-quality samples during runoff events. The Applicant reports flow data from the three surface-water monitoring stations from June 15, 2010, to October 11, 2011, with a break during the respective winter when the monitoring stations were removed to prevent their freezing (Strata, 2012a).

The results of the surface-water monitoring indicate that, where the streams enter the Ross Project area (SW-2 and SW-3), flow is in response to only snow-melt or precipitation events (i.e., ephemeral) (Strata, 2011a). The Little Missouri River, downstream from the proposed Ross Project boundary (SW-1), has flow for an extended period of the year but not all of the

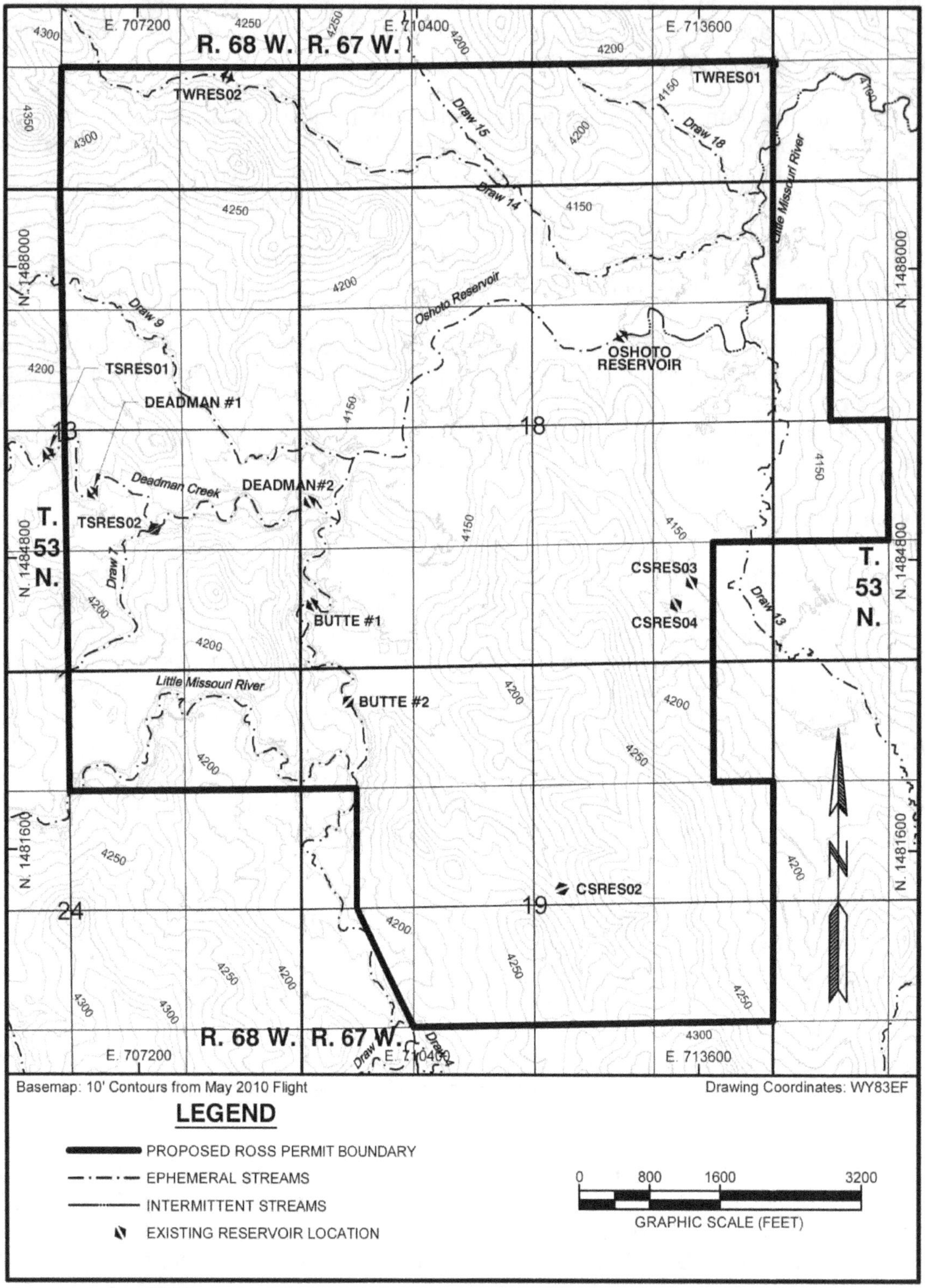

Source: Strata, 2012a.

Figure 3.12

Surface-Water Monitoring Stations at Ross Project Area

> **What are the types of streams at the Ross Project area?**
>
> **Perennial Streams:** A perennial stream is a stream or part of a stream that flows continually during all of the calendar year as a result of ground-water discharge or surface runoff.
>
> **Intermittent Streams:** An intermittent stream is a stream or part of a stream where the channel bottom is above the local water table for some part of the year, but which is not a perennial stream.
>
> **Ephemeral Streams:** An ephemeral stream is a stream which flows only in direct response to a single precipitation event in the immediate watershed or in response to a single snow-melt event, and which has a channel bottom that is always above the prevailing water table.

year and is, thus, intermittent. The Applicant compared the average daily flow observed at SW-1 to the water-surface elevation in Oshoto Reservoir (Strata, 2011a); the comparison suggests a correlation between the increased flow in the Little Missouri River downstream of Oshoto Reservoir and the amount of head in the Reservoir. This would indicate that some of the flow could be attributed to the stored capacity in Oshoto Reservoir.

All streams within the Ross Project area, including the Little Missouri River and Deadman Creek, are classified by WDEQ/Water Quality Division (WQD) as 3B streams (WDEQ/WQD, 2001). A Class 3B stream is defined by the WDEQ/WQD as an intermittent or ephemeral stream with a designated use of "aquatic life other than fish." Uses such as drinking water and fisheries are excluded in a Class 3B stream. Approximately 64 km [40 mi] downstream of the Ross Project, the Little Missouri River becomes a class 2ABWW stream at its confluence with Government Canyon Creek; at this point, the River becomes protected as a drinking water source (2AB) and warm-water (WW) fishery.

There are no long-term stream-flow records for flows within or adjacent to the Ross Project; therefore, an U.S. Army Corps of Engineers' (USACE) Hydrologic Engineering Center (HEC)-hydrologic modeling system (HMS) model was developed by the Applicant to estimate the peaks and volumes of floods for various recurrence intervals (Strata, 2011a). The resulting inundation boundaries are shown on Figure 3.13. Measured peak flows during a 2-year, 24-hour storm event in May 2011 were less than predicted by the model, suggesting that the predicted model flows are conservatively high (Strata, 2012a).

Surface-Water Quality

Data from water-quality analyses of samples obtained from the Ross Project surface-water monitoring stations in 2009 and 2010 are provided in the Applicant's *Environmental Report* (ER) and *Technical Report* (TR) (see Figure 3.12) (Strata, 2011a; Strata, 2011b). Due to reasons ranging from the Applicant's not having a landowner's permission to no-flow conditions (i.e., there was no water flowing or the water was frozen), the number of quarters in which the monitoring stations were sampled ranges from one to six (Strata, 2011a). Water-quality analytical data from samples collected in 2011 were submitted to WDEQ/LQD and are provided in the Applicant's Responses to the RAIs issued by the NRC (Strata, 2012a). The data from 2011 are generally consistent with the 2009 and 2010 data, indicating a representative characterization of surface-water quality.

1

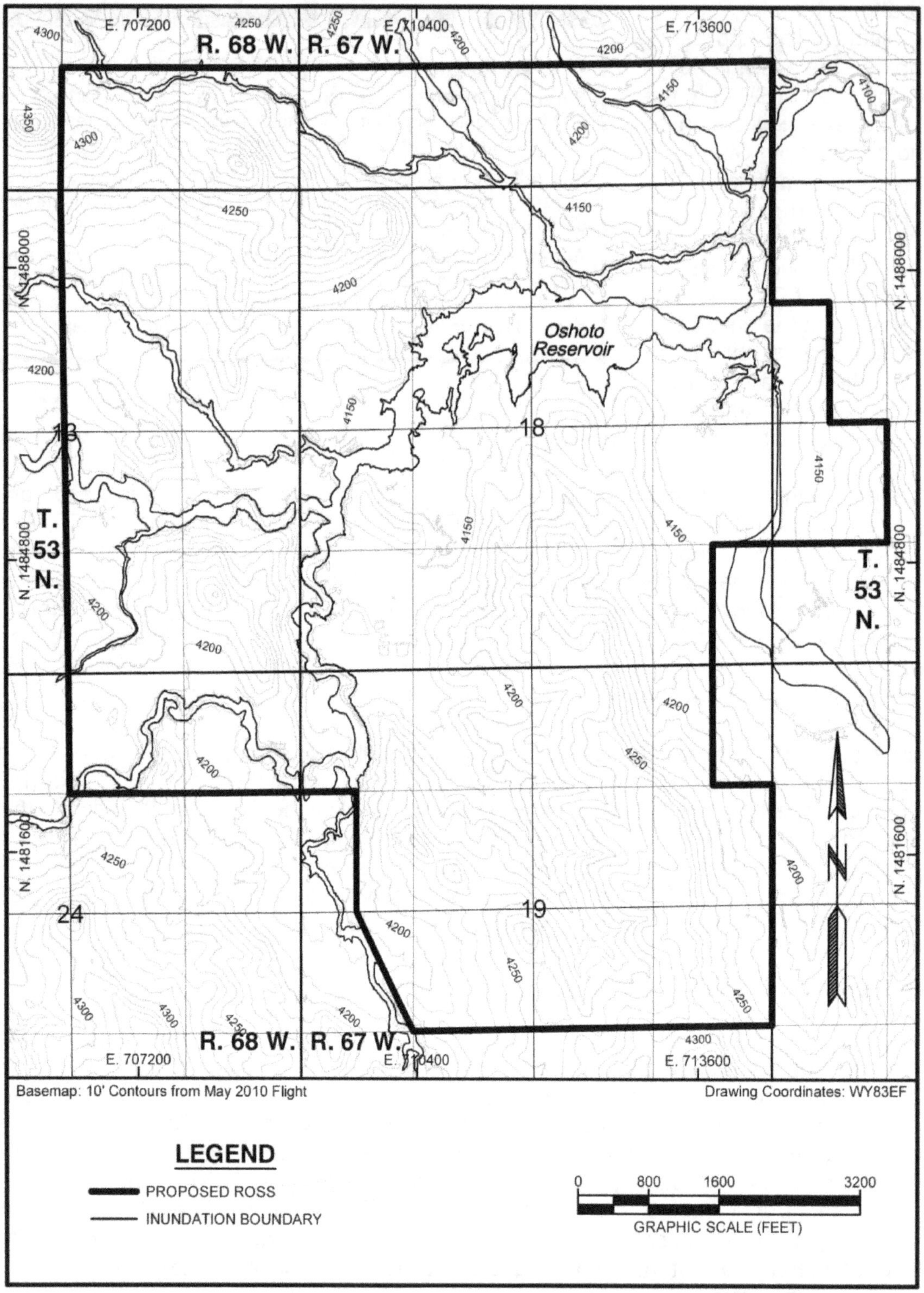

Source: Strata, 2012a.

Figure 3.13

Predicted 100-Year Flood Inundation Boundaries

The surface-water monitoring data characterizing the Little Missouri River and Deadman Creek from the first and second quarter of 2010 are summarized and described below. These data indicate that the overall water quality meets Wyoming's surface-water criteria for a Class 3B stream, which is the designation for the Little Missouri River.

- The water quality in all streams is generally consistent across the entire Ross Project area.
- The field pH measurements ranged from 7.6 – 8.9 standard units (s.u.), indicating alkaline water.
- The field measurements of dissolved oxygen ranged from 6.9 – 10.5 mg/L, indicating an intermediate to high level of oxygen in the water.
- Total salinity of the waters, expressed as total dissolved solids (TDS) concentrations, are low to moderate, ranging from 210 mg/L – 940 mg/L, and the water composition is dominated by sodium and bicarbonate.
- Iron and manganese concentrations in unfiltered samples ranged from 0.32 – 0.95 mg/L and 0.05 – 0.21 mg/L, respectively, suggesting the presence of suspended sediment in the samples.
- Dissolved metals were near or below detection limits, with the exception of iron and uranium. Iron concentrations ranged from less than 0.05 mg/L to 0.92 mg/L, with an outlier of 8.32 mg/L in the sample collected in the third quarter from Station R-5. Concentrations of dissolved uranium ranged from 0.003 – 0.02 mg/L.
- Dissolved radium-226 was less than the detection limit of 0.01 Bq/L [0.2 pCi/L]. Dissolved radium-228 was undetected (i.e., less than 0.04 Bq/L [1 pCi/L]) except for one sample obtained at Station SW2, where it was counted at 0.05 Bq/L [1.3 pCi/L].
- Gross alpha and gross beta ranged from 0.2 – 0.33 Bq/L [4 – 8.8 pCi/L) and 0. 2 – 0.41 Bq/L [6 – 11.2 pCi/L], respectively.

Other water-quality data suggest that the TDS increases downstream in the Little Missouri River and sulfate becomes the dominate anion (Langford, 1964).

The total anion/cation balances were calculated from the analyses of major ions as a quality-control check on the laboratory analyses. The balances, less than 3 percent in 31 of the 36 samples analyzed, and between 3 and 5 percent in five samples, validated the accuracy of the analyses (Strata, 2011a).

The Applicant attempted to collect water-quality samples from 11 reservoirs (see Figure 3.12) from the third quarter of 2009 through the third quarter of 2011 (i.e., quarterly) (Strata, 2011a; Strata, 2011b, Strata. 2012a). Samples were not collected when the reservoirs were dry or frozen or when the Applicant was not able to obtain the landowner's permission. These water-quality data indicate the following:

- Higher TDS corresponds to low-flow conditions in the fourth quarters of both years. TDS in samples of the reservoirs on the channels of the Little Missouri River and Deadman Creek, upstream from Oshoto Reservoir, ranged from 970 – 2,320 mg/L compared to a range of 460 – 730 mg/L in the Oshoto Reservoir and a range of 100 – 170 mg/L in the reservoir on the Little Missouri River downstream of the Oshoto Reservoir. The TDS in the reservoirs upland from the stream channels range from 110 – 1190 mg/L. Bicarbonate or carbonate (depending upon the pH) was the dominant anion in all of the waters. Sodium was the dominant cation, except in waters on the low end of the TDS range, where calcium was often the dominant cation.

- The waters in all reservoirs were alkaline, with field pH measurements generally ranging from 8 – 10 s.u.
- Field-measured dissolved oxygen ranged from 0.46 – 11.3 mg/L, suggesting seasonal low oxygen conditions.
- Similar to the streams, dissolved metals were generally at or near the laboratory detection limits, except for uranium and iron. Uranium ranged from less than 0.001 – 0.009 mg/L in all of the reservoirs except for those on Deadman Creek, where uranium concentration ranged from 0.019 – 0.087 mg/L. Detectable concentrations of dissolved iron generally corresponded to depleted dissolved oxygen levels. Measureable concentrations of total iron and manganese indicate the presence of sediment in the samples.
- The available data for radionuclides show that most of the analyses were less than the laboratory's lower limit of detection. However, detectable concentrations of lead-210, radium-226 (dissolved and suspended), dissolved radium-228, and suspended thorium-230 were detected. Gross alpha and gross beta ranged from less than 2 – 48.4 pCi/L and 3.9 – 48.5 pCi/L, respectively. The highest values of gross alpha and gross beta were measured in samples from reservoirs on Deadman Creek.

Surface-Water Uses

A search of the WSEO database of permitted surface-water rights within the Ross Project boundaries and the adjacent 3-km [2-mi] radius revealed that 43 surface-water rights existed within and adjacent to the Ross Project in 2010 (WSEO, 2006; Strata, 2011a). The search of the WSEO database indicated that nearly half of the water-right permits have been cancelled, while the remaining permits are complete, fully adjudicated, or un-adjudicated (Strata, 2011a). In addition to the permitted surface-water rights, there are at least 17 additional reservoirs within or adjacent to the Ross Project area, although none of these reservoirs was listed in the WSEO water-rights database, except for the Oshoto Reservoir (Strata, 2011a).

Surface water within the Ross Project area and surrounding 3-km [2-mi] vicinity is primarily used for livestock watering, with lesser amounts used for irrigation and industrial uses (primarily as a temporary water supply for oil- and gas-construction activities) (Strata, 2011a). Including reservoirs not listed in the WSEO database, stock reservoirs account for approximately 90 percent of the total active water rights (Strata, 2011a). Most of the stock reservoirs were constructed before 1970, and the majority are still in use today. Irrigation-water rights only account for a relatively small portion (less than 10 percent) of the surface-water rights. All of the irrigation rights were permitted 50 – 100 years ago for relatively small areas (28 ha [70 ac] or less). The one water right for Nubeth signifies the rise of uranium exploration in the late 1970s. Following this, there were some 15 temporary water-haul permits for oil- and gas-related activities from 1980 – 1991. Finally, the two most recent water rights were appropriated by the Applicant for exploration activities at the Ross Project area (Strata, 2011a).

3.5.2 Wetlands

The Federal definition of wetlands includes "those areas that are inundated or saturated by surface or ground water at a frequency and duration sufficient to support, and that under normal circumstances do support, a prevalence of vegetation typically adapted for life in saturated soil conditions. Wetlands generally include swamps, marshes, bogs, and similar areas" (33 CFR Part 328.3). Wetlands are important resources that provide habitat for aquatic fauna and flora, filter sediments and toxicants, and attenuate floodwaters.

Projects that discharge, dredge, or fill material into "Waters of the United States," a concept related to surface- and ground-water regulation which includes special aquatic sites and wetlands under the jurisdiction of the USACE, require accurate identification of wetland boundaries for Section 404 of the *Clean Water Act*-permitting process. Through the Section 404 permitting process, the USACE can authorize dredge or fill activities by issuance of a standard individual permit, regional permit, or the Nationwide Permit (NWP).

Site-specific field surveys on behalf of the Applicant were conducted at the Ross Project by WWC Engineering (WWC) staff on June 22 and 28 as well as July 8 and 21, 2010. These surveys were in accordance with the "Interim Regional Supplement to the USACE Wetlands Delineation Manual: Great Plains Region" (USACE, 2008; Strata, 2011a). These wetlands surveys were conducted to identify and to characterize the wetlands located within the Ross Project area. Existing data used in the survey included Natural Resource Conservation Service (NRCS) soil mapping, U.S. Fish and Wildlife Service's (USFWS's) National Wetlands Inventory (NWI) mapping, and aerial photography taken May 2010 (NRCS, 2010; USFWS, 2012a; Strata, 2011a).

Thirteen wetland sites were identified on the NWI maps within the Ross Project area and were investigated during the 2010 field surveys. Potential wetlands identified during the initial June survey were later visited during another survey in July to verify that wetland characteristics were present. The wetlands-survey results, photographs, and correspondence with the USACE are provided in the Applicant's ER (Strata, 2011a). All but two of the NWI areas were included in the baseline field-delineated wetlands (Strata, 2011a). The two sites not included did not have the three required characteristics for a wetland. The three criteria are: 1) hydrophytic vegetation (i.e., plants that grow in hydric soils), 2) hydric soil (i.e., soils that are commonly flooded or saturated), and 3) wetland hydrology (USACE, 2008).

Many of the potential wetland areas delineated during the 2010 field surveys were small depressions (<0.04 ha [0.1 ac]) that were in close proximity to each other but were distinct depressions separated by upland vegetation. A significant number of these small-depression areas appeared to be influenced by ground water, receiving seepage from the Lance Formation, which outcrops in the vicinity. These potential wetlands were classified according to Cowardin et al. (1979) to more accurately describe the types of potential wetlands present within the Ross Project area (Strata, 2011a). Approximately 93 percent of the potential wetlands were man-made (i.e., diked or excavated). A significant majority of these are preliminarily classified as Palustrine, Aquatic Bed, Seasonally Flooded (PABFh) or Diked. Of the areas designated as PABFh, approximately half were areas of open water. In addition, there were approximately 2.1 ha [5.1 ac] (6,750 linear m [22,130 linear ft] x an average 3-m- [10-ft]-wide channel) of "Other Waters of the U.S." identified within the Ross Project area (Strata, 2011a).

A wetlands delineation report for the Ross Project was submitted to the USACE Omaha District in Cheyenne, Wyoming, during September 2010 (Strata, 2011a). The USACE provided the Applicant a letter on December 9, 2010, that verified the following (USACE, 2010):

- The methods used to identify wetlands and other surface waters were consistent with the USACE's *Wetland Delineation Manual* and its current supplements.
- Exhibit 1 in the wetlands delineation report, entitled *Wetlands and Other Waters of the US. Delineation for the Proposed Ross ISR Project Oshoto, Wyoming* (*Wetland Map*) (dated August 23, 2010), provided an accurate depiction of the boundaries of all wetlands and other waters within the Ross Project area.
- All of the wetlands and channeled waterways identified in the delineation report are connected or adjacent to the Little Missouri River, a navigable water, and are thus likely to be Waters of the U.S. as defined in 33 CFR Part 328.

USACE's final determination of specific wetland areas would not occur until the Applicant applies for coverage for specific construction activities, such as pipeline installation and access-road stream-channel crossings. At that time, the Applicant would be required to provide a site-specific mitigation plan for its disturbance of jurisdictional wetlands (i.e., those wetlands that are under the jurisdiction of the USACE).

3.5.3 Ground Water

Regional Ground-Water Resources

The Applicant presents a description of the regional hydrogeology based upon published literature in its license application (Strata, 2011a; Strata, 2011b). The site-specific hydrogeology of the Lance Formation and the associated stratigraphy underlying the Ross Project area is not described in the GEIS; thus, detailed information is included here. Water-bearing bedrock intervals in the eastern Powder River Basin range in age from Precambrian to Paleocene (see Figure 3.7). Regionally, recharge occurs in the outcrop areas, with ground water moving away from the outcrop into the Basin. Due to the geologic dip of the units, horizons that are accessible near the Black Hills uplift are deeply buried in the Basin's center about 125 km [75 mi] west from the Ross Project area (Hinaman, 2005).

Within the northeast corner of Wyoming there are a number of water-bearing intervals tapped by municipalities and industrial users (Strata, 2011a; Langford, 1964). Below the Fox Hills aquifers, the Minnelusa Formation (210 – 270 m [700 – 900 ft] thick), and the underlying Madison Formation (90 – 270 m [300 – 900 ft] thick) are the most significant aquifers (Whitcomb and Morris, 1964). The Minnelusa and Madison aquifers are recharged at the outcrop in the area of the Black Hills uplift. Ground-water flow in all aquifers is from the recharge areas along the outcrop, westward towards the center of the Powder River Basin. Flow directions are locally modified by pumping wells. The Minnelusa Formation has received aquifer exemptions in portions of Campbell County, which allow it to be used for waste-water disposal (EPA, 1997).

The Minnelusa Formation is also an important hydrocarbon reservoir interval in the areas of the Powder River Basin that are west of the Ross Project (De Bruin, 2007). At the Ross Project area, the Minnelusa Formation is approximately 1,860 m [6,100 ft] bgs (Strata, 2011a). It is separated from the ore zone by 1,680 m [5,500 ft] of sandstone, claystone and shale, most notably the Pierre Shale which is over 600 m [2,000 ft] thick under the Ross Project area as noted in SEIS Section 3.4 (Whitcomb and Morris, 1964).

Water-supply wells in the Madison Formation have reported yields of up to 3,785 L/min [1,000 gal/min]; the Formation is an important source of drinking water for the communities of Gillette and Moorcroft. The city of Gillette operates a wellfield consisting of ten wells north of the town of Moorcroft, yielding 35,204 L/s [9,300 gal/s] from a depth of approximately 760 m [2,500 ft]. The water is piped approximately 53 km [33 mi] to Gillette and blended with locally-produced ground water from the Fort Union Formation and to a lesser degree from wells completed in the Lance and Fox Hills Formations. Other towns in the vicinity (e.g., Moorcroft, Sundance, Upton, Newcastle, and Hulett) also use the Madison Formation for municipal water supply (Strata, 2011a). In the vicinity of Gillette, the Fox Hills and Lance Formations are typically targeted by industrial users, while smaller municipalities, subdivisions, and improvement districts west of Ross Project area use wells completed within the shallower Fort Union Formation.

Local Ground-Water Resources

The detailed geologic stratigraphy and its relationship to the corresponding hydrology are illustrated in Figure 3.14. The detailed stratigraphic sequence from the land surface to the confining interval below the ore zone is, in descending order: recent, unconsolidated, surficial deposits including residual soils, colluvium, and alluvium; Lance Formation; Fox Hills Formation; and Pierre Shale (see also SEIS Section 3.4). Figure 3.14 illustrates the geophysical log and corresponding lithology obtained from type exploration drillhole No. RMR008, the location of which is shown in Figure 3.14. This particular drillhole was chosen as the “type log” by the Applicant for the Ross Project because of the clarity of the geophysical logs and the associated stratigraphic descriptions from land surface to the top of the Pierre Shale (Strata, 2011a).

Within the Ross Project there are four named aquifers existing between the land surface and the Pierre Shale. The correspondence between stratigraphic and hydrologic units, and the related nomenclature, is summarized in Table 3.4.

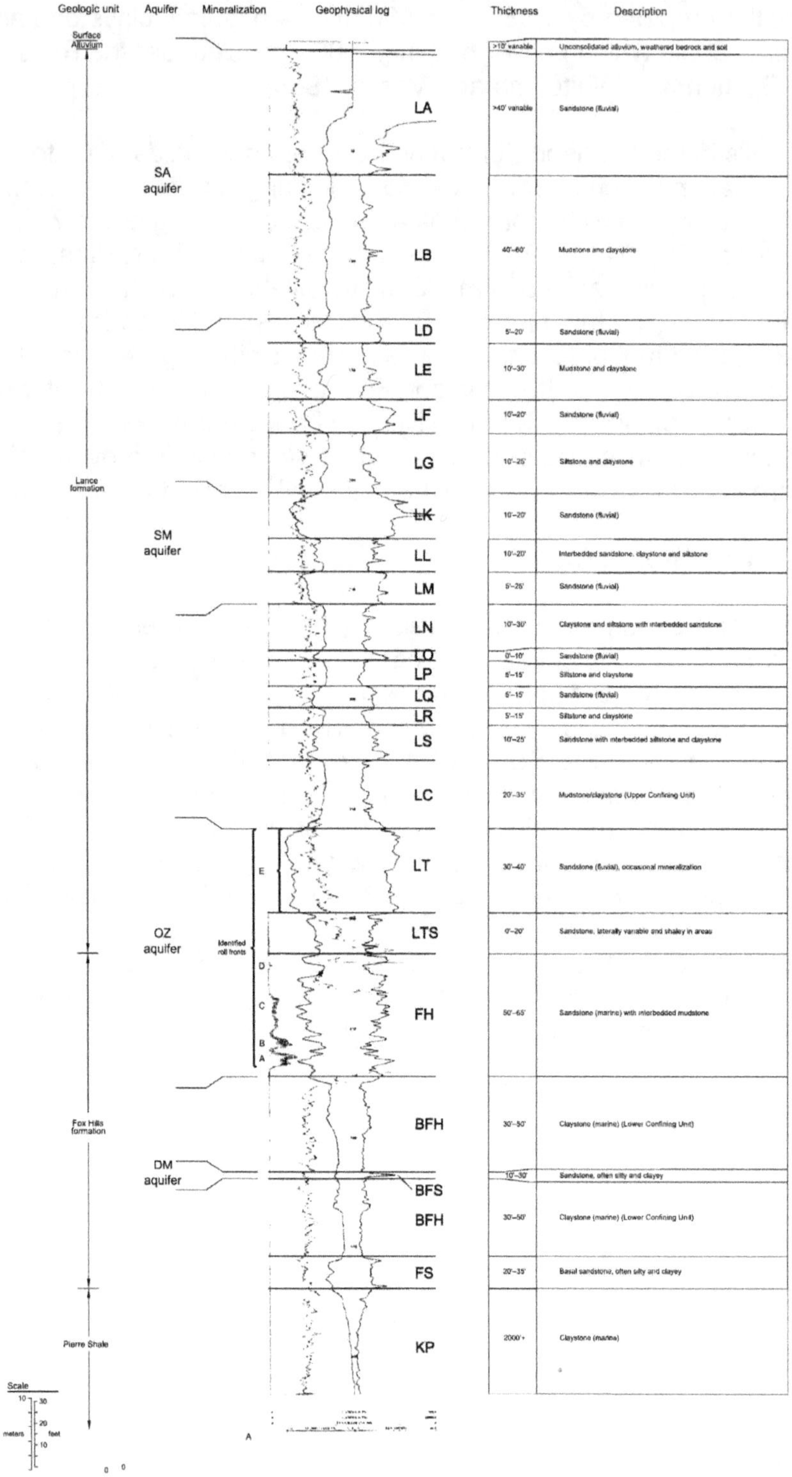

Source: Strata, 2012a.

Figure 3.14

Stratigraphic and Hydrogeologic Units at Ross Project Area

Table 3.4
Geologic Units, Stratigraphic Horizons, and Hydrologic Units of Ross Project Area

Geologic Unit	Stratigraphic Horizon	Hydrologic Unit
Lance Formation and/or Recent Alluvium/Colluvium	Qal/LA/LB	SA (Surface Aquifer)
Lance Formation	LD-LG	Lance Units (Aquitard)
	LK-LM	SM (Shallow-Monitoring Aquifer)
	LN-LS	Sandstone within Confining Unit
	LC	Upper Confining Unit
	LT-LTS	OZ (Ore-Zone Aquifer)
Fox Hills Formation	FH	
	BFH	Lower Confining Unit (Aquitard)
	BFS	DM (Deep-Monitoring Aquifer)
	BFH/FS	Sandstone within Confining Unit
Pierre Shale	KP	Regional Confining Unit (Aquitard)

Source: Strata, 2012b.

The surficial aquifer, or the SA interval, is the "water-table" aquifer within the Ross Project area. It consists of the uppermost water-bearing interval within the upper Lance Formation and the alluvium of the Little Missouri River and Deadman Creek. Ground-water levels range from near-surface in the river valleys to over 15 m [50 ft] bgs in topographically higher areas.

The sandstones of the lower Lance Formation (LT intervals) make up the upper portion of the ore zone (i.e., ore-zone [OZ] aquifer) (see Figure 3.14). The LT sands range in thickness from 9 – 12 m [30 – 40 ft] and show hydraulic continuity beneath the Ross Project area. Above the LT sands is a shale layer varying in thickness from 6 – 24 m [20 ft – 80 ft], locally called the LC interval aquitard. The Applicant designates the LC aquitard as the "upper confining unit." The LC aquitard serves as a confining unit that separates the uranium-mineralized sandstones of the FH and LT horizons and the OZ aquifer, from the water-bearing unit above (see Figure 3.14).

The water-bearing sands above the upper confining unit is referred to as the shallow-monitoring (SM) unit, or SM aquifer, and is composed of the LM- through LK-horizon sandstones. Above the SM aquifer is a sequence of thin sands, shales, and silts. Many of the thin sandstones contain water; however, these sandstones are generally discontinuous and, while they may be used locally for stock and domestic wells, they are not regionally extensive.

The Lance Formation is recharged at the outcrop and at the subcrop beneath the alluvium in the valley of the Little Missouri River and its tributaries. Natural ground-water flow would be expected to be westward from the outcrop toward the Basin.

At the Ross Project area, the thickness of the Fox Hills Formation is approximately 46 m [150 ft], with local variations of up to15 m [50 ft] or more. The Fox Hills Formation consists of an upper sandstone unit (i.e., FH horizon) and a lower sandstone unit (i.e., FS horizon) which are separated by an intervening shale, claystone, and mudstone interval (i.e., BFH horizon) containing the BFS sandstone unit (see Figure 3.14). Uranium mineralization primarily occurs within the Fox Hills Formation's sands, although in localized areas mineralization occurs within the overlying Lance Formation's (i.e., LT horizon) sandstone.

The FS and BFS sandstones represent the only water-bearing units within the lower Fox Hills Formation (see Table 3.4). Both sand units are believed to be continuous throughout the Ross Project area, although in places they are relatively thin. The BFS horizon is the nearest aquifer below the uranium-bearing sandstone (the FH horizon and also known as the ore zone) in the upper Fox Hills Formation, and in terms of uranium-recovery operations, it is referred to as the deep-monitoring (DM) interval, or the DM aquifer. It is separated from the FH sand (i.e., the ore zone) above and the FS (basal sandstone) below by a shale, claystone, and mudstone (BFH horizon). The Applicant provides potentiometric contours for the DM interval in its ER (see Figure 3.15) (Strata, 2011a).

The Pierre Shale yields very little water; it is considered regionally as a confining unit (NRC, 2009b; Whitehead, 1996). No wells are known to be completed within the Pierre Shale at the Ross Project area.

The FH horizon sandstones within the upper Fox Hills Formation contain uranium and are the primary uranium-recovery target interval for the Proposed Action. The Applicant has designated the OZ aquifer as consisting of the FH sandstones with the overlying lower Lance Formation sandstones (LT horizon). The lithologies of the ore zone range from thick-bedded, blocky sandstones to thin, interbedded sandstones, siltstones, and shales. The OZ aquifer is underlain by claystone of the Fox Hill Formation (i.e., BFH interval). Within the Ross Project area, this ore-zone interval ranges from 27 – 55 m [90 – 180 ft] thick (see Figure 3.14). Thin, silty, and clayey sandstone comprises the DM aquifer. The Applicant designates the BFH aquitard above the DM aquifer and below the ore zone as the "lower confining unit."

What terms are used to describe hydrologic characteristics?

Transmissivity: This term is used to define the flow rate of water through a vertical section of an aquifer, considering a unit width and extending the full saturated height of the aquifer under unit hydraulic gradient. Transmissivity is a function of an aquifer's saturated thickness and hydraulic conductivity.

Hydraulic Conductivity: This term represents a measure of the capacity of a porous medium to transmit water. It is used to define the flow rate per unit cross-sectional area of an aquifer under unit hydraulic gradient.

Storativity: This term is used to characterize the capacity of an aquifer to release ground water from storage in response to a decline in water levels.

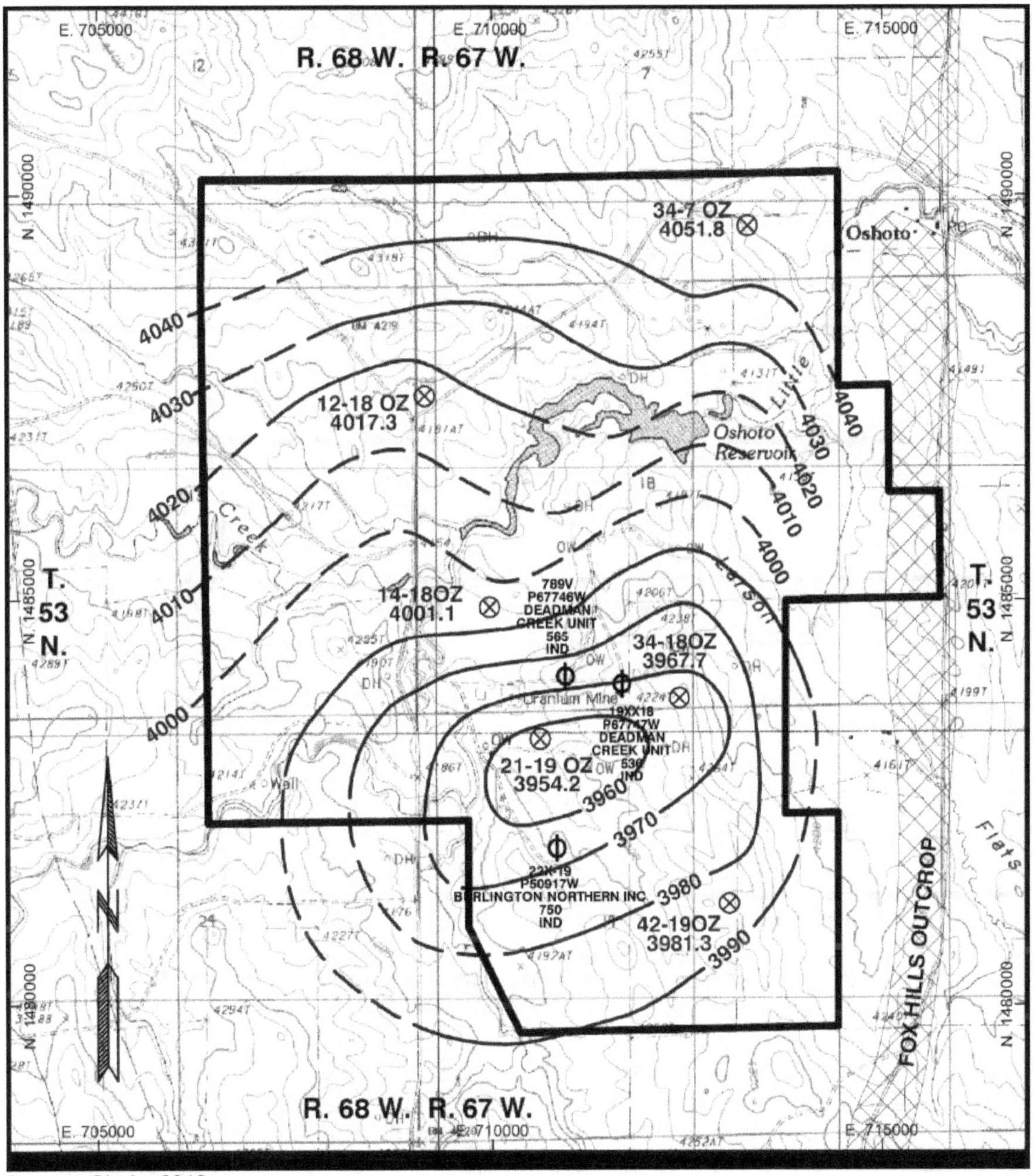

Source: Strata, 2012a.

Figure 3.15

Potentiometric Contours of Ground Water in Ore-Zone Aquifer

1

Isopachs of this structure show that it ranges in thickness from less than 3 m [10 ft] to more than 30 m [100 ft] (Strata, 2011a). Above the ore zone, the mudstone and claystone of the Lance Formation form the upper confining unit, as noted above, ranging in thickness from less than 6 m [20 ft] to more than 24 m [80 ft] (see Figure 3.14).

The FH sandstones, shales, and silts have been studied extensively through both core analysis and aquifer tests. Seven pumping tests targeting the ore zone were performed by the Applicant at six separate well clusters. Applicable methodology and testing were used and those results are shown in Table 3.5 (and additional details can be found in Strata, 2011a).

Table 3.5
Ore-Zone Aquifer Hydrogeologic Characteristics

	Transmissivity m^2/day [ft^2/day]	Hydraulic Conductivity cm/s [ft/day]	Storativity (Unitless)
Minimum	0.353 [3.80]	4.59E-05 [0.13]	4.00E-06
Maximum	34.2 [368]	2.69E-03 [7.62]	1.50E-04
Median	8.20 [88.3]	1.25E-03 [3.55]	6.10E-05
Geometric Mean	6.10 [65.6]	6.74E-04 [1.91]	4.50E-05
Average	8.15 [87.8]	1.15E-03 [3.26]	6.70E-05

Source: Addendum 2.7-F, Table 3, in Strata, 2011a.

The aquifer properties determined by the 2010 tests are comparable to results reported for previous pumping tests within the Ross Project area (Strata, 2011b).

The Applicant developed a static piezometric surface (i.e., a map showing the static water levels expressed as feet above sea level) for the ore-zone aquifer (see Figure 3.15). The ore zone's potentiometric surface shows a distinct cone of depression near the No. 21-19 well cluster that has resulted from 30 years of ground-water withdrawals by oil-field water-supply wells completed in the OZ aquifer. This pumping has changed the hydraulic gradient and the direction of ground-water flow throughout most of the Ross Project area. The potentiometric surface near the No. 34-7 well cluster, which is farthest from the oil-field water-supply wells that have been pumping for 30 years, has been least affected by such pumping. Based upon the Applicant's estimates, approximately 46 m [150 ft] of drawdown (i.e., the decline in water level) in the ore-zone aquifer has occurred in the vicinity of the No. 21-19 well cluster since pumping began in 1980 for local oil-field water-flood operations (Strata, 2011b). An updated map of the ore zone's piezometric surface prepared by the Applicant using a ground-water model provides additional detail of the drawdown associated with the withdrawals from the Merit Oil Company's (Merit's) three water supply wells (Strata, 2012b).

The Applicant also calculated horizontal gradients and vertical-head differences between the OZ, SM, and DM aquifers (Strata, 2011a). Horizontal gradients in the OZ aquifer are toward the oil-field water-supply wells, and they range from 0.009 – 0.025, with the steeper gradients being in the vicinity of the oil-field water-supply wells. Vertical-head differences between the OZ and the DM aquifers range from 6 m [20 ft] downwards in the northwestern portion of the Ross Project area to 3 m [10 ft] upwards in the area of the oil-field water-supply wells. Vertical gradients are downwards from the SM to the OZ aquifers, with head differences ranging from 15 – 46 m [50 – 150 ft].

The OZ aquifer remains a confined aquifer across the Ross Project area, with potentiometric heads ranging from approximately 46 m [150 ft] to more than 122 m [400 ft] above the top of the ore zone (Strata, 2011a). Recharge to the Fox Hills Formation and, hence, the OZ aquifer, is from precipitation along the outcrop, ground water from the subcrop beneath alluvium in the valley of the Little Missouri River and its tributaries, and from leakage from the overlying Lance Formation. Under current conditions, discharge is to the oil-field water-supply wells.

Continuous measurement of water levels for the period April to October 2010 were recorded by the Applicant in six monitoring wells completed in the OZ aquifer and are presented graphically by the Applicant in its TR (Strata, 2011b). The hydrograph for Well 34-7OZ, which is located farthest from the oil-field water-supply wells, displays the least variation. The variability in the ore-zone-well hydrographs is a function of the well locations relative to the oil-field water-supply wells in Sections 18 and 19. The wells located closest to this area (Wells 21-19OZ, 34-18OZ, 14-18OZ, and 42-19OZ) display water-level fluctuations that are related to pumping of the water-supply wells. Pumping starts and stops that occurred in late June though early July 2010 are apparent on hydrographs from these wells. A rapid water-level rise (over 4.6 m [15 ft] in Well 21-19OZ) in late September 2010 was attributed to a temporary cessation of pumping. This was followed by a rapid decline in the water level, which was interpreted as an indication of resumption of pumping.

Other than the aquifer testing that took place over the period above, other recorded perturbations are related to sampling events and barometric fluctuations. The barometric fluctuations are less than 0.2 m [0.5 ft]. During January through October 2010, the hydrograph for Well 34-7OZ showed a steady increase of approximately 0.6 m [2 ft]. The cause of this increase has not been identified; similar patterns have not been seen in other ore-zone well hydrographs. The hydrograph for Well 12-18OZ varies within a range of approximately 0.76 m [2.5 ft]. Most of the water-level changes are interpreted as responses to barometric pressure changes. However, fluctuations in the late June though early July time period coincide with pumping-related water-level changes observed in the group of four wells discussed above.

The shale, claystone, and mudstone interval, the BFH horizon and lower confining unit, separates the DM aquifer from the FH horizon. This low-permeability unit ranges in thickness from less than 3 m [10 ft] to 24 m [80 ft]. Vertical hydraulic conductivities for this interval are expected to be comparable to that of the Pierre Shale (i.e., 2×10^{-7} cm/s [5×10^{-4} ft/day] or less), based on their similar lithologies.

Pumping tests were performed on six well clusters with pumping from the OZ aquifer and monitoring of the SA, SM, and DM aquifers. No effects from pumping were measured in any of the overlying SA or SM wells. Water levels in two of the six underlying DM wells (Nos. 14-18DM and 34-18DM) declined slightly during pumping. The lower confining unit is 9 – 15 m [30 ft – 50 ft] thick in the portions of the Ross Project area where these wells are located. The response of the DM-completed wells has been interpreted by the Applicant as being due to vertical leakage across the lower confining unit via drillholes that are in close proximity to the pumping-test well cluster that have not yet been located and plugged. Prior to the Applicant's conducting the aquifer test at Well 12-18, all exploration drillholes in the vicinity of that well cluster were located and plugged, and no response of the DM-aquifer well was observed during that pumping test.

Communication between the OZ and DM aquifers in locations where the lower confining unit has been breached was demonstrated by: 1) the responses observed in the DM zone for two pumping tests, where old exploration drillholes had not been plugged and 2) the similarities in the potentiometric heads in the DM, OZ and SM aquifers in the vicinity of the oil-field water-supply wells, which are completed in both the OZ and DM intervals. To prevent communication between aquifers during uranium-recovery operation, the Applicant proposes to actively locate and plug all exploration drillholes prior to beginning wellfield operations. The Applicant proposes to actively locate and plug all exploration wells prior to beginning wellfield operation.

Ground-Water Quality

The Applicant has compiled regional water-quality data listed in the USGS's NWIS from 16 wells located in Crook and Campbell Counties that were completed in the Lance and Fox Hills aquifers (Strata, 2011a; USGS, 2012b). Data from these wells show a water quality of the Lance and Fox Hills aquifers that is slightly alkaline (i.e., median pH of 8.4) with a median TDS of 1,130 mg/L, with sodium and bicarbonate as the dominant dissolved species.

The water quality of shallow ground water from alluvial deposits on the Lance Formation is dominated by sodium, sulfate, and bicarbonate with moderate levels of TDS of approximately 1,200 – 1,400 mg/L (Langford, 1964). Rankl and Lowry (1990) noted that the water quality in the aquifer sequence through the Lance and Fox Hills Formations depends upon the stratigraphy and varies according to well depth. As well depths increase from 30.5 – 152 m [100 – 500 ft], TDS in the waters decrease sharply due to declining concentrations of calcium, magnesium, and sulfate. Water from wells at depths of 152 m [500 ft] or greater are dominated by bicarbonate and sodium.

The deep-injection-well UIC Class I permit application for the Ross Project contains estimates of water quality in deeper formations, from the Minnelusa through the Cambrian Formations (WDEQ/WQD, 2011). The Minnelusa, Deadwood, and Flathead Formations are expected to have TDS concentrations greater than 10,000 mg/L, while the Madison Formation likely has a TDS concentration around 1,000 mg/L in the vicinity of the Ross Project area.

To comply with the requirements of 10 CFR Part 40, Appendix A, Criterion 7, the Applicant has collected pre-licensing baseline ground-water-quality data from the site characterization of the Ross Project area. These data originate from three sources: 1) the Applicant's own baseline site-characterization monitoring network at the Ross Project and the respective analytical data; 2) existing water-supply-wells sampling and analysis data; and 3) historical data from the former Nubeth operation (Nuclear Dynamics, 1978). The first source of ground-water quality data is the Applicant's own ground-water-monitoring network which it constructed in 2009 and 2012 and which consists of six monitoring-well clusters and four piezometers (Strata, 2011a). Each well cluster includes four monitoring wells targeting the OZ aquifer and the aquifer units above the ore zone (SA and SM) and below the ore zone (DM) (see Figure 3.14). The Applicant provided construction details of the wells and methods used for ground-water sampling in its ER (Strata, 2011a). The four piezometers in the SA were installed in the portion of the Ross Project area proposed for the central processing plant (CPP) and surface impoundments (Strata, 2011a).

Analytical data from the 2010 quarterly samples are provided in the Applicant's ER and TR (Strata, 2011a; Strata, 2011b). Water-quality data from samples collected in 2011 and submitted to WDEQ/LQD are provided in information received subsequently from the Applicant (Strata, 2012a). The data from 2011 are generally consistent with the 2009 and 2010 data, indicating a representative characterization of ground-water quality. The data are summarized in the following paragraphs.

The average concentrations of the major cations and anions, in addition to the median field measurements of pH and average dissolved-oxygen measurements, are presented on the next page in Table 3.6. Dissolved solids (TDS) in the ground water at the Ross Project area are predominately bicarbonate-sulfate-sodium, which differs from typical ground water described in the GEIS, which is the bicarbonate-sulfate-calcium type. The pH conditions of greater than 8.5 are consistent with bicarbonate water, and dissolved oxygen levels of less than 5 mg/L suggest low-oxygen conditions. These two parameters are typical of uranium-bearing aquifers (NRC, 2009b).

The water quality data indicates distinctive water quality in each aquifer unit, i.e., the SA, SM, OZ, and DM. The distinctive water quality is made possible by the stratigraphic layers between the aquifer units that prevent vertical movement of water between the units. Average values of TDS in Strata's ground-water baseline monitoring network range from 730 mg/L in the SA to 1574 mg/L in the OZ. Ground-water from piezometers in the SA show that the TDS increases sharply with increasing distance from the Little Missouri River (Strata, 2011a).

The effects on Strata's pre-licensing baseline water quality from Nubeth can be evaluated by comparing the Strata's data with baseline data reported by Nuclear Dynamics (1978). The data from Strata (2011a, 2012a) include all four aquifer units. Nuclear Dynamics (1978) reports data from only the ore zone and the aquifer above the ore zone which is likely equivalent to the SM. The comparison shows that the TDS in the SM and OZ have decreased since 1978 (see also SEIS Section 5.7.2).

Table 3.7 summarizes the concentrations of metals, radiological parameters, ammonium, and fluoride measured by Strata in the aquifer units. With a few exceptions, the 1978 mean values are within the range of values reported by Strata (2011a, Strata, 2012a). Strata's pre-licensing baseline concentrations of arsenic, radium-226, and gross beta are slightly lower in the ore zone than was measured in 1978 (Table 3.7). Strata's concentrations of cadmium, lead and nickel are slightly lower in both the ore zone and the aquifer above the ore zone than in 1978.

Table 3.6 Average Concentrations of Major Cations and Anions in Ground Water from the Ore-Zone (OZ) Aquifer and Aquifers Above (SM & SA) and Below (DM) the Ore Zone†							
		Ross Project Monitoring-Well Data (Strata, 2011a; Strata, 2012a)				Nubeth Data (Nuclear Dynamics, 1978)	
Constituent	Units	Surficial Aquifer (SA)	Shallow-Monitoring Aquifer (SM)	Ore-Zone Aquifer (OZ)	Deep-Monitoring Aquifer (DM)	Ore Zone	Above Ore Zone
Bicarbonate	mg/L	339	449	583	295	592	653
Calcium	mg/L	21	2	6	3	6.2	6
Carbonate	mg/L	N/A	98	26	103	22	17
Chloride	mg/L	29	4	7	491	10	6
Magnesium	mg/L	13	<1**	2	<1**	2.7	2.7
Potassium	mg/L	12	15	6	19	3.2	3.9
Sodium	mg/L	224	417	545	520	622	592
Sulfate	mg/L	172	318	602	31	715	567
Total Dissolved Solids (TDS)	mg/L	730	1145	1574	1321	1629	1498
Dissolved Oxygen (DO)	mg/L	3.2	3.9	2.8	4.7	N/A***	N/A***
pH	Std. Units	8.6	9.15	8.7	9.4	8.8	8.6

Source: Strata, 2011a; Strata, 2012a; Nuclear Dynamics, 1978.

Notes:

† All values are mean concentrations, except for pH from Strata data, which is the median value, and pH reported in Nubeth, which is a mean value.

† Shading indicates a value greater than WDEQ and U.S. Environmental Protection Agency (EPA) Water Quality Standards.

* 34 percent of the 32 reported concentrations were below the detection limit, which precluded calculation of an average or median value; minimum and maximum values for carbonate concentration in mg/L were less than 5 and 218 mg/L, respectively, for this dataset.

** "<" = "Less than," where the value following the "<" is the detection limit.

***N/A = Not available.

Table 3.7
Summary of Water Quality of Ground Water from the Ore-Zone (OZ) Aquifer and Aquifers Above (SA & SM) and Below (DM) the Ore Zone[†]
2009, 2010, and 2011 Data from Ross Project Monitoring Wells Nos. 12-18, 14-18, 21-19, 34-7, 34-18, and 42-19
(As reported in Strata, 2011a; Strata, 2012a.)

		Ross Project Monitoring-Well Data								Nubeth Data	
		SA		SM		OZ		DM		Ore Zone	Above Ore Zone
Constituent*	Units	Min	Max	Min	Max	Min	Max	Min	Max	Mean	Mean
Ammonia	mg/L	<0.1**	0.6	<0.1	2.8	<0.1	0.8	<0.1	3.9	0.73	0.53
Arsenic	mg/L	<0.005	<0.005	<0.005	0.023	<0.005	<0.005	<0.005	0.014	0.11	<0.005
Barium	mg/L	<0.5	<0.5	<0.5	<0.5	<0.5	<0.5	<0.5	<0.5	<0.01	<0.10
Boron	mg/L	<0.1	0.3	0.2	0.8	0.3	0.6	0.3	1	0.32	0.6
Cadmium	mg/L	<0.002	<0.002	<0.002	<0.002	<0.002	<0.002	<0.002	<0.002	0.003	0.004
Chromium	mg/L	<0.01	<0.01	<0.01	<0.01	<0.01	<0.01	<0.01	<0.01	<0.01	<0.01
Copper	mg/L	<0.01	<0.01	<0.01	0.02	<0.01	<0.01	<0.01	<0.01	<0.01	0.01
Fluoride	mg/L	0.1	0.8	0.8	2.1	0.2	1.3	0.8	1.6	N/A**	N/A
Iron	mg/L	<0.05	0.66	<0.05	0.21	<0.05	0.69	<0.05	0.4	0.09	0.074
Lead	mg/L	<0.02	<0.02	<0.02	<0.02	<0.02	<0.02	<0.02	<0.02	0.04	0.037
Mercury	mg/L	<0.001	<0.001	<0.001	<0.001	<0.001	<0.001	<0.001	<0.001	<0.00004	0.00003
Manganese	mg/L	<0.02	0.36	<0.02	0.88	<0.02	0.06	<0.02	0.37	0.012	0.014
Molybdenum	mg/L	<0.02	0.07	<0.02	0.05	<0.02	<0.02	<0.02	0.06	<0.003	<0.005
Nickel	mg/L	<0.01	<0.01	<0.01	<0.01	<0.01	<0.01	<0.01	<0.01	0.017	0.016
Selenium	mg/L	<0.005	0.008	<0.005	0.017	<0.005	0.009	<0.005	0.03	0.003	<0.005
Silver	mg/L	<0.003	0.006	<0.003	0.011	<0.003	<0.003	<0.003	0.005	<0.003	<0.003
Uranium	mg/L	<0.001	0.007	<0.001	0.004	0.005	0.109	<0.001	0.003	0.073	0.004
Vanadium	mg/L	<0.02	<0.02	<0.02	0.02	<0.02	<0.02	<0.02	<0.02	<0.003	<0.003
Zinc	mg/L	<0.01	1.32	<0.01	0.03	<0.01	0.02	<0.01	0.09	0.011	0.016
Radium-226	pCi/L	<0.2	0.5	<0.2	3.7	0.6	12.1	<0.2	0.7	22	0.06
Radium-228	pCi/L	<1	1.8	<1	12.27	<1	1.6	<1	2.2	N/A	N/A
Gross Alpha	pCi/L	<6	13.8	<7	12.2	<5	222	<14	28.3	98	1.4
Gross Beta	pCi/L	<8	17.6	<8	319***	<8	46.8	<20	41	97	3.2

1 Source: Strata, 2011a; Strata, 2012a; Nuclear Dynamics, 1978.
2 Notes on next page.

Notes for Table 3.7:

† Analytical results are presented as minimum and maximum values for each constituent; the number of measurements that are less than the detection level precludes calculation of mean concentrations.

† Shading indicates a value greater than WDEQ and EPA Water Quality Standards.

*All constituents reported as dissolved concentrations (i.e., the samples were filtered), except ammonia and fluoride.

** "<" = "Less than," where the value following the "<" value is the detection limit.
"N/A" = Datum not available.

***319 appears to be an anomalous value; the next lowest value is 42.5.

The similarity between the pre-licensing baseline concentrations in the ore zone and aquifer above the ore zone suggests that Nubeth did not alter the baseline water quality. Table 3.8 presents the WDEQ and EPA water-quality standards for constituents that were present in Strata's that were found to exceed the standards in Strata's pre-licensing baseline data (WDEQ/WQD, 2005b; 40 CFR Part 41). Concentrations of constituents that exceed the standards are indicated by shading in Tables 3.6 and Table 3.7.

Table 3.8
Water-Quality Standards Exceeded in Ground Water at the Ross Project (Pre-Licensing Baseline)

Water-Quality Constituent	Units	WDEQ Class I Domestic	WDEQ Class II Agriculture	EPA Primary MCL	EPA Secondary MCL
Ammonia	mg/L	0.5	N/A*	N/A	N/A
Arsenic	mg/L	0.05	0.1	0.01	N/A
Boron	mg/L	0.75	0.75	N/A	N/A
Chloride	mg/L	250	100	N/A	250
Iron	mg/L	0.3	5	N/A	0.3
Manganese	mg/L	0.05	0.2	N/A	0.05
Selenium	mg/L	0.05	0.02	0.05	N/A
Sulfate	mg/L	250	200	N/A	250
Total Dissolved Solids (TDS)	mg/L	500	2000	N/A	500
Uranium	mg/L	N/A	N/A	0.03	N/A
Radium-226 + 228	pCi/L	5	5	5	N/A
Gross Alpha	pCi/L	15	15	15	N/A

Source: WDEQ/WQD, 2005b.

Notes:

* N/A = Not applicable.

Per the WDEQ/LQD Hydrology Guideline No. 8 and NRC Regulatory Guide 4.14, the water-quality data produced by the Applicant and used to compare with the water-quality standards are dissolved concentrations except for ammonium, chloride, fluoride, sulfate, and TDS (WDEQ/LQD, 2005b; NRC, 1980).

Typical of uranium-bearing aquifers described in the GEIS (NRC, 2009b), the average TDS of each aquifer unit associated with Ross Project area exceed EPA's respective Secondary drinking water maximum contaminant levels (MCLs) of 500 mg/L, but they are within all the upper limits set by WDEQ for Class II Agriculture and Livestock Classes of Use (see Tables 3.6 and 3.8) (WDEQ/WQD, 2005b). The two upper aquifers, SA and SM, contain lower TDS than the lower units, and the OZ aquifer contains the highest average TDS.

Comparison of the metals, radiological parameters, ammonium, and fluoride to EPA's MCLs for drinking water and WDEQ standards are provided in Tables 3.7 and 3.8. Ammonia was measured in all four aquifer units at concentrations greater than WDEQ's standard for domestic use, 0.5 mg/L. Iron and manganese are present in all four aquifer units in concentrations greater than WDEQ's standard for domestic use and EPA's secondary MCL for drinking water. Arsenic was measured at concentrations greater than EPA's primary drinking water standard in the SM and DM but less than WDEQ's standard for domestic use. Boron was present at concentrations greater than the WDEQ standard for domestic use in the SM and DM. Uranium and radium-226 were present in the OZ at concentrations greater than the standards (see Table 3.8). Gross alpha exceeded the standards in the OZ and DM aquifer units.

As part of its ground-water sampling and analysis efforts, the Applicant identified 29 currently operable water-supply wells within the Ross Project area and the surrounding 2-km (1.2-mi) area (Strata, 2011a). These wells included two industrial wells, 12 domestic wells, and 15 stock wells. These well locations are shown in the Applicant's ER (Strata, 2011a).

The two industrial wells, completed at depths of 163 m and 229 m [536 ft and 750 ft], were permitted in the early 1980s and provide water for enhanced oil recovery (EOR). Water used in EOR is injected into the oil-bearing rock to displace oil from the rock, thus allowing the oil to be pumped to the surface. Well No. 19X-18 was originally used by Nubeth as a recovery well for its research and development activities, before being converted to a water-supply well for the nearby oil production. The Applicant's review of the well permit reports listed in the WSEO database during 2010 determined general information about each well (WSEO, 2006; Strata, 2011a). Completion depths of permitted stock wells range from 10 – 93 m [40 – 304 ft]. Domestic wells are generally deeper than the stock wells, ranging from 46 – 180 m [150 – 600 ft]. The limited information available on these wells precluded a determination of which aquifer was supplying water to the domestic wells.

The water-supply wells were sampled in consecutive quarters in 2009 and 2010 with the same methods established for the monitoring wells (Strata, 2011a). The results of the water-quality analyses are provided in the Applicant's ER (Strata, 2011a; Strata, 2011b). Comparison between the measured water quality and WDEQ's standards and EPA's drinking- water standards are also provided in the Applicant's ER (WDEQ/WQD, 2005b; 40 CFR Part 141; Strata, 2011a). As described below for each type of well, these analyses showed that the local water supply's contaminants generally exceeded EPA's drinking water standards and often exceeded Wyoming's less stringent quality standards for agricultural use.

Domestic Wells

TDS in samples from the domestic wells consistently exceeded the Wyoming Class I (Domestic) use and the EPA Secondary MCL standards. Sulfate exceeded the Wyoming Class I, the Wyoming Class II and the EPA Secondary MCL standards in 7 of the 13 wells sampled. Gross alpha in excess of the Wyoming Class I and Class II standards, as well as the EPA Primary MCL of 0.55 Bq/L [15 pCi/L], was measured in samples from 4 of the 13 domestic wells. The Wyoming Class I and the EPA Secondary MCL iron standards were exceeded in two of the wells.

Industrial Wells

Samples from the industrial wells exceeded the Wyoming Class II standard and the EPA Secondary MCL standards for TDS and sulfate. The Wyoming Class II and the EPA MCL standards were exceeded in Well No. 19XX18 for radiological parameters: uranium, radium-226+228, and gross alpha. The gross-alpha standard was also exceeded in samples from Well No. 22X-19.

Stock Wells

The water quality of stock wells is variable. TDS often ranged from 370 to 1,610 mg/L, often exceeding the EPA Secondary MCL standard, but also consistently less than the Wyoming Class II use standard of 2,000 mg/L. Sulfate, ranging from 28 to 679 mg/L, often exceeded the Wyoming Class II and the EPA Secondary MCL standards. Gross alpha exceeded both the Class II standard and the MCL in 7 of the 15 stock wells. Selenium exceeded the Wyoming Class II and the EPA Primary MCL standards in one well.

Ground-Water Uses

In order to assess historical and current ground-water use, ground-water rights and unregistered water wells were investigated by the Applicant within the Ross Project area and the surrounding 3.2-km [2-mile] vicinity. Sources of data included WSEO-registered wells, landowner interviews, and field investigations (WSEO, 2006). The search revealed 119 ground-water rights and unregistered wells. The locations and uses of these wells are summarized in the Applicant's ER (Strata, 2011a). Historical ground-water use began with the first domestic and livestock well in 1918. From approximately 1918 – 1977, ground water was used primarily for domestic and livestock consumption, with lesser amounts of water used for irrigation.

In 1977, Nubeth permitted 14 monitoring and industrial-use wells associated with its research and development operation. In addition, between 1980 and 1991, many industrial and miscellaneous wells associated with oil and gas production were permitted in and around the Ross Project area. These include three wells within the Ross Project area itself (Nos. P50917W, P67746W and P67747W) that are currently used as water-supply wells for EOR operations (i.e., water flooding) (Strata, 2011a). In 1981, International Minerals & Chemical Corporation (IM&CC) permitted five pits (Nos. P58895W, P58896W, P58899W, P58902W and P58905W) for dewatering and dust suppression associated with bentonite mining. According to WSEO records, the water rights were cancelled prior to 2001 at the request of IM&CC.

Between 1991 and 2009, the only ground-water rights that have been filed within the Ross Project and surrounding areas are for domestic and livestock use. In 2009, the Applicant obtained ground-water rights for its pre-licensing baseline monitoring wells. The historical ground-water use within the Ross Project area is summarized in Table 3.9.

Table 3.9
Historical Ground-Water Use within
Three Kilometers [Two Miles] of Ross Project Area

Use	Number of Wells	Percent of Total Use	Appropriation Dates
Domestic Only	5	4	1943 – 1995
Domestic and Stock	15	13	1918 – 2003
Domestic, Stock, and Irrigation	1	<1	1972 – 1972
Stock Only	34	29	1933 – 2010
Stock and Irrigation	1	<1	1961 – 1961
Monitoring	39	33	1977 – 2010
Industrial or Miscellaneous	24	20	1977 – 1991
TOTAL	119	100	1918 – 2010

Source: Strata, 2011a.

Within the Ross Project area, ground-water use follows a similar pattern to that observed within the 3.2-km [2-mile] surrounding vicinity, except that historical use has been livestock only (no domestic or irrigation use). More recent uses include monitoring-well use as well as industrial uses associated with Nubeth and with water supply for oil and gas operations. Most of the ground-water rights represented in Table 3.9 have been cancelled or are no longer active. Current ground-water use is limited to four livestock wells, the Applicant's regional pre-licensing baseline monitoring wells, and three industrial wells (i.e., water supply for oil and gas production). The stock wells are completed at total depths ranging from 39 – 81 m [128 – 265 ft], which are considerably above the ore-zone aquifer. The currently operating, industrial water wells are completed at total depths of 163 – 229 m [536 – 750 ft]. Together, these wells withdraw an average of approximately 1.9 L/s [30 gal/s] from the ore-zone aquifer.

3.6 Ecology

The Proposed Action is located within the Powder River Basin of the Northwest Great Plains ecoregion. As described in the GEIS, this area is characterized by rolling prairie and dissected river breaks surrounding the Powder, Cheyenne, and Upper North Platte Rivers (NRC, 2009b). Vegetation within this region is composed of sagebrush and mixed-grass prairie dominated by blue grama (*Bouteloua gracilis*), western wheatgrass (*Pascopyrum smithii*), needle-and-thread grass (*Stipa comata*), rabbitbrush (*Chrysothamnus sp.*), fringed sage (*Artemisia frigida*), and other forbs, shrubs, and grasses (NRC, 2009b).

The Applicant has conducted a number of ecological studies of the proposed Ross Project area to address the guidelines indicated in NUREG–1569, including the identification of important species and their relative abundance, and to meet the applicable Wyoming requirements (NRC, 2003). These studies included vegetation and wildlife surveys conducted on the Ross Project area in late 2009 and 2010 (Strata, 2011a).

3.6.1 Terrestrial Species

3.6.1.1 Vegetation

The Applicant conducted pre-licensing baseline vegetation and wetland surveys during 2009 and 2010, in accordance with State and Federal guidelines (Strata, 2011a). The spatial distribution of the vegetation types within the Ross Project area are shown in Figure 3.16. The vegetation mapped at the Ross Project area included upland grassland, sagebrush shrubland, pastureland, hayland, reservoir/stock pond, wetland, disturbed land, cropland, and wooded draw. No threatened or endangered plant species have been documented on the Ross Project area.

Each vegetation community was investigated by the Applicant to establish a baseline in support of the Proposed Action. In terms of diversity, the sagebrush-shrubland vegetation type exhibited the highest total number of individual plant species recorded in 2010, followed by the upland-grassland and pasture-land vegetation types (see Table 3.10).

Table 3.10
Species Diversity by Vegetation Type at Ross Project Area

Species Type	Number of Individual Plant Species Recorded		
	Sagebrush Shrubland	**Upland Grassland**	**Pastureland**
Perennials			
Grass	16	16	9
Grass-like	2	2	0
Forb	28	27	6
Subshrub	4	4	1
Full Shrub	5	1	1
Succulent	1	1	0
Subtotal	**56**	**51**	**17**
Annuals			
Grass	2	2	0
Forb	7	3	1
Subtotal	**9**	**5**	**1**
TOTAL	**65**	**56**	**18**

Source: Strata, 2011a.

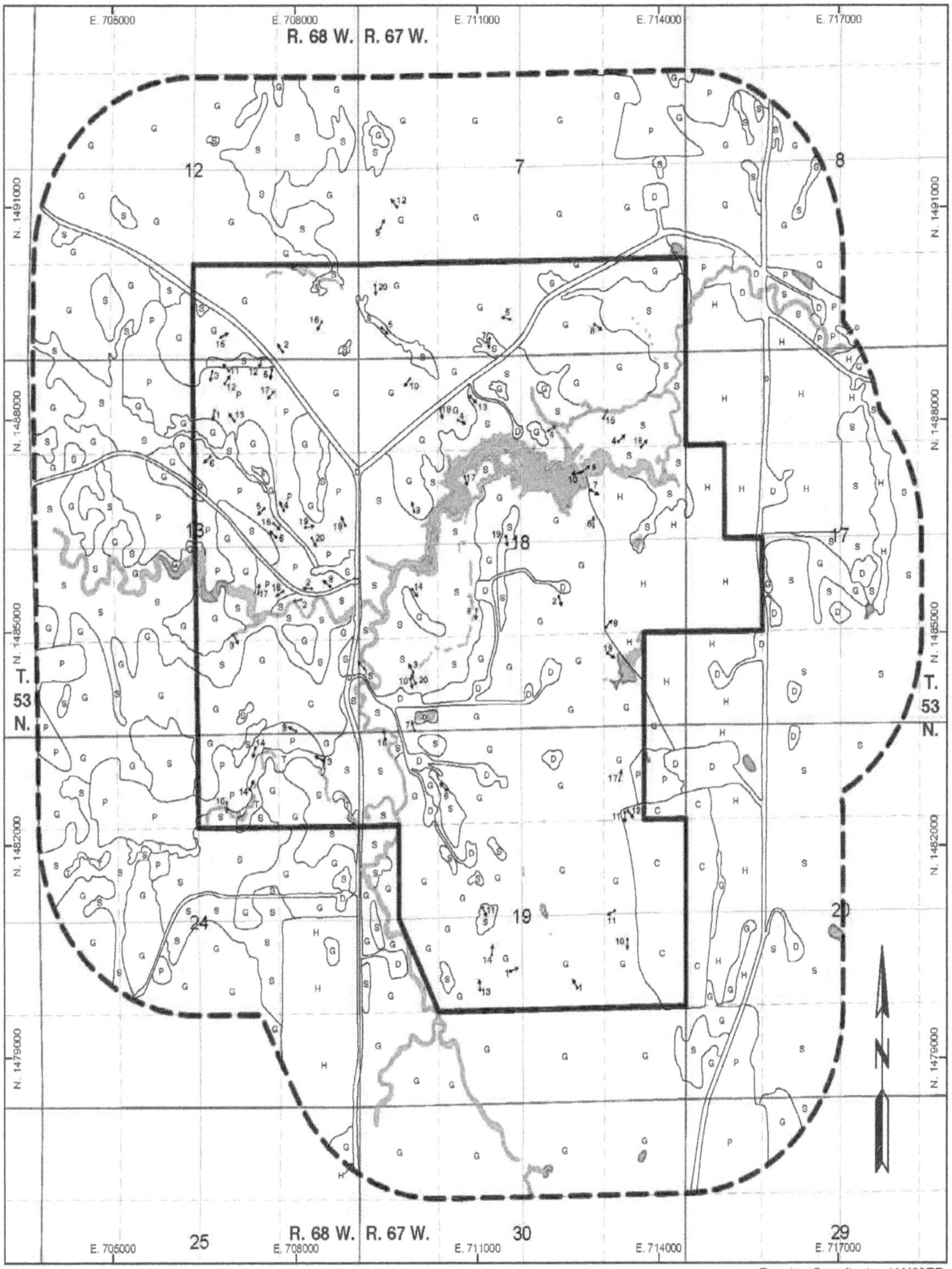

Source: Strata, 2012a.

1

Figure 3.16

Baseline Vegetation at Ross Project Area

Several species of designated and prohibited noxious weeds listed by the *Wyoming Weed and Pest Control Act* were identified on the Ross Project area. These species included field bindweed (*Convolvulus arvensis*), perennial sow thistle (*Sonchus arvensis*), quackgrass (*Agropyron repens*), Canada thistle (*Cirsium arvense*), houndstongue (*Cynoglossum officinale*), leafy spurge (*Euphorbia esula*), common burdock (*Arctium minus*), Scotch thistle (*Onopordum acanthium*), Russian olive (*Eleagnus angustifolia*), and skeletonleaf bursage (*Ambrosia tomentosa*). These weed species may be locally abundant in small areas, especially around the Oshoto Reservoir and along the Little Missouri River and Deadman Creek, but they were not common throughout the entire area of the Ross Project.

Selenium-indicator species identified on the Ross Project area in 2010 included two-grooved milkvetch (*Astragalus bisulcatus*), woody aster (*Xylorhiza glabriuscula*), and stemmy goldenweed (*Haplopappus multicaulis*); however, these indicator species were not abundant. Little larkspur (*Delphinium bicolor*), locoweed (*Oxytropis sericea* and *Oxytropis lambertii*), and meadow deathcamas (*Zigadenus venenosus*) are poisonous plants that were observed on the Ross Project area in limited numbers (locoweed is only poisonous for cattle). Cheatgrass (*Bromus tectorum*), although not a State-listed noxious weed, was abundant in some areas within the Ross Project area (Strata, 2011a).

3.6.1.2 Wildlife

Habitat Description

Background information on terrestrial vertebrate wildlife species in the vicinity of the Ross Project area was obtained from several sources, including records from the WGFD, BLM, and USFWS as well as from GEIS Section 4.4.5 (NRC, 2009b). Previous site-specific data for the Ross Project area and its surrounding environs were obtained from those same sources and Nubeth's *Environmental Report Supportive Information* (ND Resources, 1977). In addition, the Applicant completed site-specific wildlife surveys from November 2009 through October 2010 to establish one year of baseline site-characterization data (Strata, 2011a). Over 140 different species were noted during these surveys or documented by other sources, e.g. WGFD (see Table 3.11). The surveys also focused on the Applicant obtaining information regarding bald eagles' winter roosts; however, all nesting raptors, threatened and endangered species, the BLM's Sensitive Species (BLMSS), and the USFWS's "Migratory Bird Species of Management Concern in Wyoming" (SMC) (also known as "Migratory Birds of High Federal Interest") were included in the survey procedures. Surveys were also conducted on the Ross Project area for swift fox, breeding birds, and northern leopard frogs. In addition to those species that were targeted, others were noted when observed.

Table 3.11
Wildlife Species Observed on or near Ross Project Area

Scientific Name	Common Name
Mammals	
Sylvilagus audubonii	Desert Cottontail
Lepus californicus	Black-tailed Jackrabbit
Lepus townsendii	White-tailed Jackrabbit

Table 3.11 Wildlife Species Observed on or near Ross Project Area (Cont.)	
Scientific Name	**Common Name**
Mammals (*Continued*)	
Tamias minimus	Least Chipmunk
Spermophilus tridecemlineatus	Thirteen-lined Ground Squirrel
Cynomys ludovicianus	Black-tailed Prairie Dog
Sciurus niger	Eastern Fox Squirrel
Thomomys talpoides	Northern Pocket Gopher
Dipodomys ordii	Ord's Kangaroo Rat
Castor Canadensis	Beaver
Peromyscus maniculatus	Deer Mouse
Neotoma cinerea	Bushy-tailed Woodrat
Microtus Oochrogaster	Prairie Vole
Ondatra zibethicus	Muskrat
Erethizon dorsatum	Porcupine
Canis latrans	Coyote
Vulpes vulpes	Red Fox
Procyon lotor	Raccoon
Mustela frenata	Long-tailed Weasel
Taxidea taxus	Badger
Mephitis mephitis	Striped Skunk
Felis concolor	Mountain Lion
Felis rufus	Bobcat
Cervus elaphus	American Elk
Odocoileus hemionus	Mule Deer
Odocoileus virginianus	White-tailed Deer
Antilocapra americana	Pronghorn
Birds	
Branta canadensis	Canada Goose
Cygnus buccinator	Trumpeter swan
Cygnus columbianus	Tundra Swan
Anas strepera	Gadwall
Anas americana	American Wigeon
Anas platyrhynchos	Mallard
Anas discors	Blue-winged Teal
Anas crecca	Green-winged Teal
Anas cyanoptera	Cinnamon Teal
Anas clypeata	Northern Shoveler
Anas acuta	Northern Pintail
Aythya valisineria	Canvasback
Aythya americana	Redhead
Aythya collaris	Ring-necked Duck
Aythya affinis	Lesser Scaup

Table 3.11 Wildlife Species Observed on or near Ross Project Area (Cont.)	
Scientific Name	**Common Name**
Birds (*Continued*)	
Bucephala albeola	Bufflehead
Oxyura jamaicensis	Ruddy Duck
Podilymbus podiceps	Pied-billed Grebe
Podiceps auritus	Horned Grebe
Podiceps nigricollis	Eared Grebe
Pelecanus erythrorhynchos	White Pelican
Phalacrocorax auritus	Double-crested Cormorant
Ardea herodias	Great Blue Heron
Cathartes aura	Turkey Vulture
Pandion haliaetus	Osprey
Haliaeetus leucocephalus	Bald Eagle
Circus cyaneus	Northern Harrier
Accipiter striatus	Sharp-shinned Hawk
Accipiter cooperii	Cooper's Hawk
Buteo swainson	Swainson's Hawk
Buteo jamaicensis	Red-tailed Hawk
Buteo regalis	Ferruginous Hawk
Buteo lagopus	Rough-legged Hawk
Aquila chrysaetos	Golden Eagle
Falco sparverius	American Kestrel
Falco mexicanus	Prairie Falcon
Centrocercus urophasianus	Greater Sage-grouse
Tympanuchus phasianellus	Sharp-tailed grouse
Meleagris gallopavo	Wild Turkey
Porzana carolina	Sora Rail
Fulica americana	American Coot
Charadrius vociferous	Killdeer
Recurvirostra americana	American Avocet
Actitis macularia	Spotted Sandpiper
Bartramia longicauda	Upland Sandpiper
Gallinago delicata	Wilson's Snipe
Phalaropus tricolor	Wilson's Phalarope
Larus californicus	California Gull
Larus argentatus	Herring Gull
Chlidonias niger	Black Tern
Columba livia	Rock Pigeon
Zenaida macroura	Mourning Dove
Bubo virginianus	Great Horned Owl
Asio flammeus	Short-eared Owl
Chordeiles minor	Common Nighthawk
Ceryle alcyon	Belted Kingfisher

Table 3.11 Wildlife Species Observed on or near Ross Project Area (Cont.)	
Scientific Name	**Common Name**
Birds (*Continued*)	
Picoides villosus	Hairy Woodpecker
Colaptes auratus	Northern Flicker
Contopus sordidulus	Western Wood-Pewee
Sayornis saya	Say's Phoebe
Tyrannus verticalis	Western Kingbird
Tyrannus tyrannus	Eastern Kingbird
Eremophila alpestris	Horned Lark
Tachyceneta bicolor	Tree Swallow
Tachycineta thalassina	Violet-green Swallow
Stelgidopteryx serripennis	Northern Rough-winged Swallow
Riparia riparia	Bank Swallow
Hirundo pyrrhonota	Cliff Swallow
Hirundo rustica	Barn Swallow
Cyanocitta cristata	Blue jay
Pica pica	Black-billed Magpie
Corvus brachyrhynchos	American Crow
Corvus corax	Common Raven
Parus atricapillus	Black-capped Chickadee
Sitta canadensis	Red-breasted Nuthatch
Salpinctes obsoletus	Rock Wren
Troglodytes aedon	House Wren
Sialia currucoides	Mountain Bluebird
Turdus migratorius	American Robin
Oreoscoptes montanus	Sage Thrasher
Toxostoma rufum	Brown Thrasher
Sturnus vulgaris	European Starling
Lanius ludovicianus	Loggerhead Shrike
Vermivora celata	Orange-crowned Warbler
Dendroica petechia	Yellow Warbler
Dendroica coronate	Yellow-rumped Warbler
Geothlypis trichas	Common Yellowthroat
Wilsonia pusilla	Wilson's Warbler
Spizella passerine	Chipping Sparrow
Spizella breweri	Brewer's Sparrow
Pooecetes gramineus	Vesper Sparrow
Chondestes grammacus	Lark Sparrow
Calamospiza melanocorys	Lark Bunting
Ammodramus savannarum	Grasshopper Sparrow
Junco hyemalis	Dark-eyed Junco
Calcarius mccownii	McCown's Longspur
Agelaius phoeniceus	Red-winged Blackbird

Table 3.11 Wildlife Species Observed on or near Ross Project Area (Cont.)	
Scientific Name	**Common Name**
Birds (*Continued*)	
Sturnella neglecta	Western Meadowlark
Xanthocephalus xanthocephalus	Yellow-headed Blackbird
Euphagus cyanocephalus	Brewer's Blackbird
Quiscalus quiscula	Common Grackle
Molothrus ater	Brown-headed Cowbird
Icterus bullockii	Bullock's Oriole
Carpodacus mexicanus	House Finch
Carduelis pinus	Pine Siskin
Passer domesticus	House Sparrow
Amphibians	
Ambystoma tigrinum	Tiger Salamander
Pseudaris triseriata maculate	Boreal Chorus Frog
Rana pipiens	Northern Leopard Frog
Reptiles	
Phrynosoma douglassi brevirostre	Eastern Short-horned Lizard
Sceloporus graciosus graciosus	Northern Sagebrush Lizard
Chelydra serpentina serpentina	Common Snapping Turtle
Chrysemys picta belli	Western Painted Turtle
Crotalus viridis viridis	Prairie Rattlesnake
Pituophis melanoleucas sayi	Bullsnake
Thamnophis elegans vagrans	Wandering Garter Snake
Fish	
Ameiurus melas	Black Bullhead
Lepomis cyanellus	Green Sunfish
Lepomis macrochirus	Bluegill
Catastomus commersoni	White Sucker

Source: Strata, 2011a.

Mammals

Pronghorn antelope (*Antilocapra americana*), mule deer (*Odocoileus hemionus*), and white-tailed deer (*O. virginianus*) were the only big-game species that were observed on the Ross Project area during the 2009 and 2010 surveys (Strata, 2011a). American elk (*Cervus elaphus*) have been recorded in the area by the WGFD; however, none were observed during the Applicant's surveys. No crucial big-game habitats or migration corridors are recognized by the WGFD at the Ross Project or the surrounding 1.6-km [1-mi] vicinity.

Pronghorn antelope and mule deer are common but not abundant on the Ross Project area. Pronghorn herds were most often observed in sagebrush-shrubland and upland-grassland habitats, and the mule deer frequented the sagebrush-shrubland habitat (Strata, 2011a). Both species used haylands and cultivated fields in the area. White-tailed deer were not abundant, but they were observed in the riparian habitats and on the cultivated fields within and near the Ross Project area. Pronghorn antelopes' use of the Ross Project and surrounding areas has been classified by the WGFD as year long, and mule deer use within the areas as winter and year long. White-tailed deer and elk use has been classified by the WGFD as out of their normal range. The Ross Project is located within the WGFD North Black Hills pronghorn-herd unit, the Powder River and Black Hills mule deer-herd units, and the Thunder Basin and Black Hills white-tailed deer-herd units. The Ross Project area is not within a specific elk-herd unit, but it is included in the WGFD designated area referred to as "Hunt Area 129" (Strata, 2011a).

A variety of small- and medium-sized mammals could potentially be present on the Ross Project area. These mammals include a variety of predators and furbearers, such as coyote (*Canis latrans*), red fox (*Vulpes vulpes*), raccoon (*Procyon lotor*), bobcat (*Lynx rufus*), badger (*Taxidea taxus*), beaver (*Castor canadensis*), and muskrat (*Ondatra zibethicus*). Prey species that were observed included rodents (e.g., mice, rats, voles, gophers, ground squirrels, chipmunks, prairie dogs), jackrabbits (*Lepus spp.*), and cottontails (*Sylvilagus spp.*). These species are cyclically common and widespread throughout the vicinity, and they are important food sources for raptors and other predators. Each of these prey species was either directly observed during Strata's field surveys or was known to exist through the presence of burrow formation or of droppings. Jackrabbit and cottontail sightings were common.

While black-tailed prairie dogs (*Cynomys ludovicianus*) are listed as occurring in the general area of the Ross Project, no black-tailed prairie-dog colonies (important as habitat for black-footed ferrets) were located within the 1.6-km [1-mi] survey area. Other mammal species, such as the striped skunk (*Mephitis mephitis*), porcupine (*Erethizon dorsatum*), and various weasels (*Mustela spp.*) inhabit sagebrush grassland and riparian communities, and these species were recorded within the Ross Project area during the Applicant's wildlife surveys. No bat species were observed during the baseline surveys. There are no records of prior use of the Ross Project by swift fox (*Vulpes velox*), and none were observed during the 2009 or 2010 surveys.

Birds

Suitable habitat for several raptor species occurs at the Ross Project area and within the 1.6-km [1-mi] vicinity surrounding it. Several raptor species were observed during the wildlife surveys; these included the bald eagle, red-tailed hawk (*Buteo jamaicensis*), golden eagle (*Aquila chrysaetos*), ferruginous hawk (*Buteo regalis*), Swainson's hawk (*Buteo swainsoni*), northern harrier (*Circus cyaneus*), American kestrel (*Falco sparverius*), Cooper's hawk (*Accipiter cooperii*), Sharp-shinned hawk (*Accipiter striatus*), rough-legged hawk (*Buteo lagopus*), great horned owl (*Bubo virginianus*), and short-eared owl (*Asio flammeus*). Turkey vultures (*Cathartes aura*) and prairie falcons (*Falco mexicanus*) have also been recorded on the Ross Project area, but they were not seen during the Applicant's surveys.

In the vicinity of the Ross Project area, nests were observed for the ferruginous, red-tailed, and Swainson's hawks (Strata, 2011a). The only nest observed within the Project area itself was a Swainson's hawk's nest, which was observed to be inactive during the 2010 survey year. A total of seven intact nesting sites were observed within 1.6 km [1 mi] of the Ross Project area.

The wild turkey (*Meleagris gallopavo*), Greater sage-grouse (*Centrocercus urophasianus*), sharp-tailed grouse (*Tympanuchus phasianellus*), and mourning dove (*Zenaida macroura*) were observed at the Ross Project area by the Applicant. Mourning doves were recorded during the spring and summer months.

The Greater sage-grouse (*Centrocercus urophasianus*) is listed as a Federal candidate species and a Wyoming Species of Concern (WSOC) in Wyoming (75 FR 13090; WGFD, 2005a) (see SEIS Section 3.6.1.4, below). Potential sage-grouse habitat is present at the Ross Project area (upland grassland, sagebrush shrubland, pastureland, hayland, and reservoir/stock pond). Two leks, which is where male sage grouse congregate for competitive mating displays, have been recorded within several miles of the Ross Project. Leks assemble before and during the breeding season on a daily basis; the same group of males meet at traditional locations each season. However, the Ross Project area is not located in a region currently designated as a sage-grouse core area.

Breeding-bird surveys were conducted within the Ross Project area in four habitat types: upland grassland, sagebrush shrubland, pastureland/hayland, and wetland/reservoir. Twenty-seven species were recorded during the 2010 breeding-bird surveys. The Wetland/Reservoir habitat produced the greatest species diversity, with 19 species observed. The upland grassland habitat had the fewest species, with six species observed.

Natural aquatic habitats on the Ross Project occur at the Oshoto Reservoir and along the Little Missouri River. During the Applicant's wildlife surveys, 17 waterfowl and 8 shorebird species were observed. In these categories, the horned grebe (*Podilymbus podiceps*) and upland sandpiper (*Bartramia longicauda*) are the only USFWS's SMC observed within or near the Ross Project area.

3.6.1.3 Reptiles, Amphibians, and Aquatic Species

During the Applicant's baseline wildlife surveys in 2009 and 2010, the eastern short-horned lizard (*Phrynosoma douglassi brevirostre*) and northern sagebrush lizard (*Sceloporus graciosus graciosus*) were often observed. Other reptiles observed in the area included the bullsnake (*Pituophis cantenifer*), wandering garter snake (*Thamnophis elegans vagrans*), and the prairie rattlesnake (*Crotalus viridis viridis*).

Water is a limiting factor for wildlife on the Ross Project area, where only one stream flows occasionally; the Oshoto Reservoir is the major water feature within the Ross Project area. All other natural drainages are categorized as intermittent or ephemeral (see SEIS Section 3.5.1). The lack of deep-water habitat and perennial water sources decreases the potential for many aquatic species to exist. Three aquatic or semi-aquatic amphibian species and two aquatic reptiles were recorded during the Applicant's baseline surveys: the tiger salamander (*Ambystoma tigrinum*), boreal chorus frog (*Pseudacris triseriata*), northern leopard frog (*Rana pipiens)*, common snapping turtle (*Chelydra serpentine*), and western painted turtle (*Chrysemys picta*). All five species were heard and/or seen in the Oshoto Reservoir, Little Missouri River, or near stock reservoirs. All five species are common to the Ross Project and the vicinity as a whole. No egg masses were identified during the egg-mass surveys completed in early June 2010. The reason for their absence could have been that recent high winds could have broken up the egg masses and dispersed the individual eggs. During walking surveys along shorelines and riparian areas in August 2010, the leopard frog appeared to be quite common—over 500 individual adults were counted—while the chorus frog was uncommon.

The Applicant also conducted fish sampling from the Oshoto Reservoir in September 2010, under a WGFD Chapter 33 collection permit, as part of its establishing pre-licensing baseline radiological conditions for the Ross Project. The dominant fish population in the Oshoto Reservoir included black bullheads (*Ameiurus melas*) and green sunfish (*Lepomis cyanellus*); white suckers (*Catastomus* commeroni) and bluegill (*Lepomis macrochirus*) were also present. The sample fish from this population were stunted in size for their ages; high reproductive rates and limited predation leads to over-population and stunted growth. The Oshoto Reservoir and the other water bodies within the Ross Project area are not considered viable sport fisheries (see SEIS Section 3.2.2).

3.6.1.4 Protected Species

The Ute ladies'-tresses orchid (*Spiranthes diluvialis*) is Federally-listed as threatened. The species is a perennial, terrestrial orchid that occurs in Colorado, Idaho, Montana, Nebraska, Utah, Washington, and Wyoming. Within Wyoming, this orchid inhabits moist meadows with moderately dense but short vegetative cover. As noted in Fertig (2000), this species is found at elevations of 1,280 – 2,130 m [4,200 – 7,000 ft], though no known populations occur in Wyoming above 1,680 m [5,500 ft]. This species was not located during the Applicant's vegetation surveys, and it is not known to occur on or in the vicinity of the Ross Project area.

The blowout penstemon (*Penstemon haydenii*) is Federally listed as endangered, although it is not included on the list for Crook County. However, it is on the list for neighboring Campbell County, and the Applicant therefore evaluated the potential for the blowout penstemon to occur in the Ross Project area. This species is found exclusively in sparsely vegetated, early successional sand dunes or blowout areas at elevations of 1,786 – 2,268 m [5,860 – 7,440 ft] (Fertig, 2008). The Ross Project does not have sand-dune habitat, and it is outside of the elevation range in which this species is typically found. This species was not identified during Strata's vegetation surveys; appropriate habitat was not identified; and it is not known to occur on or in the vicinity of the Ross Project.

The black-footed ferret (*Mustela nigripes*) is a Federally listed endangered species, which inhabits prairie-dog colonies. A black-footed ferret survey was not required by USFWS requirements, because black-footed ferrets live exclusively in prairie-dog colonies, which are not present on or within 1.6 km [1 mi] of the Ross Project area (Strata, 2011a).

The bald eagle (*Haliaeetus leucocephalus*) was delisted from Federal threatened status in 2007, but it is still protected under the *Bald and Golden Eagle Protection Act* and the *Migratory Bird Treaty Act*. Potential habitat for bald eagle nesting and roosting activities is quite limited within the Ross Project because of the lack of trees. Bald eagles were observed from the Ross Project area during wildlife surveys that took place November and December of 2009 and January through September of 2010 (Strata, 2011a). No nests were observed, however, and the bald eagle is considered to be a winter migrant to the area.

The Greater sage-grouse (*Centrocercus urophasianus*) is Federally listed as a Candidate species, as a State of Wyoming's Species of Concern (WSOC), and as a BLMSS. On March 5, 2010, the USFWS published a finding in the FR stating that listing of the species was warranted but precluded by higher priority listing actions (75 FR 13909). The Governor of Wyoming issued Executive Order (EO) 2010-4 in August 2010 which sets out 12 provisions for oil- and gas-resource operations within core and noncore population areas to protect the species at the

State level (State of Wyoming, 2011). The WGFD published *Recommendations for Development of Oil and Gas Resources Within Important Wildlife Habitats* and the Wyoming Field Office of the BLM issued an instructional memorandum on March 5, 2010, which supplements the BLM's 2004 *National Sage-Grouse Habitat Conservation Strategy*, to be consistent with the Governor's Executive Order (EO) (WGFD, 2010; BLM, 2004; BLM, 2010a). The WGFD guidance was again updated in April 2010.

The Greater sage-grouse inhabits open sagebrush plains in the western U.S. and is found at elevations of 1,200 – 2,700 m [3,937 – 8,858 ft], corresponding with the occurrence of sagebrush habitat (69 FR 933). The Greater sage-grouse is a mottled brown, black, and white ground-dwelling bird that can be up to 0.6 m [2 ft] tall and 76 cm [30 in] in length (69 FR 933). Breeding habitat, referred to as leks (see SEIS Section 3.6.1.2), and stands of sagebrush surrounding leks are used by sage-grouse in early spring and are particularly important habitat because the birds often return to the same leks and nesting areas each year. Leks are generally more sparsely vegetated areas such as ridgelines or disturbed areas adjacent to stands of sagebrush habitat.

Two sage-grouse leks are known to occur within 3 km [2 mi] of the Ross Project area. The Oshoto Lek (Sections 28 and 29, T53N, 67W) and the Cap'n Bob Lek (Section 32, T53 N, R67W) have been identified; no other sage-grouse leks were identified during the wildlife surveys. Details of sage-grouse mating activities for these leks are summarized in Table 3.12. A ground survey of the Oshoto and Cap'n Bob leks were conducted by the Applicant on two days in April 2010. On the Cap'n Bob lek, a total of two males and one female were observed on one day, and two males were observed on the second day; no sage-grouse were observed at the Oshoto Lek during the survey. No broods or brood-rearing areas were identified during the Applicant's 2010 survey. In addition, no sage-grouse wintering areas were identified on the Ross Project area (Strata, 2011a).

Threats to this species' survival include habitat loss, agricultural practices, livestock grazing, hunting, and land disturbances from energy and mineral development as well as the oil and gas industry (Sage-Grouse Working Group, 2006). Although the two leks described earlier were recorded near the Ross Project, the Project area is not located within a designated sage-grouse core area. Additionally, although sharp-tailed grouse were observed on the Ross Project area during only the 2009 winter survey, they are considered year-long residents of the Project area.

Table 3.12
Summary of Sage-Grouse Activity in Oshoto and Cap'n Bob Leks

Year of Survey Activity	Oshoto	Cap'n Bob
1985	6 males	No information
1988	0	"
1988	0	"
1991	0	"
1994	0	"
1997	0	"
2000	0	"
2001	5 males	"
2004	2 males	"
2007	0	10 males
2007	0	10 males
2010	0	2 males 1 female
2010	0	2 males

Source: Strata, 2011a.

The mountain plover (*Charadrius montanus*) is Federally proposed as threatened and is a Wyoming Species of Greatest Conservation Need. The species is a small bird approximately 17.8 cm [7 in] in height with light brown and white coloring. The mountain plover is a native of the short-grass prairie and is found in open, dry shrubland, or agricultural fields with short vegetation and bare ground. Prairie dogs and other burrowing animals provide highly suitable habitat for the mountain plover.

Mountain plover breeding habitat includes the western Great Plains and Rocky Mountain states extending from the Canadian border to northern Mexico (75 FR 37353). The prime breeding and nesting period for the mountain plover is from April 10 through July 10 (BLM, 2007a). In Wyoming, the greatest concentration of mountain plovers is found in the south central part of the state, but they can be found in every county (Andres 2009; UW, 2010). This bird is often found in areas with heavy grazing and landscapes with excessive surface disturbance. USFWS originally proposed this species as threatened on February 16, 1999 (64 FR 7587); the proposal was withdrawn on September 9, 2003, but it was reinstated on June 29, 2010 (68 FR 53083; 75 FR 37353). This species was not observed during either the 2009 or 2010 wildlife surveys (Strata, 2011a).

Table 3.13 lists species that occur in Crook County and that are Federally listed under the *Endangered Species Act* (ESA), State-listed under the *Final Comprehensive Wildlife Conservation Strategy for Wyoming*, or are listed as a BLMSS.

Table 3.13
Species of Concern in Crook County and at Ross Project Area

Common Name *Scientific Name*	USFWS Species of Management Concern (Level)[1]	BLM Sensitive Species	Wyoming Species of Concern Status[2]	Observed on the Ross Project Area
Mammals				
Hayden's Shrew *Sorex haydeni*			NSS4*	
Vagrant Shrew *Sorex vagrans*			NSS3*	
Long-eared Myotis *Myotis evotis*		Yes	NSS2*	
Northern Myotis *Myotis septentrionalis*			NSS2*	
Little Brown Myotis *Myotis lucifugus*			NSS3*	
Long-legged Myotis *Myotis volans*			NSS2*	
Fringed myotis *Myotis thysanodes*		Yes	NSS2*	
Hoary Bat *Lasiurus cinereus*			NSS4*	
Silver-haired Bat *Lasionycteris noctivagans*			NSS4*	
Big Brown Bat *Eptesicus fuscus*			NSS3*	
Black-tailed Prairie Dog *Cynomys ludovicianus*			NSS3*	Yes
Plains Pocket Gopher *Geomys bursarius*			NSS4*	
Olive-backed Pocket Mouse *Perognathus fasciatus*			NSS3*	
Silky Pocket Mouse *Perognathus flavus*			NSS3*	
Western Harvest Mouse *Reithrodontomys megalotis*			NSS3*	
Prairie Vole *Microtus ochrogaster*			NSS3*	Yes
Sagebrush Vole *Lemmiscus curtatus*			NSS4*	
Swift Fox *Vulpes velox*		Yes	NSS4*	

Table 3.13
Species of Concern in Crook County and on the Ross Project Area
(*Continued*)

Common Name *Scientific Name*	USFWS Species of Management Concern (Level)[1]	BLM Sensitive Species	Wyoming Species of Concern Status[2]	Observed on the Ross Project Area
***Mammals* (*Continued*)**				
Black-footed Ferret *Mustela nigripes*			NSS1*	
Birds				
Waterfowl and Shorebirds				
Trumpeter swan *Cygnus buccinator*		Yes	NSS2	Yes
Northern Pintail *Anas acuta*			NSS3	Yes
Canvasback *Aythya valisineria*			NSS3	Yes
Redhead *Aythya americana*			NSS3	Yes
Lesser Scaup *Aythya affinis*			NSS3	Yes
Horned Grebe *Podiceps auritus*	Yes (NL)			Yes
Western Grebe *Aechmophorus occidentalis*			NSS4	
American Bittern *Botauosus lentiginosus*	Yes (I)		NSS3	
Great Blue Heron *Ardea herodias*			NSS4	Yes
Black-crowned Night-Heron *Nycticorax nycticorax*			NSS3	
White-faced Ibis *Plegadis chihi*		Yes	NSS3	
Sandhill Crane *Grus canadensis*			NSS3	
Mountain Plover *Charadrius montanus*	Yes (I)	Yes	NSS4*	
Upland Sandpiper *Bartramia longicauda*	Yes (I)		NSS4	Yes
Marbled Godwit *Limosa fedoa*	Yes (NL)			
Long-billed Curlew *Numenius americanus*	Yes (I)	Yes	NSS3*	

Table 3.13
Species of Concern in Crook County and on the Ross Project Area
(*Continued*)

Common Name *Scientific Name*	USFWS Species of Management Concern (Level)[1]	BLM Sensitive Species	Wyoming Species of Concern Status[2]	Observed on the Ross Project Area
Raptors				
Bald Eagle *Haliaeetus leucocephalus*	Yes (I)		NSS2	Yes
Northern Goshawk *Accipiter gentilis*		Yes	NSS4*	
Swainson's Hawk *Buteo swainsoni*			NSS4	Yes
Ferruginous Hawk *Buteo regalis*	Yes (I)	Yes	NSS3*	Yes
Golden Eagle *Aquila chrysaetos*	Yes (III)			Yes
Merlin *Falco columbarius*			NSS3*	
Peregrine Falcon *Falco peregrinus*	Yes (I)		NSS3*	
Prairie Falcon *Falco mexicanus*	Yes (III)			Yes
Burrowing Owl *Athene cunicularia*	Yes (I)	Yes	NSS4	
Short-eared Owl *Asio flammeus*	Yes (I)		NSS4	Yes
Upland Game				
Greater Sage-grouse *Centrocercus urophasianus*		Yes	NSS2	Yes
Other				
White Pelican *Pelecanus erythrorhynchos*			NSS3	Yes
Franklin's Gull *Larus pipixcan*			NSS3	
Forster's Tern *Sterna forsteri*			NSS3	
Black Tern *Chlidonias niger*			NSS3	Yes
Black-billed Cuckoo *Coccyzus erythropthalmus*	Yes (II)			
Yellow-billed Cuckoo *Coccyzus americanus*	Yes (II)	Yes	NSS2*	

Table 3.13
Species of Concern in Crook County and on the Ross Project Area
(*Continued*)

Common Name *Scientific Name*	USFWS Species of Management Concern (Level)[1]	BLM Sensitive Species	Wyoming Species of Concern Status[2]	Observed on the Ross Project Area
Other (*Continued*)				
Lewis's Woodpecker *Melanerpes lewis*	Yes (II)		NSS3*	
Willow Flycatcher *Empidonax traillii*	Yes (II)		NSS3	
Pinyon Jay *Gymnorhinus cyanocephalus*	Yes (IV)			
Pygmy Nuthatch *Sitta pygmaea*			NSS4*	
Sage Thrasher *Oreoscoptes montanus*	Yes (II)	Yes	NSS4*	Yes
Loggerhead Shrike *Lanius ludovicianus*	Yes (II)	Yes		Yes
Dickcissel *Spiza americana*	Yes (II)		NSS4	
Brewer's Sparrow *Spizella breweri*	Yes (I)	Yes	NSS4	Yes
Sage Sparrow *Amphispoza belli*	Yes (I)	Yes	NSS4	
Lark Bunting *Calamospiza melanocorys*	Yes (II)		NSS4	Yes
Baird's Sparrow *Ammodramus bairdii*	Yes (I)	Yes		
Grasshopper Sparrow *Ammodramus savannarum*	Yes (II)		NSS4	Yes
McCown's Longspur *Calcarius mccownii*	Yes (I)		NSS4	Yes
Chestnut-collared Longspur *Calcarius ornatus*			NSS4	
Bobolink *Dolichonyz oryzivorus*			NSS4	
Cassin's Finch *Carpodacus cassinii*	Yes (IV)			
Amphibians				
Tiger Salamander *Ambystoma tigrinum*			NSS4*	Yes
Plains Spadefoot *Scaphiopus bombifrons*			NSS4*	

Table 3.13 Species of Concern in Crook County and on the Ross Project Area (*Continued*)				
Common Name *Scientific Name*	**USFWS Species of Management Concern (Level)[1]**	**BLM Sensitive Species**	**Wyoming Species of Concern Status[2]**	**Observed on the Ross Project Area**
***Amphibians* (*Continued*)**				
Great Plains Toad *Bufo cognatus*			NSS4*	
Boreal Chorus Frog *Pseudaris triseriata maculate*			NSS4*	Yes
Bullfrog *Rana catesbeiana*			NSS4*	
Northern Leopard Frog *Rana pipiens*		Yes	NSS4*	Yes
Reptiles				
Northern Sagebrush Lizard *Sceloporus graciosus graciosus*			NSS4*	Yes
Western Painted Turtle *Chrysemys picta belli*			NSS4*	Yes
Prairie Rattlesnake *Crotalus viridis viridis*			NSS3*	Yes
Plains Hognose Snake *Heterondon nasicus nasicus*			NSS4*	
Bullsnake *Pituophis melanoleucas sayi*			NSS4*	
Wandering Garter Snake *Thamnophis elegans vagrans*			NSS4*	
Eastern Yellowbelly Racer *Coluber constrictor flaviventris*			NSS4*	

Source: Strata, 2011a.
Notes: See next page.

Notes for Table 3.13::

[1] *USFWS Level:*

- Level I (Conservation Action): Species clearly needs conservation action.
- Level II (Monitoring): The action and focus for the species is monitoring (M). Declining population trends and habitat loss are not significant at this point.
- Level III (Local Interest): Species that Wyoming Partners In Flight may recommend for conservation action that are not otherwise high priority but are of local interest (LI).
- Level IV (Not Considered Priority): Additional species of concern, but not considered a priority species.

[2] *WGFD Status:*

- NSS1: 1996 Nongame Bird and Mammal Plan Species of Special Concern with a Native Species Status of 1.
- NSS2: 1996 Nongame Bird and Mammal Plan Species of Special Concern with a Native Species Status of 2.
- NSS3: 1996 Nongame Bird and Mammal Plan Species of Special Concern with a Native Species Status of 3.
- NSS4: 1996 Nongame Bird and Mammal Plan Species of Special Concern with a Native Species Status *of 4.*

**Species listed wholly or in part due to absence of data.*

The Wyoming Field Office of the USFWS also uses the SMC list for conducting reviews related to non-coal, surface-disturbance projects. Thirty-two birds on the WSOC list were identified on this list for the Ross Project area (see Table 3.13). Surveys for avian WSOC, including sage-grouse, bald eagle, and mountain plovers, were conducted in 2009 and 2010 for the Ross Project area. Table 3.14 lists the avian WSOCs that were observed on the Ross Project area during the Applicant's 2009 and 2010 baseline surveys (Strata, 2011a), including their primary nesting habitats and historical occurrence in the general Ross Project vicinity.

In addition to the species previously discussed above, 20 bird species on the U.S. Fish and Wildlife Service's (USFWS's) SMC list could potentially be present within the Ross Project area. Of these 20 bird species, 7 have been observed within or near the Ross Project (see Table 3.13). Ten non-raptor or non-game bird species on the BLMSS list could potentially occur within the Ross Project. Of the ten bird species, four have been observed on or near the Ross Project area (see Table 3.14). Thirty-two non-raptor or non-game bird species on the WSOC list could potentially be present within the Ross Project area. Of the 32 bird species, 15 have been actually observed on or near the Project area (see Tables 3.13 and 3.14).

Table 3.14 Avian Species of Concern Observed at Ross Project Area		
Common Name *Scientific Name*	Primary Nesting Habitat(s)[1]	Status[2]
Level 1 Species of Concern/Conservation Needed		
Bald Eagle *Haliaeetus leucocephalus*	Montane Riparian, Plains/Basin Riparian	Uncommon year-long resident
Ferruginous Hawk *Buteo regalis*	Shrub Steppe and Short-Grass Prairie	Summer uncommon resident
Upland Sandpiper *Bartramia longicauda*	Short-Grass Prairie	Summer uncommon resident
Short-eared Owl *Asio flammeus*	Short-grass Prairie and Meadows	Common year-long resident
Brewer's Sparrow *Spizella breweri*	Shrub Steppe and Mountain-Foothills Shrub	Common summer resident
McCown's Longspur *Calcarius mccownii*	Shrub steppe and short-grass prairie	Common summer resident
Level 2 Species of Concern/Continued Monitoring Recommended		
Sage Thrasher *Oreoscoptes montanus*	Shrub Steppe	Common summer resident
Loggerhead Shrike *Lanius ludovicianus*	Shrub Steppe	Common summer resident
Lark Bunting *Calamospiza melanocorys*	Shrub Steppe and Short-Grass Prairie	Abundant summer resident
Grasshopper Sparrow *Ammodramus savannarum*	Shrub Steppe and Short-Grass Prairie	Common summer resident
Level 3 Species of Concern/Species of Local Interest		
Golden Eagle *Aquila chrysaetos*	Specialized (Cliffs)	Common year-long resident
Prairie Falcon *Falco mexicanus*	Specialized (Cliffs)	Common year-long resident

Sources: USFWS, 2011, and USGS, 2011.

3.7 Meteorology, Climatology, and Air Quality

3.7.1 Meteorology

The region of the Ross Project area is characterized by hot summers and cold winters, and rapid temperature fluctuations are common. The Rocky Mountains (the "Rockies") have a great influence on the climate. As air crosses the Rockies from the west, much moisture is lost on the windward sides of the Mountains, and the air becomes warmer as it descends on the eastern slopes (NRC, 2009b). The Ross Project area is located in this semi-arid area (Strata, 2011a).

The closest National Weather Service (NWS) station with a long recording period is Gillette Airport, which is located 56 km [35 mi] southwest of the Ross Project (Strata, 2011a). As the GEIS noted, there is a NWS station in Crook County, at Colony, Wyoming (72 km [45 mi] northeast of the Ross Project) (NRC, 2009b). This station, however, ceased operation in 2008. In addition, the Applicant has installed a site-specific meteorology station in 2010, where meteorology data has been collected every month since the station went online (Strata, 2011a).

Temperature

As described in the GEIS, the northwest Great Plains region has summer nights that are normally cool, even though daytime temperatures can be very warm. Winters can be quite cold; however, warm spells during winter months are common. The average temperatures for the two NWS stations in the vicinity of the Ross Project area, Colony and Gillette Airport, are shown in Table 3.15, in addition to the information collected by the Applicant in 2010 (NRC, 2009b; NWS, 2011; Strata, 2011a).

Table 3.15
Average, Minimum, and Maximum Temperatures in Ross Project Vicinity

Station	Average Temperature °C [°F]	Average Minimum Temperature °C [°F]	Average Maximum Temperature °C [°F]
Ross Project[1]	8.9 [48]	- 4.3 [24.3]	23.9 [75]
Gillette Airport[2]	8.1 [46.5]	N/A	N/A
Colony[3]	8.3 [47]	- 5.3 [22.5]	22.4 [72.3]

Source: Strata, 2011a; NRC, 2009b; NWS, 2011.

Notes: N/A = Data not available.
1 = Monitoring period 2010
2 = Monitoring period 1902 – 2009
3 = Monitoring period 1971 – 2000

At the Gillette Airport station, the warmest month of the year is July, with an average temperature of 23.6 °C [74.5 °F] (Strata, 2011a). The coldest month is December, with an average temperature of -4.7 °C [23.6 °F]. This trend was also observed at the Ross Project's meteorology station, with an average July temperature of 23.1 °C [73.6 °F] and an average December temperature of -4.7 °C [23 °F] for 2010.

Wind

The average wind speed at the Gillette Airport station is 16.9 km/hr [10.5 mi/hr], with an average maximum wind speed from 2000 – 2009 of 77 km/hr [48 mi/hr] (Strata, 2011a). The highest winds were recorded in January through March, with the lowest speeds from July through September. As shown on the wind rose for the Ross Project area, the prevailing wind direction in the fall and winter is north/northwest (as shown in Figure 3.17), whereas in the spring and summer, the winds are generally from the southeast. The highest wind speeds tend to occur from the north-northwest.

During the 12 months of monitoring at the Applicant's meteorology station in 2010, the average annual wind speed was 18.5 km/hr [11.5 mi/hr], ranging from a minimum wind speed of 0 km/hr [0 mi/hr] to a maximum wind speed of 73.4 km/hr [45.6 mi/hr]. More southerly winds were recorded at the Ross Project than at the Gillette Airport station (as shown in Figure 3.18); however, as at Gillette Airport, the highest wind speeds are from the northwest.

Precipitation

The Ross Project area and the surrounding area receive relatively little rainfall, with average annual precipitation ranging from 25 – 38 cm [10 – 15 in]. The region receives an average annual snowfall of 127 – 152 cm [50 – 60 in]. At the Gillette Airport station, between 2005 – 2009, the average annual precipitation was measured at 30.5 cm [12 in] (Strata, 2011a). Approximately one-half of the precipitation is associated with spring snows and thunderstorms. May is the wettest month, with more than 5 cm [2 in] of precipitation, while January is the driest month, with average precipitation of approximately 1.3 cm [0.5 in] or less (Strata, 2011a).

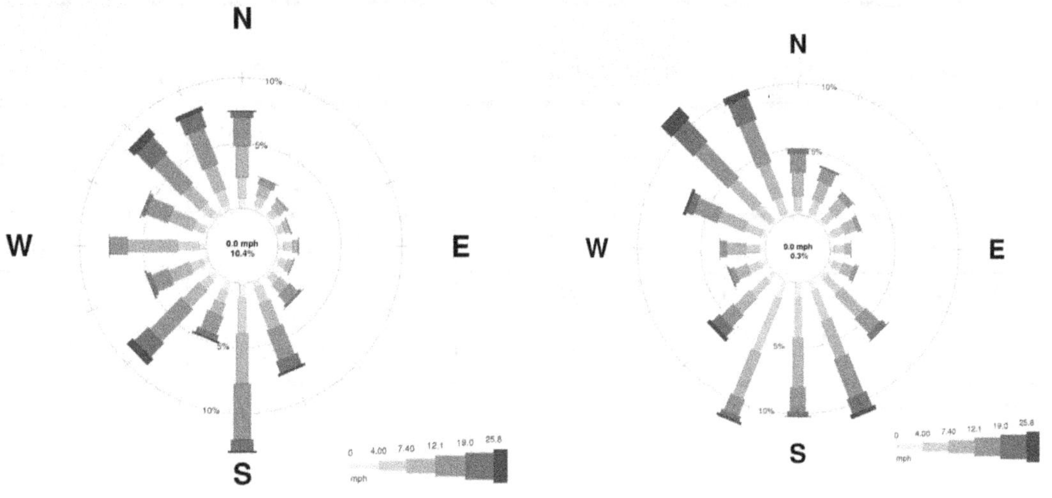

Source: Strata, 2012a.

Figures 3.17 and 3.18

Gillette Airport Wind Rose (Left)
Ross Project Area Wind Rose (Right)

At the Applicant's onsite meteorology station, the total precipitation measured in 2010 was 24.8 cm [9.8 in], compared to 32.5 cm [12.8 in] for the same period at the Gillette Airport station (Strata, 2011a). The difference in precipitation during 2010 was primarily due to the fact that Gillette Airport received 6.4 cm [2.5 in] more in the month of May than the Ross Project. Otherwise, the monthly precipitation data are very similar.

Evaporation

As with the majority of the western U.S., the evaporation rate in northeastern Wyoming exceeds the rate of precipitation. As discussed in the GEIS, evaporation rates in the region range from 102 – 127 cm/yr [40 – 50 in/yr] (NRC, 2009b). An evaporation pan was installed at the Ross Project's meteorology station in June 2010; however, data are available from only June through late October 2010, because the gauge was removed to prevent its freezing. At the Gillette Airport station, evaporation in 2010 varied from slightly more than 10 cm [4 in] in April to almost 25 cm [10 in] in July and August. For the period of time the evaporation pan operated at the Ross Project, similar rates were observed (Strata, 2011a).

Atmospheric Stability Classification and Mixing Height

Atmospheric stability classification and mixing height are environmental variables that influence the ability of the atmosphere to disperse air pollutants. The stability class is a measure of atmospheric turbulence, and mixing height characterizes the vertical extent of contaminants mixing in the atmosphere. The nearest upper-air data available from the NWS are from Rapid City, South Dakota, approximately 170 km [106 mi] southeast of the Ross Project (Strata, 2011a). However, Rapid City is approximately 1,700 m [5,577 ft] lower in elevation than the Ross Project, and it is on the other side of the Black Hills. Therefore, the data are likely not representative of conditions at the Ross Project area.

Stability-class information was collected using the Applicant's meteorological station, which demonstrated that the class distributions were predominantly neutral approximately 62 percent of the time. Other calculated conditions were Stability Class D (17 percent) and Class E (Strata, 2011a). The classification that results in the least vertical mixing (Class F) was approximately 4.7 percent at the Ross Project area, while Classes A through C ranged from 3 percent to 6.7 percent (Strata, 2011a).

Average annual mixing heights were not reported, although Wyoming has provided statewide mixing heights to be used in dispersion modeling (see Table 3.16) (Strata, 2011a).

Table 3.16
Statewide Mixing Heights for Dispersion Modeling

Stability Class	Mixing Height (m [ft])
Class A	3,450 [11,319]
Class B	2,300 [7,546]
Class C	2,300 [7,546]
Class D	2,300 [7,546]
Class E	10,000 [32,808]
Class F	10,000 [32,808]

Source: Strata, 2011a.

Stability classes E and F are given an arbitrarily high number by the WDEQ/Air Quality Division (AQD) to indicate an absence of a distinct boundary in the upper atmosphere.

3.7.2 Climatology

On a larger scale, climate change is a subject of national and international interest. The recent compilation of the current scientific understanding in this area by the U.S. Global Change Research Program (GCRP), a Federal advisory committee, was considered in preparation of this SEIS (GCRP, 2009). Average temperatures in the U.S. have risen more than 1.1 °C [2 °F] over the past 50 years and are projected to rise more in the future. During the period from 1993 – 2008, the average temperature in the Great Plains increased by approximately 0.83 °C [1.5 °F] from 1961 to 1979 baseline temperatures (GCRP, 2009). The projected change in temperature over the period from 2000 – 2020, which encompasses the period that the Ross Project would be licensed, ranges from a decrease of approximately 0.28 °C [0.5 °F] to an increase of approximately 1.1 °C [3.4 °F]. Although the GCRP did not incrementally forecast a change in precipitation by decade, it did project a change in spring precipitation from the baseline period (1961 – 1979) to the next century (2080 – 2099). For the region in Wyoming where the Ross Project is located, the GCRP forecast a 10 – 15 percent increase in spring precipitation (GCRP, 2009).

The EPA has determined that potential changes in climate caused by greenhouse gases (GHG) emissions endanger public health and welfare based on a body of scientific evidence assessed by the GCRP as well as the National Research Council (74 FR 66496). The Administrator of the EPA has issued an endangerment finding based on a technical support document compiled by these scientific organizations. This endangerment finding specifies that, while ambient concentrations of GHG emissions do not cause direct adverse health effects (such as respiratory issues or toxic effects), public health risks and impacts can result indirectly from changes in climate. Based on the EPA's determination, the NRC recognizes that GHGs may have an effect on climate change. In Memorandum and Order CLI-09-21, the Commission provided guidance to NRC staff to consider carbon dioxide and other GHG emissions in its *National Environmental Policy Act* (NEPA) reviews (NRC, 2009a). GHG emissions, as projected for the Ross Project, are considered as an element of the air-quality impacts evaluation in this SEIS; GHG emissions are discussed in SEIS Section 5.

3.7.3 Air Quality

As described in GEIS Section 3.4 (NRC, 2009b), all of the NSDWUMR is classified as an attainment area for all the primary criteria pollutants under the National Ambient Air Quality Standards (NAAQS) (NRC, 2009b). (The EPA sets NAAQS for air pollutants considered harmful to public health and the environment [40 CFR Part 50]. Some states, such as Wyoming, also set their own Ambient Air Quality Standards,

What is an air-quality attainment area?

The attainment status of an area refers to whether or not its air quality "attains" the National Ambient Air Quality Standards (NAAQS) for specific air pollutants. That is, an attainment area is a particular geographic area where the respective concentrations of primary (or "criteria") air pollutants meet the health-based NAAQS for the corresponding primary air pollutants. If the area persistently exceeds the NAAQS for one or more primary air pollutants, it is classified as being in "non-attainment" for the particular air pollutant(s) that exceed(s) the respective NAAQS standard. The Powder River Basin is an attainment area for PM_{10}.

such as the Wyoming Ambient Air Quality Standards [WAAQS].) Primary NAAQS are established to directly protect public health, and secondary NAAQS are established to protect public welfare by safeguarding against environmental and property damage. As discussed in GEIS Section 3.4.6, the NAAQS defines acceptable ambient-air concentrations for six common nonradiological particulate and gaseous air pollutants (i.e., primary or criteria pollutants): nitrogen oxides (as NO_2), ozone (O_3), sulfur oxides (as SO_2), carbon monoxide (CO), lead (Pb), and particulate matter (less than 10 and 2.5 μm in diameter [PM_{10} and $PM_{2.5}$]). In particular, most of the Powder River Basin, where significant coal mining activities are ongoing, and which includes the Ross Project area, is currently designated an attainment area for all pollutants (Strata, 2011a).

As noted above, states may develop standards that are more strict than or that supplement the NAAQS. The WDEQ/AQD has submitted a draft revision of its own WAAQS to the appropriate State boards. These revisions would result in Wyoming's adding one-hour NO_2 and SO_2 standards and revoking the current 24-hour and 1-hour standards for SO_2 of the existing WAAQS to be identical with NAAQS (see Table 3.17). The Wyoming-specific annual (arithmetic mean) PM_{10} standard of 50 $\mu g/m^3$, which is required for short-term modeling of surface coal mine emissions, will be retained. Some primary and secondary NAAQS are presented in Table 3.17 (WDEQ/AQD, 2010).

The air quality in the vicinity of the Ross Project area is currently in compliance with the NAAQS for all primary air pollutants, including particulates (i.e., fugitive dusts) and combustion-engine gaseous emissions.

Table 3.17
National and Wyoming Ambient Air Quality Standards

Criteria Pollutant	National Primary Standards	Wyoming Primary Standards	Averaging Time	Secondary Standards
Carbon Monoxide	9 ppm (10,000 $\mu g/m^3$)	9 ppm (10,000 $\mu g/m^3$)	8 Hours†	N/A*
	35 ppm (40,000 $\mu g/m^3$)	35 ppm (40,000 $\mu g/m^3$)	1 Hour†	N/A
Nitrogen Dioxide	0.053 ppm (100 $\mu g/m^3$)	0.05 ppm (100 $\mu g/m^3$)	Annual Arithmetic Mean	Same as Primary
	0.100 ppm (187 $\mu g/m^3$)	0.100 ppm (187 $\mu g/m^3$)	1 Hour	N/A
Particulate Matter (10-µm Diameter) (PM_{10})	150 $\mu g/m^3$	150 $\mu g/m^3$	24 Hours	Same as Primary
	N/A	50 $\mu g/m^3$	Annual Arithmetic Mean	N/A
Particulate Matter (2.5-µm Diameter) ($PM_{2.5}$)	12.0 $\mu g/m^3$	12.0 $\mu g/m^3$	Annual Arithmetic Mean	Same as Primary
	35 $\mu g/m^3$	35 $\mu g/m^3$	24 Hours[a]	Same as Primary
Ozone	0.08 ppm (157 $\mu g/m^3$)	0.08 ppm (157 $\mu g/m^3$)	8 Hours[b]	Same as Primary
Sulfur Oxides	N/A	*23 ppm (Will Revoke) 60 $\mu g/m^3$*	Annual Arithmetic Mean	N/A
	N/A	*100 ppm (Will Revoke) 260 $\mu g/m^3$*	24 Hours†	N/A
	75 ppm 200 $\mu g/m^3$	*75 ppm (Will Add) 200 $\mu g/m^3$*	1 Hour	N/A
	N/A	0.5 ppm (1,300 $\mu g/m^3$)	3 hours†	0.5 ppm (1,300 $\mu g/m^3$)

Source: Modified from EPA's "National Ambient Air Quality Standards (NAAQS)," as of October 2011.

Notes:

† Not to be exceeded more than once per year.

* N/A = Not applicable.

[a] To attain this standard, the 3-year average of the 98th percentile of 24-hour concentrations at each population-oriented monitor within an area must not exceed 35.0 $\mu g/m^3$ (effective December 18, 2006).

[b] To attain this standard, the 3-year average of the fourth highest daily maximum 8-hour average ozone concentrations measured at each monitor within an area over each year must not exceed 0.08 ppm.

Italics: Standard is in the rulemaking process in Wyoming. The intention is for WAAQS to reflect NAAQS, while retaining the State annual-average PM_{10} standard of 50 $\mu g/m^3$.

3.7.3.1 Particulates

"Particulates" refers to particles that are suspended in the air. Some particles are large enough to be seen (e.g., smoke and wind-blown dust), while others are too small to be visible. Agriculture, forestry, transportation, wind, and fire all contribute airborne particulates to the atmosphere. The NAAQS and WAAQS specify the allowable concentration of airborne particulates of 10 microns in diameter or smaller, or "PM_{10}," to 150 µg/m^3 [9.4 x 10^{-9} lb/ft^3] over 24 hours (see Table 3.17). Wyoming has a supplemental annual (arithmetic mean) PM_{10} standard of 50 µg/m^3 [3.1 x 10^{-9} lb/ft^3] that is averaged over the year (WDEQ/AQD, 2010). The NAAQS also limits allowable concentrations of airborne particles that are 2.5 microns in diameter or smaller ($PM_{2.5}$). Based on the pre-operational background data collected by the Applicant, three radionuclide particulates of interest (natural uranium, Ra-226 and Th-230) are found at concentrations at or below the minimum analytical detection limit and one radionuclide particulate (Pb-210) is found at concentrations just above the minimum analytical detection limits. The detected Pb-210 particulate levels are consistent with the background radon flux as Pb-210 is a progeny of the radon-222 decay.

The eastern portion of the Powder River Basin has an extensive network of PM_{10} monitoring stations that are operated by the mining industry because of the density of the coal mines in the region. There are five surface coal mines within approximately 48 km [30 mi] of the Ross Project area. PM_{10} compliance with the NAAQS and WAAQS 24-hour standards at these five mines (and, by inference, at the Ross Project area) has been consistently demonstrated by these stations (Strata, 2011a); However, there have been three small excursions over the 24-hour PM_{10} at the mines that were determined to be due to high wind conditions. There are also monitoring stations operated by the WDEQ/AQD in the cities of Sheridan, Gillette, Arvada, and Wright, where particulates are generally measured as PM_{10}.

The WDEQ/AQD operates a $PM_{2.5}$ particulate sampler at the Buckskin Mine, about 48 km [30 mi] west of the Ross Project area. Ambient air-quality monitoring data from 2005 – 2009 from the Buckskin Mine show that the average $PM_{2.5}$ ranged from 5.1 – 6.2 µg/m^3 [3.2 – 3.9 x 10^{-10} lb/ft^3], about one-third the annual mean $PM_{2.5}$ standard of 15 µg/m^3 [9.4 x 10^{-10} lb/ft^3]. No excursions above the 24-hour standard of 5 µg/m^3 were recorded at the Mine. The data indicate that particulates from highway and non-road-construction vehicles comprise approximately 28 percent of the total PM_{10} and $PM_{2.5}$ particulate emissions.

As discussed in GEIS Section 3.4.6, prevention of significant deterioration (PSD) requirements identify maximum allowable increases in concentrations for particulate matter for areas designated as in attainment. Different increment levels are identified for different classification areas, with Class I areas having the most stringent requirements. The nearest Class I areas to the Ross Project area is the Northern Cheyenne Indian Reservation (in Montana) and Wind Cave National Park (South Dakota); these areas are 130 km [80 mi] and 160 km [100 mi] from the Ross Project area, respectively. The other sensitive areas are the Class II Devils Tower and the Class II Cloud Peak Wilderness Area. These areas are approximately 16 km [10 mi] and 130 km [80 mi], respectively, from the Ross Project area (Strata, 2011a).

3.7.3.2 Gaseous Emissions

Existing regional air pollutants are known to include gaseous emissions, such as NO_2 and O_3, which have been extensively monitored near the Ross Project area and in the Powder River

Basin since 1975 (Strata, 2011a). See Table 3.17, which presents both the respective NAAQS and WAAQS gaseous-emission standards. Radon is a gaseous air emission which is described further in SEIS Section 3.12.1 under ***Air***. Based on the pre-operation background sampling, the radon concentrations in air through the Ross Project ranges from 0.5 to 2.0 pCi/L with a resultant exposure between 9.2 to 38.2 mrem. These values are consistent with expected background levels for radon in air overlying mineralized environments (Strata, 2011a).

Air-quality monitoring for gaseous emissions within the Powder River Basin includes measuring ozone (as O_3) and nitrous oxides (as NO_2) at two WDEQ/AQD stations, the closest of which is 29 km [18 mi] from the Ross Project area. A Wyoming Air Resources Monitoring System (WARMS), which is operated by the BLM, monitors sulfur- and nitrogen-oxide concentrations near Buffalo, Sheridan, and Newcastle. Nitrogen oxides (as NO_2) are also monitored by the WDEQ at the Thunder Mountain Basin National Grassland monitoring station, 29 km [18 mi] west of the Ross Project area as well as at private monitoring stations at the Belle Ayr and Antelope coal mines (see SEIS Section 5.2). All of these monitoring stations routinely indicate that the annual mean NO_2 emissions are well below the NAAQS and WAAQS.

Ozone is also monitored in the Powder River Basin which is considered an ozone attainment area. Although no violations of the ozone standard have occurred in the area, the levels reported by these nearby air-quality monitoring stations are sometimes close to the respective ozone standard.

PSD requirements also incorporate gaseous-emission standards (e.g., for NO_2, SO_2, and O_3) for maximum allowable increases in concentrations for areas designated as in attainment. As discussed above, Class I areas have the most stringent requirements; Class I areas nearest to the Ross Project area are listed above in SEIS Section 3.7.3.2.

3.8 Noise

As described in GEIS Section 3.4.6, eastern Wyoming is predominantly rural and undeveloped, except for the heavily mined Powder River Basin. Rural areas tend to be quiet, and natural phenomena, such as wind, rain, insects, and livestock, tend to contribute the most to background noise. The unit of measure used to represent sound-pressure levels is the decibel (dB) (and on the A-weighted scale, dBA or A-weighted decibel). dBA is a measure designed to simulate human hearing by placing less emphasis on lower frequency noises, because the human ear does not perceive sounds at low frequencies in the same manner as sounds at higher frequencies. In the undeveloped rural areas of Wyoming, the existing background ambient noise levels range from 22 decibels (dB) on calm days up to 38 dB, depending upon factors such as wind and traffic (NRC, 2009b).

How Is sound measured?

The human ear responds to a wide range of sound pressures. The range of sounds people normally experience extends from low to high pressures by a factor of 1 million. Sound is commonly measured using decibels (dB). Another common sound measurement is the A-weighted sound level (dBA). The equivalent sound level is expressed as an A-weighted sound level over a specified period of time—usually 1 or 24 hours. The A-weighting measures different sound frequencies and the variation of the human ear's response over the frequency range. Higher frequencies receive less A-weighting than lower ones.

> **What is noise?**
>
> Sound waves are characterized by frequency and measured in hertz (Hz). Noises that are perceptible to human hearing range from 31 to 20,000 Hz. Audible sounds (those that can be heard) range from about 60 dB at a frequency of 31 Hz to less than about 1 dB between 900 and 8,000 Hz. dBAs assume a human receptor to a particular noise-producing activity.

It should be noted that noise levels lessen with increasing distance from the respective source. Noise from a line source, such as a highway, is reduced by approximately 3 dB per doubling of distance. For example, road noise at 15 m [49 ft] from a highway is reduced by 3 dB at 30 m [98 ft] and further reduced by an additional 3 dB at 60 m [197 ft]. For point sources, such as equipment, compressors, and pumps, the reduction factor with distance is greater, at approximately 6 dB per doubling of distance.

The land uses in the Ross Project area (see Section 3.2) include livestock grazing, oil production, crop production, ordinary transportation, recreation, and wildlife habitat. Existing ambient noise levels at the Ross Project area were measured by the Applicant to establish pre-licensing baseline conditions at the residences located on New Haven Road and 11 residences in a 3-km [2-mi] vicinity of the Ross Project. Future site-specific noise levels associated with uranium-recovery activities would be measured against these baseline conditions to identify relative increases in noise levels.

The baseline noise study specifically studied the two nearest residences to the Ross Project. The first nearest residence is 210 m [690 ft] from the Ross Project's boundary and approximately 762 m [2,500 ft] from the location of the CPP in the Proposed Action. The second residence is 255 m [835 ft] from the boundary and 1,707 m [5,600 ft] from the proposed location of the CPP. Because these residences are so close to the Ross Project area, they bound the upper range of noise for all four of the residences next to the Ross Project area, where all of the residences are located within 0.48 km [0.3 mi] of the Ross Project's boundary (Strata, 2011a). The noise levels at these two residences averaged 35.4 dBA and 37.4 dBA, depending upon simultaneous factors such as wind speed, traffic volume, vehicular speed, and the type of load being transported (Strata, 2011a).

Truck traffic, in particular bentonite hauling from the Oshoto bentonite mine 5 km [3 mi] north of the Ross Project area and, less frequently, livestock hauling, are the main contributors to existing traffic noise on D and New Haven Roads. According to the U.S. Department of Transportation (USDOT), typical noise levels at road speeds ranging from 80 – 113 km/hr [50 – 70 mi/hr] are 62 – 68 dBA (passenger automobiles), 74 – 79 dBA (medium trucks), and 80 – 82 dBA (heavy trucks) (USDOT, 1995). Posted speed limits for D Road, which passes adjacent to the Ross Project area, are 88 km/hr [55 mi/hr] for automobiles and 72 km/hr [45 mi/hr] for trucks. Peak noise levels attributed to truck traffic have been measured at 80 – 90 dBA (Strata, 2011a). A passing truck hauling bentonite registered 73.4 dBA at the residence on New Haven Road.

In a separate noise study, the Applicant collected baseline measurements at the Applicant's Field Office for an entire week; the data yielded an average day-night noise level (l_{dn}) of 41.6 dBA overall, with no variance between weekday and weekend measurements (Strata, 2011a). The l_{dn} is the A-weighted equivalent noise level for a 24-hour period that includes a noise level at nighttime that is 10 dBA lower than the daytime noise level. Nighttime hours are considered to be from 10 p.m. to 7 a.m. (EPA, 1978).

The Wyoming Department of Transportation (WYDOT) has defined Noise Abatement Criteria (NAC) that take into account land use, because different land-use areas are sensitive to noise in different ways (NACs are used for impact determinations only). The WYDOT procedures consider a person to be affected by traffic noise from highways when existing or future sound levels approach or exceed the NAC, or when expected future sound levels exceed existing sound levels by 15 dBA. In addition, the sound characteristics of noise can affect the acceptability of noise levels to receptors and the acceptability of noise levels is increased when the noise is familiar and routine (WYDOT, 2011). There are no NACs for undeveloped land. The exterior of residential structures would be considered affected by highway traffic above 67 dBA $L_{eq(h)}$ (i.e., equivalent continuous noise level).

Ambient noise levels in larger communities would be expected to be similar to other urban areas (i.e., approximately 50 – 78 dBA). However, the nearest cities to the Ross Project are all quite distant from the Ross Project area and are, thus, not expected to be affected by the noise levels at the Ross Project (nor, conversely, affect the noise levels from the Ross Project). For example, Casper, Wyoming, which has a population of 55,000 and is 225 km [140 mi] away from the Ross Project area (USCB, 2010), and smaller communities, such as Hulett and Moorcroft, which are located 22 km [14 miles] and 35 km [22 miles] away from the Ross Project area, respectively, are too distant to contribute to the noise environment at the Ross Project area.

3.9 Historical, Cultural, and Paleontological Resources

Both NEPA and the *National Historic Preservation Act of 1966* (NHPA), as amended, require Federal agencies to consider the effects of their undertakings on historical and cultural properties. The historic preservation review process is outlined at regulations promulgated by the Advisory Council on Historic Preservation in 36 CFR Part 800. Historical properties are resources eligible for listing in the National Register of Historic Places (NRHP) and may include sites, buildings, structures, districts, or objects. Amendments to Section 101 of the NHPA in 1992 explicitly allowed properties of traditional religious and cultural importance to be eligible for inclusion on the NRHP (and the Wyoming Register of Historic Places). Eligible properties generally must be at least 50 years old and possess criteria of eligibility as defined in 36 CFR Part 60.4; these criteria include: 1) association with significant events in the past, 2) association with the lives of persons significant in the past, 3) embodiments of distinctive characteristics of type, period, or construction, or 4) yield or be likely to yield important information. Historical properties must also possess integrity, defined as the ability of a property to convey its significance (NPS, 1997a).

NEPA established the responsibility of the Federal government to employ all practicable means to preserve important historical, cultural, and natural aspects of national heritage. Implementing regulations for Section 106 provide guidance on how NEPA and Section 106 processes can be coordinated (at Section 800.8[a]) and set forth the manner in which the NEPA process and its documentation can be used to comply with Section 106 (Section 800.9[c]). The NHPA regulations also address the Federal government's responsibility to identify historical and cultural properties and assess the effects of a given Federal undertaking on those properties (Sections 800.4 through 800.5).

As a Federal undertaking, the issuance of an NRC source and byproduct material license for the Ross Project has the potential to affect historic properties located on, in, beneath, or near the Ross Project area. The NRC is required, in accordance with the NHPA, to make a reasonable effort to identify historic properties in the area of potential effect (APE) for the Project. The APE is defined by the Ross Project site boundary and its immediate environs, which may be impacted by the Ross Project construction, operation, aquifer restoration, and decommissioning activities. If historic properties are known to be present, the NRC is required to assess the effects of its issuing a license for uranium-recovery operations on identified properties and to resolve any adverse impacts to those properties.

Several additional statutes and EOs apply to Federal land managed by the BLM, most notably the *Native American Graves Protection and Repatriation Act* (NAGPRA) and the *Archaeological Resources Protection Act* (ARPA). NAGPRA is applicable to burials found on BLM-managed lands, and in that context provides for the protection of Native American remains, funerary objects, sacred objects, or objects of cultural patrimony, and their repatriation to affiliated Native American Tribes following a consultation process between Tribes and the land managing federal agency. ARPA regulates the permitting of archaeological investigations on public land, including those managed by BLM. The State of Wyoming also has a statute pertaining to archaeological sites and human remains, entitled *Archaeological Sites* (Wyoming Statute Ann. §36-1-114, et seq.). The Wyoming State Historic Preservation Office (SHPO) administers and is responsible for oversight and compliance review for Section 106 of the NHPA and NAGPRA as well as compliance with other Federal and State historic-preservation statutes and regulations. The Wyoming SHPO and the Wyoming State Office of the BLM have entered into a programmatic agreement that describes the manner in which the two entities would interact and cooperate under the BLM's National Programmatic Agreement.

3.9.1 Cultural Context of Ross Project Area

The following information is provided as an aid to the reader to understand the Ross Project area in terms of potential prehistoric and historic events that would reasonably be expected to have occurred and that would have left behind artifacts (archaeological resources) of interest to present-day archeologists, paleontologists, and present-day Native American Tribes of this area.

The Ross Project area is within a portion of Wyoming inhabited by aboriginal hunting and gathering people for more than 13,000 years. Throughout the prehistoric past, this area was used by highly mobile hunters and gatherers who exploited a wide variety of resources. The immense expanse of grassland in the Plains region was home to vast herds of bison, also known as buffalo. Exploitation of this resource by indigenous groups structured the Northwest Plains culture area. Fur traders, explorers, and military men were the first Euro-Americans to enter the region and encounter the mounted Indians of the region. These bison-dependent people and their way of life were eventually displaced by permanent farming and ranching settlement.

3.9.1.1 Prehistoric Era

Past research activities within the Northwestern Plains culture area have defined a sequence of cultural periods that provide a general context for identification and interpretation of archaeological resources within the proposed Ross Project area. This chronology for the

Northwestern Plains was developed from the work of Frison (1991; 2001) with age ranges provided in years Before Present (B.P.):

- Paleoindian period (13,000 – 7,000 years B.P.)
- Early Archaic period (7,000 – 5,000 years B.P.)
- Middle Archaic period (5,000 – 4,500 to 3,000 years B.P.)
- Late Archaic period (3,000 – 1,850 years B.P.)
- Late Prehistoric period (1,850 – 400 years B.P.)
- Protohistoric period (400 – 250 years B.P.)
- Historic period (250 – 120 years B.P.)

The most-recent two cultural periods, about which more is known, are more thoroughly discussed in a separate section below.

The Paleoindian period includes various complexes (Frison, 1991; Frison, 2001). Each of these complexes is correlated with a distinctive projectile point style derived from generally large, lanceolate and/or stemmed point morphology. The Paleoindian period is traditionally thought to be synonymous with the "big game hunters" who exploited megafauna such as bison and mammoth (Plains Paleoindian groups), although evidence of the use of vegetal resources has been noted at a few Paleoindian sites (foothill-mountain groups).

The Early Archaic period projectile point styles reflect the change from large lanceolate types that characterized the earlier Paleoindian complexes to large side- or corner-notched types. Subsistence patterns reflect exploitation of a broad spectrum of resources, with a much-diminished use of large mammals.

The onset of the Middle Archaic period has been defined on the basis of the appearance of the McKean Complex as the predominant complex on the Northwestern Plains around 4,900 years B.P. (Frison, 1991; Frison, 2001). McKean Complex projectile points are stemmed variants of the lanceolate point. These projectile point types continued until 3,100 years B.P. when they were replaced by a variety of large corner-notched points (e.g., Pelican Lake points) (Martin, 1999, as cited in Strata, 2011a). Sites dating to this period exhibit a new emphasis on plant procurement and processing.

The Late Archaic period is generally defined by the appearance of corner-notched dart points. These projectile points dominate most assemblages until the introduction of the bow and arrow around 1,500 years B.P. (Frison, 1991). This period witnessed a continual expansion of occupations into the interior grassland and basins, as well as the foothills and mountains.

The Late Prehistoric period is marked by a transition in projectile point technology around 1,500 years B.P. The large corner-notched dart points characteristic of the Late Archaic period are replaced by smaller corner- and side-notched points for use with the bow and arrow. Approximately 1,000 years B.P., the entire Northwestern Plains appears to have suffered an abrupt collapse or shift in population (Frison, 1991). This population shift appears to reflect a narrower subsistence base focused mainly on communal procurement of pronghorn antelope and bison.

3.9.1.2 Protohistoric/Historic Periods

The Protohistoric period witnessed the beginning of European influence on prehistoric cultures of the Northwestern Plains. Additions to the material culture include, most notably, the horse and European trade goods, including glass beads, metal, and firearms. Projectile points of this period include side-notched, tri-notched, and un-notched points, with the addition of metal points. Introduction of the horse on the southern Plains in the 1600s spread northward to other Tribes, and mounted buffalo hunters became the classic Plains culture known in the period of Euro-American contact. New diseases also spread across the continent with the first arrival of Europeans, affecting Native peoples even before the physical appearance of the newcomers.

The Plains Tribes shared a basic commonality of style in their material culture, with regional and Tribal variation. This material culture was strongly characterized by its dependence on bison. Bison played a part in all aspects of physical life by providing food, clothing, shelter, tools, and fuel (dung), as well as embodying a spiritual force (DeMallie, 2001). The need to follow the seasonal movements of bison herds resulted in seasonal variation in residential patterns. Summer encampments of large groups gathered to hunt, using cooperative hunting techniques such as driving a herd over a cliff (buffalo jump sites) or into a corral at the bottom of a slope or a cut bank.

Extended family and village groups moved along with the herds, hauling their belongings and portable dwellings to new encampments. Originally, long, low, multiple-family tents, the classic Plains teepee built on a foundation of supporting poles, developed following the adoption of the horse (DeMallie, 2001). Extended families were organized in nomadic bands or semi-sedentary villages, each independent but sharing the same language and culture, with the size of their aggregations determined by ecological factors. Communal hunting needed for the bison hunts gave way to smaller, scattered social groups that were optimal at other times. The need for horse pasturage also limited the size and duration of residential groups. Smaller Tribes stayed together more of the year, but large Tribes might only congregate for summer hunts. The largest Tribes, such as the Blackfoot and Crow, might rarely gather in a single place and tended toward more lasting divisions that can be viewed as separate Tribes with their own territories and linguistic distinctions (DeMallie, 2001).

Plains groups shared a fundamental belief in the power inherent in all living beings. This power was accessible to individuals in dreams and visions but was particularly useful to medicine men and priests, whose more heightened understanding and experience of power gave them a special role in the ritual life of Plains communities. Sacred power was acquired by individuals through vision seeking during a retreat and accompanied by fasting and prayer while awaiting the appearance of spiritual beings in a special form, sometimes an animal that embodied a teaching and protective spirit (DeMallie, 2001).

During the historic period, the Plains Tribes came under duress from the effects of a rapidly changing world. As soldiers, settlers, bison hunters, and other Tribal nations pushed westward, epidemic diseases ravaged the native populations, and the dislocation of conflict increased, leading to changing demographic patterns and a breakdown of traditional systems of food gathering and inter-group exchange patterns. As missionaries came onto the Plains they professed belief systems that conflicted with, and sometimes even forbade, native traditional rites related to a life view that often mingled the spirit and physical worlds. The influx of trading post goods, the shift in hunting patterns, and the loss of access to the seasonal migrations of prey produced a distorting effect that challenged native life. Cultural transformation was rapid, and was characterized by a long period of hostilities with the white settlers and disagreements among various Tribal entities regarding the course of action in the face of encroachment. Eventual resolution of conflict came through military means and treaties that established the present-day reservation system.

The only Tribal reservation in Wyoming is the Wind River Indian Reservation, located approximately 273.6 km [170 mi] southwest of the Ross Project. The Crow and Northern Cheyenne Indian Reservations in Montana (approximately 160 and 146 km [100 and 91 mi] northwest, respectively) and the Pine Ridge Indian Reservation in South Dakota (approximately 185 km [115 mi] southeast) are the other Tribal reservation communities nearest the proposed Ross Project site. A review of the literature indicates that Devils Tower, which is called *Mato Tipila* by some Native Americans which means "Bear Lodge" (other names for Devils Tower include: Bear's Tipi, Home of the Bear, Tree Rock and Great Gray Horn) (NPS, 2012), (located approximately 18 km [10 mi] from the Ross Project) is a sacred area for several Plains Tribes (Hanson and Chirinos, 1991, as cited in Strata, 2011a). According to the U.S. National Park Service (NPS), over 20 Tribes have potential cultural affiliation with Devils Tower. Six Tribes (Arapaho, Crow, Lakota, Cheyenne, Kiowa, and Shoshone) have historical and geographical ties to the Devils Tower area (NPS, 1997b). Many Native American Tribes of the northern Plains refer to Devils Tower in their legends and consider it a sacred site.

3.9.1.3 Historic Era

The historical context of the Ross Project area includes several themes common to all of northeastern Wyoming. The earliest cumulative historic impact was associated with intermittent exploration, fur trapping, gold seeking, and military expedition, circa 1810s – 1870s. This era was followed by large-scale stock raising (1870s – 1900s). The dry-land farming/homesteading movement was the most substantial historic expansion, occurring from the 1910s – 1930s. The Great Depression resulted in the government assistance programs of the mid- to late-1930s, which affected the settlement patterns of this region. Post-war ranching (1945 to present) is the latest historic theme. Crook County, where the Ross Project is situated, was formed in 1875 and named for Brigadier General George Crook, a commander during the Indian Wars.

Although Euro-Americans began to pass through Wyoming in the early 1800s, these visits were limited to government expeditions of discovery and various British and American fur trapping brigades. Beginning in the 1840s, emigrants of the "great western migration" passed along the Oregon-California Trail along the Platte River and through South Pass heading for lands in Oregon, California, and the Salt Lake Valley, but few if any stayed on in the region. As the lands in the west became more populated and the cattle industry made its way into Wyoming in the 1860s, the region began to attract its own settlers.

The Texas Trail, which operated from 1876 – 1897, was used to move cattle as far north as Canada. Most of the early cattle herds passed through Wyoming and were used to establish Montana's ranching industry. As cattlemen recognized the value of Wyoming's grassland, several large cattle ranches were established and flourished until the devastating blizzards in the winter of 1886-1887. The close of the cattle baron era provided an opening for Wyoming's sheep industry. Several large ranches, including the 4J and G-M, were established in the Gillette area south of the proposed Ross Project; however, the industry experienced steady declines in the 1900s (Massey, 1992; Rosenberg, 1991, as cited in Ferguson, 2010). The dry-land farming movement of the late 19th and early 20th centuries had a profound effect on the settlement of northeastern Wyoming during the years around World War I. The most intensive period of homesteading activity in northeastern Wyoming occurred in the late 1910s and early 1920s. Promotional efforts by the State and the railroads, the prosperous war years for agriculture in 1917 and 1918, and the Stock Raising Act of 1916 with its increased acreage (but lack of mineral rights) all contributed to this boom period. It soon became evident, however, that dry-land farming alone would not provide a living and farmers began to increase their livestock holdings (Ferguson, 2010).

A severe drought in 1919 followed by a severe winter, along with a fall in market prices in 1920, forced out many small holders. During the 1920s the size of homesteads in Wyoming nearly doubled while the number of homesteads decreased, indicating the shift to livestock raising (LeCompte and Anderson 1982, as cited in Strata, 2011a). A period of drought began in 1932, leading to Federal drought relief programs. In April of 1932, the Northeast Wyoming Land Utilization Project began repurchasing the sub-marginal homestead lands and making the additional acres of government land available for lease. Two million acres within five counties, including about 226,624 ha [560,000 ac] of Federally-owned lands, were included in the Thunder Basin Project (LA-WY -1) to alter land use and to relocate settlers onto viable farmland (Resettlement Administration, 1936, as cited in Ferguson, 2010).

During the development program to rehabilitate the range, impounding dams were erected, wells were repaired, springs developed, and homestead fences removed while division fences were constructed for the new community pastures. The government paid former farmers to remove homesteads and their efforts were so successful that almost no trace remains. The remaining subsidized ranches were significantly larger and provided a stabilizing effect on the local economies. The Thunder Basin Grazing Association, the Spring Creek Association, and the Inyan Kara Grazing Association were formed to provide responsible management of the common rangeland.

Uranium was first discovered in Wyoming in 1918 near Lusk. Nuclear Dynamics and Bethlehem Steel Corporation formed the Nubeth Joint Venture (Nubeth) to develop new uranium recovery districts in the western U.S. with specific attention focused on northeastern Wyoming's Powder River Basin (Strata, 2011a). The initial discovery of uranium near Oshoto was made by Albert Stoick during an over-flight of the area. This was followed by macroscopic sampling efforts and then regional exploration work by the Nubeth Joint Venture (Nubeth) (Buswell, 1982, as cited in Strata, 2011a). Nubeth received a Wyoming Department of Environmental Quality/Land Quality Division (WDEQ/LQD) License to Explore (No. 19) in August 1976 and an NRC license in April 1978 (No. SUA-1331). The Nubeth research and development facility was constructed and operated from August 1978 through April 1979. No precipitation of a uranium product took place, however, and all recovered uranium was stored as a uranyl carbonate solution. All final approvals for Nubeth's decommissioning were granted by the NRC and WDEQ by 1986 (Strata, 2011a).

3.9.2 Historical Resources

Buildings and Structures

No buildings or structures eligible for the NRHP or Wyoming State Register were identified within the Ross Project area (Ferguson, 2010). An earthen structure in the Ross Project area, the Oshoto Dam, did not meet the criteria for eligibility for listing in the NRHP (48 CFR Part 2157). The original dam has been rebuilt numerous times because of flood damage, most recently in 2005, and is considered to be essentially a reconstruction rather than the original dam.

Archaeological Sites

A Class III Cultural Resource Inventory (Class III Inventory) was conducted in support of the Ross Project in April 2010 and July 2010 (Ferguson, 2010). The Inventory included a pedestrian survey in transects of 30-m [102-ft] intervals throughout the Ross Project area. Subsurface exposures such as cut banks, anthills, rodent burrows, roads ruts, and cow tracks were examined. Shovel probes were placed at the discretion of the surveyors, primarily in locations where artifacts or features were located or where soil had accumulated. The Inventory focused on landforms where intact sites might be expected, such as intact, stable terraces and their margins as well as areas of exposure (Ferguson, 2010). In November 2011, a geophysical investigation consisting of a magnetometer survey was conducted at several sites within the Ross Project Area and additional shovel tests were conducted in May 2012 and June 2012.

In preparation for the Class III Inventory, a Class 1 Inventory (i.e., a records search) was conducted for the Ross Project area in 2010; this search included the records of the Wyoming Cultural Records Office (WYCRO), the WYCRO online data base, and the BLM's Newcastle Field Office (Ferguson, 2010).

The records search showed that, prior to the 2010 Class III Inventory, no substantial block inventory (i.e., survey) had been conducted in the Project area. Small-scale investigations, including two associated with power lines and buried telephone cables as well as a drilling-pad and access-road survey, have been conducted in the Ross Project area. Only one survey, an inventory for a linear buried telephone cable in Section 13, identified one prehistoric campsite, 48CK1603. Avoidance of this campsite was recommended as a result. The campsite lies on both State of Wyoming and private land, and it was described as "bisected" by D Road (Ferguson, 2010).

During the Applicant's Class III Inventory for the Ross Project, 24 new sites and 21 isolated finds were recorded. Twenty-three of the recorded sites are prehistoric camps, and one is a historic-period homestead. The 24 sites along with the previously identified 48CK1603 are listed in Table 3.18. Paleontological materials, believed to be out of context, were found at two of the sites. These two sites produced projectile points that represent Middle Archaic and Late Archaic periods; other fragments found indicate Late Prehistoric-period occupation. Twenty-one isolates were also recorded during the Inventory. All but two of these are prehistoric artifacts; the two historic isolates are trash scatters. In addition to the sites identified during the Class III Inventory, the potential exists for deeply buried sites to be found within the Ross Project area because of its propitious location near the headwaters of the Little Missouri River.

Fifteen sites identified for the Ross Project have been recommended by the Applicant as eligible for the NRHP (Ferguson, 2010). These are: Nos. 48CK1603, 48CK2073, 48CK2075, 48CK2076, 48CK2078, 48CK2079, 48CK2080, 48CK2081, 48CK2082, 48CK2083, 48CK2085, 48CK2089, 48CK2090, 48CK2091, and 48CK2092. All of these sites are considered eligible under Criterion D of the NRHP, because they are likely to yield information important to our knowledge of prehistory. Collectively or individually, the sites have the potential to yield important information about the occupations at the headwaters of the Little Missouri River and possibly to add to the understanding of the prehistoric cultural relationships between the Little Missouri River region and the Powder River Basin. Two of the sites, Nos. 48CK2083 and 48CK2091, also provide temporal information (Ferguson, 2010).

In general, the Class III Inventory considered that sites located on intact terrace settings, where site preservation was sufficient for research purposes, were recommended as eligible. The remaining nine sites, where landforms lacked soil development and surfaces were eroded or deflated, were not considered likely to retain additional research potential. The NRC staff is in the process of consulting with the Applicant, interested Tribes, and Wyoming SHPO to evaluate the archaeological sites identified during the Applicant's Class III Inventory.

3.9.3 Cultural Resources

Implementing regulations for NHPA, specifically 36 CFR Part 800.4l(a)(1), require the NRC to determine and document the respective APE in consultation with the Wyoming SHPO and the Tribal Historic Preservation Offices (THPOs) (36 CFR Part 800). The definition of an APE is defined in 36 CFR Part 800.16(d) as the geographic area or areas within which an undertaking may directly or indirectly cause alterations in the character or use of historic properties, if any such properties exist (36 CFR Part 800). The APE is influenced by the scale and nature of an undertaking, and it may be different for different types of effects caused by the undertaking.

The APE for the Ross Project area would include all lands where construction, operation, aquifer restoration, and decommissioning activities are proposed. This would include associated staging areas and new access roads in addition to the actual footprint of ground disturbance. In addition, the APE for the Ross Project would need to take into account additional areas where potential effects to traditional cultural properties (TCPs) are identified.

3.9.3.1 Culturally Significant Locations

No Native American heritage, special interest, or sacred sites have been formally identified or recorded to date that are directly associated with the Ross Project area. The geographic position of the Project area between mountains considered sacred by various Native American cultures (the Big Horn Mountains to the west, the Black Hills and Devils Tower to the east), however, creates the possibility that existing, specific locations could have special religious or sacred significance to Native American groups.

3.9.3.2 Tribal Consultation

According to Executive Order (EO) No. 13175, *Consultation and Coordination with Indian Tribal Governments*, the NRC is encouraged to "promote government-to-government consultation and coordination with Federally-recognized Tribes that have a known or potential interest in existing licensed uranium-recovery facilities or applications for new facilities" (NRC, 2009b). Although the NRC, as an independent regulatory agency, is explicitly exempt from the Order, NRC remains committed to its spirit. The agency has demonstrated a commitment to achieving the Order's objectives by implementing a case-by-case approach to interactions with Native American Tribes. NRC's case-by-case approach allows both NRC and the Tribes to initiate outreach and communication with one another.

As part of its obligations under Section 106 of the NHPA and the regulations at 36 CFR 800.2(c)(2)(B)(ii)(A), the NRC must provide Native American Tribes "a reasonable opportunity to identify its concerns about historic properties, advise on the identification and evaluation of historic properties and evaluation of historic properties, including those of religious and cultural importance, articulate its views on the undertaking's effects on such properties, and participate in the resolution of adverse effects." Tribes that have been identified as potentially having concerns about actions in the Powder River Basin include the Assiniboine and Lakota (Montana), Blackfoot, Blood (Canada), Crow, Cheyenne River Lakota, Crow Creek Lakota, Devil's Lake Lakota, Eastern Shoshone, Flandreau Santee Dakota, Kootenai and Salish, Lower Brule Lakota, Northern Arapaho, Northern Cheyenne, Oglala Lakota, Pigeon (Canada), Rosebud Lakota, Sisseton-Wahpeton Dakota, Southern Arapaho, Southern Cheyenne, Standing Rock Lakota, Three Affiliated Tribes, Turtle Mountain Chippewa, and Yankton Dakota (NPS, 2010). On February 9, 2011, the NRC staff formally invited 24 Tribes (see SEIS Section 1.7.3.2) to participate in the Section 106 consultation process for the proposed Ross Project. The NRC staff invited the Tribes to participate as consulting parties in the NHPA Section 106 process and sought their assistance in identifying Tribal historic sites and cultural resources that may be affected by the proposed action.

SEIS Section 1.7.3.2 describes in detail the consultation activities undertaken by NRC with Tribal governments. At this time, the NRC staff is coordinating with interested Tribes to conduct a survey of the Ross Project area to identify sites of religious and cultural significance to Tribes. Correspondence and other documents related to the NRC's Section 106 Tribal consultation efforts are listed in Appendix A.

Table 3.18 Historic and Cultural Properties Identified within the Ross Project Area		
Smithsonian Number	**Preliminary NRHP Eligibility Recommendation[a]**	**Cultural Affiliation/Site Type**
48CK1603	Eligible	Prehistoric campsite
48CK2070	Not eligible	Prehistoric artifact and possible stone ring
48CK2071	Not eligible	Prehistoric campsite
48CK2072	Not eligible	Late prehistoric campsite
48CK2073	Eligible	Prehistoric campsite
48CK2074	Not eligible	Prehistoric campsite
48CK2075	Eligible	Unknown prehistoric camp site
48CK2076	Eligible	Prehistoric stone feature; Historic cans
48CK2077	Not eligible	Prehistoric campsite
48CK2078	Eligible	Unknown prehistoric camp site; historic debris
48CK2079	Eligible	Unknown prehistoric camp site
48CK2080	Eligible	Unknown prehistoric camp site
48CK2081	Eligible	Unknown prehistoric camp site
48CK2082	Eligible	Unknown prehistoric camp site
48CK2083	Eligible	Late Archaic Prehistoric campsite
48CK2084	Not eligible	Prehistoric campsite
48CK2085	Eligible	Unknown prehistoric camp site
48CK2086	Not eligible	Prehistoric campsite
48CK2087	Not eligible	Unknown cairn
48CK2088	Not eligible	Historic homestead (Maros Homestead)
48CK2089	Eligible	Prehistoric campsite

Table 3.18 Historic and Cultural Properties Identified within the Ross Project Area (Continued)		
Smithsonian Number	Preliminary NRHP Eligibility Recommendation[a]	Cultural Affiliation/Site Type
48CK2090	Eligible	Unknown prehistoric camp
48CK2091	Eligible	Middle Archaic camp
48CK2092	Eligible	Unknown prehistoric camp
48CK2093	Not eligible	Prehistoric lithic scatter

[a] The eligibility recommendations reflected in this table are those provided by the Applicant's consultant as reflected in the Class III survey report. However, the NRC staff's review of the Applicant's eligibility recommendations for the identified sites is ongoing. Therefore, for the purposes of this NEPA document, those sites that the applicant has recommended as not eligible will be treated as eligible.

3.10 Visual and Scenic Resources

What are the objectives for the visual resource classes?

Class I: To preserve the existing character of the landscape. This class provides for natural ecological changes; however, it does not preclude very limited management activity. The level of change to the characteristic landscape should be very low and must not attract attention.

Class II: To retain the existing character of the landscape. The level of change to the characteristic landscape should be low. Management activities may be seen, but should not attract the attention of the casual observer. Any changes must repeat the basic elements of form, line, color, and texture found in the predominant natural features of the characteristic landscape.

Class III: To retain partially the existing character of the landscape. The level of change to the characteristic landscape should be moderate. Management activities may attract attention but should not dominate the view of the casual observer. Changes should repeat the basic elements found in the predominant natural features of the characteristic landscape.

Class IV: To provide for management activities that require major modifications of the existing character of the landscape. The level of change to the characteristic landscape can be high. These management activities may dominate the view and be the major focus of viewer attention. However, every attempt should be made to minimize the impact of these activities through careful location, minimal disturbance, and repeating the basic elements.

The Ross Project area is located in a landscape of gently rolling topography and large, open expanses of upland grassland, pasture- and haylands, sagebrush shrubland, and intermittent riparian drainages. Intermittent streams are fed by ephemeral drainages that seasonally drain the adjacent uplands. A mountainous landscape east of the Ross Project can be seen; this landscape includes Devils Tower and the Missouri Buttes.

To quantify visual and scenic resources on the land it administers, the BLM has established an evaluation methodology that defines the visual and scenic quality of land through a Visual Resource Inventory (VRI). The VRI process provides a means for determining visual values. The VRI consists of a scenic-quality evaluation, sensitivity-level analysis, and a delineation of distance zones. Based on these three factors, BLM-administered lands are placed into one of four VRI classes.

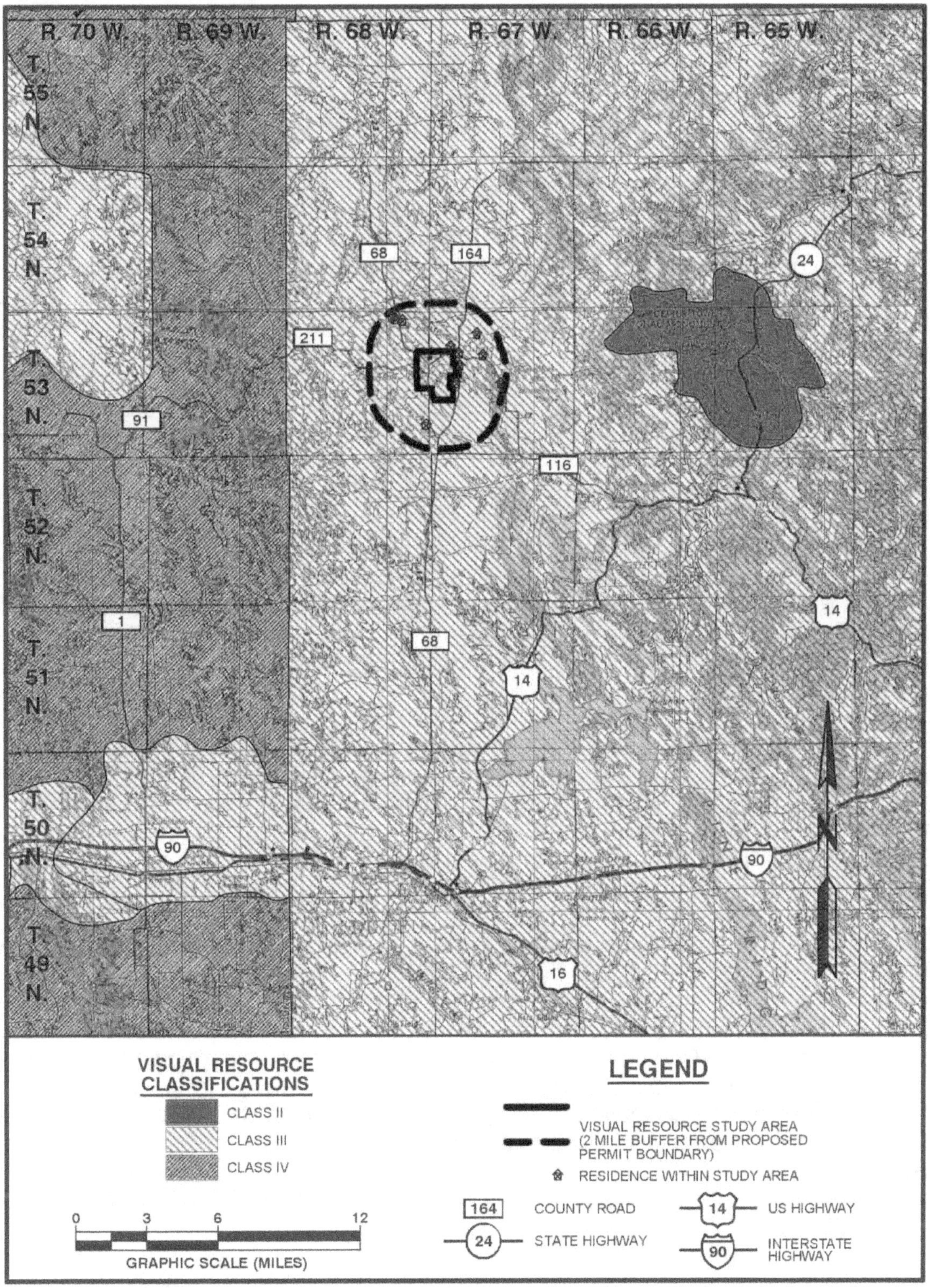

Source: BLM, 2000; BLM, 2001.

Figure 3.19

Regional Visual Resources Management Classifications

These classes represent the relative value of the visual resources.

Classes I and II are designated as the most valued, Class III represents a moderate value, and in Class IV, visual resources are of the least value. The VRI classes provide the basis to assess visual values during the resource management planning (RMP) process conducted for all BLM-administered lands (see Figure 3.19) (BLM, 2010b). The VRI classes are considered in addition to other land uses, such as livestock grazing, recreational pursuits, and energy development when the BLM establishes its Visual Resource Management (VRM) classes during the RMP process. All public lands must be placed into one of the four VRM classes. VRM classes may or may not reflect the VRI classes, depending upon other resource considerations (i.e., a VRI Class II area could be managed as a VRM Class III, or vice versa). The text box above describes the VRM classes and the BLM objectives for each visual classification (BLM, 2007c).

The regional visual and scenic resources in the vicinity of the Ross Project area are described below, and the following section describes Ross Project-specific visual and scenic resources.

3.10.1 Regional Visual and Scenic Resources

The NSDWUMR is located within the Great Plains physiographic province, adjacent to the southern end of the Black Hills (NRC, 2009b). The northeastern corner of Wyoming, within which the Ross Project is located, is managed by the BLM's Newcastle Field Office. Most of the surrounding area is categorized as VRM Class III, but there are some Class II areas located around Devils Tower and the Black Hills National Forest, along the Wyoming-South Dakota border (see Figure 3.19).

Five areas of visually managed land are located within 32.2 km [20 m] of the Ross Project area, including Devils Tower (16 km [10 mi]) and the Missouri Buttes to the east of the Ross Project. Thunder Basin National Grassland (9.10 km [6 mi]) to the west and south, Keyhole State Park (18 km [11 mi]) to the southeast, and Black Hills National Forest (64 km [40 mi]) to the east (Strata, 2011a). These monuments, parks, and forests in the general vicinity of the Ross Project are indicated in Figure 3.20 (Strata, 2011a).

President Theodore Roosevelt established Devils Tower as a national monument on September 24, 1906. The Monument rises 386 m [1,267 ft] above the Belle Fourche River and is visible for at least 16 km [10 mi], as it is visible from the Ross Project area. Devils Tower and the surrounding countryside of pine forest, woodlands, and grassland attract visitors from around the world. The 545-ha [1,350-ac] park allows climbing, hiking, backpacking, and picnicking. Recreational climbing at Devils Tower has increased significantly in recent years. In 1973, there were approximately 312 climbers; currently, there are approximately 5,000 to 6,000 climbers a year (NPS, 2008). As noted above, the BLM VRM classification for Devils Tower is Class II. Beginning in 1995, climbers have enacted a voluntary closure, or a "no climbing period," for the entire month of June as an act of respect for Native American cultural values (NPS, 2008) (see SEIS Section 3.9.1.2).

The Black Hills National Forest (VRM Class II) encompasses streams, lakes, reservoirs, canyons and gulches, caves, varied topography, and vegetation, all of which provide habitat for an abundance of wildlife (Strata, 2011a). Keyhole State Park (VRM Class III) is home to a variety of wildlife. Keyhole Reservoir is the primary attraction to the Park and provides visitors

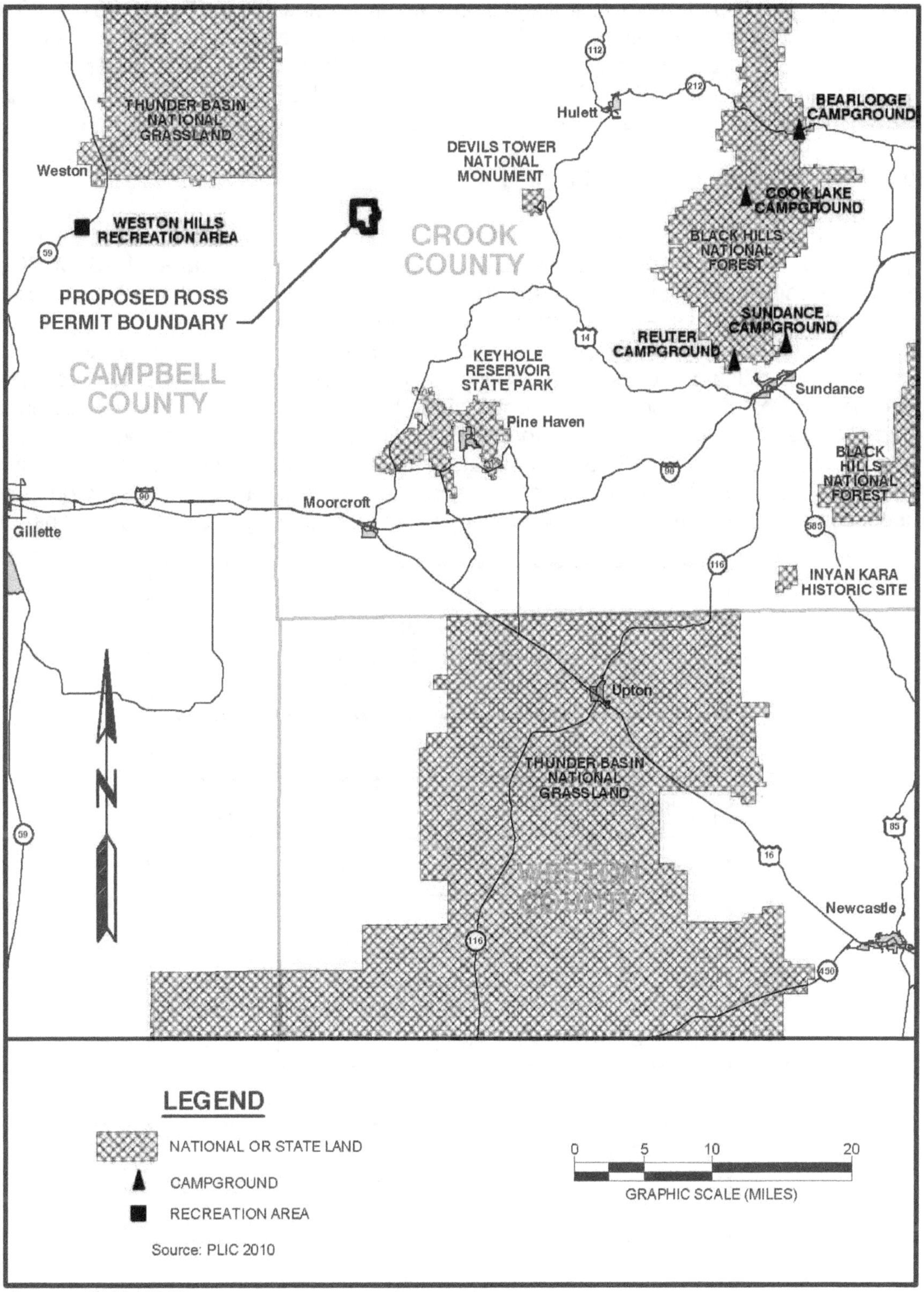

Source: PLIC, 2010,
as shown in Strata, 2012a.

Figure 3.20

Roads, National Parks, National Monuments, and Forests in Vicinity of Ross Project Area

many recreational opportunities including fishing, camping, and hiking (Strata, 2011a). The Thunder Basin National Grassland (VRM Class IV) also provides many opportunities for recreation, including fishing, hiking, and bicycling. Lush, green pastures at the Grassland provide abundant wildlife habitat. The U.S. Forest Service (USFS) manages the Grassland to conserve the natural resources of grass, water, and wildlife habitats (Strata, 2011a).

3.10.2 Ross Project Visual and Scenic Resources

The Applicant conducted a site-specific scenic-quality inventory and evaluation of the Ross Project area in October 2010, using the BLM VRI methodology (see Figure 24) (BLM, 2010b). The scenic-quality evaluation for the visual-resource study area was evaluated based on the key factors of landform, vegetation, water, color, influence of adjacent scenery, scarcity, and cultural modifications. The average scenic-quality index for the Ross Project area was determined by a rating of the scenic quality of four individual aspects (the cardinal compass points) viewed from a high point in the center of the Ross Project. The individual scores were averaged to get a scenic-quality score for the entire Ross Project area. The scenic-quality evaluation presented in Table 3.19 shows that the visual-resource evaluation rating calculated for the Ross Project area is a 10.5 out of a possible 32. More detailed information on the Ross Project scenic-quality inventory and evaluation, including photos, can be found in Appendix B.

Table 3.19
Scenic-Quality
Inventory and Evaluation
(Arithmetic Average of Four Views)

Key Factor	Score
Landform	2.00
Vegetation	3.00
Water	0.50
Color	2.50
Influence of Adjacent Scenery	1.25
Scarcity	2.00
Cultural Modifications	-0.75
TOTAL	**10.50**

The BLM VRM classifications for the lands within and near the Ross Project area are shown on Figure 3.19 (BLM, 2000; BLM, 2001). The land west of the Ross Project is located in Campbell County and is categorized as VRM Class IV, while the land surrounding the Ross Project in Crook County to the east is categorized as VRM Class III. The areas studied for visual resources include the Ross Project and the 3.2-km [2-mi] surrounding vicinity. Thus, this visual-resources area is located entirely within Crook County, and it is consequently categorized as VRM Class III. The level of change allowed by the BLM to the characteristic landscape in Class III management areas would be moderate (BLM, 2010b).

No developed parks or recreational areas are located within the Ross Project and the 3.2-km [2-mi] area around the Project (Strata, 2011a). Within these areas, there are 11 residences in addition to storage tanks; pump jacks; small maintenance buildings; public and private roads and road signage; utilities and poles (power and other utility lines); agricultural features (fences, livestock, stock tanks, and cultivated fields), and environmental-monitoring installations are prominent in the immediate foreground, and they are often noticeable in foreground views by the casual observer.

Of the 11 residences within the study area, 4 residences have unobstructed views to the Ross Project area where the uranium-recovery facility and wellfields would be constructed, and they are in close proximity to the Ross Project in general. The closest residence is 210 m [690 ft] from the Project boundary. Of the 11 residences, 8 are located to the east of the Project area with views to the east (e.g., Devils Tower) and 3 of the 11 residences are northwest of the Ross Project area. Figure 3.21 indicates the areas where the Ross Project facility (i.e., CPP and surface impoundments) would be visible, and Figure 3.22 indicates the potential areas where light pollution from the Ross Project could impact. Photographs used to document the visual-resource study are included in Appendix B.

3.11 Socioeconomics

The Ross Project's region of influence (ROI) is defined as the area within which the Ross Project's socioeconomic impacts and benefits are reasonably anticipated to be concentrated. The Ross Project would be located in Crook County, but it is close enough to the Campbell County line that both counties are within this area of potential impacts. The ROI extends approximately 57 miles to the eastern boundary of Crook County, 41 miles to the northern boundary of Crook County, 115 miles to the western boundary of Campbell County, and 121 miles to the southern boundary of Campbell County. The ROI includes all of the towns and unincorporated areas within Crook County, in which the Project's facility and wellfields would be located and, therefore, would benefit from mineral-production tax revenues. It also includes adjacent Campbell County, which hosts the nearest, largest urban area (i.e., Gillette) and is, consequently, a potential source of labor, services, and materials to support the Ross Project.

3.11.1 Demographics

In Campbell County, Gillette, Wyoming, is the nearest urban area to the Ross Project; it is approximately 53 km [33 mi] to the southwest of the Project. Gillette would likely serve as a regional logistics hub as well as a source of personnel and supplies for the Ross Project (Strata, 2011a). Moorcroft, Wyoming, is approximately 35 km [22 mi] from the Ross Project area and could be a source of personnel as well as a place of residence for Project staff (Strata, 2011a).

Table 3.20 presents the 2000 and 2010 population data for the potentially affected jurisdictions in the ROI. The population in Crook County was 7,083 persons as of 2010, having increased 20.3 percent over 2000 levels (USCB, 2012). The population in Campbell County was 46,133

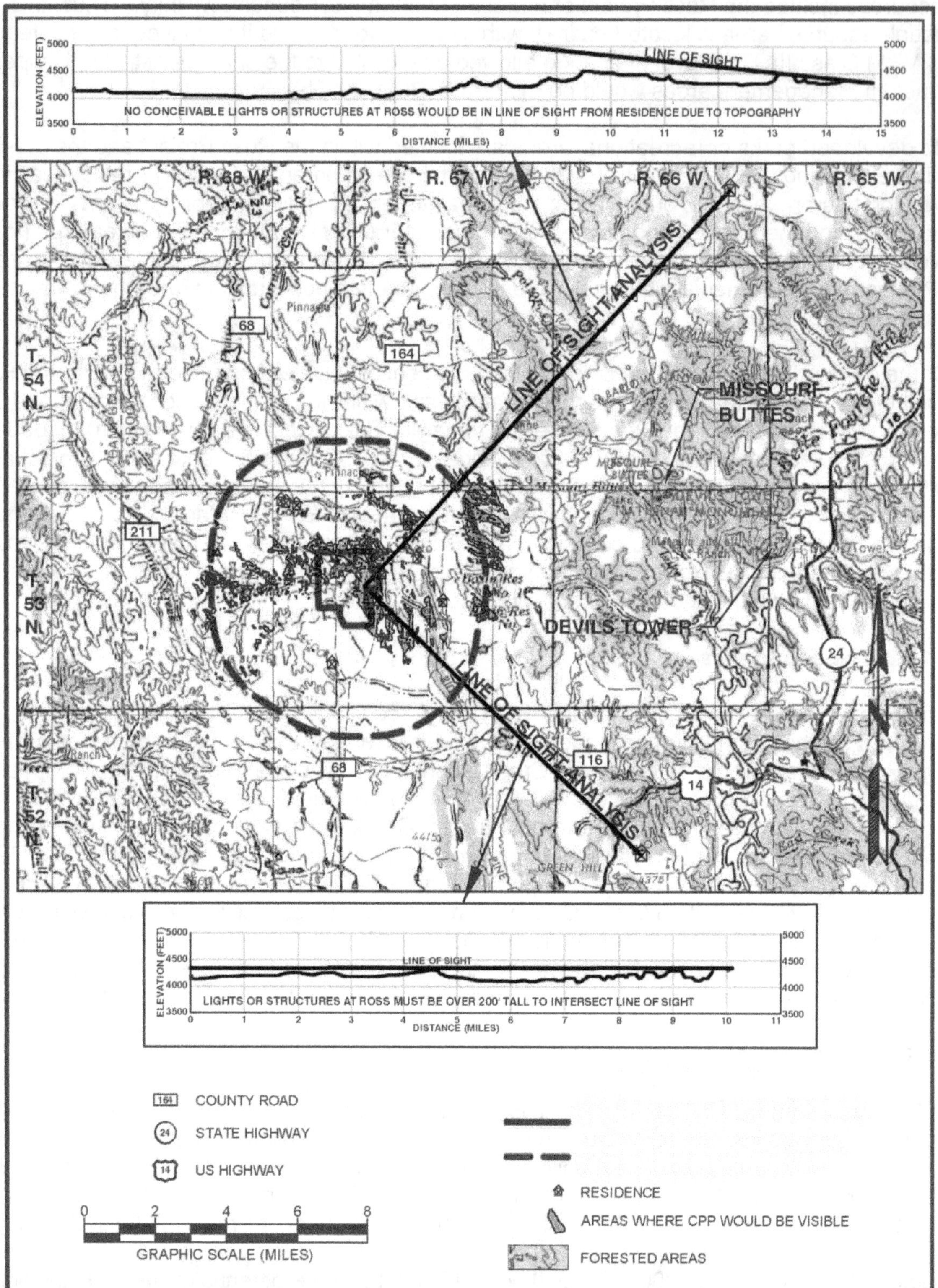

Source: Strata, 2012a.

Figure 3.21

Viewshed Analysis of Ross Project Area

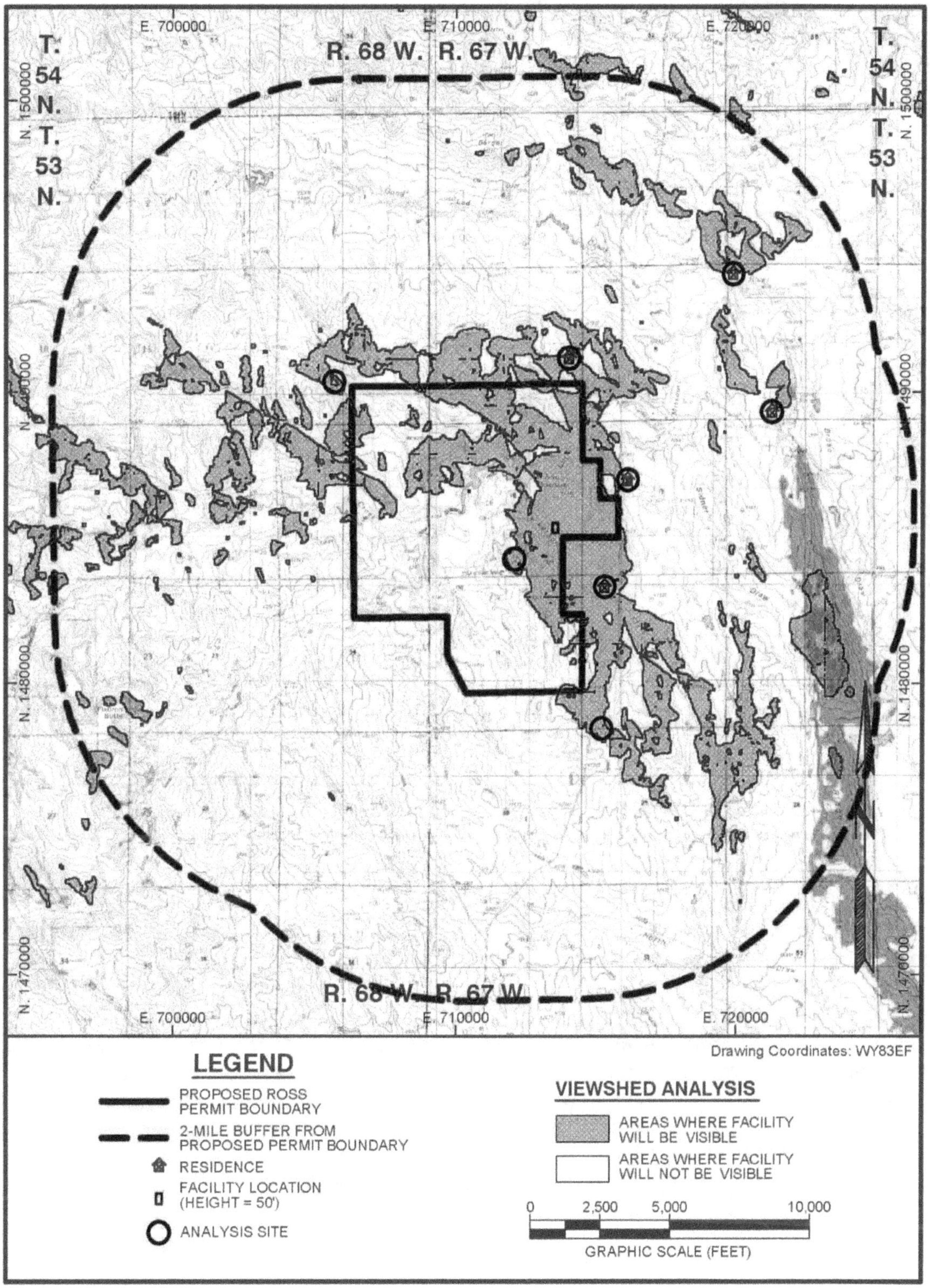

Source: Strata, 2012a.

Note: Prior to construction of the Ross Project, baseline monitoring for potential light pollution would be conducted at eight sites.

Figure 3.22

Light-Pollution Study Area

persons as of 2010, having increased 36.9 percent over 2000 levels. In contrast, population of Wyoming as a whole increased only 14.1 percent between 2000 and 2010. Crook County is the third least populous county in Wyoming, whereas Campbell County is the third most populous.

Table 3.20
Populations in Crook County, Campbell County, and Wyoming
2000 and 2010

Jurisdiction	2000	2010	Change	Total Change (percent)	Annual Average Change (percent)
Crook County	5,887	7,083	1,196	20.3%	1.9%
Hulett	408	383	-25	-6.1%	-0.6%
Moorcroft	807	1,009	202	25.0%	2.3%
Pine Haven	222	490	268	120.7%	8.2%
Sundance	1,161	1,182	21	1.8%	0.2%
Campbell County	33,698	46,133	12,435	36.9%	3.2%
Gillette	19,646	29,087	9,441	48.1%	4.0%
Wright	1,347	1,807	460	34.1%	3.0%
TOTAL ROI	39,585	53,216	13,631	34.4%	3.0%
TOTAL WYOMING	493,782	563,626	69,844	14.1%	1.3%

Source: USCB, 2012

Between 2000 and 2010, Gillette grew by 48.1 percent, faster than the county as whole and much faster than the entire State. This is largely attributable to the growth in the energy sector, conventional oil and gas, coal mining, and power plant construction.

The population of Campbell County is younger than the Wyoming average, has more people per household, more households with individuals under 18 years of age, fewer households with individuals over 65 years of age, and slightly more female householders with no husband present and with their own children under 18 years old (USCB, 2012). Conversely, the population of Crook County is older than the Wyoming average with a higher median age, smaller percentage of households with individuals under 18 years of age, and a higher percentage of households with persons 65 years of age or older.

3.11.2 Income

Per capita personal income in Crook County was $45,843 per person in 2009 and was $49,986 per person in Campbell County (USBEA, 2011). By comparison, per capita income in Wyoming was $49,887 and $40,936 in the U.S. (USBEA, 2011). Based upon the population characteristics discussed above, total personal income in the two-county area was $2.6 billion. Per capita income in Crook and Campbell counties grew at an average annual rate of 3.9 percent over the 2000 to 2009 period (USBEA, 2011). In contrast, per capita income in Wyoming grew at a slightly lower rate of 3.4 percent per year, while the rate of growth in the U.S. over the same period was only 0.8 percent.

Average earnings per job in Crook County were $35,371 in 2009, having increased 2.9 percent annually since 2000. Average earnings per job in Campbell County are almost twice as high as in Crook County and were $64,612 in 2009, having increased 2.9 percent annually since 2000. In contrast, earnings per job State-wide were $46,831 and $52,358 in the U.S. for the same period.

3.11.3 Housing

As of 2010, there were 18,955 housing units in Campbell County (USCB, 2012). Of these, 1,783 were vacant housing units, representing an overall vacancy rate of 9.4 percent (USCB 2012). Of the 1,783 vacant units, 689 of the vacant units were for rent. In contrast, there were only 3,595 housing units in Crook County in 2010. Of these, 674 were vacant housing units, for an overall vacancy rate of 18.7 percent. Of the vacant units, only 54 vacant units were for rent.

Homeownership rates in the two Counties are high by state and national standards. Owner-occupied units accounted for 73.3 percent of all occupied units in Campbell County and 79.3 percent of all occupied units in Crook County (USCB, 2012). Homeownership for the State is 69.2 percent of the population, compared to the entire U.S. where homeownership is 65.1 percent of the population.

3.11.4 Employment Structure

Wyoming State Data

In October 2009, the seasonally-adjusted unemployment rate in Wyoming reached 7.4 percent for the first time since September 1987. Unemployment rates have been on the decline since that time, with the August 2011 rate reported at 5.5 percent (BLS, 2011; WDWS, 2011a).

State-wide employment grew 6.5 percent between the years 2000 and 2010 and stood at 273,313 employed persons in 2010 (WDWS, 2011a). By August 2011, employment was 296,424 persons, up from 277,625 persons in August 2010.

Trade, transportation, and utilities employment represent the largest employment sector in Wyoming, with 24.0 percent of employed persons as of 2010 (WDWS, 2011a), comparable to the U.S. average of 23.0 percent. State-wide employment in the natural resources and mining sector amounted to 13.4 percent of all employment, significantly higher than the U.S. average of 1.7 percent.

Crook and Campbell County Data

Employment in Crook County over the past decade has typically been in the 3,000 to 3,400 range, with peak employment registered at 3,404 persons in 2008 (WDWS, 2011a). Average annual employment in 2010 was 3,284 persons. The August 2011 monthly level is currently at 3,475 persons, down slightly from the August 2010 level of 3,527 persons. Unemployment rates in Crook County have been typically low by national standards, ranging from 2.7 percent to 4.3 percent over the 2000 to 2007 period, but subsequently rose to 5.8 percent in both 2009 and 2010 (BLS, 2011). The unemployment rate as of August 2011 stood at a slightly reduced level of 5.0 percent, representing 175 unemployed persons at this time.

In contrast to Crook County, employment in Campbell County over the past decade has typically been in the 20,000 to 28,000 range, with peak employment registered at 28,492 persons in 2009 (WDWS, 2011a). Employment dropped slightly in 2010 to 27,531 persons and August 2011 levels are currently at 25,542 persons, up slightly from the comparable period in 2010, but still down from 2010 averages.

Unemployment rates in Campbell County also have been typically low by national standards, ranging from 2.0 percent to 3.7 percent over the 2000 to 2008 period, but subsequently rose to 5.5 percent in 2009 and 6.0 percent in 2010 (BLS, 2011). The unemployment rate as of August 2011 stood at a reduced level of 4.4 percent, representing 11,166 unemployed persons at this time.

3.11.5 Finance

The State of Wyoming does not levy a personal or corporate income tax, nor does Wyoming impose a tax on intangible assets such as bank accounts, stocks, or bonds (Strata, 2011a). In addition, Wyoming does not assess any tax on retirement income earned and received from another state. Revenues to the State of Wyoming come from three sources: taxes on mineral production, earnings on investments, and general-fund revenues. Taxes on mineral production include property taxes on the assessed value of production, severance taxes, royalties on production of State-owned minerals, and the State's share of Federal mineral royalties. General-fund revenues include sales (at 4 percent) and use taxes, charges for sales and services, franchise taxes, and cigarette taxes. The third source of State revenues is earnings from the Wyoming Permanent Mineral Trust Fund and pooled investments.

Cities and counties receive revenues in the form of property taxes as well as local sales and use taxes up to 2 percent, including special assessments such as capital-facilities taxes and revenue sharing from the State. Local governments are responsible for collection of property taxes, which are the primary source of funding for public schools and for municipalities, counties, and other local government units. Although Crook County has a slightly higher average mill levy than Campbell County, the mill levy is applied to a much lower evaluation, thus the property taxes raised in Crook County amounted to only a little more than 4 percent of those raised in Campbell County in FY 2010 (Strata, 2011a).

3.11.6 Education

Kindergarten through 12th grade (K-12) public schools in Wyoming are generally organized at the county or sub-county level by school district. Campbell and Crook counties each have one public school district. Campbell County School District operates 16 elementary schools, 2 junior high schools, 2 high schools, and 1 combined junior/high school (Strata, 2011a). Crook County operates a single K-12 school, 2 elementary schools, 2 secondary (grades 7-12) schools, and 1 high school (grades 8-12).

Campbell County has higher school attendance rates than Wyoming as a whole in all grade levels, except college or graduate school (Strata, 2011a). The student-teacher ratio is 19.6 to 1 (Campbell County School District, 2012). Crook County is below the State average at the nursery and preschool ages as well as at the kindergarten and college/graduate school levels, but well above the State average at the elementary (grades 1 – 8) and high-school levels. The student-teacher ratio is 11 to 1 (Education.com, Inc., 2012).

Wyoming also has seven community-college districts. The Northern Wyoming Community College District consists of the main campus in Sheridan, a satellite campus in Gillette, and outreach centers in Buffalo, Kaycee, and Wright. The Gillette campus is the closest post-secondary school to the Ross Project area (Strata, 2011a).

3.11.7 Health and Social Services

Campbell County Memorial Hospital is the principal health-care provider in northeast Wyoming and offers a full range of health services, including emergency room and outpatient surgery services (Strata, 2011a). It is located approximately 65 miles from the Ross Project area. The Heptner Radiation Oncology Center was completed in 2002, and an expansion of medical oncology services was completed in 2008 to form the Cancer Care Center at Campbell County Memorial Hospital. An approximately 560 m^2 [6,000-ft^2] expansion of the Emergency Department was completed in 2009 and an extensive laboratory was completed in late 2009. The laboratory project included the first full chemistry automation line in Wyoming. A $68-million expansion project on the Hospital began in June 2009, with construction of a 3.5 level, 294-space parking structure adjacent to the main entrance of the Hospital. Construction began on a three-level Hospital addition, capable of supporting three additional levels, in 2010. In addition to the Hospital, Campbell County also has outpatient and walk-in clinics, surgery and rehabilitation centers, and numerous senior-residence facilities.

The Crook County Medical Services District consists of a hospital and clinic located in Sundance, as well as clinics located in Moorcroft and Hulett. The District also provides a long-term-care facility attached to the hospital in Sundance (Strata, 2011a).

Sundance, Moorcroft, and Hulett have an ambulance service to cover each town and surrounding areas. Each service has Emergency Medical Technician (EMT) Intermediates, EMT Basics, and Emergency Medical Responders (EMRs) serving on their teams. Of these, Moorcroft is closest to the Ross Project area.

A community survey of needs and services was published in June 2010 by the Campbell County CARE Board. The primary purpose of this needs assessment was to better understand the needs of people who are living in poverty in Campbell County. This survey showed that both low-income clients and agencies ranked, in order, the following services as the most highly rated needs of the County:

- Emergency services
- Housing
- Health

- Nutrition/food
- Employment and training

3.12 Public and Occupational Health and Safety

The existing pre-licensing baseline radiological conditions at the Ross Project area are discussed below.

3.12.1 Existing Site Conditions

As required by 10 CFR Part 40, Appendix A, Criterion 7, the Applicant has conducted one year of pre-licensing, pre-operational baseline radiological monitoring of the Ross Project area. It began its monitoring activities in August 2009. The resulting monitoring data establish the Ross Project area's baseline characteristics prior to NRC licensing. This site-characterization monitoring was developed and implemented in accordance with the following NRC guidelines:

- NRC Regulatory Guide 4.14, *Radiological Effluent and Environmental Monitoring at Uranium Mills*, Revision 1 (NRC, 1980).
- NRC Regulatory Guide 3.46, *Standard Format and Content of License Applications, Including Environmental Reports, For In Situ Uranium Solution Mining*, Section 2.9 ("Radiological Background Characteristics") (NRC, 1982a).
- NRC Regulatory Guide 3.8, *Preparation of Environmental Reports for Uranium Mills* (NRC, 1982b).
- NUREG–1569, *Standard Review Plan For In Situ Leach Uranium Extraction License Applications* (NRC, 2003).

These pre-licensing baseline radiological data represent the condition of the Ross Project area prior to development or construction of any Ross Project facility, wellfields, or any other structural improvements. These data would support future assessments of any environmental impacts that could occur as a result of the Ross Project's construction, operation, and decommissioning, including accidental releases. That is, for most resource areas, the site-characterization data collected by the Applicant would be used to compare and contrast any data collected during the operation of the Ross Project as well as post-operational data collected later.

In the case of ground-water resources, however, additional post-licensing, pre-operational data would be collected (i.e., after the NRC license has been issued, but before actual uranium recovery in a wellfield is initiated, as would be required by the NRC license). This post-licensing, pre-operational data set, which would be established for each wellfield prior to uranium recovery in that wellfield, would serve as a benchmark for the Applicant to determine whether an excursion has occurred (i.e., by way of the upper control limits (UCLs) established for that particular wellfield) and whether the ground water in a wellfield has been restored to the respective target values. These further sampling and analysis activities are discussed in SEIS Sections 2.1.1.1 and 3.5.3.

As discussed in SEIS Section 3.5.3, results from ground-water site-characterization samples can be compared to the specific regulatory standards published by the EPA and the WDEQ/WQD. However, most of the analytical results discussed in this section cannot be compared easily to existing standards because the standards are specified in units other than the reported laboratory units. That is, for example, gross alpha results are reported in picoCuries/volume (pCi/L) [Bq/L] or pCi/kg [Bq/kg] (i.e., in liquid or solid matrices, respectively). This unit is a measurement of the radioactivity in a sample (such as ground water or soil). However, the units of radiation-dose standards are specified in radiation dose/unit time (Sievert or millirem [Sv or mrem]/unit time), and pCi/L or pCi/kg concentrations cannot be straightforwardly converted to mrem/unit time, which is a standard for a human's radiation dose, without extensive modeling (including the conversion to a Total Effective Dose Equivalent [TEDE] which is one of the units used in radiation-protection regulations) (see SEIS Section 4.13). The NRC staff has taken the pre-licensing baseline data supplied by the Applicant and reviewed the modeling that the Applicant performed to determine the respective total radiation dose currently present at the Ross Project area, given the radioactivity-concentration values included in Strata's license application (Strata, 2011b; Strata, 2012b). The modeling and the pre-operational monitoring results performed by the Applicant indicate that the existing conditions at the Ross Project area do not exceed any radiation-dose guidelines or standards in the applicable regulations.

How are potential radiation exposures and doses calculated?

Radiation dose estimates are quantified in units of either **Sievert** or **rem** and are often referred to in either milliSievert (mSv) or millirem (mrem) where 1,000 mSv = 1 Sv and 1,000 mrem = 1 rem (Sv = 100 rem). These units are used in radiation protection to quantify the amount of damage to human tissue expected from a dose of ionizing radiation.

Person-Sv (or person-rem) is a metric used to quantify population radiation dose (also referred to as collective dose). It represents the sum of all estimated doses received by each individual in a population and is commonly used in calculations to estimate latent cancer fatalities in a population exposed to radiation.

Radiation dose is a measure of the amount of ionizing energy that is deposited in a human body. Ionizing radiation is a natural component of the environment and ecosystem, and members of the public are exposed to natural radiation continuously. Radiation doses to the general public occur as a result of the radioactive materials found in the Earth's soils, rocks, and minerals (including those in the Ross Project area). For example, radon-222 (Ra-222) is a radioactive gas that escapes into ambient air from the decay of uranium (and its progeny radium-226), which is found in most soils and rocks. Naturally occurring low levels of uranium and radium are also found in drinking water and foods. Cosmic radiation from space is another natural source of radiation. In addition to these natural sources, there are also artificial or human-made sources that contribute to the radiation dose the general public routinely receives. For example, medical diagnostic procedures using radioactive materials and x-rays are the primary human-made source of radiation the general public experiences. For comparison, the National Council for Radiation Protection estimates the average dose to the public from all natural radiation sources (terrestrial and cosmic) is 3.1 millisieverts (mSv) [310 millirem (mrem)] per year. In Wyoming, this figure is approximately 3.15 mSv/year [316 mrem/yr] (NRC, 2009b).

Pre-Licensing Baseline Radiological Conditions

Table 3.21 presents the range (i.e., the minimum and maximum values) of selected pre-licensing baseline data for the some of the radiological parameters required by the NRC's Regulatory Guide 4.14 (Strata, 2011b; NRC, 1980). Individual reported values for the various radiological parameters can be found in the Applicant's TR (Strata, 2011b).

Pre-Licensing Baseline Sample Matrices, Locations, and Results

The Applicant's pre-licensing baseline environmental-monitoring program was conducted under rigorous sampling-and-analysis procedures and quality-control methods (Strata, 2011b). During the Applicant's environmental monitoring efforts, local ground and surface waters were sampled and analyzed as were samples of sediments, vegetation, air, wildlife, and fish. Direct gamma ("γ") radiation was also measured. The pre-licensing baseline monitoring program included the Applicant's obtaining samples of the following matrices at the specified locations and having the samples analyzed for the radiological parameters shown in Table 3.21. The range of the values obtained by laboratory analysis of these samples is presented in Table 3.21 as well.

Surface Water

The surface waters at the Ross Site were sampled by the Applicant at 14 locations. These locations included both the Oshoto Reservoir and two creek samples (one each from Deadman Creek and the Little Missouri River) during June 2010. Ten other water reservoirs in the Lance District were sampled as well. Three locations on the Ross Site are set up to automatically collect samples during any significant runoff events, although none occurred during the monitoring period (Strata, 2011b). In addition, intermittent and ephemeral surface-water channels were sampled when water was present. Figure 3.14 shows these locations.

Ground Water

Ground-water samples were collected during the Applicant's pre-licensing baseline site-characterization efforts at the Ross Project area. The samples were collected at six locations within the Ross Project area using monitoring wells screened from various horizons within the Lance/Fox Hills aquifer, on-site and nearby privately owned water supply wells. The results of all ground-water samples are more fully discussed in SEIS Section 3.5.3. Note that for samples where metals, including uranium, were to be analyzed, these samples were filtered, yielding "dissolved" concentrations in the data reported. This methodology is described in SEIS Section 3.5.3.

As discussed in the Applicant's license application and in SEIS Section 3.5.3, several ground-water samples exceeded radiological criteria specified by the EPA for its MCLs, and some exceeded more than one of the criteria. The three MCLs are:

- Uranium = 30 μg/L
- Radium-226+228 = 5 pCi/L [0.19 Bq/L]
- Gross Alpha = 15 pCi/L [0.56 Bq/L]

Table 3.21 Range of Analytical Results of Pre-Licensing Baseline Samples* All Sample Matrices					
Matrix	**Type**	**Selected Parameters**	**Range (If Any) of Results**** (Minimum to Maximum)	**Units**	**Any Samples Greater than Detection Limit?**
Water					
	Surface Water[†,††]				
		Lead-210	<1 – 1.46	pCi/L	Yes
		Polonium-210	<1**	pCi/L	No
		Radium-226	<0.02 – 0.46	pCi/L	Yes
		Radium-228	<1 – 1.52	pCi/L	Yes
		Thorium-230	<0.2	pCi/L	No
		Uranium[a]	<0.001 – 0.089	mg/L	Yes
		Gross Alpha	<2 – 48.7	pCi/L	Yes
	Ground Water[†,††]				
		SA Zone			
		Lead-210	<1	pCi/L	No
		Polonium-210	<1	pCi/L	No
		Radium-226	<0.2 – 0.5	pCi/L	Yes
		Radium-228	<0.1 – 1.8	pCi/L	Yes
		Uranium	<0.001 – 0.007	mg/L	Yes
		Gross Alpha	<6 – 13.8	pCi/L	Yes
		SM Zone			
		Lead-210	<1 – 1.34	pCi/L	Yes
		Polonium-210	<1	pCi/L	No
		Radium-226	<0.2 – 3.7	pCi/L	Yes
		Radium-228	<0.1 – 1.3	pCi/L	Yes
		Uranium	<0.001 – 0.004	mg/L	Yes
		Gross Alpha	<7 – 12.2	pCi/L	Yes
		Ore Zone			
		Lead-210	<1 – 4.89	pCi/L	Yes
		Polonium-210	<1 – 22.9	pCi/L	Yes
		Radium-226	0.6 – 12.1	pCi/L	Yes
		Radium-228	<0.1 – 1.4	pCi/L	Yes
		Uranium	0.005 – 0.109	mg/L	Yes
		Gross Alpha	<5 – 222	pCi/L	Yes

Table 3.21
Range of Analytical Results of Pre-Licensing Baseline Samples*
All Sample Matrices
(*Continued*)

Matrix	Type	Selected Parameters	Range (If Any) of Results** (Minimum to Maximum)	Units	Any Samples Greater than Detection Limit?
		DM Zone			
		Lead-210	<1 – 1.16	pCi/L	Yes
		Polonium-210	<1	pCi/L	No
		Radium-226	<0.2 – 0.7	pCi/L	Yes
		Radium-228	<0.1 – 2.2	pCi/L	Yes
		Uranium	<0.001 – 0.013	mg/L	Yes
		Gross Alpha	<14 – 28.3	pCi/L	Yes
	Piezometers in SA Zone				
		Lead-210	<1	pCi/L	No
		Polonium-210	<1	pCi/L	No
		Radium-226	<0.2 – 0.53	pCi/L	Yes
		Radium-228	<0.01 – 2.5	pCi/L	Yes
		Uranium	<0.01 – 0.264	mg/L	Yes
		Gross Alpha	<8.44 – 218	pCi/L	Yes
Soil					
	Surface and Subsurface Soils				
		Lead-210	<0.2 – 2.0 ± 0.7	pCi/g	Yes
		Radium-226	<0.005 – 14.4 ± 2.0	pCi/g	Yes
		Thorium-230	<0.2 – 1.29 ± 0.59	pCi/g	Yes
		Uranium	<0.01 – 2.80	mg/kg	Yes
		Gross Alpha	<1 – 3.6 ± 1.7	pCi/g	Yes
	Sediments				
		Lead-210	<1 – 471 ± 6.1	pCi/g	Yes
		Radium-226	0.8 ± 0.1 – 1.5 ± 0.1	pCi/g	Yes
		Thorium-230	0.39 ± 0.14 – 371 ± 58	pCi/g	Yes
		Uranium	0.876 – 2.24	mg/kg	Yes
		Gross Alpha	1.1 ± 0.4 - 2.8 ± 0.6	pCi/g	Yes

Table 3.21
Range of Analytical Results of Pre-Licensing Baseline Samples*
All Sample Matrices
(*Continued*)

Matrix	Type	Selected Parameters	Range (If Any) of Results** (Minimum to Maximum)	Units	Any Samples Greater than Detection Limit?
Air					
	Particulates				
		Lead-210	6.25×10^{-8} – 1.14×10^{-5}	pCi/L	Yes
		Radium-226	<Detection Limits[d]	pCi/L	No
		Thorium-230	<Detection Limits – 9.74×10^{-8}	pCi/L	Yes
		Uranium	$<1.16 \times 10^{-8}$ – 9.41×10^{-9}	pCi/L	Yes
	Radon				
		Average Radon[b]	0.3 ± 0.04 – 2.0 ± 0.13	pCi/L	Yes
Vegetation					
	Grazing Vegetation				
		Lead-210	3.9 ± 0.5 – 264 ± 19.1	pCi/L	
		Polonium-210	0.225 ± 0.51 – 23.4 ± 7.2	pCi/L	
		Radium-226	1.12 ± 0.08 – $1{,}530 \pm 0.4$	pCi/L	
		Thorium-230	<0.2 – 89.5 ± 16.4	pCi/L	
		Uranium	0.0017 – 8.99	mg/kg	
	Wetland Vegetation				
		Lead-210	9.07 ± 4.1 – 43.1 ± 6.1	pCi/L	
		Polonium-210	1.87 ± 1.7 – 5.88 ± 2.8	pCi/L	
		Radium-226	0.3 ± 0.1 – 11.4 ± 0.5	pCi/L	
		Thorium-230	<0.2 – 3.9 ± 1.5	pCi/L	
		Uranium	0.0005 – 0.0019	mg/kg	
	Hay[c]				
		Lead-210	122 ± 13	pCi/L	
		Polonium-210	7.61 ± 4.1	pCi/L	
		Radium-226	123 ± 1.1	pCi/L	
		Thorium-230	0.83 ± 0.20	pCi/L	
		Uranium	3.10	mg/kg	

Table 3.21
Range of Analytical Results of Pre-Licensing Baseline Samples*
All Sample Matrices
(*Continued*)

Matrix	Type	Selected Parameters	Range (If Any) of Results** (Minimum to Maximum)	Units	Any Samples Greater than Detection Limit?
	Vegetable[c]				
		Lead-210	2.95 ± 4.9	pCi/L	
		Polonium-210	2.55 ± 1.8	pCi/L	
		Radium-226	<0.05	pCi/L	
		Thorium-230	0.40 ± 0.90	pCi/L	
		Uranium	0.0001	mg/kg	
Animal					
	Livestock (Beef)[c]				
		Lead-210	3.12 ± 4.8	pCi/L	
		Polonium-210	<1.0	pCi/L	
		Radium-226	0.288 ± 0.05	pCi/L	
		Thorium-230	<0.2	pCi/L	
		Uranium	<0.001	mg/kg	
	Wildlife (Deer)[c]				
		Lead-210	13.0 ± 7.5	pCi/L	
		Polonium-210	3.68 ± 3.75	pCi/L	
		Radium-226	1.8 ± 1.5	pCi/L	
		Thorium-230	7.6 ± 4.2	pCi/L	
		Uranium	<0.001	mg/kg	
	Fish[c]				
		Lead-210	60.4 ± 93.6	pCi/L	
		Polonium-210	<1.0	pCi/L	
		Radium-226	175 ± 15	pCi/L	
		Thorium-230	0.6 ± 0.6	pCi/L	
		Uranium	0.0160	mg/kg	
Direct Gamma					
	Gamma Survey		5.3 – 25.3 ± 1.54	μR/hr	
	TLD Exposure[d]		17.3 – 30.1	mrem/day	

Source: Strata, 2011b.
Notes: See next page.

Notes for Table 3.21:

- * As suggested by NUREG-4.14.
- ** “<” = “Less than,” where the value following the “<” value is the detection limit.
- † Results also discussed in SEIS Sections 3.5.1 and 3.5.3, Water Quality.
- †† All metals concentrations in water matrices reported as dissolved concentrations (i.e., the samples were filtered).
- [a] All uranium concentrations were obtained by wet-chemistry analysis, not isotope speciation by alpha or gamma spectrometry.
- [b] Averages are radon concentrations taken over three months at each monitoring station.
- [c] One sample only.
- [d] Averages taken from approximately three-month exposures of thermo luminescent dosimeters (TLDs) at each monitoring station. Each value is the “Environmental Dose,” where the Environmental Dose is the Reported Dose (i.e., recorded by the TLD) minus the Transit Dose (i.e., dose received by TLD while in transit to laboratory).

Monitoring Wells and Piezometers

Six well clusters were used by the Applicant to sample ground water quarterly in 2010 (Strata, 2011b). An additional four piezometers in the CPP area were also used quarterly beginning in May 2010 (a piezometer is a device that measures the pressure [more precisely, the piezometric head] of ground water at a specific location.) As described in SEIS Section 2.1.1.1, the six well clusters allowed access to four different ground-water systems in the SA, SM, OZ, and DM zones.

Drinking Water Wells

Twenty-nine local drinking water wells were also sampled quarterly, beginning in July 2009. Some of these samples could not always be obtained because some of the wells were either inaccessible during winter or non-functioning (Strata, 2011b).

Sediments

The sediments at Oshoto Reservoir as well as those at the three surface-water monitoring stations were sampled in August 2010 (Strata, 2011b). Two cups of sediment were sampled for each location and analyzed for Uranium, Ra-226, Th-230, Pb-210, and gross alpha.

Soil

Soil samples at the Ross Project area were obtained from 39 locations; each location was sampled at three depths (i.e., 0-30, 30-60, and 60-100 cm [0-11.8, 11.8-23.6, and 23.6-39.4 in]) (Strata, 2011b). Figure 3.23 indicates the locations of soil sampling activities. These include the three nearest residences, Strata's Oshoto Field Office, the potential locations of the surface impoundments and the CPP, and locations over the major ore bodies where production and recovery wells could be located.

1

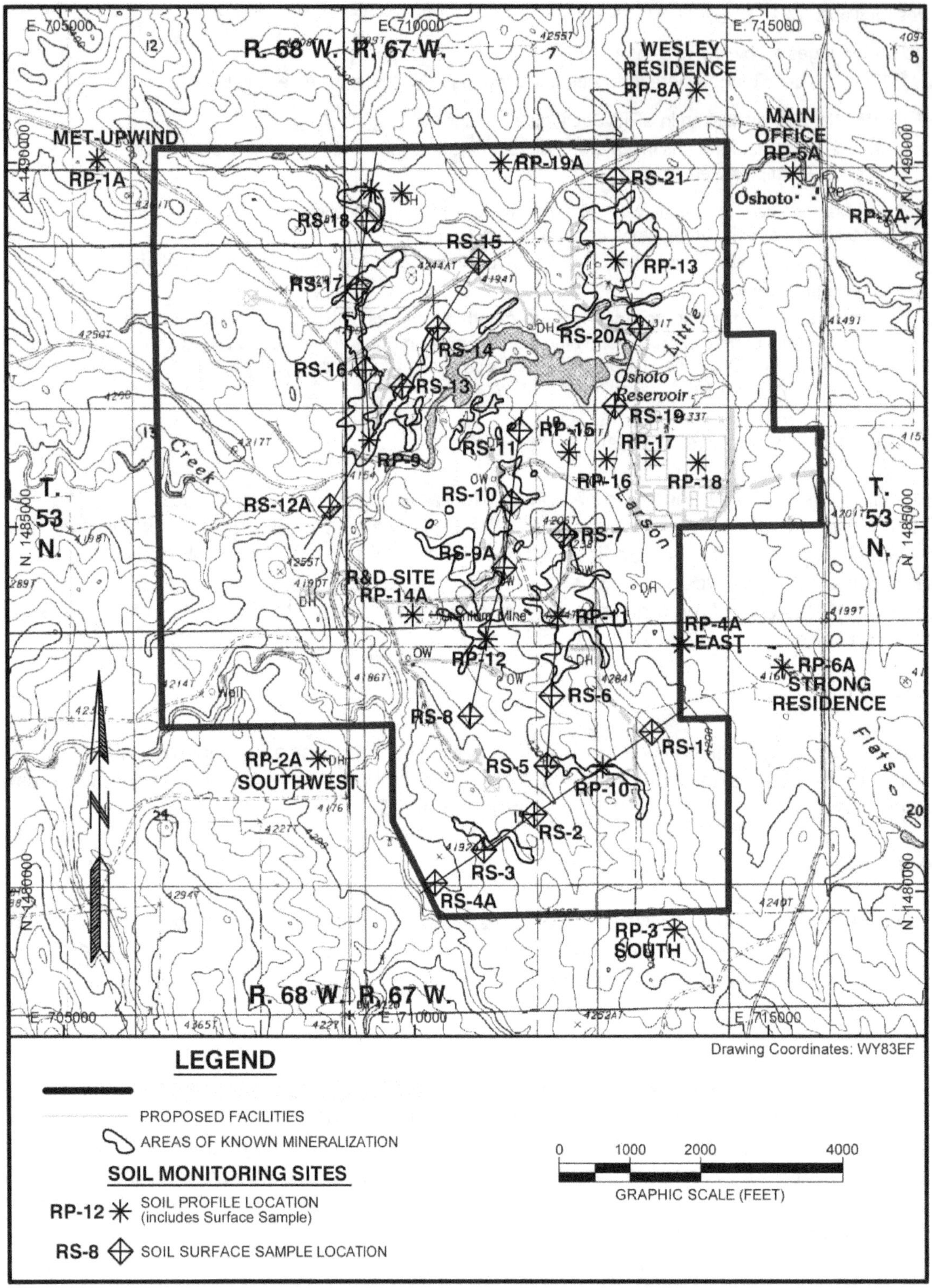

Source: Strata, 2011b.

Figure 3.23

Soil Sampling Locations at Ross Project Area

Air

Particulates

Samples of airborne particulates (e.g., dust) were collected by the Applicant at the six air-sampling stations shown in Figure 3.24. Five of these stations commenced operation in January 2010; the sixth began operating in November 2010. The filters at each air-sampling location were collected weekly and then later composited for analysis (i.e., the filters from each sampling station were composited with the filters from only that respective station, the filters having been collected weekly over an entire quarter for a total of approximately 13 filters per composite sample) (Strata, 2011b).

Radon

Seventeen radon-sampling locations were established by the Applicant, and the results at each were collected quarterly beginning in January 2010; two of these stations were established in mid-2010, resulting in fewer samples. The radon (i.e., a potential gaseous emission) samplers are situated at each of the particulate-sampling locations as well as in the proposed CPP and surface-impoundment areas, the four nearest residences, the former research and development site that had been explored by Nubeth, and over two ore bodies that have been identified for potential uranium recovery (Strata, 2011b).

Vegetation

Vegetation at the Ross Project area was sampled by the Applicant in cooperation with the neighboring landowners after a field study to determine the best vegetation-sample locations was conducted in 2010. Eleven vegetation samples were ultimately collected at downwind locations and near the potential locations of the CPP and surface impoundments as well as along the major ore bodies in the mid- to late summer of 2010.

Animals

Livestock

Beef from locally raised cattle were sampled in cooperation with local landowners. Because horses are not raised in the area for human consumption, no horse-meat samples were obtained. A single beef sample was collected in July 2010 (Strata, 2011b).

Wildlife

Based on the wildlife surveys discussed in SEIS Section 3.6, the only wildlife potentially hunted at or near the Ross Project area for human consumption are deer and pronghorn antelope. One deer-meat sample was obtained from a local landowner who had hunted the deer in the Project's vicinity during the 2010 hunting season (Strata, 2011b).

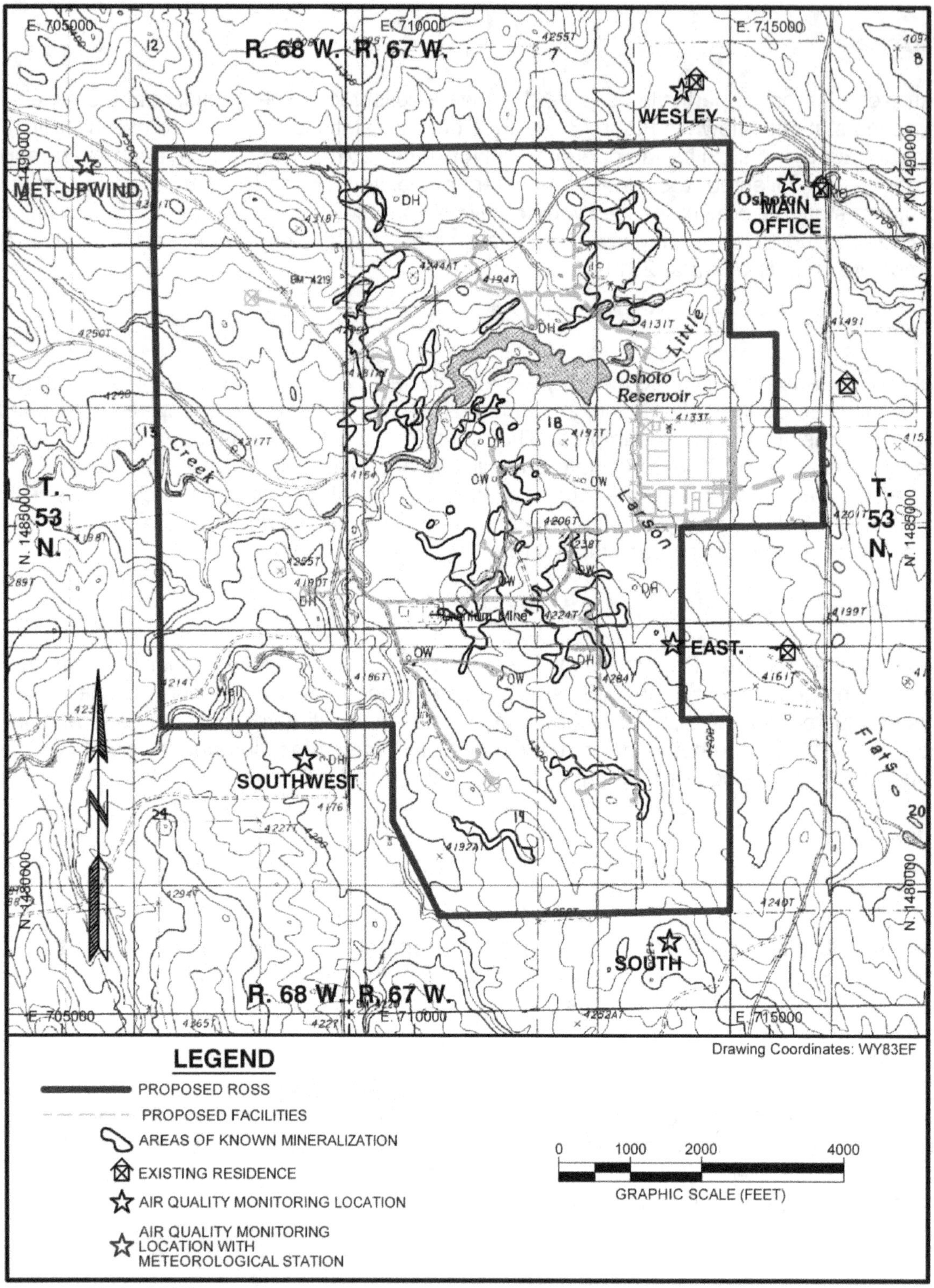

Source: Strata, 2011b.

Figure 3.24

Air-Particulate Sampling Stations at Ross Project Area

Fish

A single composite sample from 99 fish that were caught at the Oshoto Reservoir was collected. Although it is reported by local landowners that fish from the Reservoir are not consumed by humans (Strata, 2011b), this sample was nonetheless submitted for analysis in September 2010.

Direct (Gamma) Radiation

Gamma Field Survey

A field survey performed by a contractor for the Applicant was conducted during July 19 through 22, 2010. During this survey, a total of 80,833 points were surveyed for gamma radiation (Strata, 2011a). In addition, ten soil samples were obtained for an evaluation of the potential relationship between radiation levels and radium concentrations in the corresponding soils (Strata, 2011b). The survey was performed according to the *Multi-Agency Radiation Survey and Site Investigation Manual* (MARSSIM) (NRC, 2000), which is the generally accepted methodology for gamma field surveys.

Long-Term Gamma Study

A long-term study to measure long-term gamma radiation by thermo-luminescent dosimeters (TLDs) was implemented by the Applicant at the same time the radon monitoring stations were established. Ultimately, a total of 17 TLDs (and 2 controls) would be installed around the Ross Project area to measure quarterly gamma exposures.

3.12.2 Public and Occupational Health and Safety

The exposure of members of the public to hazardous chemical is regulated by the EPA and by the State of Wyoming under a variety of statutes and regulations. The NRC, however, has the statutory responsibility, under the *Atomic Energy Act* (AEA), to protect public and occupational health and safety with respect to radioactive materials and radiation exposures. NRC regulations at 10 CFR Part 20 specify annual radiation dose limits to members of the public of 1 mSv [100 mrem] TEDE and 0.02 mSv [2 mrem] per hour from any external radiation sources (see SEIS Section 3.12.1 for a discussion of the units of radiation dose) (10 CFR Part 20). The existing public and occupational health and safety concerns that exist at the Ross Project area today, where it currently presents minimal chemical and radiation exposures, are discussed below.

3.12.2.1 Public Health and Safety

A factor in any assessment of risks to public health and safety, including both chemical and radiation exposures, is the proximity of potentially impacted populations and the nearest receptors. As described in SEIS Section 3.2, the Ross Project area is located in a sparsely populated area of western Crook County (Strata, 2011a). The nearest community is Moorcroft, Wyoming, 35 km [22 miles] to the south, with an estimated population of approximately 1,000 persons. The unincorporated town of Oshoto which is adjacent to the Ross Project area has only a very small population (approximately 50 persons). There are no residences on the proposed Ross Project area; however, within 3 km [2 mi], there are 11 residences with approximately 30 residents. The nearest residence to the Ross Project's boundary is approximately 210 m [690 ft] away, and the nearest residence to the CPP is about 762 m [2,500 ft] away (see SEIS Sections 3.2 and 3.8).

In addition, access to the Ross Project by non-local members of the public is very limited because much of it is privately owned land; there are few public roads that enter the area; and there are no actual public attractions or recreational activities within the Ross Project area or its immediate environs. Moreover, as described in SEIS Section 3.12.1, the hazardous substances known to be present at the Ross Project area are crude oil, associated oil-contaminated water and trash, propane and methanol, and, potentially, polychlorinated biphenyls (PCBs) (Strata, 2011a). Thus, there are very limited non-radiological public health and safety concerns at the Ross Project area because there are: 1) few close residential receptors, all of whom are located offsite; 2) few, if any, members of the public who can access the Project area; and 3) very few hazardous materials are present.

With respect to the existing radiological hazards that are present at the Ross Project area, the same limitations exist as described above for nonradiological hazards: few nearby residents, no public access, and few sources of radiation exposure. The pre-licensing, site-characterization results presented in Table 3.21 indicate exposures to only common background radiation as described in SEIS Section 3.12.1. Soil results presented in Table 3.21 indicate the radionuclide concentrations in soils that are naturally occurring, including the decay products (i.e., progeny) of the naturally occurring uranium, thorium, and radon. The surface- and ground-water pathways, as described above (see SEIS Section 3.12.1), yield little if any radiation exposure to those receptors located offsite because the analytical results of surface- and ground-water samples indicate concentrations of radionuclides that are essentially at or below the respective detection limits and/or below regulatory guidelines. Finally, animal samples indicate limited concentrations of naturally occurring radionuclides. Thus, there are very limited public health and safety concerns at the Ross Project area as it is currently characterized.

3.12.2.2 Occupational Health and Safety

<u>Nonradiological</u>

Occupational health and safety (i.e., industrial safety) is regulated by the State of Wyoming under the Occupational Safety and Health Administration Program. However, occupational health and safety hazards within the Ross Project area are limited by the existing land uses, which are primarily grazing, agriculture, and oil production (see SEIS Section 3.2). Known occupational health and safety concerns include common physical health and safety hazards as well as, potentially, exposures to hazardous substances. Occupational exposures could include normal, industrial, airborne hazardous substances associated with servicing equipment (e.g., vehicles); fugitive dust generated by agricultural activities and by access road use during well-drilling activities; and various chemicals used in agriculture or during oil extraction.

A common type of occupational hazard includes injuries and illnesses. According to the Wyoming Department of Workforce Services (WDWS), the most common lost-day injuries among mineral-extraction workers, including oil-production workers (currently the only type of consistent occupational worker present at the Ross Project area), were from strains and sprains that often resulted from slips, trips, falls, or lifting. The Bureau of Labor Statistics (BLS) compiles annual reports of incidence rates of nonfatal occupational injuries and illnesses by industry and case types. The most recent reports include data from 2009 and 2010. For the category "uranium-radium-vanadium ore mining," annual average employment is given as 1,000 and 900 in 2009 and 2010, respectively. For both years, no total recordable cases either during work or not during work were reported (BLS, 2009; BLS, 2010).

Radiological

The occupational standard promulgated by the NRC is 50 mSv [5 rem] for TEDE over the entire human body (other limits pertain to exposures other than whole body). In addition, all radiation exposures are to be limited to "as low as reasonably achievable" (ALARA). However, only a few pre-construction activities are currently taking place at the Ross Project area—activities such as drillhole plugging and abandonment, monitoring well installation, and environmental monitoring sample collection by the Applicant's personnel. As the pre-licensing baseline data demonstrate (Strata, 2011a), little radioactivity is available to come into contact with these personnel at the Ross Project area today. As a result, there is currently only a small occupational exposure to radiation (i.e., there are few personnel to be exposed and few sources of radioactivity that yield measureable doses).

3.13 Waste Management

Few wastes are currently generated at the Ross Project area, either liquid or solid. Those that are generated are described below.

3.13.1 Liquid Waste

Sources of liquid waste generated at the Ross Project area currently include uranium-exploration drilling, monitoring wells drilling and development, and oil-production facilities (Strata, 2011a).

Drilling the many exploration drillholes on the Ross Project generates drilling fluids and muds (i.e., cuttings). These wastes are classified as technologically enhanced, naturally occurring radioactive materials (TENORM); they are defined by EPA as "[n]aturally occurring radioactive materials that have been concentrated or exposed to the accessible environment as a result of human activities such as manufacturing, mineral extraction, or water processing" (EPA, 2008). Drilling wastes (i.e., fluid, muds, cuttings) are collected and disposed of by the Applicant in onsite excavated pits, or mud pits, that are dug for this specific purpose pursuant to the various EPA regulations governing TENORM, such as those in 40 CFR Part 192. They are allowed to evaporate and dry, and then the dried pits are reclaimed according to WDEQ/LQD requirements, usually within one construction season.

Drilling fluids and muds similar to those created during uranium-exploration drilling are also generated during the Applicant's drilling of its preconstruction monitoring wells and drillholes that it is using to support its license application to the NRC (Strata, 2011a). These fluids are contained and evaporated in mud pits the same as those above, which are constructed adjacent to the drilling pads (Strata, 2011b). An average of 23, 000 liters [6,000 gallons] of ground water along with 12 m^3 [15 yd^3] of drilling muds, are produced during the development and sampling of monitoring wells (Strata, 2011b).

Ground water has also been produced during well tests conducted to characterize aquifer properties (Strata, 2011a). This TENORM water is discharged under a temporary WYPDES Permit No. WYG720229 (WDEQ/WQD, 2011).

Crude oil and water used in its production could be present at the three oil-producing wells on the Ross Project area. These wastes are categorized by EPA as "special wastes" and are exempt from the Federal hazardous waste regulations under Subtitle C of the *Resource Conservation and Recovery Action* (RCRA).

3.13.2 Solid Waste

Few solid wastes are currently generated at the Ross Project area; no AEA-regulated wastes are currently generated. The solid wastes currently generated include predominantly miscellaneous trash from the existing agricultural and oil-production activities that currently take place at the Project area. Agricultural wastes are either disposed of at private landfills or at the local state-permitted landfill in Moorcroft; no private landfills have been identified at the Ross Project area (Strata, 2011a).

Oil-production solid wastes, such as rags contaminated by oil, propane, or methanol, are "special wastes" according to EPA regulations (i.e., they are generated in the production of crude oil) and are exempted from the EPA's hazardous waste regulations under Subtitle C of RCRA (Strata, 2011a). There is one existing stockpile of discarded oil-production tubing that has been identified on the Ross Project area.

3.14 References

10 CFR Part 40. Title 10, "Energy," *Code of Federal Regulations*, Part 40, "Domestic Licensing of Source Material," Appendix A, "Criteria Relating to the Operation of Uranium Mills and the Disposition of Tailings or Wastes Produced by the Extraction or Concentration of Source Material from Ores Processed Primarily for their Source Material Content." Washington, DC: Government Printing Office. 1985, as amended.

33 CFR Part 328. Title 33, "Navigation and Navigable Waters," *Code of Federal Regulations*, Part 328, "Definition of Waters of the United States." Washington, DC: Government Printing Office. 1986, as amended.

36 CFR Part 60. Title 36, "Parks, Forests, and Public Property," *Code of Federal Regulations*, Part 60, "National Register of Historic Places." Washington, DC: Government Printing Office. 1966, as amended.

36 CFR Part 800. Title 36, "Parks, Forests, and Public Property," *Code of Federal Regulations*, Part 800, "Protection of Historic Properties." Washington, DC: Government Printing Office. 2004, as amended.

40 CFR Part 50. Title 40, "Protection of the Environment," *Code of Federal Regulations*, Part 50. "National Primary and Secondary Ambient Air Quality Standards." Washington, DC: U.S. Government Printing Office.

40 CFR Part 61. Title 40, "Protection of the Environment," *Code of Federal Regulations*, Part 61, "National Emission Standards for Hazardous Air Pollutants," Subpart B, "National Emission Standards for Radon Emissions from Underground Uranium Mines." Washington, DC: Government Printing Office. 1989, as amended.

40 CFR Part 141. Title 40, "Protection of the Environment," *Code of Federal Regulations*, Part 141, "National Primary Drinking Water Regulations." Washington, DC: Government Printing Office. 1975, as amended.

40 CFR Part 143. Title 40, "Protection of the Environment," *Code of Federal Regulations*, Part 143, "National Secondary Drinking Water Regulations." Washington, DC: Government Printing Office. 1970, as amended.

40 CFR Part 192. Title 40, "Protection of the Environment," *Code of Federal Regulations*, Part 192, "Health and Environmental Protection Standards for Uranium and Thorium Mill Tailings." Washington, DC: Government Printing Office. 1983, as amended.

68 FR 53083. *Federal Register.* Volume 68, Page 53083. "Withdrawal of Proposed Rule to List the Mountain Plover as Threatened."

74 FR 66496. *Federal Register.* Volume 74, Page 66496. "Endangerment and Cause or Contribute Findings for Greenhouse Gases Under Section 202(a) of the Clean Air Act."

(US)ACE (U.S. Army Corps of Engineers). *Interim Regional Supplement to the Corps of Engineers Wetland Delineation Manual: Great Plains Region.* ERDC/EL TR-08-12. Washington, DC: USACE. March 2008. Agencywide Documents Access and Management System (ADAMS) Accession No. ML13024A331

(US)ACE. Letter from Paige Wolken, Department of the Army, Corps of Engineers, Omaha District, Wyoming Regulatory Office to Tony Simpson, Strata Energy, Inc. December 9, 2010.

Algermissen, S. T., D. M. Perkins, P. C. Thenhaus, S. L. Hanson, and B. L. Bender. *Probabilistic Estimates of Maximum Acceleration and Velocity in Rock in the Contiguous United States.* U.S. Geological Survey Open File Report 82-1033, Scale 1:7,500,000. Washington, DC: U.S. Geological Survey (USGS). 1982. ADAMS Accession No. ML13022A452.

Andres, Brad A. and Kelli L. Stone. *Conservation Plan for the Mountain Plover* (Charadrius montanus). Version 1.0. Manomet, MA: Manomet Center for Conservation Sciences. 2009. ADAMS Accession No. ML13022A455.

(US)BEA (U.S. Bureau of Economic Analysis, USDOC). *Regional Economic Information System. Employment, earnings per job, per capita personal income estimates, Crook County, Campbell County, State of Wyoming, and United States.* Washington, DC: USDOC. 2011.

BLM (Bureau of Land Management, USDOI). *Visual Resource Contrast Rating.* Manual 8431-1. Washington, DC: USBLM. 1986. ADAMS Accession No. ML13011A379.

BLM. *Record of Decision and Approved Resource Management Plan for Public Lands Administered by the Newcastle Field Office.* BLM/WY/PL-00/027+1610. Newcastle, WY: USBLM. September 2000. ADAMS Accession No. ML13016A384.

BLM. *Approved Resource Management Plan for Public Lands Administered by the Bureau of Land Management, Buffalo Field Office.* Buffalo, WY: USBLM. April 2001.

BLM. *Final Environmental Impact Statement and Proposed Plan Amendment for the Powder River Basin Oil and Gas Project.* Volumes 1 – 4. WY-072-02-065. Buffalo, WY: USBLM. January 2003.

BLM. *Task 2 Report for the Powder River Basin Coal Review, Past and Present and Reasonably Foreseeable Development Activities.* Prepared for the BLM Wyoming State Office, BLM Casper Field Office, and BLM Montana Miles City Field Office. Fort Collins, CO: ENSR Corporation. July 2005a. ADAMS Accession No. ML13016A451.

BLM. *Task 1A Report for the Powder River Basin Coal Review, Current Air Quality Conditions.* Prepared for the BLM Wyoming State Office, BLM Casper Field Office, and BLM Montana Miles City Field Office. Fort Collins, CO: ENSR Corporation. September 2005b.

BLM. *Task 3A Report for the Powder River Basin Coal Review – Cumulative Air Quality Effects.* Prepared for the BLM Wyoming State Office, BLM Casper Field Office, and BLM Montana Miles City Field Office. Fort Collins, CO: ENSR Corporation. February 2006.

BLM. *Final Report, Mountain Plover* (Charadrius montanus) *Biological Evaluation.* Cheyenne, WY: USBLM, USFWS, U.S. Forest Service (USFS). March 2007a. ADAMS Accession No. ML13016A463.

BLM. *Potential Fossil Yield Classification (PFYC) System for Paleontological Resources on Public Lands.* Instruction Memorandum No. 2008-009. Washington, DC: USBLM. October 2007b.

BLM. *Visual Resource Management.* Manual 8400. Washington, DC: USBLM. 2007c. ADAMS Accession No. ML101790184.

BLM. *Update of Task 3A Report for the Powder River Basin Coal Review, Cumulative Air Quality Effects for 2015.* Prepared for the BLM Wyoming State Office, BLM Casper Field Office, and BLM Montana Miles City Field Office. Fort Collins, CO: ENSR Corporation. October 2008a.

BLM. *Rawlins Resource Management Plan, Final Environmental Impact Statement.* Rawlins, WY: USBLM. 2008b.

BLM. *Update of Task 3A Report for the Powder River Basin Coal Review, Cumulative Air Quality Effects for 2020.* Prepared for the BLM Wyoming State Office, BLM Casper

Field Office, and BLM Montana Miles City Field Office. Fort Collins, CO: ENSR Corporation. December 2009a.

BLM. *Gunnison and Greater Sage-Grouse Management Considerations for Energy Development Instruction* (*Supplement to National Sage-Grouse Habitat Conservation Strategy*). Memorandum No. 2010-071. Washington, DC: USBLM. 2010a.

BLM. *Visual Resource Inventory*. Manual H–8410–1. Washington, DC: USBLM. 2010b. ADAMS Accession No. ML13014A644.

(US)BLS (U.S. Bureau of Labor Statistics). "Incidence rates of non-fatal occupational injuries and illnesses by industry and case types." At http://www.bls.gov/iif/oshwc/osh/os/ostb2435.pdf (as of June 25, 2012). 2009. ADAMS Accession No. ML13014A700.

(US)BLS. "Incidence rates of non-fatal occupational injuries and illnesses by industry and case types". 2010. At http://www.bls.gov/iif/oshwc/osh/os/ostb2813.pdf (as of June 25, 2012). ADAMS Accession No. ML13014A709.

(US)BLS. *Local Area Unemployment Statistics, Original Data Value, Campbell and Crook Counties, WY, 2001 to 2011*. Washington, DC: USDOL. 2011. ADAMS Accession No. ML13017A248.

Buswell, M. D. *Subsurface Geology of the Oshoto Uranium Deposit, Crook County, Wyoming*, M.S. Thesis, South Dakota School of Mines and Technology. Rapid City, SD: South Dakota School of Mines and Technology. May 1982.

Campbell County School District (Campbell County School District No. 1.) *Facts About Campbell County School District*. Gillette, WY: Campbell County School District. 2012. ADAMS Accession No. ML13011A252.

Case, James C. and J. Annette Green. *Earthquakes in Wyoming*. Information Pamphlet 6. Laramie, WY: Wyoming State Geological Survey (WSGS). 2000. ADAMS Accession No. ML13017A259.

Case, James C., Robert Kirkwood, and Rachel N. Toner. *Basic Seismological Characterization for Crook County, Wyoming*. Laramie, WY: WSGS. 2002. ADAMS Accession No. ML13023A274.

(US)CB (U.S. Census Bureau). "The United States Census Bureau." 2010. At www.census.gov (as of June 25, 2012).

(US)CB. "State and County QuickFacts, Wyoming, Crook County, Campbell County." 2012. At http://quickfacts.census.gov/qfd/states/56000.html (as of June 25, 2012).

Connor, Carol W. *The Lance Formation—Petrography and Stratigraphy, Powder River Basin and Nearby Basins, Wyoming and Montana*. U.S. Geological Survey Bulletin 1917-I. Washington, DC: USGS. 1992. ADAMS Accession No. ML13022A443.

Cowardin, Lewis M., Virginia Carter, Francis C. Golet, and Edward T. LaRoe. *Classification of Wetlands and Deepwater Habitats of the United States.* Prepared for the U.S. Fish and Wildlife Service (USFWS), U. S. Department of the Interior (U.S. Department of Interior [DOI]). Jamestown, ND: Northern Prairie Wildlife Research Center. 1979.

(US)DA (U.S. Department of Agriculture). *2007 Census of Agriculture.* Washington, DC: USDA. 2009. ADAMS Accession No. ML13014A730.

De Bruin, R. H. *Oil and Gas Map of the Powder River Basin, Wyoming.* Wyoming Geological Survey, MS-51-CD, Scale 1:350,000. Cheyenne, SD: Wyoming Geological Survey. 2007.

DeMallie, Raymond J. "Introduction" in *Plains* (Raymond J. DeMallie, Volume Editor) in *Handbook of the American Indian*, Vol. 13 (William C. Sturtevant, General Editor). Washington, DC: Smithsonian Institution. 2001.

Dodge, Harry W. Jr. and J. Dan Powell. *Stratigraphic and Paleoenvironment Data for the Uranium-bearing Lance and Fox Hills Formations, Crook and Northern Weston Counties, Northeastern Wyoming.* U.S. Geological Survey Open File Report 75-502. Washington, DC: USGS. 1975. ADAMS Accession No. ML13023A235.

Dodge, Harry W. Jr. and C. W. Spencer. "Thinning of the Fox Hills Sandstone, Crook County, Wyoming – A Possible Guide to Uranium Mineralization" in *Short Papers of the U.S. Geological Survey, Uranium-Thorium Symposium* (John S. Campbell, Editor). U.S. Geological Survey Circular 753. Washington, DC: USGS. 1977. ADAMS Accession No. ML13023A250.

(US)DOI (U.S. Department of the Interior). *Bureau of Land Management, National Sage-Grouse Habitat Conservation Strategy.* Washington, DC: USDOI. November 2004. ADAMS Accession No. ML13016A323.

(US)DOT (U.S. Department of Transportation). *Highway Traffic Noise Analysis and Abatement Policy and Guidance.* Washington, DC: Noise and Air Quality Branch, Office of Environment and Planning, Federal Highway Administration, USDOT. June 1995. ADAMS Accession No. ML13023A373.

Dunlap, C. M. "The Lewis, Fox Hills and Lance Formations of Upper Cretaceous Age in the Powder River Basin, Wyoming" in *Thirteenth Annual Field Conference Guidebook, Powder River Basin.* Casper, WY: Wyoming Geological Association (WGA). 1958.

Education.com, Inc. "Crook County School District No. 1 Student-Teacher Ratio." 2012. Redwood City, CA: Education.com, Inc. 2012. ADAMS Accession No. ML13011A266.

(US)EPA (U.S. Environmental Protection Agency). "Approval of the Non-Significant Revision of the State of Wyoming Underground Injection Control Program Comprised by the Aquifer Exemption of the Minnelusa Formation underlying Portions of Campbell County, WY." Letter to Mr. Dennis Hemmer, Director, Wyoming Department of Environmental Quality, from Kerrigan G. Clough, Assistance Regional Administrator, Office of Pollution, Tribal and Assistance. December 1997. ADAMS Accession No. ML13024A059.

(US)EPA. "Protective Noise Levels: Condensed Version of EPA Levels Document." EPA Report No. 550/9-79-100. ML13015A552. Washington DC: EPA. November 1978.

Ferguson, D. *A Class III Cultural Resource Inventory of Strata Energy's Proposed Ross ISR Uranium Project, Crook County, Wyoming* (Redacted Version). Prepared for Strata Energy, Inc., Gillette, Wyoming. Butte, MT: GCM Services, Inc. 2010.

Fertig, Walter. "Status Review of the Ute ladies tresses (*Spiranthes diluvalis*) in Wyoming" in *Wyoming Natural Diversity Database*. Laramie, WY: University of Wyoming. 2000. ADAMS Accession No. ML13023A267.

Fertig, Walter. "State Species Abstract: *Penstemon Haydenii*, Blowout Penstemon" in *Wyoming Natural Diversity Database*. Laramie, WY: University of Wyoming. 2008.
Frison, George C. *Prehistoric Hunters of the High Plains*. Second Edition. San Diego, CA: Academic Press, Inc. 1991.

Frison, George C. "Hunting and Gathering Tradition: Northwestern and Central Plains" in *Plains* (Raymond J. DeMallie, Volume Editor) in *Handbook of the American Indian*, Vol. 13 (William C. Sturtevant, General Editor). Washington, DC: Smithsonian Institution. 2001.

(US)FWS (U.S. Fish and Wildlife Service, USDOI). *Black-Footed Ferret* (Mustela nigripes), *Five-Year Review: Summary and Evaluation.* Pierre, SD: USFWS. November 2008. ADAMS Accession No. ML13024A061.

(US)FWS. "National Wetlands Inventory." At http://www.fws.gov/wetlands/ (as of June 25, 2012). April 2012a.

(US)FWS. "Species of Concern, Birds of Conservation Concern." At http://www.fws.gov/wyominges/Pages/Species/Species_SpeciesConcern/BirdsConsvConcern.html (as of June 25, 2012). February 2012b.

(US)FWS. "Migratory Bird Species of Management Concern in Wyoming." At http://www.fws.gov/migratorybirds/Current BirdIssues/Management/BMC.html (as of June 25, 2012). 2012c.

(US)FS (U.S. Forest Service, USDOI). *Final Environmental Impact Statement for the Northern Great Plains Management Plan.* May 2001. At http://www.fs.fed.us/ngp/plan/feis_chpts_appx.htm (as of January 8, 2013). ADAMS Accession No. ML13029A319.

(US)GCRP (U.S. Global Change Research Program). "Regional Climate Impacts, Great Plains." 2009. At http://www.globalchange.gov/publications/reports/scientific-assessments/us-impacts/regional-climate-change-impacts/great-plains (as of April 12, 2012). ADAMS Accession No. ML13017A256.

(US)GS (U.S. Geological Survey, USDOI). "Bird Checklists of the United States, Birds of Northeastern Wyoming." August 2006. At http://www.npwrc.usgs.gov/resource/birds /chekbird/r6/newyomin.htm (as of June 25, 2012).

(US)GS. "2009 Earthquake Probability Mapping." November 2010. At http://geohazards.usgs.gov/eqprob/2009/ (as of June 25, 2012).

(US)GS. "Quaternary Fault and Fold Database of the United States." February 2011. At http://earthquake.usgs.gov/hazards/qfaults/ (as of June 25, 2012).

(US)GS. "USGS Surface-Water Data for the Nation (Historical Monthly and Annual Flow Statistics from Gaging Stations 06334000, 06334500, 06335500, 06336000, and 06337000)" April 2012a. At http://waterdata.usgs.gov/nwis/sw/ (as of June 25, 2012).

(US)GS. "Water Quality Samples for Wyoming." April 2012b. At http://nwis.waterdata. usgs.gov/wy/nwis/qwdata (as of June 25, 2012[b]).

Hinaman, Kurt. *Hydrogeologic Framework and Estimates of Ground-Water Volumes in Tertiary and Upper Cretaceous Hydrogeologic Units in the Powder River Basin, Wyoming.* U S. Geological Survey Scientific Investigations Report 2005-5008. Prepared in Cooperation with the U.S. Bureau of Land Management. Reston, VA: USGS. 2005. ADAMS Accession No. ML13011A278.

Langford, Russell H. "The Chemical Quality of the Ground Water" in Harold A. Whitcomb and Donald A. Morris, *Ground-Water Resources and Geology of Northern and Western Crook County, Wyoming.* U.S. Geological Survey Water-Supply Paper 1698. Prepared in Cooperation with the State Engineer of Wyoming. Washington, DC: Government Printing Office. 1964.

Lisenbee, Alvis L. "Tectonic History of the Black Hills Uplift" in *Thirty-Ninth Field Conference Guidebook, Eastern Powder River Basin – Black Hills.* Casper, WY: WGA. 1988. ADAMS Accession No. ML13023A327.

Merewether, Edward Allen. *Stratigraphy and Tectonic Implications of Upper Cretaceous Rocks in the Powder River Basin, Northeastern Wyoming and Southeastern Montana.* U.S. Geological Survey Bulletin 1917. Denver, CO: U.S. Geological Survey, Information Services. 1996. Accession No. ML13011A281.

ND Resources. Nubeth Joint Venture, Environmental Report Supportive Information to Application for Source Material License In-Situ Solution Mining Test Site Sundance Project Crook County, Wyoming. December 1977.

(US)NPS (U.S. National Park Service, USDOI). "How is Devil's Tower a Sacred Site to American Indians?" August 1994. At http://www.nps.gov/deto/historyculture/sacredsite.htm (as of June 25, 2012).

(US)NPS. *How to Apply the National Register Criteria for Evaluation.* National Register Bulletin No. 15. Washington, DC: USNPS. 1997a. Accession No. ML13024A164.

(US)NPS, Jeffery R. Hanson, and Sally Chirino. *Ethnographic Overview and Assessment of Devils Tower National Monument*, Wyoming. Paper 4. Washington, DC: USNPS. 1997b.

(US)NPS. *Final General Management Plan/Environmental Impact Statement, Devils Tower National Monument, Crook County, WY.* At http://www.nps.gov/deto/parkmgmt/devils-tower-general-management-plan.htm (as of January 8, 2013). November 2001.

(US)NPS. "Standing Witness: Devils Tower National Monument A History." By J. Rodgers. http://www.nps.gov/history/history/online_books/deto/history/index.htm (as of February 24, 2013). 2008.

(US)NPS. Devils Tower National Monument Frequently Asked Questions. http://www.nps.gov/deto/faqs.htm (as of January 31, 2013). December 2012.

(US)NRC (U.S. Nuclear Regulatory Commission). *Radiological Effluent and Environmental Monitoring at Uranium Mills*, Revision 1. Regulatory Guide 4.14. Washington, DC: NRC. 1980. ADAMS Accession No. ML003739941.

(US)NRC. *Standard Format and Content of License Applications, Including Environmental Reports, For In Situ Uranium Solution Mining.* Regulatory Guide 3.46. Washington, DC: NRC. 1982a. ADAMS Accession No. ML003739441.

(US)NRC. *Preparation of Environmental Reports for Uranium Mills, Revision 2.* Regulatory Guide 3.8. Washington, DC: NRC. October 1982b. ADAMS Accession No. ML003740211.

(US)NRC. *Final Environmental Impact Statement To Construct and Operate the Crownpoint Uranium Solution Mining Project, Crownpoint, New Mexico.* NUREG–1508. Washington, DC: NRC. February 1997. ADAMS Accession No. ML062830096.

(US)NRC. *Standard Review Plan for In Situ Leach Uranium Extraction License Applications, Final Report.* NUREG–1569. Washington, DC: NRC. June 2003. ADAMS Accession No. ML032250177.

(US)NRC. *In the Matter of Duke Energy Carolinas [and] In the Matter of Tennessee Valley Authority, Memorandum and Order.* CLI-09-21. November 3, 2009a. ADAMS Accession No. ML093070690.

(US)NRC. *Generic Environmental Impact Statement for In-Situ Leach Uranium Milling Facilities*, Volumes 1 and 2. NUREG–1910. Washington, DC: NRC. 2009b. ADAMS Accession Nos. ML091480244 and ML091480188.

NRCS (Natural Resources Conservation Service, U.S. Department of Agriculture). "Soil survey geographic database for Crook County, Wyoming." 2010. At http://soildatamart.nrcs.usda.gov/download.aspx?Survey=WY011&UseState=WY (as of April 12, 2012).

Nuclear Dynamics. *Quarterly Report, Summary of Water Quality Program.* Source Material License No. SUA-1331, Docket No. 40-8663. 1978. ADAMS Accession No. ML12135A358.

NWS (National Weather Service, National Oceanic and Atmospheric Administration, U.S. Department of Commerce [DOC]). "Wyoming Climate Summaries" for "Colony, Wyoming

(481905).” At http://www.wrcc.dri.edu/cgi-bin/cliMAIN.pl?wycolo (as of June 25, 2012). 2012. ADAMS Accession No. ML13024A342.

Parker, Patricia and Thomas King. *Guidelines for Evaluating and Documenting Traditional Cultural Properties.* National Register Bulletin No. 38. Washington, DC: U.S. National Park Service (USNPS). 1998. ADAMS Accession No. ML13011A335.

Rankl, James G. and Marlin E. Lowry. *Ground-water-flow Systems in the Powder River Structural Basin, Wyoming and Montana.* Water Resources Investigations Report No. 85-4229. Washington DC: USGS. 1990. ADAMS Accession No. ML13024A320

Robinson, Charles S., William J. Mapel, and Maximilian H. Bergendahl. *Stratigraphy and Structure of the Northern and Western Flanks of the Black Hills Uplift, Wyoming, Montana, and South Dakota.* U.S. Geological Survey Professional Paper 404. Washington, DC: USGS. 1964. ADAMS Accession No. ML13024A325.

Sage-Grouse Working Group (Northeast Wyoming Sage-Grouse Working Group). “Northeast Wyoming Sage-Grouse Conservation Plan.” Cheyenne, WY: Wyoming Department of Game and Fish (WDGF). August 15, 2006. ADAMS Accession No. ML13016A281.

Strata (Strata Energy, Inc.). *Ross ISR Project USNRC License Application, Crook County, Wyoming, Environmental Report, Volumes 1, 2 and 3 with Appendices.* Docket No. 40-09091. Gillette, WY: Strata. 2011a. ADAMS Accession Nos. ML110130342, ML110130344, and ML110130348.

Strata. *Ross ISR Project USNRC License Application, Crook County, Wyoming, Technical Report, Volumes 1 through 6 with Appendices.* Docket No. 40-09091. Gillette, WY: Strata. 2011b. ADAMS Accession Nos. ML110130333, ML110130335, ML110130314, ML110130316, ML110130320, and ML110130327.

Strata. *Ross ISR Project USNRC License Application, Crook County, Wyoming, RAI Question and Answer Responses, Environmental Report, Volume 1 with Appendices.* Docket No. 40-09091. Gillette, WY: Strata. 2012a. ADAMS Accession No. ML121030465.

Strata. *Ross ISR Project USNRC License Application, Crook County, Wyoming, RAI Question and Answer Responses, Technical Report, Volumes 1 and 2 with Appendices.* Docket No. 40-09091. Gillette, WY: Strata. 2012b. ADAMS Accession Nos. ML121020357 and ML121020361.

(US)WRDS (U.S. Water Resources Data System) and U.S. Department of Homeland Security. “Earthquakes in Wyoming.” 2010. At http://www.wrds.uwyo.edu/wrds/wsgs/hazards/quakes/quake.html (as of June 25, 2012).

UW (University of Wyoming, Department of Geology and Geophysics). “Collection of Fossil Vertebrates Database.” No Date. At http://paleo.gg.uwyo.edu (as of June 25, 2012). ADAMS Accession No. ML13024A346.

UW. "Mountain Plover Data Compilation" in *Wyoming Natural Diversity Database*. 2010. At http://www.uwyo.edu/wyndd/data-dissemination/priority-data-comp/mtn-plover/ (as of June 25, 2012).

Whitcomb, Harold A. and Donald A. Morris. *Ground-Water Resources and Geology of Northern and Western Crook County, Wyoming*. U.S. Geological Survey Water-Supply Paper 1698. Washington DC: Government Printing Office. 1964.

Whitehead, R. L. "Ground Water Atlas of the United States, Montana, North Dakota, South Dakota, Wyoming." HA 730-I. 1996. At http://pubs.usgs.gov/ha/ha730/ch_i/index.html (as of June 25, 2012).

WDAI (Wyoming Department of Administration and Information). "Wyoming Population Estimates and Forecasts." June 2010. At http://eadiv.state.wy.us/pop/ (as of June 25, 2012).

Wyoming (State of Wyoming, Office of the Governor). *Greater Sage-Grouse Core Area Protection*. EO No. 2011-5. Cheyenne, WY: State of Wyoming Executive Department. June 2, 2011. ADAMS Accession No. ML13015A702

WDEQ/AQD (Wyoming Department of Environmental Quality/Air Quality Division). "Air Quality," *Rules and Regulations,"* Chapter 2. Cheyenne, WY: WDEQ/AQD. 2010.

WDEQ/LQD (Wyoming Department of Environmental Quality/Land Quality Division). *Topsoil and Overburden*. Guideline 1. Cheyenne, WY: WDEQ/LQD. November 1994. ADAMS Accession No. ML13024A064.

WDEQ/LQD. "Noncoal In Situ Mining," *Rules and Regulations*, Chapter 11. Cheyenne, WY: WDEQ/LQD. 2005a.

WDEQ/LQD. "Hydrology, Coal and Non Coal." Guideline 8. Cheyenne, WY: WDEQ/LQD. March 2005b.

WDEQ/WQD (Wyoming Department of Environmental Quality/Water Quality Division). "Wyoming Surface Water Classification List." At http://deq.state.wy.us/ wqd/watershed/surface standards/Downloads/Standards/2-3648-doc.pdf. 2001. ADAMS Accession No. ML13016A388.

WDEQ/WQD. "Groundwater Sampling for Metals: Summary." ML13015A691. At http://deq.state.wy.us/wqd/groundwater/downloads/CBM/Sampling%20 Procedures%20for%20Metals%20April6.doc%5E.pdf (as of April 26, 2005). 2005a.

WDEQ/WQD (Wyoming Department of Environmental Quality/Water Quality Division). "Water Quality," *Rules and Regulations*, Chapter 8. Cheyenne, WY: WDEQ/WQD. 2005b.

WDEQ/WQD. "Wyoming Surface Water Quality Standards," *Rules and Regulations*, Chapter 1 . Cheyenne, WY. WDEQ/WQD. 2007. ADAMS Accession No. ML13024A069.

WDEQ/WQD. *Strata Energy, Inc. – Ross Disposal Injection Wellfield, Final Permit 10-263, Class I Non-hazardous, Crook County, Wyoming.* Cheyenne, WY: WDEQ/WQD. April 2011. ADAMS Accession No. ML111380015.

WDWS (Wyoming Department of Workforce Services). *Wyoming Labor Force Trends.* Laramie, WY: Research and Planning Division, WDWS. 2011a.

WDWS. "Wyoming Nonagricultural Wage and Salary Employment, Final Benchmark 1990-2009 - Preliminary Benchmark 2010." 2011b. At http://wydoe.state.wy.us/lmi/ces/naanav9002.htm.

WDWS (Wyoming Department of Workforce Services). "Labor and Employment," *Rules and Regulations*, Chapter 11. Cheyenne, WY: WDWS. 2012.

WGFD. *A Comprehensive Wildlife Conservation Strategy for Wyoming.* Cheyenne, WY: WDGF. July 2005a. ADAMS Accession No. ML13024A147.

WGFD (Wyoming Department of Game and Fish). *Avian Species of Special Concern, Native Species Status List.* Buffalo, WY: WDGF. January 2005a.

WGFD. *Recommendations for Development of Oil and Gas Resources Within Important Wildlife Habitats.* Version 6.0. Cheyenne, WY: WDGF. March 2010. ADAMS Accession No. ML13016A480.

WSEO (Wyoming State Engineer's Office). "Water Rights Database." 2006. At http://seo.state.wy.us/wrdb/index.aspx (as of June 25, 2012).

WYDOT (Wyoming Department of Transportation). *WYDOT Vehicle Miles Book.* Cheyenne: WY: WYDOT. 2011.

4 ENVIRONMENTAL IMPACTS AND MITIGATION MEASURES

4.1 Introduction

As discussed in this Supplemental Environmental Impact Statement (SEIS) Sections 2 and 3, the Generic Environmental Impact Statement (GEIS) evaluated the potential environmental impacts of in situ recovery (ISR) projects in four distinct geographic regions, including the Nebraska-South Dakota-Wyoming Uranium Milling Region (NSDWUMR), where the proposed Ross Project area is located. Four project phases were evaluated in the GEIS for each of the geographic regions (i.e., construction, operation, aquifer restoration, and decommissioning). The activities that would occur during the four project phases at the Ross Project and their timeframes are described in SEIS Section 2. Because of the similarities between the ISR projects examined in the GEIS and the proposed Ross Project, many of the conclusions found in the GEIS can be used to identify and rate the relative impacts of the Proposed Action in this SEIS. However, if the results of the GEIS's impact analyses indicated a wide range of impacts on a particular resource area (e.g., from SMALL to LARGE), then that resource area was evaluated in greater detail within this site-specific SEIS.

The information that has been used to perform these site-specific impact analyses has been obtained from the license-application documents submitted by the Applicant to the U.S. Nuclear Regulatory Commission (NRC) in 2011 as well as subsequent information provided by the Applicant in 2012 (Strata, 2011a; Strata, 2011b; Strata, 2012a; Strata, 2012b). The NRC staff has compiled related information from publicly available sources as well (see SEIS Section 2.1). All of this information has allowed the NRC to perform site-specific assessments of the environmental impacts of the proposed Ross Project facility and wellfields, as needed, and to evaluate the measures that would successfully mitigate those impacts.

NRC established a standard of significance for its analyses of environmental impacts during the conduct of its environmental reviews, as described in the NRC guidance NUREG–1748 (NRC, 2003). This standard is summarized as follows:

SMALL: The environmental effects are not detectable or are so minor that they will neither destabilize nor noticeably alter any important attribute of the resource considered.

MODERATE: The environmental effects are sufficient to alter noticeably, but not destabilize, important attributes of the resource considered.

LARGE: The environmental effects are clearly noticeable and are sufficient to destabilize important attributes of the resource considered.

This section of the SEIS analyzes the four lifecycle phases (i.e., construction, operation, aquifer restoration, and decommissioning) of the proposed Ross Project, consistent with the analytical approach used in the GEIS (NRC, 2009). This assessment is conducted for the Proposed Action and the two Alternatives (the No-Action and North Ross Project Alternatives). The impacts are organized by the environmental resource and management areas commonly examined for the satisfaction of the *National Environmental Policy Act* (NEPA) requirements. These areas include:

- Land Use
- Transportation
- Geology and Soils
- Water Resources (Surface and Groundwaters)
- Ecology
- Air Quality
- Noise
- Historical, Cultural, and Paleontological Resources
- Visual and Scenic Resources
- Socioeconomics
- Environmental Justice
- Public and Occupational Health and Safety (Nonradiological and Radiological)
- Waste Management

The respective mitigation measures that would moderate the identified environmental impacts are also discussed in this section for each resource and management area. Many types of mitigation measures can be considered when any particular resource or management area's impacts are evaluated. Some of the mitigation measures that are described in this section of the SEIS include:

- Permit and License Requirements
- Regulatory Requirements and Standards
- Facility Design Criteria and Modifications
- Process and System Adjustments
- Engineering and Management Techniques
- Best Management Practices (BMPs)
- Standard Operating Procedures (SOPs)
- Management and Operating Plans
- Training Prerequisites
- Scheduling and Phasing Variations

The respective environmental impacts and associated mitigation measures identified and evaluated in this section are also summarized in Section 8, Summary of Environmental Impacts and Mitigation Measures, in Table 8.1.

4.2 Land-Use Impacts

The Proposed Action could impact local land use during all phases of the Project's lifecycle. Potential land-use impacts could result from land disturbance during, especially, the Ross Project's construction and decommissioning; from grazing and access restrictions; and from competing access for mineral rights. These potential impacts could be greater in the areas where there are higher percentages of private landownership. As shown in Table 2.1, the surface owners of the Ross Project area include private owners (553 ha [1,367 ac]), the State of Wyoming (127 ha [314 ac]), and the U.S. Bureau of Land Management (BLM) (16 ha [40 ac]). At the end of operation, final site reclamation would occur during decommissioning, and all lands would be returned to their current land use. These current land uses include livestock grazing, crop agriculture, and wildlife habitat. Detailed discussion of the potential environmental impacts to land use during construction, operation, aquifer restoration, and decommissioning and site restoration for the proposed Ross Project are provided in the following sections.

4.2.1 Alternative 1: Proposed Action

Alternative 1 consists of four phases: construction, operation, aquifer restoration, and decommissioning of an ISR uranium-recovery facility and wellfields.

4.2.1.1 Ross Project Construction

The GEIS identified potential land-use impacts during construction resulting from land disturbances and site-access restrictions that could limit other grazing, mineral extraction, or recreational activities (NRC, 2009). As discussed in GEIS Section 4.4.1, potential impacts to most aspects of land use from the construction of an ISR facility would be SMALL (NRC, 2009). This is because the amount of area disturbed by the construction would be small in comparison to the available lands; the majority of the site would not be fenced; potential conflicts over mineral access would be expected to be negotiated and agreed upon; only a small portion of the available land would be restricted from grazing; and the open spaces for hunting and off-road vehicle access would be minimally impacted by the fencing associated with the ISR facility. The GEIS defined land-use impacts to be SMALL when they ranged from 50 – 750 ha [120 – 1,880 ac] (NRC, 2009).

What are mineral rights, oil rights, and drilling rights?

Rights may be conferred to remove minerals, oil, or sometimes water that may be present on and under some land. In jurisdictions supporting such rights, they may be separate from other rights to the land. The rights to develop minerals, and the purchase and sale of those rights, are contractual matters that must be agreed between the parties involved.

Construction-phase activities during the Proposed Action would include construction of buildings, other auxiliary structures, and surface impoundments; wells, wellfields, and pipelines; and transportation and utility infrastructure (e.g., roads and lighting). The Applicant estimates that construction activities would disturb a total of 113 ha [280 ac] of land, which represents 16 percent of the Ross Project area. The impacts on specific areas of the Proposed Action by construction activities are summarized in Table 4.1.

Table 4.1
Summary of Land Disturbance during Construction of Proposed Action

Activity	Total Area Impacted by Proposed Action (ha [ac])	Total Area Impacted in the Year Preceding Proposed Action Operation (ha [ac])	Primary Current Use
Central Processing Plant	22 [55]	22 [55]	Dryland crop production Pasture
Wellfield Modules	65 [160]	14 [85]	Livestock grazing Oil and gas production
Access Roads	12 [30]	5 [12]	Livestock grazing
Deep-Injection Wells	2 [5]	1 [3]	Livestock grazing
Pipelines	6 [15]	2 [5]	Various
Utilities	6 [15]	2 [5]	Various
TOTAL	~ 113 [280]	~ 47 [116]	

Source: Strata, 2011a.

The Applicant would mitigate short-term impacts resulting from construction activities by phasing its activities and limiting the amount of land disturbance at any one time; promptly restoring and reseeding disturbed areas; coordinating efforts with the oil-production company currently operating within the Ross Project area (i.e., Merit Oil Company [Merit]); using existing roads wherever possible; following existing topography during access-road construction to minimize the need to cut and fill; minimizing secondary and tertiary access-road widths; and locating access roads, pipelines, and utilities in common corridors. In addition, the Applicant would establish surface-use agreements with surface owners/lessees to mitigate and/or to compensate for their temporary loss of use in areas which are currently used for livestock grazing or crop production. Cultivated fields would be specifically avoided, where possible, during facility construction and wellfield installation.

As shown in Table 2.1, of the 16 ha [40 ac] of BLM surface-administered land within the Ross Project area, 0.5 ha [1.3 ac] would be disturbed by the Proposed Action. This disturbance would take place during the construction phase. The Applicant would restrict hunting during the life of the Project in order to protect workers. Hunting and recreation are not major land use activities within the Ross Project area and there is no public access to BLM lands, therefore impacts would be minimal.

All of the construction activities at the Ross Project would result in temporary, short-term impacts, with the current use restored following construction, except for the area where the central processing plant (CPP) and surface impoundments (i.e., the facility) would be constructed. The use of the Ross Project lands, however, would be restored after all uranium-recovery activities have ceased. The area of surface disturbance the Applicant estimates for the Proposed Action is less than that identified in the GEIS, and no site-specific impacts have been identified for the Proposed Action that would change the magnitude of the impacts identified by the GEIS (NRC, 2009). Thus, the land-use impacts resulting from the Ross Project would be SMALL.

4.2.1.2 Ross Project Operation

The primary land-use impact during the Ross Project's operation would be due to the Applicant's installing additional wellfields and its operating the processes and circuits located in the CPP; however, these impacts are generally the same as those addressed in the construction-phase analysis above. Additionally, the affected area would be reclaimed over the longer term.

As during the construction phase, the Applicant would reduce ongoing impacts to livestock grazing by fencing less than 12 percent of the Ross Project area at any one time, including the CPP and wellfields, during active operation of the Ross Project. In addition, the Applicant would continue to work with Merit, as discussed above, so as not to impact its oil-recovery operation.

No further land-use impacts have been identified for the Ross Project beyond those identified in the GEIS. Thus, the land-use impacts resulting from the operation of the Proposed Action would be SMALL.

4.2.1.3 Ross Project Aquifer Restoration

Land use impacts during aquifer restoration would be similar to those during construction, as they could involve temporary access restrictions, and are expected to be SMALL according to GEIS Section 4.4.1 (NRC, 2009). The impacts to land use during the Proposed Action's aquifer-restoration phase would be similar to those during the construction and operation phases, and they are consistent with the GEIS. These impacts could involve temporary access restrictions, but they are expected to be few. Mitigation measures during the Proposed Action's aquifer-restoration phase would be identical to those identified for its construction and operation. Therefore, the land-use impacts resulting from aquifer-restoration activities at the Ross Project would be SMALL.

4.2.1.4 Ross Project Decommissioning

As discussed in GEIS Section 4.4.1, land-use impacts would temporarily increase during decommissioning and related site restoration of an ISR facility due to the additional equipment that would be used for dismantling and removal of wellfields, pipelines, and other wellfield components as well as the demolition of the processing plant itself and any surface impoundments. In addition, the reclamation of the site would involve heavy equipment and significant earth disturbance. However, these short-term impacts would not be greater than those experienced during the construction phase. Therefore, the GEIS concluded that the land-use impacts that result from the decommissioning an ISR facility would be SMALL (NRC, 2009).

During decommissioning, the Ross Project area would be returned to its approximate preconstruction state, including surface topography and drainage patterns. All roads and wellfields would be removed and reclaimed, unless exempted by the request of a landowner. Topsoil would be salvaged and redistributed on disturbed areas to a depth approximately equal to pre-licensing baseline conditions. Additional subsoil would be ripped as needed to minimize soil compaction prior to revegetation. Revegetation would be completed in accordance with an approved restoration plan, which would be required as part of Strata's Permit to Mine, and a seed mix approved by WDEQ/Land Quality Division (LQD) and the landowners would be used. Seeding would be conducted by either drill or broadcast methods, as appropriate. Once vegetation has been re-established (and all radioactive materials have been removed), the Project area would be released for unrestricted use and would no longer require a license from the NRC.

The land-use impacts resulting from the decommissioning of the Proposed Action would be SMALL and the site's restoration would ameliorate all land-use impacts caused by earlier phases of the Proposed Action.

4.2.2 Alternative 2: No Action

Under the No-Action Alternative, the Ross Project would not be licensed and the land would continue to be available for other uses. Although limited construction activities could occur, the 113 ha [280 ac] of land surface potentially disturbed during the Proposed Action would remain mostly undisturbed. No pipelines would be laid and no additional access roads would be constructed. The Applicant could continue with some preconstruction activities, such as abandonment of exploration drillholes and the collection of environmental monitoring data, but these activities would have little land use impact.

The current land uses of natural-resource extraction and livestock grazing would continue with no access restrictions within the Ross Project area. Impacts to current land uses from the continued oil-production activities could also occur from accidental breaks or failures in equipment and infrastructure; however, these impacts are no different than would occur whether or not the Proposed Action were to be licensed, constructed, or operated. There would be no impact from activities associated with construction and operation of the Proposed Action under the No-Action Alternative.

Under the No-Action Alternative, there would also be no impacts due to aquifer-restoration or decommissioning activities at the Ross Project area, because no wells would have been installed nor wellfields developed for uranium recovery. Thus, there would be no impact to the current land uses. There would be no impact to land use from decommissioning activities because the Ross Project would not have been licensed, constructed, or operated. No buildings would require decontamination and dismantling; no topsoil would need to be reclaimed; and no land would need to be revegetated. The land-use impacts of the No-Action Alternative would be SMALL.

4.2.3 Alternative 3: North Ross Project

Under Alternative 3, the North Ross Project would generally be the same as the Proposed Action, except that the facility (i.e., the CPP, associated buildings, and auxiliary structures as well as the surface impoundments) would be located to the north of where it would be located in the Proposed Action, as described in SEIS Section 2.1.3. This north-site facility would be located about 900 m [3,000 ft] northwest of that the Proposed Action. Construction activities would still disturb an approximate total of 113 ha [280 ac] of land, which represents 16 percent of the total Ross Project area. The impacts from each activity would be approximately the same as those summarized in Table 4.1, except that construction of the surface impoundments at the north site could require additional engineering, while the containment barrier wall (CBW) would not need to be constructed.

For Alternative 3, the CPP would not be located in an area of dry-land crop agriculture or pasture. Therefore, Alternative 3 would cause less impact to land use if the CPP and surface impoundments were to be constructed at the north site. Nonetheless, there would be an increased loss of wildlife- and livestock-grazing opportunities during the construction and operation phases of Alternative 3, just as in the Proposed Action; these impacts would result from the construction of access roads and installation of wells, pipelines, and utilities. The total land area disturbed would be essentially the same (approximately 113 ha [280 ac]). During Alternative 3's operation and decommissioning as well as during the restoration of the underlying aquifer, this Alternative's impacts would be the same as those discussed earlier for the Proposed Action, because the area of land-use disturbance would generally be the same. Finally, because the impacts to land use would generally be the same in Alternatives 1 and 3, the mitigation measures for Alternative 3 would be the same, as would be their effectiveness, as those described for Alternative 1. Based upon this analysis, the land-use impacts resulting from Alternative 3 would be SMALL.

4.3 Transportation

The Proposed Action could impact transportation during all phases of the Project's lifecycle. Transportation impacts would result from workers commuting to and from the Ross Project area;

visitors, such as regulatory agency personnel, travelling to and from the Project; from shipments to the Ross Project area of supplies, materials, and chemicals used during the uranium-recovery and milling processes; from shipments of other materials including uranium-loaded ion-exchange (IX) resins from future satellite areas within the Lance District (which are considered in SEIS Section 5, Cumulative Impacts) and/or other offsite ISR or waste-water treatment facilities (i.e., toll milling); and shipments of yellowcake and wastes from the Ross Project area to other, offsite facilities such as a uranium-conversion facility. Transportation impacts could also include increased fugitive dust that would be released during the increased traffic, increased traffic accidents, increased noise, and increased incidental wildlife or livestock mortalities, compared to current area conditions. Fugitive-dust impacts are evaluated as air-quality impacts and public and occupational health impacts in SEIS Sections 4.7 and 4.13, respectively. Noise impacts are evaluated in SEIS Section 4.8. Wildlife and livestock mortalities are evaluated as potential ecological impacts in SEIS Section 4.6. Detailed discussion of the other potential environmental impacts from Project-related transportation to and from the Ross Project area during construction, operation, aquifer restoration, and decommissioning is provided in the sections below.

4.3.1 Alternative 1: Proposed Action

Alternative 1, the Proposed Action, consists of four phases: construction, operation, aquifer restoration, and decommissioning of a uranium-recovery facility and wellfields. During the Proposed Action, transportation impacts for all phases of the Ross Project would result from the increased traffic on roads compared to current (2010) levels (see Figure 4.1); these traffic increases are summarized in Table 4.2.

Table 4.2
Estimated Number of Workers and Traffic Volumes for Ross Project

Project Phase	Average No. Daily Workers	Traffic	
		Passenger Vehicles per Day	Trucks per Day
Construction	200	400	24
Operation	60	120	16
Aquifer Restoration	20	40	12
Decommissioning and Site Restoration	90	180	10

Source: Strata, 2011a.

Note: Vehicle counts are to and from the Ross Project (two one-way trips per vehicle per day) and each assume that each worker would be in a separate passenger vehicle.

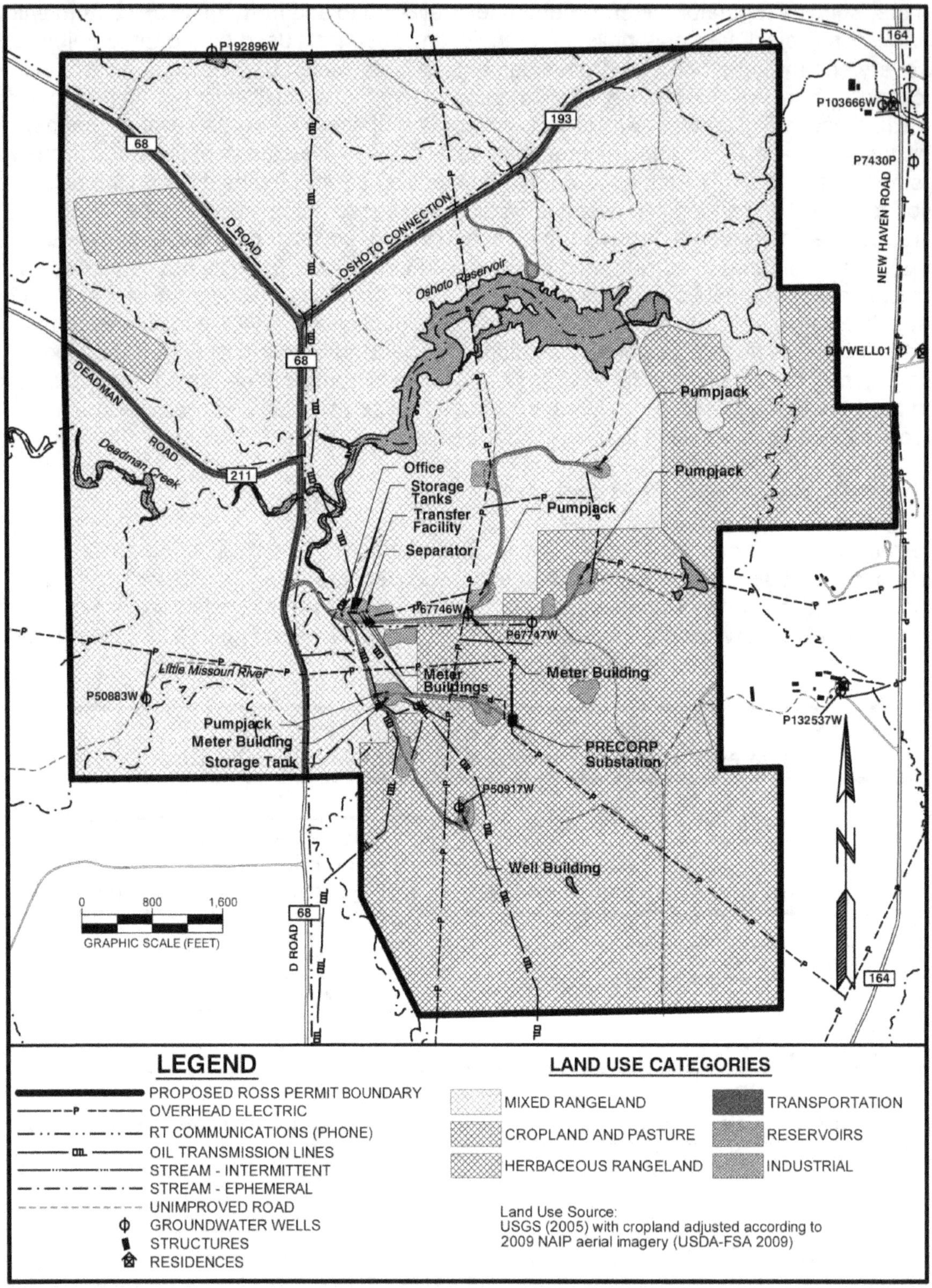

Source: Strata, 2012a.

Figure 4.1

Ross Project Design Components to be Decommissioned and Land Uses to be Restored

4.3.1.1 Ross Project Construction

As described in GEIS Section 4.4.2, the increase in daily traffic on most roads that would be used for construction-supply transport and workforce commutes would not be significant and, therefore, traffic-related impacts would be SMALL (NRC, 2009). Roads with the lowest average annual daily traffic volumes, such as local county roads, would have higher (i.e., MODERATE) potential impacts, particularly when the ISR facilities are experiencing peak employment (NRC, 2009). The limited duration of construction activities (i.e., 12 – 18 months), suggests that impacts would be of short duration in many areas where such a facility would be sited.

The highest traffic volumes resulting from the proposed Ross Project would occur during the construction phase of the Proposed Action because of the relatively large workforce (i.e., 200 persons) and the frequent material and equipment shipments. The increased traffic is expected to be 400 passenger cars and 24 trucks per day, which, compared to 2010 levels, represents a traffic increase of approximately 400 percent on the New Haven Road south of the Ross Project area, which would be the workers' primary route to the Project area (Strata, 2011a). This volume is higher than that assumed in the GEIS (NRC, 2009). This significant increase in traffic could result in more traffic accidents as well as wear and tear on the road surfaces. It is expected that additional road-maintenance activities would be needed. Due to the increased projected traffic volumes on the local and county roads between I-90 and the Ross Project area, the construction impacts would be MODERATE to LARGE with respect to the traffic levels and the road-surface wear and tear.

The increase in traffic on I-90 itself would be approximately 10 percent when compared to 2010 volumes. This increase to traffic on the interstate-highway system would be SMALL, and such impacts would mostly be related to increased traffic volume. However, the Interstate-highway system has been built to accommodate additional capacity and, therefore, the resulting impacts, if any, would be minor.

As noted above, traffic impacts to local roads are expected to be greatest during the Proposed Action's construction, and the Applicant identifies the following expected mitigation measures (Strata, 2011a):

- Improve signage, including speed-limit signs, on D and New Haven Roads.
- Implement a policy to enforce speed limits for Strata employees and contractors. The Applicant and Crook County have already executed a Memorandum of Understanding (MOU) that specifies the activities that Strata would undertake to assist with speed-limit controls, among other requirements (Strata, 2011d).
- Perform a safety analysis of the county roads where increased traffic would occur. Potential enhancements could include a decreased truck speed on D and New Haven Roads, or the assignment of "daytime headlight sections" to increase safety.
- Perform routine assessments of road conditions. The MOU between the Applicant and Crook County also includes a maintenance agreement to address road-maintenance needs.
- Explore a coalition with other companies operating heavy trucks on the county roads (e.g., the haulers of bentonite from the nearby mine) to provide additional assistance to Crook County for safety and maintenance needs.

- Work with Crook County to upgrade some portions of the roads by adding gravel to specially identified sections.
- Evaluate the feasibility of an employee carpooling program, or a park-and-ride system, in Gillette or Moorcroft. Alternatives could also include a van-pool system.

These mitigation measures would substantially reduce the transportation impacts associated with the Proposed Action's construction; with mitigation, the impacts of transportation would be SMALL to MODERATE.

4.3.1.2 Ross Project Operation

As discussed in GEIS Section 4.4.2, during the operation phase at an ISR facility, the facility-related traffic volume would be unlikely to generate any significant environmental impacts above those expected during the construction phase. Dust, noise, and possible incidental wildlife- or livestock-mortality impacts on or near a facility's access roads could continue to occur. The GEIS concluded that the potential impacts from transportation during facility operation could range from SMALL to MODERATE (NRC, 2009).

> **What are "best management practices"?**
> Best management practices (BMPs) are techniques, methods, processes, activities, or incentives that are effective at delivering a particular outcome. BMPs can also be defined as efficient and effective ways of meeting a given objective based on repeatable procedures that have proven themselves over time, such as specific standard operating procedures (SOPs). Well-designed BMPs combine existing managerial and scientific knowledge with knowledge about the resource being protected. The Wyoming Department of Environmental Quality (WDEQ) defines best practicable technology as a "technology-based process determined by WDEQ as justifiable in terms of existing performance and achievability (in relation to health and safety) which minimizes, to the extent safe and practicable, disturbances and adverse impacts of the operation on human or animal life, fish, wildlife, plant life and related environmental values." (WDEQ, 2007, as cited in NRC, 2009b).

The GEIS also assessed the potential for accidents and their consequences when the accidents involve the transportation of hazardous chemicals and radioactive materials. The GEIS recognized the potential for high consequences from a severe accident involving transportation of hazardous chemicals in a populated area. The GEIS stated that the probability of such accidents is low because of the small number of shipments, comprehensive regulatory controls, and the ISR facility operator's use of best management practices (BMP). For radioactive material shipments (for example, yellowcake product, loaded IX resins, or radioactive wastes), compliance with transportation regulations would be expected to limit radiological risk during normal ISR operations. The GEIS concluded there would be a low radiological risk in the unlikely event of an accident. The use of emergency-response protocols would help to mitigate the consequences of severe accidents that involve the release of radioactive materials. This SEIS reviews the radiological consequences of such accidents in Section 4.13.1 (NRC, 2009).

During the operation phase, increased traffic over that in 2010 would be present due to employee traffic; shipments of process chemicals, loaded IX resins, yellowcake, and vanadium; and shipments of solid, hazardous, and radioactive wastes to and from the CPP and/or wellfields. These shipments are included in the truck count in Table 4.2. Potential impacts to other resources could again occur during uranium-recovery operation, as discussed earlier. Impacts to local roads would be less significant during operation than during construction due to the lower traffic associated with facility and wellfield operation, although the traffic on these roads would still be double that in 2010 (Strata, 2011a). In total, the increase in anticipated traffic during the Ross Project's operation phase is significant when compared to current levels, although there are low and manageable risks associated with yellowcake, process-chemical, and waste transportation. Consequently, the transportation impacts during the operation phase would be less significant than during construction and would nonetheless be SMALL to LARGE. However, the magnitude of these impacts would be mitigated by the same measures used during the construction phase. Thus, with mitigation, transportation impacts would be SMALL to MODERATE.

GEIS Section 4.2.2.2 as cited by GEIS Section 4.4.2.2 evaluated yellowcake transportation, and assumed shipment volumes would range from 34 – 145 yellowcake shipments per year. The Applicant estimates that there would be 75 shipments of yellowcake per year from the Ross Project based on the maximum annual production rate (i.e. including yellowcake produced from toll milling), which is within the range of the GEIS analysis (Strata, 2011a). The GEIS indicated that 145 yellowcake shipments per year from a single ISR facility could result in 0.04 and 0.003 cancer deaths per year, depending on the amount of yellowcake released during a transportation accident (NRC, 2009). To minimize the risk of an accident involving yellowcake transport associated with the Proposed Action, the material would be transported in accordance with U.S. Department of Transportation (USDOT), Wyoming Department of Transportation (WYDOT), and NRC regulations, managed as a "low-specific activity" (LSA) material, and shipped on exclusive-use vehicles. Only properly licensed and trained drivers would transport LSA materials. Should a transportation accident occur, the NRC concluded that the consequences of such accidents would be limited because the Applicant would develop emergency-response protocols for yellowcake and other transportation accidents. Also, shipping companies would ensure their personnel receive proper emergency-response training. Emergency-response protocols would include communication equipment and emergency-spill cleanup kits on each vehicle and at the shipping and receiving facilities (Strata, 2011a). Based on this analysis, the impacts due to a potential accident involving the transportation of yellowcake during the operations phase of the proposed Ross Project would be SMALL.

The Applicant estimates that approximately four bulk-chemical, fuel, and other supply and material deliveries would be made per day throughout the operation phase of the Proposed Action (Strata, 2011a). This number of shipments is greater than the daily number of chemical-supply shipments considered in GEIS Section 4.4.2 (estimated at approximately one per day); however, these shipments would be made in accordance with the applicable USDOT hazardous-materials-shipping requirements and spill response would be similar to the response for yellowcake shipments. The Applicant conducted an analysis, using the injury rate of 4.3×10^{-7} per mile, to determine the risk of an injury to a member of the general public that could result from a transportation accident involving the shipment of anhydrous ammonia. The applicant found that these shipments could result in 0.002 injuries per year. The NRC staff reviewed the Applicant's analysis and verified that reasonable input parameters were used. Chemical shipments would be conducted safely and the probability of an accident involving these shipments would be SMALL. As described in GEIS Section 4.4.2.2 and 4.2.2.2, the likelihood of an incident in a populated area would be small, given the precautions that would be taken with hazardous chemical shipments. Therefore, the potential environmental impacts of accidents involving chemical transportation during Ross Project operations would be SMALL.

The CPP is designed to process more yellowcake than is expected to be recovered at the proposed Ross Project (Strata, 2011a). The Applicant indicates that it proposes to accept uranium-loaded IX resins from other ISR operations as well as, potentially, those from offsite domestic-sewage facilities as noted in SEIS Section 2.1.1. The Applicant would expect to receive four shipments of resin per day. GEIS Section 4.2.2.2 as cited in GEIS Section 4.4.2.2 concluded that the potential radiological impacts of IX resins shipments would be lower than the risks from yellowcake shipments based on the less concentrated nature of the resins; the uranium being chemically bound to the resins, which would limit dispersion in the event of a spill; and the small transport distance relative to yellowcake shipments. Although the number of shipments proposed by the Applicant is higher than the one truck per day assumed in the GEIS, the other three factors evaluated in the GEIS would ensure that the probability of an accident that involves uranium-loaded IX resins would be small. Compliance with the applicable NRC and USDOT regulations for shipping IX resins would also reduce the risk of accidents involving these shipments. Therefore, the environmental impacts of accidents involving shipments of IX resins during Ross Project operations would be SMALL

The vanadium extracted by the Applicant in the CPP's vanadium circuit is considered a hazardous material by USDOT and would be shipped in sealed transport vehicles to an offsite processing facility (see SEIS Section 2.1.1) (Strata, 2011a) in accordance with USDOT regulations. It is anticipated that there would be 45 shipments of vanadium from the Ross Project each year. Due to the low number of shipments, the probability of an accident involving vanadium shipments would also be small. Because of the less hazardous nature of vanadium as compared with yellowcake, the environmental impacts of accidents involving shipments of vanadium would be SMALL.

The operation of the Proposed Action would also generate radioactive wastes. These would be shipped in 208-L [55-gal] drums inside sealed roll-off containers in accordance with applicable USDOT regulations. Only five such waste shipments are anticipated during a year; given the infrequent nature of these shipments, they do not represent a significant impact to local traffic conditions or a significant increased risk of accidents. Thus, the impacts of the shipment of radioactive wastes to traffic would be SMALL. Other solid wastes would be transported to a local municipal landfill in Moorcroft, Sundance, and/or Gillette. The Applicant estimates that one trip per week would be required to remove solid waste from the Ross Project. This number would represent a SMALL impact to the local roads, both in terms of traffic volume and impacts to local road maintenance. Finally, the Applicant anticipates that there would be one shipment of hazardous wastes from the Ross Project each month. The hazardous waste is expected to include used oil, oil-contaminated soil, oily rags, used batteries, expired laboratory reagents, fluorescent light bulbs, spent solvents, and degreasers. Given the low number of shipments, this represents a SMALL impact to the local traffic and the local roads and would have SMALL environmental impacts in the case of an accident due to the small volumes generated at the Ross Project.

To mitigate transportation impacts, many of the mitigation measures instituted during the Ross Project's construction would continue during operation. Additional mitigation measures would be implemented for the shipment of materials, such as yellowcake, uranium-loaded IX resins, and vanadium as well as solid, hazardous, and radioactive wastes. Two mitigation measures that would address all such shipments would be 1) coordination with local emergency-response personnel, and 2) the requirement that only appropriately licensed transporters would be used. The Applicant would develop a protocol, or a SOP, to provide ongoing training to local emergency-response personnel, including EMTs, firefighters, and municipal and county law-enforcement personnel. For each type of material, specific information would be provided about the physical and chemical characteristics of the substances being shipped, the related hazards, the potential exposure pathways, and appropriate spill-response, containment, and cleanup procedures. This training would be ongoing and would include updates on a routine schedule or as new substances are transported to or from the Ross Project. All shipments would be made by appropriately licensed transporters in accordance with USDOT and WYDOT hazardous-material regulations and requirements.

The release of a radioactive material as a result of a transportation-related incident would prompt the activities described in USDOT's hazardous-materials regulations at 49 *Code of Federal Regulations* (CFR) 171, Subpart B, "Incident Reporting [and] Notification." Among other activities, these regulations require immediate notice of certain incidents, preparation of detailed incident reports, submission of examination reports, and assistance with investigations and special studies. Should an accident occur that results in a release of any yellowcake or other radioactive materials to the environment, the Applicant would perform a post-cleanup radiological survey of the affected area to ensure that there are no long-term hazards associated with the released radioactive material or of spill-response and cleanup activities.

4.3.1.3 Ross Project Aquifer Restoration

As discussed in GEIS Section 4.4.2, the potential transportation impacts during aquifer restoration would be equal to or less than the potential impacts during ISR facility operation (NRC, 2009). At the Ross Project, the number of uranium-recovery workers, and therefore the number of personal vehicles, would decline significantly during aquifer restoration from the construction and operation phases (from 200 to 60 to 20 workers). Thus, the potential transportation impacts discussed above for the Ross Project's construction and operation would be reduced due to the anticipated smaller traffic volume during this phase of the Project.

Yellowcake, vanadium, and uranium-loaded IX-resin shipments could remain the same if the CPP continues to process uranium-loaded IX resins during the Ross Project wellfield's aquifer restoration. The shipments of process chemicals would similarly depend upon whether the CPP would continue to process loaded resins after the Ross Project's wellfields are no longer engaged in uranium recovery. Should the CPP continue to process loaded IX resins, there would not be a reduction in worker commuting as discussed above.

However, the impacts would be similar to those during uranium-recovery operation at the Ross Project, and these would be expected to be SMALL to MODERATE due to the workforce of 60 or 20 workers. Mitigation measures implemented during aquifer restoration at the Ross Project would be identical to those implemented during its construction and operation phases.

4.3.1.4 Ross Project Decommissioning

During ISR facility decommissioning, the GEIS concluded that transportation impacts as a result of worker commutes would steadily decrease, but initially there would be a large increase in decommissioning-phase workers. GEIS Section 4.4.2 also concluded that, based on the concentrated nature of yellowcake when shipped, the longer distance of the yellowcake shipments when compared to waste shipments, and the number of shipments when compared to byproduct waste shipments, the potential radiological risks from transportation accidents involving byproduct waste shipments during decommissioning would be bounded by the yellowcake transportation risks during operations. Overall, according to GEIS Section 4.4.2, transportation impacts would be SMALL (NRC, 2009).

During the decommissioning phase of the Ross Project, the Applicant expects that the workforce would initially increase to approximately 90 workers (up from 20 workers during aquifer restoration). Traffic on the local roads would thus increase over that of the aquifer-restoration phase, but it would still be less than half of that expected during the Proposed Action's construction phase. Fuel shipments would increase due to the operation of heavy equipment during decommissioning activities. Little or no yellowcake or vanadium would be shipped during the decommissioning phase; however, Project decommissioning would result in an increase in shipments of radioactive and other solid wastes. The Applicant estimates that the frequency of radioactive-waste shipments would increase from the approximately 5 per year during the operation and aquifer-restoration phases, to between 100 – 200 shipments per year during the decommissioning phase (Strata, 2011a). These shipments would still be relatively infrequent compared to passenger vehicular traffic, and they would have only a small impact on traffic volume. Solid-waste shipments are expected to increase from approximately one per week during operation and aquifer restoration to about two per week during decommissioning. Hazardous-waste shipments are expected to remain unchanged at approximately one per month throughout all four Proposed Action phases.

As anticipated in the GEIS, the potential radiological risks associated with transportation accidents involving byproduct waste shipments during decommissioning at the Ross Project would be bounded by the risks associated with transporting yellowcake during operations. The GEIS assumed that the distance between the yellowcake conversion facility and the proposed project would be greater than the distance between the waste disposal facility and the proposed project. Consistent with the GEIS, the distance from the Ross Project area to the conversion facility that would accept the yellowcake is 2,029 km [1,260 mi] whereas the byproduct waste would travel between 378 km [235 mi] to 1,610 km [1,000 mi] to a disposal facility. The GEIS also assumed that there would be up to 145 yellowcake shipments per year and 300 total byproduct material shipments during decommissioning (based on 4,593 m^3 [6,008 yd^3] of byproduct material generated during decommissioning and each shipment containing 15 m^3 [20 yd^3] of byproduct material), which would result in more yellowcake shipments than byproduct material shipments overall. The Applicant estimates that there would be 75 shipments of yellowcake per year during operations and 3,823 m^3 [5,000 yd^3] of byproduct material generated during decommissioning (250 total shipments of byproduct material during decommissioning), which would also result in more yellowcake shipments than byproduct material shipments overall.

Potential transportation impacts would be less during decommissioning than those occurring during construction; however, they would be still be SMALL to MODERATE due to the increased workforce required for decommissioning (approximately 90 workers). Mitigation measures implemented during the Proposed Action's decommissioning would be identical to those that would be implemented during all of the other phases of the Ross Project.

4.3.2 Alternative 2: No Action

Under the No-Action Alternative, the Ross Project would not be licensed and the land would continue to be available for other uses. However, traffic volumes and patterns would likely increase from the 2010 pre-licensing baseline conditions noted in SEIS Section 3.3 because additional residences could be expected to be built near the Ross Project over time. The Applicant has projected that volumes would increase approximately 2 percent per year, even without the Ross Project's construction and operation (Strata, 2011a). There would be no transportation of materials of any kind to or from the Ross Project to support uranium-recovery activities. There would be no transportation of either radioactive or solid wastes from the Proposed Action because the Ross Project would neither be licensed nor constructed and operated. The current transportation activities to support ongoing oil production and bentonite mining would be the same. In addition, the Applicant would continue with some preconstruction activities, such as abandonment of exploration drillholes and collection of environmental data. These activities are similar to those currently occurring at the Ross Project area, and, although short-term increases in activity could occur, these impacts would be SMALL.

4.3.3 Alternative 3: North Ross Project

Under Alternative 3, the North Ross Project would generally be the same as the Proposed Action, except that the facility (i.e., the CPP, associated buildings, and auxiliary structures as well as the surface impoundments) would be located to the north of where it would be located in the Proposed Action, as described in SEIS Section 2.1.3. This change in facility location would cause a change in the impacts to local roads as compared to current conditions, because additional roads would be used that would not be used during the Proposed Action at the south site—most notably, the Oshoto Connection and the D Road north of the D Road/New Haven Road intersection (see Figure 2.1 in SEIS Section 2). There would likely be less localized impact to the New Haven Road, as it is anticipated that the majority of the traffic from the Proposed Action would access the Ross Project area by travelling D Road to the Oshoto Connection (Strata, 2011a). Although this change would minimize impacts to the New Haven Road, it would nevertheless cause a corresponding increase in impacts to the D Road and the Oshoto Connection as both roads are similarly constructed and maintained. Since the total traffic counts would remain the same during all phases of Alternative 3 as those for the Proposed Action, the transportation impacts would be the same as those described earlier for Alternative 1, SMALL to LARGE. As the same mitigation measures discussed for the Proposed Action would be employed for Alternative 3, the resulting transportation impacts would be SMALL to MODERATE.

4.4 Geology and Soils

4.4.1 Alternative 1: Proposed Action

Alternative 1, the Proposed Action, consists of four phases: construction, operation, aquifer restoration, and decommissioning of a uranium-recovery facility and wellfields.

4.4.1.1 Ross Project Construction

As described in GEIS Section 4.2.3 and 4.4.3, the principal impacts to geology and soil during construction would result from disturbance of soil and surficial bedrock by construction activities. These activities include the Applicant's clearing ground or topsoils, eliminating the vegetation that is present; cutting, filling, and grading the ground surface, preparing it for the construction of the CPP, surface impoundments, access roads, utility corridors, and wellfields; excavating and backfilling trenches for pipelines and other subsurface design components; and excavating the mud pits, CBW, and flood-control diversion channel (NRC, 2009). As the GEIS noted, the impacts on geology and soils from construction activities depend upon local topography, surface bedrock geology (i.e., the rock immediately below the soil), and soil characteristics. The GEIS concluded that, with the implementation of appropriate BMPs, the impacts on geology and soils would be SMALL, if less than 15 percent of an ISR project's area would be affected. As described earlier in SEIS Section 4.2, approximately 113 ha [280 ac] of land, or about 16 percent of the Ross Project area, would be disturbed during the lifecycle of the project (Strata, 2011b). This area is slightly larger than that identified in the GEIS; thus, a site-specific analysis is provided here.

Geology

Construction activities are not expected to encounter bedrock, except for localized impacts to the surficial bedrock by construction of the CBW. The wall would be a 0.7-m- [2-ft] wide barrier of a soil-bentonite mixture extending from the surface to at least 0.7 m- [2 ft-] into bedrock. The impacts of the CBW's construction would be SMALL, due to the relatively small and localized effects on the bedrock below it.

The impacts from the Applicant's drilling and developing injection, recovery, and monitoring wells as well as installing the deep-injection wells are discussed in SEIS Section 4.5.

Soils

The impacts on soils would occur largely during the construction phase of the Proposed Action, when most of the ground disturbance takes place. Potential soils impacts include soils loss (by wind and water erosion), soils compaction, increased salinity, soils-productivity loss, and soils contamination. Surface-disturbing activities would expose the soils and subsoils at the Ross Project area and would temporarily increase the potential for soil loss because of wind and water erosion. As described in SEIS Section 3.4.2, the soils in the Ross Project area have a moderate to severe potential to be affected by wind erosion. One soil type, Vona fine sandy loam—which makes up less than 3 percent of the entire Ross Project area—has a severe potential for wind erosion. Water-erosion hazards range from negligible to moderate for the soil types found within the Ross Project area.

Soils at the Ross Project also have the potential to become compacted, particularly during construction activities where heavy equipment is being operated. Soil compaction could result in a decrease in water infiltration, thereby increasing runoff. To decrease the potential for compaction, existing roads would be used where possible; secondary access-road widths would be minimized, and a one-way-in/one-way-out policy would be implemented by the Applicant to access wellfields. Compacted soils would be further addressed in the decommissioning plan (DP) that the Applicant would be required to submit to the NRC (Strata, 2011a).

During preconstruction activities, the Applicant has been employing various methods of soil reclamation, according to landowner preference. These methods have included Strata's "ripping" compacted soil with the teeth of a grader, loosening compacted soil with a disk, or simply replacing topsoil and reseeding. These techniques would continue to be refined and coordinated with WDEQ/LQD and the respective landowners during the Proposed Action.

Saline soils are very susceptible to soil loss. Saline soils were not found on the Ross Project during the Applicant's soil surveys. However, the use of magnesium chloride for dust control could increase the salinity of the local soils (Strata, 2011a). If magnesium chloride were to be used on access roads for fugitive-dust control or if a salt and sand mixture were to be used for traction on primary access roads during the winter, the Applicant would sample the soils beneath and adjacent to access roads for salinity during the Proposed Action's decommissioning phase. Any salt-impacted soils would be removed at that time.

Loss of topsoil and disturbance of soils could affect the soils' structure and microbial activity. In turn, these changes could reduce soil productivity. Based upon the total anticipated disturbance area of 113 ha [280 ac] and the average topsoil depth of 0.53 m [1.7 ft], the volume of topsoil stockpiled during the life of the Proposed Action is estimated to be up to approximately 600,000 m^3 [800,000 yd^3] (Strata, 2011b). This estimate could be conservatively high because most of the wellfields and access roads would be located outside of the 100-year floodplain at the Ross Project area, where topsoil would be thinner than average. This volume of topsoil would generally not be removed from access roads to and from the wellfields, and much of the topsoil would be replaced promptly after removal for pipeline and utility-corridor trenching.

To mitigate the potential loss of top soil as well as soil productivity, topsoils would be salvaged and stockpiled for wellfield-decommissioning and site-restoration activities. Sequential wellfield decommissioning is anticipated by the Applicant; once a wellfield is depleted, it would be decommissioned and the field's wells properly abandoned. This decommissioning would occur as each wellfield is taken out of service; it would not be delayed until the end of the entire Ross Project's lifecycle.

The Applicant proposes to locate a relatively large topsoil stockpile near the CPP (see Figure 2.5 in SEIS Section 2.1.1) (Strata, 2011a). Any topsoils that are stripped before the construction of roads and drilling pads in the wellfields would be stockpiled in nearby piles, typically spaced approximately 600 m [2,000 ft] apart along access roads to minimize the soil compaction, fugitive dust, combustion gases, and noise associated with long topsoil hauls.

Related mitigation measures designed to minimize soil loss, and to diminish fugitive dust (see SEIS Section 4.7.1.1) would include the Applicant: 1) constructing topsoil stockpiles on the leeward side of hills, where possible; 2) constructing topsoil stockpiles away from ephemeral-stream channels or any other flood-prone areas; 3) avoiding construction within areas susceptible to flooding; 4) minimizing the disturbance of surface-water drainages (i.e., roads and pipelines would cross drainages perpendicular to the flow direction [as described in SEIS Section 3.4.2]); 5) wetting exposed soils during construction to minimize soil loss from wind erosion; 6) employing sediment-control BMPs, such as silt fences, sediment logs, and straw-bale check dams in all disturbed areas; 7) implementing additional sediment-control BMPs for topsoil stockpiles, including seeding and installing a perimeter ditch and water-collection sump to trap storm water and sediment; and 8) restoring and reseeding disturbed areas as quickly as possible, typically within a single construction season (Strata, 2011a; WDEQ/LQD, 2005). Many of these BMPs are consistent with those identified by NRC in the GEIS in Section 7.4 and are commonly used at other ISR facilities (Strata, 2011a; NRC, 2009b).

To minimize soil-productivity impacts, the Applicant would use corresponding BMPs including several of the mitigation measures identified above to prevent soil loss. These BMPs include the Applicant 1) protecting topsoil stockpiles from wind and water erosion; 2) seeding topsoil stockpiles during inactive periods with an appropriate perennial seed mix; 3) redistributing topsoil and applying a permanent seed mix approved by WDEQ/LQD during the Proposed

Action's decommissioning phase; and 4) using information gathered from reference areas over the long term to perform statistical, quantitative, and qualitative comparisons approved by WDEQ/LQD.

Although the subsurface would be exposed during the Applicant's excavation of mud pits and pipeline trenches, the primary area of subsoil disturbance would be where the CPP and surface impoundments are to be constructed. The subsoils there would be disturbed by the cut, fill, and grading activities necessary to create a relatively level site and by the excavations for the surface impoundments, CBW, and flood-control diversion channel. The quantity of excess subsoils generated from construction of the CPP and surface-impoundment area is estimated to be approximately 60,000 m^3 [80,000 yd^3]. This material could be used to provide a slightly elevated and relatively level primary access road, or it could be stored in a subsoil stockpile separate from the topsoil stockpiles.

During the Proposed Action's construction, additional potential soil impacts could occur from the introduction of drilling fluids and muds to the soils near the recovery, injection, and monitoring wells. However, the volume of these drilling fluids would be small, and these fluids and muds would be contained within the mud pits excavated near each drillhole's drilling pad. Other potential soil impacts could also occur from spills and leaks of fuel or lubricants from heavy-construction equipment and passenger vehicles that would be operated during construction of the Ross Project. However, such spills and leaks would be contained and cleaned up immediately if they were to occur. Oil- or lubricant-contaminated soil would be disposed offsite in an appropriately permitted facility.

During construction, up to five Class I deep-injection wells would be installed in aquifers approximately 2,000 m [6,400 ft] below ground surface (bgs). These wells would be used for the disposal of process solutions, including brine and excess permeate. The Applicant's drilling of these wells and their completion and testing would be governed by the applicable Underground Injection Control (UIC) Class I Permit from WDEQ (WDEQ/WQD, 2011). Thus, the surface and subsurface area disturbed by these particular wells would be very limited.

Therefore, the potential impacts of the Proposed Action's construction to soils would be SMALL.

4.4.1.2 Ross Project Operation

As described in GEIS Section 4.4.3, the potential impacts to geology and soils during the operation of an ISR facility could include: soil loss due to surface-water runoff and erosion; soil compaction as described above; increased soil salinity due to the use of magnesium chloride for dust control; soil contamination caused by spills and leaks of lixiviant, as the solution moves through pipelines between the wellfields and the CPP; transportation accidents, which could involve liquids; other accidental spills and leaks associated with waste management; and changes to the uranium-bearing formations as a result of the disposal of brine and other liquid byproduct wastes in UIC Class I deep-injection wells. The GEIS concluded that the impacts on geology and soils from an ISR operation would be SMALL.

<u>Geology</u>

During uranium-recovery operation, the lixiviant dissolves the uranium-mineral coatings on the sandstones in the targeted ore zone; this geochemical change in the rock would result in mineralogical changes to the ore zone, but it would not affect the rock matrix nor rock structure. The thickness and depth of the ore zone at the Ross Project are similar to the ore zones evaluated in the GEIS (NRC, 2009). The GEIS concluded that it is unlikely that geochemical alteration of the ore zone would result in any compression or subsidence that would be translated to the ground surface.

Based upon historical uranium-recovery operations in the NSDWUMR, reactivation of geologic faults would not be anticipated (NRC, 2009b; Strata, 2011b). As established in SEIS Section 3.4.4, earthquake activity in the area of the Ross Project is very low. Potential impacts associated with increased earthquake risk because of the operation of injection wells would be avoided by Applicant's maintaining the injection pressure at a level that does not exceed the fracture pressure of the receiving rock formation, as specified in the WDEQ/Water Quality Division (WQD) permit. See SEIS Section 2.1 for a related discussion of how in situ uranium recovery is different than hydrofracking.

The potential impacts from the operation of the Proposed Action to Ross Site geology would be SMALL.

Soils

During the operation of the Proposed Action, potential impacts from soil loss would be minimized by proper design and operation of surface-runoff features and implementation of BMPs, as described for those during construction. Soil compaction would be minimal during the Proposed Action's operation, due to low density of roads across the Ross Project area. Mitigation measures to minimize soil compaction and to diminish increases in soils salinity would be the same as those identified for the construction phase of the Proposed Action. The potential for a release of yellowcake or IX resin during a transportation accident has been determined by NRC to be small; however, the magnitude of the impacts of this type of accident is described in SEIS Section 4.2 (NRC, 2009).

In the event of releases of process solutions from pipelines, module buildings, process vessels, or surface impoundments, the process-control system described in SEIS Section 2.1.1.2 would quickly alert an operator, who could then take action including a full shutdown of the leaking components as well as initiating immediate containment and cleanup. As noted in the GEIS, during 1996, the operator of the Crow Butte Uranium Project in Dawes County, Nebraska, logged 27 spill incidents of process solutions, with volumes ranging from 45 – 65,000 L [12 – 17,305 gal] (NRC, 2009). This potential for soil contamination at the Ross Project would be minimized by the Applicant: 1) adhering to the NRC and WDEQ design criteria for uranium-recovery facilities; 2) designing successful spill-containment and leak-detection systems; 3) training employees on monitoring process parameters and recognizing potential upset conditions before spills or leaks occur; 4) training employees on inspection SOPs, spill-control BMPs, and a storm water pollution prevention plan (SWPPP); 5) frequently inspecting waste-management systems and effluent-control systems; and 6) training all employees on spill detection, containment, and cleanup procedures (Strata, 2011a). Additional information on the excursion-monitoring and spill-detection systems incorporated into the design of the Ross Project is presented in SEIS Section 2.1.1 of this SEIS.

The design criteria for the Proposed Action include leak-detection capability in each wellfield module building, where an alarm inside the CPP would signal the on-duty operator that a spill has occurred. (The CPP would be staffed 24 hours a day.) In addition, routine, weekly inspections of wellfield module buildings and wellheads would be conducted by Strata personnel. Such inspections would ensure that all piping and equipment, wellheads, and valve manholes are visually inspected (Strata, 2011b). Other wellfield leak-detection monitoring and control measures would include the continuous measurement of flows and pressures for injection and recovery trunk lines and feeder lines as well as the presence of leak-detection sensors in valve manholes and in the protective box around each wellhead. In addition, all pipelines would have been hydrostatically tested before they were buried, and the Applicant would institute a monitoring program for leaks and other abnormalities as required by the NRC license (Strata, 2011b).

To minimize the potential for subsurface pipeline leaks, the WDEQ/WQD requirements for potable-water stream crossings would be incorporated into the design and construction of all pipeline stream crossings. These requirements include the Applicant: 1) providing a minimum of 0.6 m [2 ft] of soil cover (at the Ross Project, 1.2 – 1.8 m [4 – 6 ft] would typically be used) over the respective pipelines to guard against damage from livestock and to protect them from freezing; 2) using pipes with flexible, watertight joints, such as polyvinyl chloride (PVC) or high-density polyethylene (HDPE); and 3) installing accessible isolation valves at both ends of water crossings so that the section could be isolated for testing or repair.

Two levels of engineering controls would also minimize potential impacts to soils from the unintended release of process solutions within the CPP itself. The first level of protection is the primary containment accomplished by pipelines, vessels, and surface impoundments, all of which would be tested for leaks during construction. The second level of protection is the secondary containment that is provided by curbs, berms, and sumps for all chemical-storage tanks, process vessels, and all piping and equipment inside the CPP building (Strata, 2011a).

The design and operation of the surface impoundments would also minimize the likelihood of liquid releases. The surface impoundments would include a double-liner and leak-detection system, and they would be operated so as to maintain sufficient reserve capacity to permit the Applicant to transfer the contents of a surface-impoundment cell to another in the event of a leak, in order to facilitate repair or replacement. To minimize the likelihood of releases, impoundment embankments would be monitored and inspected weekly by the Applicant in accordance with NRC-approved inspection protocols (Strata, 2011a).

Further, to minimize the potential impacts of soil contamination, such as short-term, elevated concentrations of radiological parameters and other associated chemical constituents above baseline levels, the Applicant would be required to establish immediate spill detection, response, containment, and cleanup protocols and SOPs (NRC, 2009) by its NRC license. For example, immediate spill response could include the Applicant shutting down the leaking pipeline, recovering as much of the spilled fluid as possible, and collecting samples of the impacted soils for comparison of constituent-concentration values (e.g., uranium, radium, and other indicators) to baseline conditions. Soils contaminated by spills or leaks would be removed in accordance with Criteria 6(6) of Appendix A to 10 CFR Part 40, which requires that soil concentrations not exceed background concentrations by more than 0.2 Bq/g [5 pCi/g] of radium-226, averaged over the first 15 cm [5.9 in] below the ground's surface. Analytical tests would be required to demonstrate that no such residual contamination exists. Baseline concentrations have been established by the Applicant through its pre-licensing, site-characterization monitoring program (see SEIS Section 3.12.1), and additional determination of background values would have been established by a post-licensing, pre-operational monitoring program prior to major Ross Project construction. Soils contaminated by spills or leaks would be properly disposed of at an offsite properly licensed and permitted disposal facility (Strata, 2011a).

The NRC's monitoring requirements would specify that licensees must report designated types and volumes spills to the NRC within 24 hours (NRC, 2009). These spills include those that cause unplanned contamination that meets the criteria at 10 CFR Part 40.60 as well as those spills that could cause public or occupational exposures that exceed the limits established in 10 CFR Part 20, Subpart M (see SEIS Section 4.13). Additional reporting requirements could be imposed by the State or by NRC license conditions. The spill response requirements would be defined in the NRC license. All of these spill-response protocols would be implemented if other liquid radioactive or chemical materials or wastes, or if solid radiologically and/or chemically contaminated materials or wastes, were to be spilled or dispersed.

Potential impacts to the soils at the Ross Project would be mitigated by the Applicant's implementation of BMPs and other spill-related procedures, plans, and programs that would be required in the NRC license. As noted above, all contaminated soils and sediments would be removed and disposed of according to the requirements of the 10 CFR Part 40, Appendix A. These mitigation measures would substantially minimize the impacts to the soils and sediments of the Ross Project area; these impacts would be SMALL.

4.4.1.3 Ross Project Aquifer Restoration

As described in GEIS Section 4.4.3, aquifer restoration would impact the geology of the deep-injection aquifers similarly to the operation of an ISR facility. With respect to ore-zone and soils, the potential for accidental spills and leaks would be similar, but less than, those described for the operation phase. Lixiviant would not be used during aquifer restoration so there would not be potential impacts to geology from dissolution of uranium and other constituents from the ore zone. As the quality of ground water from the exempted aquifer improves during restoration, the potential impacts of process-solution spills or leaks from pipes and pumps would decrease compared to potential impacts during operations. The GEIS determined that the potential impacts to geology and soils would be SMALL.

The potential impacts to Ross Project geology and soils associated with aquifer restoration at the Ross Project would be similar, but less, than those associated with its operation. The relative magnitude of impacts would be less because the concentrations of radionuclides, metals, and TDS in the water moving through the pipes, pipelines, and injection and recovery wells would be lower during aquifer restoration than during uranium-recovery operation. Also, there would less transport of uranium-bearing solutions and fewer shipments of yellowcake or vanadium; thus, less potential for spills and leaks than during operation. As previously described for the operation phase of the Ross Project, impacts to soils resulting from spills would be concentrations of radionuclides and other chemical constituents above established baseline or background values, but these elevated concentrations would be eliminated upon spill cleanup. Thus, the potential impacts of the Proposed Action's aquifer restoration to geology and soils would be SMALL.

4.4.1.4 Ross Project Decommissioning

GEIS Section 4.4.3 described the activities associated with the decommissioning of an ISR facility, including decontamination of surfaces, dismantling of process components and associated structures, demolishing buildings and other structures, removal of buried pipelines, and plugging and abandonment of wells and wellfield components (NRC, 2009). The GEIS determined that most of the impacts to geology and soils during the decommissioning phase would be short-term and SMALL. In fact, because the goal of decommissioning and site restoration is to restore, to the extent practical, the environment to preconstruction conditions through activities such as redistributing, seeding, and contouring soil that would have been stockpiled during the earlier phases of the Ross Project, the overall long-term impacts to geology and soils would be SMALL (NRC, 2009).

<u>Geology</u>

The potential impacts to the geology of the ore zone at the Proposed Action would depend upon the density of plugged and abandoned drillholes and wells. At the end of the life of the Ross Project, the wellfields (whether recently operated or decommissioned some time ago) would contain approximately 3,000 drillholes and wells; these would include those drillholes from Strata's ore-zone delineation efforts and geotechnical investigations, ground-water monitoring wells used for site characterization, the injection and recovery wells from uranium-recovery activities, and Nubeth Joint Venture (Nubeth) drillholes and wells. This would represent an average density of approximately 4.3 wells/ha [1.7 wells/ac]. All of these drillholes and wells would be properly abandoned by the Applicant with concrete or a similar material. Each drillhole and well would be required to be filled with a concrete plug up to 15 cm [6 in] in diameter, through the entire depth of the drillhole or well (WDEQ/LQD, 2005). The density of these concrete plugs is not great enough to alter the geology of the ore zone nor the surrounding stratigraphy. As described in SEIS Section 2.1.1.1, well-abandonment records would be maintained onsite at the Ross Project until termination of its NRC license. The impacts to ground water from improperly abandoned drillholes and wells are discussed in SEIS Section 4.5.
Potential impacts to the geology of the deep-injection aquifers (i.e., the Flathead and Deadwood Formations) would also be similar, but have less magnitude, than during the aquifer-restoration phase, because there would be only minimal volumes (less than 38 L/min [10 gal/min]) of liquid byproduct wastes injected into the Class I deep-disposal wells during the decommissioning phase.

The surficial bedrock would be affected locally by the actions necessary to breach the CBW to re-establish aquifer flow. The potential impacts from these relatively small and local effects on bedrock beneath the CBW would be SMALL as would all impacts related to geology.

<u>Soils</u>

The potential impacts to Ross Project area soils during the decommissioning of the Proposed Action would result from activities associated with land reclamation and site restoration, including the excavation and cleanup of contaminated soils. These decommissioning impacts would be similar to those resulting from construction of the Proposed Action. The BMPs, SOPs, and other mitigation measures described earlier for the construction and operation phases would continue to be implemented. Thus, the potential impacts from decommissioning activities to the local soils would be SMALL.

4.4.2 Alternative 2: No Action

Under the No-Action Alternative, the Ross Project would not be licensed and the land would continue to be available for other uses. However, until that decision is made by the NRC, the impacts of soils compaction and soils loss by heavy equipment and vehicular traffic across the Ross Project area could occur during the Applicant's continuing conduct of: 1) different types of surveys (e.g., continued ecological surveys); 2) boring of exploration and geotechnical drillholes; 3) drilling and monitoring of all types of ground-water wells; 4) locating and abandoning Nubeth drillholes and wells; and 5) installing and observing surface-water and meteorological monitoring stations.

As of August 2011, the Applicant had drilled and then plugged approximately 612 holes it installed during site characterization, geotechnical investigation, and ore-zone delineation; an additional 51 were also drilled and are now used as pre-licensing site-characterization ground-water monitoring wells. The Applicant has also located and properly abandoned 55 Nubeth drillholes. Under the No-Action Alternative, the 51 drillholes would need to be responsibly abandoned by the Applicant, plugging the full depth of the drillhole or well with concrete. However, the potential impacts of all of these preconstruction and current activities would be short-term, and the related traffic over the Ross Site area would be low density and minimal. Thus, neither the geology nor the soils would sustain significant impacts; the impacts to the geology and soils as a result of Alternate 2 would be SMALL.

4.4.3 Alternative 3: North Ross Project

Under Alternative 3, the North Ross Project would generally be the same as the Proposed Action, except that the facility (i.e., the CPP, associated buildings, and auxiliary structures as well as the surface impoundments) would be located to the north of where it would be located in the Proposed Action, as described in SEIS Section 2.1.3. The geology and soils at the north site are similar, but there are a few important differences. The most important difference is that the north site slopes to the southeast at a grade of 5 – 15 percent, where the slope at the Proposed Action's facility location, the south site, has a significant percentage of ground surface with less than 1 percent slope. Given that the cells in the surface impoundments have approximate dimensions of 75 m x 165 m [250 ft x 540 ft], significant additional grading would be necessary to construct the surface impoundments at the north site as compared to the south site's location. Also, given that the use of above-grade embankments (to minimize the volume of release during a catastrophic failure) should be minimized from engineering and environment-protection points of view, then the maximum depth of excavation to create each impoundment at the north site would be on the order of 4 – 12 m [13 – 40 ft], with an impoundment depth of 4.6 m [15 ft] and slopes of 5 – 15 percent. It is estimated that the north site would require the grading of an additional 0.4 – 1.2 ha [1 – 3 ac] to accommodate the sloping site.

The additional construction effort associated with these deeper cuts and larger disturbed areas would result in greater soils impacts than those resulting from Alternative 1, the Proposed Action. In addition, these deep cuts would likely encounter bedrock within 1.5 – 7.5 m [5 – 25 ft], increasing the cost and complexity of the construction activities. Embankments could reduce the depths of the excavations, but they would increase the volume of a potential release of process solutions and other liquid byproduct wastes if a catastrophic release were to occur.

Another important difference between Alternative 3 and the Proposed Action is that the north site is not underlain by shallow ground water and, thus, a CBW would not be required. As a result, the soils loss and soils compaction associated with construction of the CBW at the south site would not occur under Alternative 3.

The potential impacts to geology and soils from construction of Alternative 3 would be SMALL and similar to the Proposed Action. In addition, the potential impacts to geology and soils from the operation and aquifer restoration of Alternative 3 would be the same as those of the Proposed Action and would be SMALL.

Alternative 3 would also result in similar impacts to the geology and soils of the Ross Project area during the Proposed Action's decommissioning, except for activities associated with the decommissioning of the surface impoundments. The larger surface impoundments would require larger areas of recontouring and revegetation during site restoration, which would result in a marginally greater potential for the soils loss and soils compaction. However, the impacts to the surficial geology and soils as a result of the Applicant's cutting through the CBW to re-establish aquifer flow in the Proposed Action would be eliminated. In total, the potential impacts to geology and soils during the decommissioning of Alternative 3 would be SMALL.

4.5 Water Resources

The Proposed Action could impact water resources, both surface and ground waters, during all phases of the Project's lifecycle. As discussed in Section 3.2.4, surface and ground waters in the Ross Project area are currently used for livestock and wildlife watering, crop irrigation, and enhanced oil recovery (EOR). The largest water right within 3 km [2 mi] of the Ross Project area is Permit No. P6046R for the Oshoto Reservoir with a permitted capacity of 21 ha-m/yr [173 ac-ft/yr]. The Applicant proposes to convert this water right for use at the Ross Project (Strata, 2011a). The Applicant would have the option of providing alternative sources of water to supply the EOR operation. This section describes the potential impacts to water resources and the corresponding mitigation measures the Applicant proposes throughout the Proposed Action's lifecycle as well as those of the two other Alternatives.

4.5.1 Alternative 1: Proposed Action

Alternative 1 consists of four phases: construction, operation, aquifer restoration, and decommissioning of an ISR uranium-recovery facility and wellfields.

The Ross Project has the potential to impact quantity and quality of surface and ground waters to varying degrees during each phase of the project. The Applicant intends to use local water for the construction of the facility and wellfields, operation of the Proposed Action, and its aquifer restoration, and decommissioning phases. Consumptive ground-water use results from the Applicant injecting 1.25 percent less water than is withdrawn during uranium-recovery operation. Non-production surface- and ground-water use for domestic needs, dust control, and agricultural irrigation is provided in the Table 4.3.

Table 4.3 Estimated Non-Production Water Use					
		Typical Water Usage (L/min [gal/min])			
Type of Use	Source	Construction	Operation	Restoration	Decommissioning
Domestic	Ground Water	3.4 [0.9]	7.2 [1.9]	6.1 [1.6]	6.8 [1.8]
Dust Control	Surface Water	27.2 [7.2]	13.6 [3.6]	13.6 [3.6]	27.2 [7.2]
Irrigation	Ground Water	0.4 [0.1]	0.4 [0.1]	0.4 [0.1]	0.4 [0.1]
Construction	Surface Water	60.2 [15.9]	31.4 [8.3]	0.0	31.4 [8.3]
	TOTAL	91.2 [24.1]	52.6 [13.9]	20.1 [5.3]	48.8 [12.9]

Source: Modified From Strata, 2012a.

The Applicant anticipates that ground water from the shallow-monitoring (SM) zone would be used for domestic purposes and agricultural irrigation, while surface water from either the Oshoto Reservoir or the Little Missouri River would be used for road and construction dust control. Although the GEIS Section 4.4.4.1 did not address consumptive use of surface water, and it assumed that all required water uses would be provided by ground water, the analysis of impacts to ground water and surface water is nonetheless applicable due to the fact that process water from ground water is the largest component of Ross Project water use.

In addition, the Applicant proposes BMPs consistent with those identified by NRC as commonly employed at ISR facilities and that are summarized in GEIS Section 7.4 (Strata, 2011a; NRC, 2009b).

4.5.1.1 Ross Project Construction

Surface Water

As described in GEIS Sections 4.2.4.1.1 and 4.4.4.1.1, the potential impacts to surface waters that could result from the construction of the Proposed Action include land clearance and disturbance for buildings and auxiliary structures as well as the surface impoundments, wellfields, pipelines, access roads, and utilities; stream-channel disturbance for limited periods and minor wetland encroachment. In addition, spills and leaks of fuels and lubricants as well as the discharge of well-drilling fluids from installation, development, and testing of wells could potentially impact surface-water quality. The potential for these impacts would be mitigated through proper planning, thoughtful design, sound construction methods, permit requirements, and BMPs as described in GEIS Section 7.4 (NRC, 2009). The GEIS considered that changes to stream flow (from land grading and other topographic changes) and to natural drainage patterns would be mitigated or restored after the ISR facility's construction phase is complete. Additionally, while impacts from incidental spills into surface water drainages could occur, they would be expected to be only temporary. The quality of storm water discharged during the construction phase would be controlled by permits from cognizant regulatory authorities. The GEIS concluded that potential impacts to surface water during the construction phase of an ISR facility would be expected to be SMALL to MODERATE, depending upon site-specific conditions.

The Applicant intends to use approximately 88 L/min [23 gal/min] of surface water from either the Oshoto Reservoir or the Little Missouri River for dust control during construction (see Table 4.3). This equates to an annual use of 4.6 ha-m [37 ac-ft/yr], significantly less than the currently permitted annual appropriation for Oshoto Reservoir of 21 ha-m [173 ac-ft/yr]. Thus, the potential impacts of the Proposed Action's construction to surface-water quantity would be SMALL.

Suspended-sediment concentrations in storm water at the Ross Project area could be increased due to vegetation removal and soil disturbance during construction of the Proposed Action. The Applicant estimates that 45 ha [110 ac], or 7 percent of the Ross Project area, would be disturbed by the end of the construction phase (Strata, 2011a). Given this disturbance, the Applicant would need to obtain a "Large Construction General WYPDES Storm Water Permit" from WDEQ/WQD for the Proposed Action. Under this Permit, the Applicant would be required to implement a SWPPP to address storm-water runoff during construction activities. The SWPPP would describe the nature and sequence of construction activities, identify potential sources of pollution, and describe the BMPs that must be used, including erosion and sediment controls (e.g., silt fences, sediment logs, and/or straw-bale check dams) and operational controls (e.g., housekeeping, signage, and/or hydrocarbon storage requirements).

In addition, the construction of a single well (injection, projection, or monitoring) would generate a quantity of drilling fluid estimated at 22,700 L [6,000 gal] and about 11 m^3 [15 yd^3] of drilling muds. In total 1,500 – 2,000 wells would be drilled and the wastes generated could potentially impact water quality. However, the wells would be drilled at different times throughout the Project. The drilling fluids and muds would be contained in a mud pit constructed near the well that is being installed to prevent discharge to surface water. These wastes would then be evaporated and dried over time.

Other potential surface-water impacts could occur from leaking fuel or lubricants from heavy construction equipment and passenger vehicles that would be operated during the construction phase of the Proposed Action. Any such leaks of equipment fluids would be mitigated by the Applicant locating construction activities away from surface-water features, when possible, and rapidly responding to leaks by properly sealing the equipment as needed and by containing and cleaning up the leakage.

Stream channels within the Ross Project would be potentially impacted when crossed by roads, pipelines, and utilities. The Applicant estimates that three stream-channel crossings would be constructed and one existing stream-channel crossing would need to be rehabilitated during the construction phase of the Proposed Action. In addition, there are several instances where tertiary roads would access wellfields and would cross ephemeral drainages. To mitigate impacts, these channel crossings would consist of unconstructed, two-track roads that would be constructed away from drainages where possible; ephemeral channel crossings would involve minimal land disturbance, and they would not be used during flow events. In the instances where it is necessary to cross a stream channel, the crossing would be made perpendicular to the channel and would include a culvert capable of passing the runoff resulting from a 10-year, 24-hour precipitation event. Sediment load would be mitigated by sediment-control BMPs. Pipeline crossings would be constructed in the same corridor as road crossings where possible to minimize disturbance. The impacts to surface-water flow from construction activities across a stream channel would also be minimized by the Applicant routing flow around active construction activities, storing the water in temporary sediment surface impoundments, or passing the water through sediment-control measures prior to discharge (Strata, 2011a). Given the site-specific mitigation measures to be implemented by the Applicant, the potential impacts of the Proposed Action's construction to surface-water quality would be SMALL.

The Applicant has applied for a permit with the U.S. Army Corps of Engineers (USACE) through USACE's Section 404 of the *Clean Water Act* (CWA) permitting process that, if granted, would authorize dredge and fill activities and require the Applicant to mitigate the disturbance of wetlands. While the impacts to surface water could have MODERATE impacts before mitigation (NRC, 2009), the Section 404 permit would establish conditions that could mitigate such impacts. The Applicant anticipates that it would be required to operate in accordance with a Nationwide Permit (NWP) for specific construction activities.

The Ross Project area hosts approximately 26 ha [65 ac] of potential wetlands mostly situated along the Little Missouri River and adjacent to the Oshoto Reservoir (Strata, 2011a). Construction of the Proposed Action would have the potential to impact up to 0.8 ha [2 ac] of wetlands. Prior to disturbing any USACE-verified wetlands, the Applicant would apply for coverage under a USACE permit for specific construction activities such as pipeline installation and access-road stream-channel crossings. For example, the Permit application would require the Applicant to provide a site-specific mitigation plan for construction-related disturbance of jurisdictional wetlands (i.e., wetlands regulated by the USACE under Section 404 of the CWA).

Depending upon the nature of the anticipated wetlands disturbance, mitigation could include the Applicant re-establishing temporarily disturbed wetlands in place, enhancing other existing wetlands, or constructing additional wetland areas for circumstances where the disturbance would be long term. Mitigation measures would ensure that the Proposed Action does not result in a net loss of wetlands. Thus, while the impacts to wetlands could have MODERATE impacts before mitigation (NRC, 2009), a USACE CWA permit would establish conditions that could mitigate such impacts to wetlands. The potential impacts of the Proposed Action's construction to wetlands consequently would be SMALL.

Ground Water

As stated in GEIS Section 4.2.4 and 4.4.4, potential impacts to ground water during an ISR facility's construction are primarily from consumptive water use and contamination caused by: drilling fluids and muds during injection, recovery, and monitoring well drilling; and fuel and lubricant spills and leaks from construction equipment. It is further noted in the GEIS that ground-water use during an ISR facility's construction phase would be limited, and that ground water would be protected by implementing BMPs such as spill-prevention and spill-cleanup protocols. A limited amount of drilling fluids and muds would be introduced into the environment during well installation. Because of the limited nature of construction activities and the implementation of BMPs to protect shallow ground water, the GEIS concluded that construction impacts on ground water would be SMALL (NRC, 2009).

Although construction of the CBW during the Proposed Action is not part of the typical ISR design considered in the GEIS, the analysis of impacts to ground water provided in the GEIS are applicable because the effects of the CBW on shallow ground water are localized and the presence of the CBW would not affect the surrounding ground water.

In the following sections, potential impacts and mitigation measures are considered for three aquifer units: 1) The unconfined shallow (near-surface) aquifers; 2) the confined aquifers hosting the ore-zone (OZ) as well as those above and below the ore zone (the shallow-monitoring [SM] and the deep-monitoring [DM]); and 3) the deeper aquifers below the DM aquifer.

Shallow Aquifers

Potential impacts to the quantity of water in the shallow aquifers during construction of the Proposed Action would be caused by the quantity taken from the Oshoto Reservoir and the quantity involved in the installation of the CBW surrounding the facility (i.e., the CPP and surface impoundments). In the vicinity of the Oshoto Reservoir, the Reservoir stage (i.e., the volume of water it contains) and the shallow-aquifers water levels are closely related (Strata, 2012b). Although the Applicant anticipates an annual withdrawal of 4.6 ha-m/yr [37 ac-ft/yr] of water during construction, that volume is less than the permitted annual appropriation for the Oshoto Reservoir, 21 ha-m/yr [173 ac-ft/yr] (Strata, 2012b). Any changes in ground-water levels due to water usage from Oshoto Reservoir would be small and restricted to the area around the Reservoir.

Construction of the CBW (see SEIS Section 2.1.1.1) could impact the quantity of water in the shallow aquifer because the CBW would isolate the shallow aquifer at the Ross Project facility. Preconstruction dewatering within the facility's area would lower water levels locally in the shallow aquifer, but the normal ground-water flow regime would not be disrupted. The Applicant anticipates that the construction dewatering following installation of the CBW would be a one-time event and require little continuing maintenance. Ground-water use would be mitigated by the design of the CBW, which would prevent any leakage inside the CBW that would require removal by pumping. Thus, the potential impacts from the construction of the Proposed Action to ground-water quantity in the shallow aquifers would be SMALL.
In addition, shallow-aquifer water levels could increase slightly on the hydraulically up-gradient side of the CBW and could decrease slightly on the hydraulically down-gradient side of the CBW in response to the lower permeability of the CBW relative to the shallow aquifer. The changes in ground-water levels would be restricted to the area adjacent to the CBW (Strata, 2011a).

Potential water-quality impacts to the shallow aquifer that could occur during construction include spills or leaks from construction equipment and the introduction of drilling fluids. The potential for the shallow ground water to be impacted by drilling fluids and muds is minimal because of the small volume of fluids used, and because the fluids would be contained within a mud pit in accordance with WDEQ/LQD and EPA requirements. Impacts to ground water during well drilling would be further limited by the nature of the bentonite or polymer-based drilling additives in the drilling fluids. These additives are designed to limit infiltration in an aquifer (i.e., to a few inches) and to isolate the drillhole from the surrounding geologic materials via a wall-cake or veneer of drilling-fluid filtrate, further diminishing the potential for impacts. Thus, the potential impacts of the Proposed Action's construction to ground-water quality in the shallow aquifers would be SMALL.

Ore-Zone and Surrounding Aquifers

Ground water used for domestic uses and agricultural irrigation during the Proposed Action's construction is estimated to be 3.8 L/min [1.0 gal/min] (see Table 4.3). A water-supply well drawing water from the SM aquifer would be used to supply these needs. Based upon yields from regional baseline wells and other wells completed in the SM aquifer, ground-water modeling indicates that the aquifer could support this level of withdrawal with little drawdown (Strata, 2011b). The potential impacts of the Proposed Action's construction on the ground-water quantity available from the confined aquifers, therefore, would be SMALL.

Drilling for mineral delineation and well installation would potentially impact the SM aquifer, the OZ aquifer laterally adjacent to the ore zone, and the DM aquifer. Improperly abandoned drillholes, overly penetrating drillholes, or lack of well integrity could result in the mixing of industrial-use ground water from the OZ aquifer with the chloride-dominated ground water of the DM aquifer or the stock-water quality of the overlying SM aquifer. This mixing would be localized and any significant changes in water quality would be detected by monitoring wells.

To mitigate potential impacts to the confined aquifers from drilling, the Applicant proposes to continue to comply with WDEQ/LQD rules for well completion and drillhole abandonment (WDEQ/LQD, 2005). The Applicant would rely upon the geological model developed to determine total depths for drill holes, thus preventing over-penetration into underlying aquifers. Onsite geological and engineering supervision would continue throughout the construction phase. Wells installed for further hydrologic studies, pre-licensing baseline site characterization, and production infrastructure would pass mechanical integrity testing (MIT) prior to use (see SEIS Section 2.1.1). Consequently, the potential impacts from the Proposed Action's construction on the ground-water quality within the confined aquifers would be SMALL.

Deep Aquifers

Construction of the Ross Project would not impact the aquifers below the DM aquifer. The Flathead and Deadwood Formations would be tapped by the construction of the Class I injection well(s) discussed in SEIS Section 2.1.1, where that well(s) would be used for the disposal of brine and other byproduct liquid wastes during the Ross Project's operation, aquifer restoration, and decommissioning phases. The potential impacts of construction of the Proposed Action on the quantity and quality of ground water present within the deep aquifers would be SMALL.

4.5.1.2 Ross Project Operation

This section describes potential impacts and mitigation measures to surface and ground waters associated with operation of the Proposed Action.

Surface Water and Wetlands

As described in GEIS Sections 4.2.4 and 4.4.4, surface waters could be impacted by accidental spills during ISR operations. Spills from the CPP or wellfields as well as spills during transportation could impact storm-water runoff or contaminate shallow aquifers that are hydraulically connected to surface waters. The GEIS determined that surface-water monitoring and spill response would limit the impacts of potential surface spills to SMALL; however, impacts of spills to surface waters that are connected to shallow aquifers would be SMALL to MODERATE, depending upon the specifics of an incident. Activities posing potential impacts to surface waters from uranium-recovery operation would be regulated by Federal agencies. According to the GEIS, the Applicant's use of BMPs, and implementation of required mitigation measures would moderate the impacts of the Proposed Action's operation from MODERATE to SMALL, depending upon local conditions.

The Applicant estimates that approximately 45 L/min [12 gal/min] of surface water from either the Oshoto Reservoir or the Little Missouri River would be used during the Proposed Action's operation for continuing construction activities in the wellfields and for dust control (see Table 4.3). The estimated annual use of 2.4 ha-m [19 ac-ft/yr] would be significantly less than the existing, permitted annual appropriation for Oshoto Reservoir of 21 ha-m [173 ac-ft/yr]. Ground water produced from monitoring and testing wells outside the exempt (ore-zone or OZ) aquifer would be discharged according to a temporary WYPDES Permit, comparable to the permit obtained by the Applicant for development of its monitoring wells installed in 2010. This water would either infiltrate into the ground or add to the surface water in the Little Missouri River.

Flow in the Little Missouri River could potentially be affected during operation. Water from the Little Missouri River infiltrates into the OZ aquifer where the Fox Hills and Lance Formations outcrop at the ground surface east of the Ross Project area (Strata, 2011a). The Applicant's ground-water model shows that infiltration would increase by approximately 6 L/min [1.5 gal/min], decreasing the average annual discharge of the Little Missouri River by less than 0.005 percent just downstream of the Wyoming-Montana border (Strata, 2011a). Thus, no mitigation measures would be warranted for this very small volume and the potential impacts of the Proposed Action's operation on surface-water quantity would be SMALL.

Storm-water runoff from impervious surfaces, including buildings, roads, and parking areas, could result in higher water flows, channel erosion, and increased sediment concentrations in surface waters. The Applicant predicts a peak flow of 1.4 m^3/s [50 ft^3/s] during a 100-year, 24-hour storm (Strata, 2011a). This peak flow represents an increase of less than 1 percent of the peak flow in the Little Missouri River of 170 m^3/s [6,000 ft^3/s]. In addition, BMPs would be implemented by the Applicant to reduce erosion and the likelihood of increased sediment loads.

Surface-water runoff would be mitigated by the Proposed Action's storm-water control system that would route all storm water to a sediment surface impoundment sized to hold runoff from the 100-year, 24-hour runoff event. A flood-control diversion channel around the CPP and surface impoundments (i.e., the facility itself) would prevent storm water originating in the ephemeral stream channel upstream of the facility from encountering process solutions or chemicals. Mitigation measures employed by the Applicant to reduce soil erosion would also mitigate storm-water runoff across the Ross Project. Protection of wellheads and module buildings from large runoff events would typically be accomplished by placement on high ground out of the flood plain. When wells or other facility components must be placed within the 100-year-flood inundation area, appropriate engineering controls would be used to ensure safety and environmental protection. The injection, recovery, and monitoring wells would be protected from flooding by the installation of cement seals around the well casings and the use of watertight well caps.

Measures designed to mitigate the impacts from suspended sediment would be contained in the WYPDES Storm-Water Permit required by the Applicant prior to uranium-recovery operation. The Permit would include a SWPPP that describes erosion and sediment controls as well as operational controls that would be used to ensure that storm-water discharges from the Ross Project facility do not cause a violation of Wyoming's surface-water quality standards (WDEQ/WQD, 2007). Storm-water BMPs would be inspected semiannually or as required by the WYPDES Storm-Water Permit. The SWPPP would be updated as needed, such as when potential problems are identified during inspections or when there are changes in uranium-recovery operation (e.g., transition from operation to aquifer restoration). The WYPDES Storm-Water Permit would also require storm-water discharge sampling and analysis as well as compliance with a numeric effluent limit for total suspended sediment.

Release of process solutions from uranium-recovery wellheads, pipelines, module buildings, or process vessels; accidental discharge from surface impoundments; or release of yellowcake or IX resin during a transportation accident could result in surface-water contamination if the release(s) reached a surface-water body. Impacts from releases that do reach surface water(s) would be short-term, elevated concentrations of radionuclides and associated chemical constituents at levels above post-licensing, pre-operational baseline conditions. Cleanup of contaminated sediments associated with a spill would follow the same requirements as those for soil cleanup efforts (see SEIS Section 4.4.1.2). Any impacts to surface waters would decline over time as the contaminated fluids are dispersed in the surface-water body.

The potential for release of process solutions would be mitigated by the control system in place at the Ross Project which continually monitors pressure and flow. Accidental discharge from surface impoundments would be mitigated by the size and design of the impoundments and by regular inspections. Because roads would cross surface-water drainages in only a few, isolated locations, it is unlikely that a transportation accident would result in a release to any surface water. Further mitigation of impacts would be accomplished by Applicant's personnel containing and cleaning up any release before the solution could migrate to a surface-water body. Therefore, given these mitigation measures, the potential impacts of the operation of the Proposed Action on surface-water quality would be SMALL.

The potential impacts of the Proposed Action's operation to the Ross Project area's wetlands would be the same as described for the Ross Project's construction-phase impacts and the impacts would be SMALL.

Ground Water

The GEIS concluded in GEIS Sections 4.2.4 and 4.4.4 that the amounts of ground water used in routine activities such as dust suppression, cement mixing, and well drilling are small and would have a SMALL and temporary impact.

At an ISR facility, a network of buried pipelines would be used during in situ uranium recovery for transporting lixiviant between pump houses and the CPP as well as connecting injection and extraction wells to manifolds inside the header houses. The failure of pipeline fittings or valves, or well mechanical-integrity failures, in shallow aquifers could result in spills or leaks of lixiviant, which could impact water quality in the shallow aquifers. Potential environmental impacts due to spills and leaks from pipelines could be MODERATE to LARGE depending upon site-specific conditions, including whether 1) the ground water in the shallow aquifers is close to the ground surface; 2) the shallow aquifers are important sources for local domestic or agricultural water supplies; or 3) the shallow aquifers are hydraulically connected to other locally or regionally important aquifers; or 4) the shallow aquifers have either poor water quality or yields that are not economically suitable for production (NRC, 2009). The use of surface impoundments to manage process solutions generated during ISR activities could also impact shallow aquifers by failure of impoundment embankments or their liners. Potential impacts of such failures would be expected to be minimized as stated in the NRC license, where requirements such as installation of leak-detection systems, maintenance of reserve capacity, and embankment inspections would be required. Thus, the GEIS concluded that impacts of the use of surface impoundments on ground water would be SMALL (NRC, 2009).

As discussed in GEIS Sections 4.2.4 and 4.4.4, potential environmental impacts to ground-water resources in the OZ and surrounding aquifers include consumptive water use and changes to water quality (NRC, 2009). Consumptive use arises from the fact that ISR operations withdraw about 2 percent more water than is injected into the wellfields, which is referred to as "production bleed." Ground-water bleed ensures a net inflow of ground water into the wellfield to minimize the potential movement of lixiviant and its associated contaminants out of the wellfield. Bleed water is generally disposed of through a waste-water control system, and it is not re-injected into the ISR wellfields. The GEIS determined that the short-term impacts of consumptive use could be MODERATE, but temporary, if the OZ aquifer outside the exempted portion of ore zone is used locally. (Uranium-recovery requires exemption of the uranium-bearing aquifer as an underground source of drinking water and is exempted through Wyoming's UIC program administered by the WDEQ.). Therefore, the long-term consumptive-use impacts would be expected to be SMALL in most cases, depending on site-specific conditions.

The GEIS noted that water quality in the OZ aquifer would be degraded during ISR operations (NRC, 2009). A licensee would be required, by its WDEQ Permit to Mine and would be by its NRC license, to initiate aquifer-restoration activities to restore the OZ aquifer to preoperational conditions, if possible. If the aquifer cannot be returned to post-licensing, pre-operational conditions described in SEIS Section 2.1.1.1, the NRC would require that the aquifer meet the U.S. Environmental Protection Agency (EPA) maximum contaminant levels (MCLs) provided in 10 CFR Part 40, Appendix A, Table 5C or Alternate Concentration Limits (ACLs), as approved by NRC (10 CFR Part 40; NRC, 2009b). For these reasons, the NRC determined in the GEIS that potential impacts to water quality of the uranium-bearing aquifer (i.e., ore zone, production zone or unit, or mineralized zone) as a result of ISR operations would be expected to be SMALL and temporary (NRC, 2009).

In GEIS Section 4.2.4 as cited by GEIS Section 4.4.4, the potential for vertical and horizontal excursions of degraded ground water outside of the uranium-production zone (i.e., the ore zone) is discussed. The impact of horizontal excursions could be MODERATE or LARGE, if a large volume of contaminated water leaves the ore zone and moves down-gradient and impacts an area outside the ore zone which is being used for consumption (NRC, 2009). The historical record for several licensed ISR facilities indicates that excursions occur at ISR operations (NRC, 2009). Most of the excursions are horizontal and were recovered within months after detection. Vertical excursions tend to be more difficult to recover than horizontal excursions, and in a few cases, remained on excursion status for as long as eight years. The vertical excursions were traced to thinning of the confining geologic interval below the ore zone and improperly abandoned drillholes from earlier exploration activities (NRC, 2009).

To reduce the likelihood and consequences of potential excursions, the NRC requires licensees to identify preventive measures before starting ISR operations. In general, the potential impacts of vertical excursions to ground-water quality in surrounding aquifers would be SMALL if the vertical hydraulic-head gradients between the OZ aquifer and the adjacent aquifer are small; if the vertical hydraulic conductivities of the confining geologic units are low; and if the confining geologic units are sufficiently thick (NRC, 2009). Environmental impacts, however, would be expected to be MODERATE or LARGE if the confining units are discontinuous, thin, or fractured (NRC, 2009). The NRC requires assurance of the integrity of the confining units to minimize the potential impacts from horizontal excursions.

As indicated in GEIS Sections 4.2.4.2.2.3 and 4.4.4.2.2.3, the potential environmental impacts from disposal of liquid effluents into deep aquifers below ore-bearing aquifers would be SMALL, if water production from the deep aquifers is not economically feasible; if the ground-water quality from these aquifers is not suitable for domestic or agricultural uses (e.g., high salinity); and if they are confined above by sufficiently thick and continuous low-permeability layers (NRC, 2009). Under different environmental laws such as the CWA, the *Safe Drinking Water Act*, and the *Clean Air Act* (CAA), the EPA has statutory authority to regulate activities that could affect the environment. Underground injection of liquids requires a permit from the EPA or from an authorized State UIC program. As noted in SEIS Section 2.1, the WDEQ has been authorized to administer the UIC program in Wyoming.

In the following sections, the potential impacts and mitigation measures related to the Proposed Action's operation are considered for the three types of aquifers: 1) the unconfined shallow (i.e., near-surface) aquifers; 2) the confined aquifers hosting the ore zone as well as those above and below the ore zone (the SM and the DM aquifers); and 3) the deep aquifers below the DM aquifer.

Shallow Aquifers

Potential impacts from operation to ground-water quantity in the shallow aquifers would be similar to those described for the Proposed Action's construction phase and would be SMALL.

During ISR operation, the water quality throughout the Ross Project has the potential to be impacted by accidental spills or leaks from chemical-storage areas, process-solution vessels, or the surface impoundments as well as by spills and leaks of lixiviant from failure of a pipeline or a shallow break in the casing of an injection or recovery well. To reduce the risk of pipeline failure, the Applicant would hydrostatically test all pipelines prior to use and install leak-detection devices in manholes along the pipelines. The Applicant's implementation of BMPs during Ross Project operation would reduce the likelihood and magnitude of spills or leaks and facilitate expeditious cleanup.

Further, the Applicant would monitor recovery and injection pipelines and immediately shut down affected pumps if a spill or leak were detected (Strata, 2011b). The CPP would include a control room where a master control-system would allow remote monitoring and control of ISR, wellfield, and deep-well-disposal operations (Strata, 2011b). Operators would be located in the CPP's control room 24 hours a day and would use a computer-based station to command the control system.

MIT would be conducted on all Class III injection wells, recovery wells, and monitoring wells (see SEIS Section 2.1.1). Construction of all wells and their respective MIT would comply with the pertinent WDEQ/LQD regulations (WDEQ/LQD, 2005).

The Applicant would also implement spill control, containment, and cleanup measures in the CPP and surface-impoundment areas (i.e., the facility). These measures would include secondary containment for process-solution vessels and chemical storage tanks, a geosynthetic liner beneath the CPP's foundation, dual liners with a leak-detection system for the surface impoundments, and a sediment impoundment to capture storm-water runoff. In the event of a surface-impoundment leak, sufficient capacity would be reserved in the other impoundments' cells to allow the contents of the leaking cell to be rapidly transferred, minimizing the volume of the release. In addition, the ground-water levels within the CBW would be maintained below the ground-water levels in the shallow aquifer outside the CBW. This would impose inward and upward hydraulic gradients and therefore minimize the potential for contaminated ground water to migrate into the regional system. Thus, the potential impacts of the Proposed Action's operation to ground-water quality in the shallow aquifers would be SMALL.

Ore-Zone and Surrounding Aquifers

Potential impacts from the consumptive use of ground water from the ore-zone and surrounding aquifers were evaluated by the Applicant using a regional numerical model (Strata, 2011b). The conditions simulated by the Applicant were for two ISR "mine units" operating simultaneously, as described in SEIS Section 2.1.1. Details of the ISR simulations and results of the modeling are provided in Addendum 2.7-H of the Applicant's TR (Strata, 2011b).

During the production simulation, each wellfield module was estimated to operate at a maximum rate of 44 L/s [700 gal/m] or 1.10 L/s [17.5 gal/m] per well. Estimated bleed rate during production was estimated at 1.25 percent (0.55 L/s [8.75 gal/m] per module, 0.0138 L/s [0.219 gal/m] per recovery well). The ground-water sweep operation was estimated to remove 50 percent of the pore volume of the wellfield (see SEIS Section 2.1.1.2). Based upon the three-month sweep period, the estimated flow rate during sweep was 0.0827 L/s [1.31 gal/m] per recovery well. Aquifer-restoration activities were assumed to last approximately six months (actual time could vary based upon wellfield conditions). The bleed during restoration would be expected to vary depending upon whether or not aquifer restoration is occurring concurrent with uranium recovery in other wellfields. When restoration is occurring in one wellfield and uranium recovery is simultaneously occurring in another wellfield, excess bleed from the well undergoing uranium recovery would be used to offset reverse-osmosis (RO) losses within the wellfield in restoration.

The simulations assumed no changes in flow rates within the stock and domestic wells within the model area. Estimated flow rates for the oil-field water-supply wells were developed based upon average historical flow rates for the last two years of recorded flow (2008 and 2009). Three of the oil-field water-supply wells (Nos. 22X-19, 19XX, and 789V) are located immediately adjacent to Modules 2-6 and 2-7. The Applicant has been in communication with Merit, the owner of these wells, and is currently exploring alternative water sources that would allow it to suspend use of the wells before and during uranium recovery. Currently, the goal is to have the Merit wells shut off approximately two years prior to uranium recovery. Given the uncertainty associated with the future status of the Merit wells, the Applicant simulated two uranium-recovery scenarios. Scenario 1 assumed that an alternative water supply could be found, allowing the Merit wells to be taken out of operation two years prior to uranium recovery, and kept out of operation until recovery operations cease. Scenario 2 assumed that an alternative water supply source could not be located and that, during uranium-recovery operation, the Merit oil-field water-supply wells operated at the assumed 2008 – 2009 average flow rates.

The maximum modeled drawdowns for select wells in the OZ aquifer, within and adjacent to the Ross Project area, at the end of uranium-recovery operation and aquifer restoration for the two scenarios are presented in Addendum 2.7-H of the Applicant's TR (Strata, 2011b). The most significant estimated drawdown occurs in the Wesley No. TW02 well located in the SWSW Section 8, Township 53 North, Range 67 West, with 10.2 m [33.3 ft] of drawdown or 42.4 percent of the available head under Scenario 2 at the end of aquifer restoration. This well supplies water to a structure that is currently used by the Applicant as its Field Office for the Ross Project and to provide water to livestock.

Potential impacts to the SM-aquifer water quantity, because of withdrawals during uranium recovery and aquifer restoration in the ore zone, were also evaluated by the regional ground-water model (Strata, 2011b). Under the two recovery scenarios evaluated, the estimated maximum amounts of drawdown ranged from 1.5 – 5 m [5 – 15 ft] within the Ross Project area following the Proposed Action's operation and aquifer-restoration phases.

Impacts from consumptive use of ground water from the ore zone would be minimized by cessation of water withdrawals by the Merit oil-field water-supply wells. The ground-water model simulated a single operational sequence of wellfield development, recovery, and aquifer restoration. Different operational approaches could be more effective in reducing impacts, and the Applicant proposes to investigate these as wellfield installation and testing progresses.

In the event that uranium recovery at the Proposed Action prevents the full use of a well which provides water under a valid water right, the Applicant would commit to providing an alternative source of water of equal or better quality and quantity, subject to Wyoming water statute requirements.

In the regional numerical model, the model's lower boundary was the base of the ore zone/top of the lower confining unit. As a result, potential impacts to the DM aquifer were not evaluated by the model. The DM aquifer supports only one well (Merit Well No. 22X-19), and it has only limited hydraulic conductivity and yield. Thus, as the model demonstrates, the potential impacts from the Proposed Action's operation to ground-water quantity in the confined aquifers would be SMALL.

There is potential for water-quality impacts (vertical excursions) to the SM and DM aquifers from the lixiviant-fortified ground water during injection and withdrawal from the OZ aquifer, although this potential is mitigated by the natural confining units of fine-grained mudstones, siltstones, and claystones above and below the OZ aquifer (see SEIS Section 3.5).

The Applicant tested the integrity of the lower confining unit separating the OZ aquifer from the DM aquifer with a six pump tests; in two of the six tests, pumping of the OZ aquifer showed a possible response in the DM aquifer (Strata, 2011a). These responses were interpreted by the Applicant as due to improperly plugged previous exploration drillholes that have not yet been properly abandoned. Other aquifer tests by Nubeth and the Applicant recorded no response in the aquifers vertically adjacent to the ore zone. Different water qualities, observed in the OZ and DM aquifers also support the premise of hydraulic separation. Stratigraphic sections created by the Applicant from the geologic logs of the drillholes have provided further support for the continuity and integrity of the shale confining units (Strata, 2011b). The thickness of the shale unit between the OZ and the DM aquifers is generally greater than 6 m [20 ft], except for an area along the southern edge of the Ross Project area where the unit thins to about 1.5 m [5 ft]. The Applicant would continue geologic evaluation and hydrologic testing to characterize the integrity of the lower confining unit, through observations of piezometric levels in the SM and DM aquifers.

The Applicant would implement a WDEQ-approved MIT program for all injection and recovery wells to ensure casing integrity (WDEQ/LQD, 2005). Breaches to the integrity of the confining unit from old exploration drillholes would be minimized by the Applicant locating the drillholes within the wellfields and beneath the Proposed Action as well as plugging and abandoning them with low-hydraulic-conductivity materials such as cement or heavily mixed bentonite grout according to methods approved by WDEQ as described in Section 2 of this SEIS (Strata, 2011b). As of October 2010, the Applicant had located 759 of the 1,682 holes from Nubeth exploration activities and had plugged 55 of them (Strata, 2011b). The Applicant proposes to actively locate and plug all exploration drillholes prior to beginning wellfield operation.

If the Ross Project were to be licensed by the NRC, the NRC license would include a requirement that the Applicant install a ring of monitoring wells around each wellfield. The wells would allow monitoring of the SM and DM aquifers as well as the OZ aquifer around their perimeters. The ground-water model discussed in SEIS Section 3.5.3, Local Ground-Water Resources, indicates that a spacing of 122 –183 m [400 – 600 ft] between the production wellfields and perimeter monitoring-well ring would be sufficient to detect an excursion; thus, spacing between the monitoring wells is also proposed to be 122 –183 m [400 – 600 ft] (Strata, 2011b). The simulations indicated that a head change or hydraulic anomaly would rapidly become apparent in the perimeter wells before any geochemical changes in the ground water would be detected. The NRC would require an early-warning system of pressure transducers to detect anomalous hydrostatic pressure increases in the perimeter monitoring wells and sampling of monitoring wells with a semi-monthly frequency. Mitigation in the event of a vertical excursion of lixiviant-containing ground water to the SM or DM aquifers could require withdrawal and treatment of contaminated ground water from these aquifers.

During the Proposed Action's operation, the ground-water quality in the OZ aquifer would be impacted during uranium-recovery operation. The Applicant proposes to file an exemption request with WDEQ/LQD for exemption of the OZ aquifer as a source of drinking water based upon the fact that some constituents in the ground water (e.g., TDS, sulfate, ammonia, radium-226+228, and gross alpha) currently exceed applicable standards for human or livestock water consumption as shown in Tables 3.6, 3.7, and 3.8 and in SEIS Section 3.5.3 (Strata, 2011b). The uranium and vanadium in the ore zone would be oxidized and mobilized by the introduction of lixiviant into the OZ aquifer through injection wells. In addition to the uranium and vanadium, other constituents would also be mobilized, including anions, cations, and trace metals (Strata, 2011b). These impacts to the water quality of OZ aquifer within the wellfields would be short-term because aquifer restoration would be required by the NRC license to return these constituent concentrations to each wellfield's respective NRC-approved baseline (i.e., post-licensing, pre-operational) concentrations, or ACLs as approved by the NRC.

The quality of the non-exempted OZ aquifer outside the perimeter monitoring-well rings could be impacted via a lateral excursion resulting from a local wellfield imbalance. A wellfield imbalance occurs when the rate of injected lixiviant exceeds the rate of extraction by the recovery wells, resulting in a migration of lixiviant laterally away from the wellfield. The Applicant proposes a computer-based control system, staffed 24 hours a day within the CPP, to monitor injection pressures and recovery-well flow rates so that wellfield balance would be maintained (Strata, 2011a).

In the event of an operational upset, the ground-water model, integrated with injection and recovery well data, would allow for a determination of potential migration paths and assist the system's operator in making decisions on mitigating actions. The Applicant notes that the heterogeneous lithology of the sandstones produces lateral and vertical variations in permeability, with uranium mineralization concentrated in the higher-permeability sediments. Lateral migration of lixiviant would therefore be limited by the less-permeable and un-mineralized zones within the ore-zone sandstones.

Temporary impacts to water quality would result if an excursion were to occur. Typical lixiviant circulating through the ore zone would contain high concentrations of sodium, bicarbonate, chloride, and sulfate with TDS up to 12,000 mg/L and concentrations of uranium, vanadium, and radium greater than 100 mg/L (NRC, 2009; Strata, 2011a; WDEQ/WQD, 2011). As described in SEIS Section 3.5, the water qualities in the surrounding aquifers have much lower TDS, averaging 1,092 mg/L, 1,600 mg/L, and 1,268 mg/L in the SM, OZ, and DM aquifers, respectively, or about 10 percent of the TDS of the lixiviant. Preconstruction monitoring by the Applicant has shown concentrations of uranium (less than 0.004 mg/L]) and radium (less than 0.01 Bq/L [0.4 pCi/L]) in the SM and DM aquifers (Strata, 2011a). Higher concentrations of uranium (maximum value of 0.109 mg/L) and radium (maximum value of 0.44 Bq/L [12.01 pCi/L]) were measured in the ore zone (Strata, 2011a). Temporary impacts to water quality from an excursion of increased concentrations of TDS, uranium, radium, and other radionuclides as well as elements such as arsenic, selenium, and vanadium that are mobilized with the uranium would be expected.

The potential impacts of the operation of the Proposed Action to ground-water quality in the confined aquifers above and below the ore zone would, therefore, be SMALL. The short-term potential impacts of lixiviant excursions from uranium-recovery operation to the OZ aquifer outside the exempted area would be SMALL to MODERATE. Detection of excursions through the network of monitoring wells, followed by the Applicant's pumping of ground water to recover the excursion would reduce long-term potential impacts to the OZ aquifer outside the exempted portion to SMALL.

Deep Aquifers

The Applicant plans to dispose of brine and other liquid byproduct wastes into five deep wells discharging into the Flathead and Deadwood Formations, which are defined as the Formations that occur beneath the base of the Icebox Shale member of the Winnipeg Group and above the top of the Precambrian basement. There are no porous and permeable zones below the Deadwood and Flathead Formations that would make suitable injection zones. Because of the depth in the stratigraphic column at which these Formations occur and the apparent lack of oil or other hydrocarbons, there has been little exploration of these intervals and few data are available for the Ross Project area. To improve its understanding of the targeted Formations, the Applicant plans to drill one deep well for hydraulic testing as a preconstruction activity (Strata, 2011a). If the capacity in the targeted Formation for injected solutions is less than anticipated by the Applicant, more wells than five may be needed.

The UIC Class I Permit issued by the WDEQ identified the confining unit immediately above the discharge zone as consisting of approximately 16 m [52 ft] of Icebox Shale. An additional confining unit immediately above the Icebox Shale is the Red River Formation, which consists of 97 – 140 m [318 – 460 ft] of cryptocrystalline to microcrystalline impermeable dolomite. The top of the discharge zone occurs about 2,488 m [8,163 ft] below the ground surface, and the total thickness of the injection zone for the wells is estimated to be 180 m [592 ft]. In issuing the UIC Permit, the WDEQ/WQD determined that, at the depths and locations of the injection zones specified in the Permit, the use of ground water from the Flathead and Deadwood Formations is economically and technologically impractical (WDEQ/WQD, 2011).

The data that are available for the Formations targeted for deep-well injection suggest that ground water contains greater than 10,000 mg/L TDS. The estimated water quality of the brine, one liquid effluent that would be injected in the deep-injection wells, comprises the following constituent concentrations: 4,000 – 40,000 mg/L TDS; 5 – 25 mg/L uranium as U_3O_8; and 14.8 – 92.5 Bq/L [400 – 2,500 pCi/L] Ra-226. Its pH is between 6 and 9. WDEQ concluded that the liquid effluents could be suitably isolated in the deep aquifers, and they would not affect any overlying underground sources of drinking water. The deep-injection wells would be installed and tested in accordance with WDEQ/WQD Class I disposal-well standards and the UIC Permit. The Permit requires the Applicant to control effluent pressures at the wellhead to ensure that the fracture pressure of the Formation is not exceeded. Regular monitoring of the water quality of the injected brine is required by the Permit, and pH would have to meet the respective upper control limits (UCLs) to be injected (WDEQ/WQD, 2011). The Permit also prohibits injection of hazardous waste as defined by EPA and WDEQ. Thus, the potential impacts of the Proposed Action's operation to ground-water quantity and quality in the deep aquifers would be SMALL. The conditions of the UIC Permit would mitigate potential impacts, including those described above.

4.5.1.3 Ross Project Aquifer Restoration

The Proposed Action's aquifer-restoration methodology would use a combination and sequence of: 1) ground-water transfer; 2) ground-water sweep; 3) reverse osmosis (RO), permeate injection, and recirculation; (4) stabilization; and (5) water treatment and surface conveyance. The Applicant proposes to use ground-water sweep selectively (i.e., around the perimeter of the wellfield) rather than throughout the entire wellfield to minimize the consumptive use of ground water (Strata, 2011a). After the first wellfield is depleted, the Applicant would conduct aquifer restoration concurrently with operation of subsequent wellfields. Consumptive use of ground water during the aquifer-restoration phase is generally greater than during uranium-recovery operation (NRC, 2009).

Surface Water

As described in GEIS Sections 4.2.4.1.3 and 4.4.4.1.3, the activities occurring during aquifer restoration that could impact surface waters include management of waste water, permeate reinjection, storm-water runoff, and accidental spills and leaks (NRC, 2009). The GEIS concluded that the potential impacts to surface water due to the management of ground water extracted during aquifer restoration would be SMALL. An ISR operator's compliance with permit conditions, use of BMPs, and execution of mitigation measures would reduce impacts from storm-water runoff as well as accidental spills and leaks such that they would be SMALL to MODERATE, depending upon site-specific conditions.

At the Ross Project, the Applicant intends to use approximately 13.6 L/min [3.6 gal/min] of water obtained from either the Oshoto Reservoir or the Little Missouri River for dust control during aquifer restoration (see Table 4.3). The potential impacts would thus be comparable to those during the Proposed Action's construction and operation phases.

Potential increases in sediment concentrations during the Proposed Action's aquifer-restoration phase would also be comparable to its operation phase. Potential risk of surface-water contamination associated with releases of process solutions and/or waste liquids as well as spills of other materials during aquifer restoration would be comparable to the operation phase of the Proposed Action, although the concentration of uranium-bearing solutions would decline. Thus, the potential impacts of aquifer restoration to surface-water quantity and quality would be SMALL.

The potential impacts of aquifer restoration during the Proposed Action to the wetlands on the Ross Project area would be the same as discussed under the Ross Project's construction.

Ground Water

As the GEIS states in Sections 4.2.4 and 4.4.4, the potential environmental impacts on ground-water resources during aquifer restoration are related to ground-water consumptive use and waste-management practices, including liquid-effluent discharges to the surface impoundments and deep disposal of brine resulting from the RO process. In addition, aquifer restoration directly affects ground-water quality in the vicinity of the wellfield being restored (NRC, 2009). The purpose of aquifer restoration is to return the ground-water quality in the production zone (i.e., the exempted ore zone) to ground-water protection standards specified at 10 CFR Part 40, Appendix A. These standards require that the concentration of a given hazardous constituent must not exceed 1) the NRC-approved background concentration of that constituent in ground water, 2) the respective numeric value in the table included in Paragraph 5C, if the specific constituent is listed in the table and if the background level of the constituent is below the value listed, or 3) an ACL the NRC establishes for the constituent. Potential impacts are affected by the aquifer-restoration methodologies chosen, the respective severity and extent of the contamination, and the current and future uses of the ore-zone and surrounding aquifers in the vicinity of an ISR facility. Consequently, the GEIS concluded that the potential impacts of ground-water consumption during aquifer restoration could range from SMALL to MODERATE, depending on site-specific conditions.

Shallow Aquifers

Potential impacts to water quantity of the shallow aquifers during restoration would be reduced, compared to the construction and operation phases of the Proposed Action. The impact to the aquifers' water levels from consumptive use of water from the Oshoto Reservoir and the Little Missouri River would also be moderated, because of the lower-volume withdrawals from the surface-water bodies.

In addition, potential impacts to water quality would again be reduced when compared to the Proposed Action's operation because no lixiviant would be used in the injection stream and the concentration of chemicals in the recovered ground water would be significantly less than during ISR operations. The Applicant's implementation of BMPs during uranium-recovery operation would also reduce the likelihood and magnitude of spills and leaks, and thorough cleanup would be facilitated. The ground-water mitigation measures during aquifer restoration would be the same as those described for the operation of the Proposed Action. Thus, the potential impacts of aquifer restoration to ground-water quantity and quality of the shallow aquifers would be SMALL.

Ore-Zone and Surrounding Aquifers

The magnitude of potential impacts to water quantity of the OZ aquifer and the surrounding aquifers during the aquifer-restoration phase of the Proposed Action would be greater than from its operation because of the greater consumptive use of ground water (Strata, 2011a). Ground-water modeling estimates of the drawdown in the shallow-monitoring (SM) aquifer during both Ross Project operation and aquifer restoration were less than 5 m [15 ft]. The exempted OZ aquifer was predicted to experience significant drawdowns in three wells on the Ross Project area, with minor drawdowns in wells within 3.2 km [2 mi] of the Project. The conservative regional impact analysis conducted by the ground-water modeling predicts a reduction in the available head in wells used for stock, domestic, and industrial use. Although these effects would be localized and short-lived, the Applicant would commit to provide an alternative source of water of equal or better quantity and quality, subject to Wyoming water-statute requirements, in the event that aquifer-restoration operations prevent the full use of a well under a valid water right (Strata, 2011a; Strata, 2012a). Consequently, the potential impacts of the Proposed Action's aquifer-restoration phase to ground-water quantity of the confined aquifers would be SMALL to MODERATE.

The potential for excursions during aquifer restoration that would affect water quality in the aquifers vertically adjacent to the exempted OZ aquifer would be similar to those described earlier for the Proposed Action's operation. However, the magnitude of impacts would be less because the injection and recovery flow rates would be lower during aquifer restoration than during active uranium recovery and the ore-zone water quality would improve throughout active aquifer-restoration activities. The concentrations of radiological parameters and other chemical constituents in the permeate that would be injected as "clean" water to restore the exempted OZ aquifer would be lower than the pre-licensing baseline ore-zone water quality reported by the Applicant, except for radium-226 (Strata, 2011a). Dissolved radium-226 measured in the OZ aquifer has ranged from 0.03 Bq/L [0.71 pCi/L] to 0.44 Bq/L [12.01 pCi/L], and the typical radium-226 concentration is 1.1 Bq/L [30 pCi/L] (Strata, 2011a). The potential impacts of aquifer restoration to ground-water quality of the confined aquifers would be SMALL.

Deep Aquifers

The Applicant estimates that less than 860 L/d [227 gal/d] of brine and other byproduct wastes would be disposed in the Class I injection wells during aquifer restoration at the Proposed Action. Although the volume of waste injected would be greater during the aquifer-restoration phase than during the Ross Project's operation phase, the potential impacts would be similar because the injection pressures would not increase beyond the limit established by WDEQ's UIC Permit. These pressure limits would ensure that the capacity of the Class I receiving aquifer is not exceeded. The potential impacts from aquifer restoration to ground-water quality of the deep aquifers would, therefore, be SMALL.

4.5.1.4 Ross Project Decommissioning

The decommissioning activities of the Proposed Action that might impact surface water and/or ground water include the Applicant dismantling the CPP, auxiliary structures, and the surface impoundments; removing buried pipelines; excavating and removing any contaminated soil; plugging and abandoning wells using accepted practices; breaching the CBW; and restoring and revegetating all disturbed areas. Figure 4.1 indicates the components of the Proposed Action that would be in place by the end of its decommissioning.

Surface Water

As described in GEIS Sections 4.2.4 and 4.4.4, during the decommissioning phase, temporary impacts to water quality would be anticipated due to sediment loading during the excavation and removal of pipelines, drainage crossings, and other infrastructure (NRC, 2009). As the GEIS noted, an Applicant's compliance with permit conditions, its use of BMPs, and its observance of required mitigation measures would reduce decommissioning impacts to SMALL to MODERATE, depending upon site-specific conditions.

For the Proposed Action, the Applicant intends to use surface water from either the Oshoto Reservoir or the Little Missouri River for dust control and any demolition activities during the Project's decommissioning. As shown in Table 4.3, the Applicant estimates that approximately 42 L/min [11 gal/min] of surface water would be used during facility and wellfield decommissioning. This withdrawal rate is between the quantities of anticipated water use during the Proposed Action's construction and operation phases.

The primary impacts to surface water during the decommissioning of the Ross Project would be from activities associated with the removal of constructed Project components, reclamation and restoration of the land impacted during the Proposed Action, and the cleanup of any contaminated soils. These impacts would be similar to those that result from the construction of the Proposed Action. Removal of buried pipelines and the roads near stream channels during the decommissioning phase would result in temporary disturbances that could impact surface-water quality. Potential surface-water contamination could occur from spilled or leaked fuel or lubricants from construction equipment and passenger vehicles that would be operated during decommissioning activities, although the equipment would generally be located away from surface-water bodies. These potential impacts to surface-water quality would be mitigated using the same measures as implemented during the Proposed Action's construction (e.g., BMPs and spill-response protocols). The potential impacts to surface-water quantity and quality from the Ross Project's decommissioning would be SMALL.

The potential impacts to wetlands from the Proposed Action's decommissioning would be SMALL, as they would be the same as discussed under the Proposed Action's construction.

Ground Water

As described in GEIS Sections 4.2.4 and 4.4.4, the impacts to ground water during the decommissioning of an ISR facility are primarily associated with consumptive use of ground water, potential spills of fuels and lubricants, and well abandonment (NRC, 2009). Ground-water consumptive use during decommissioning activities would be less than during operation and aquifer-restoration activities. BMPs would reduce the likelihood of spills and leaks. After

ISR operations are completed and a facility is decommissioned, improperly abandoned wells could impact aquifers above the OZ aquifer by providing hydrological connections between aquifers (NRC, 2009). To ensure that this consequence does not happen at the Ross Project, all injection, recovery, and monitoring wells would be plugged and abandoned in accordance with UIC Permit requirements. The GEIS determined that implementation of BMPs and compliance with permit requirements would ensure that the potential impacts to ground water would be SMALL during decommissioning; the Proposed Action's decommissioning would include observance of these procedures and requirements.

Shallow Aquifers

During decommissioning, finger drains (see SEIS Section 2.1.1.4) would be created along the up-gradient and down-gradient sides of the CBW and backfilled with permeable material (gravel). These gravel-filled breaches in the CBW would create a highly permeable flow path through the CBW that would allow the natural flow of the shallow aquifer ground water beneath the CPP and in the immediate vicinity outside the CBW to be restored. Water levels would be monitored by the Applicant to verify that the CBW reclamation and ground-water restoration is complete. After uranium-recovery operation is complete, unidentified, improperly abandoned wells (i.e., from previous subsurface explorations not associated with the Applicant or its operations) could continue to impact aquifers above the ore-zone and adjacent aquifers by providing hydrologic connections between aquifers. The Applicant's implementation of BMPs and SOPs for the plugging and abandonment of its own wells during decommissioning of the Proposed Action would reduce the likelihood of shallow-aquifer contamination. In addition, other BMPs employed by the Applicant would reduce the likelihood and magnitude of spills and leaks during equipment and vehicular operation and would facilitate any soil or other cleanup required. Thus, the impacts to shallow aquifers during the Proposed Action's decommissioning would be SMALL.

Ore-Zone and Surrounding Aquifers

As part of the decommissioning of the Proposed Action and the concomitant land reclamation and restoration activities, all monitoring, injection, and production wells would be plugged and abandoned in accordance with the UIC Permit requirements. The wells would be filled with cement and/or bentonite and then cut off below plow depth to ensure ground water does not flow through the abandoned wells (Stout and Stover, 1997). Proper implementation of these procedures would isolate the wells from ground-water flow. Thus, the impacts to the ore-zone and vertically adjacent aquifers would be SMALL.

Deep Aquifers

The Applicant estimates that less than 38 L/day [10 gal/day] of brine and other liquid byproduct wastes would be disposed in the Class I injection wells during the decommissioning of the Proposed Action. The potential impacts to ground-water quantity and quality during decommissioning would be SMALL and less than the other phases of the Ross Project.

4.5.2 Alternative 2: No Action

Under the No-Action Alternative, the Ross Project would not be licensed and the land would continue to be available for other uses. Mud pits that could continue to be constructed at each well site to manage drilling fluids and muds would have little potential of impacting surface waters and no potential of impacting ground water. The roads across the Ross Project area would be graded, contoured, and revegetated, also leaving little potential for them to impact surface water by increasing sediments.

Similarly, although no license would be issued and no Ross Project would be constructed or operated in the No-Action Alternative, preconstruction activities would cause potential impacts. The respective impacts to ground water depend upon the density of plugged and abandoned wells and drillholes. As of August 2011, the Applicant had drilled and plugged approximately 612 holes it installed during site and geotechnical characterization; an additional 51 were drilled and are now used as site-characterization ground-water monitoring wells. The Applicant has also located and properly abandoned 55 Nubeth drillholes. Thus, the drillhole density is approximately 1 hole per 1 ha [2.5 ac]. Under the No-Action Alternative, the 51 monitoring wells, and any others that could be located, would need to be properly abandoned, where each well and drillhole would be filled with a concrete plug up to 6 inches in diameter through the entire depth of the hole. The low density of these properly plugged and abandoned wells and drillholes would not affect the ground-water flow or quality.

Thus, the potential impacts from the No-Action Alternative to surface and ground waters, relative to the existing Ross Site area and including the preconstruction activities that have already occurred, would be SMALL

4.5.3 Alternative 3: North Ross Project

Under Alternative 3, the North Ross Project would generally be the same as the Proposed Action, except that the facility (i.e., the CPP, associated buildings, and auxiliary structures as well as the surface impoundments) would be located to the north of where it would be located in the Proposed Action, as described in SEIS Section 2.1.3. The hydrology of the north site differs from that under the location of the CPP in the Proposed Action. The depth to the unconfined, shallow ground-water aquifer is greater, which would eliminate the need for a CBW. However, the north site contains two ephemeral streams rather than the one in the Proposed Action. These ephemeral drainages extend over 760 m [2,500 ft], before entering the Little Missouri River, compared to the Proposed Action where the facility is within 300 m [1,000 ft] of the Oshoto Reservoir and the Little Missouri River. Because of the drainage, the design of the facility could require that the CPP and the surface impoundments be constructed across a drainage that leads directly to the Little Missouri River. The ground's surface slopes to the southeast at a grade of 5 – 15 percent compared with a slope of less than 1 percent for the south location in the Proposed Action. Thus, the construction of the surface impoundments on the steeper slope would require a large increase in the area of disturbed land and would require that significant design and engineering considerations be addressed in order to mitigate potential impacts to surface water.

Nonetheless, most of the potential impacts of and mitigation measures for this Alternative would be the same as for the Proposed Action. Only the differences in impacts between the Proposed Action and the North Ross Project are described below.

4.5.3.1 North Ross Project Construction

Impacts to surface and ground waters during construction are expected to be generally the same as the Proposed Action, although the steeper slopes at the north site would require more engineering and construction activity. As a result, there would be a slight increase in the potential for impacts to surface and ground waters in the shallow aquifer. However, the impacts to shallow ground water in the Proposed Action, which result from the construction of the CBW and, in particular, the alteration of the surficial ground-water flow regime, would not be a consequence of this Alternative. At the north site, shallow ground-water levels are estimated to be at a depth of greater than 15 m [50 ft], within the sandstones of the LA interval of the Lance Formation (as discussed in SEIS Section 3.4); however, during high-precipitation events or after significant snowmelt, perched ground water could be present above the regional water table. If the CBW is not needed and not constructed by the Applicant, then the need for dewatering the shallow aquifer would be eliminated and thereby would reduce the consumption of ground water by a small amount.

Construction of the storm-water control system and implementation of BMPs during construction of the Alternative 3 facility would be more involved, in order to protect the two ephemeral drainages from impacts of erosion and increased sediment loads. If the Alternative 3 design required the CPP and the surface impoundments to be separated by a drainage (as shown in Figure 2.11 in SEIS Section 2.1.3), the construction of the pipeline network would also require additional construction and engineering activity. However, the BMPs during construction would minimize potential impacts to surface and ground waters from construction of Alternative 3; thus, the impact would still be SMALL.

4.5.3.2 North Ross Project Operation

Alternative 3 would result in many of the same potential impacts to surface water during its operation as the Proposed Action's. The proximity of the facility to two ephemeral drainages would increase the risk of surface-water impacts from spills and leaks, where the released material could make its way into surface water. The potential for impact to surface water would be mitigated by the distance of approximately 0.8 km [0.5 mi] to the Little Missouri River. The greater distances from the CPP to the Little Missouri River in Alternative 3, when compared to those of the Proposed Action, would also strengthen the natural mitigation of impacts from discharge of excess permeate. Operation of the wellfields during the North Ross Project would be the same as during the Proposed Action and, therefore, the potential impacts and mitigation measures associated with the wellfields would be the same. Thus, the potential impacts to surface water of Alternative 3's operation would be SMALL.

The greater thickness of the vadose (i.e., unsaturated) zone under the north site would also provide additional natural protection to the shallow ground water in the event of a release of process chemicals, recovery solutions, or liquid wastes within the CPP and surface-impoundment areas. If contaminants reached the ground water, remediation by pump-and-treat methods would be required. With the Proposed Action, ground-water levels within the CBW would be maintained lower than surrounding and underlying ground-water levels, and would thus prevent any migration of contaminants away from the CPP and surface impoundments. Because there would be no difference between the location and operation of the wellfields under Alternative 3 as compared with the Proposed Action, the potential MODERATE impacts from lixiviant excursions discussed in SEIS Section 4.5.1.2 could also occur under Alternative 3.

Therefore, the potential impacts to ground water of the operation of Alternative 3 would be SMALL to MODERATE due to the potential for lixiviant excursions.

4.5.3.3 North Ross Project Aquifer Restoration

Because the wellfields would be in the same locations in Alternative 3, this Alternative does not include any modifications to the wellfields from what was described for the Proposed Action (because they follow the subsurface uranium mineralization), the wellfields would result in the same potential impacts to ground water during Alternative 3's aquifer restoration phase as in the Proposed Action. These potential impacts would be SMALL to MODERATE, due to potential drawdowns during aquifer restoration.

4.5.3.4 North Ross Project Decommissioning

Alternative 3 would result in generally the same potential impacts to surface and ground waters during its decommissioning as would the Proposed Action, with the following exceptions: The surface-impoundment area requiring recontouring and revegetation would be larger and more extensive; thus, the potential for surface-water impacts associated with these activities would be marginally greater. Unlike with the Proposed Action, it would not be necessary to cut gravel-filled channels through a CBW, thereby eliminating the potential for the associated surface-water impacts. The potential impacts during Alternative 3's decommissioning to the surface drainages through the north site would be the same as described above for Alternative 3's operation. The potential impacts to surface and ground waters from decommissioning of Alternative 3 would be SMALL.

4.6 Ecology

The Proposed Action could impact ecological resources, including both flora and fauna during all phases of the Project's lifecycle. These impacts could include removal of vegetation from the Ross Project area; reduction in wildlife habitat and forage productivity, and an increased risk of soil erosion and weed invasion; the modification of existing vegetative communities; the loss of sensitive plants and habitats; and the potential spread of invasive species and noxious weed populations. Impacts to wildlife could include loss, alteration, or incremental fragmentation of habitat; displacement of and stresses on wildlife; and direct and/or indirect mortalities. Aquatic species could be affected by disturbance of stream channels, increases in suspended sediments, pollution from fuel spills, and habitat reduction. The potential environmental impacts to and related mitigation measures for ecological resources during the construction, operation, aquifer restoration, and decommissioning of the Proposed Action and the two Alternatives are discussed in the following sections.

4.6.1 Alternative 1: Proposed Action

Alternative 1, the Proposed Action, consists of four phases: construction, operation, aquifer restoration, and decommissioning of an ISR uranium-recovery facility and wellfields.

4.6.1.1 Ross Project Construction

As discussed in GEIS Section 4.4.5, the potential impacts to terrestrial vegetation during the construction of ISR facilities could include removal of vegetation from ISR facility sites (and the associated reduction in wildlife habitat and forage productivity and the increased risk of soil erosion and weed invasion), the modification of existing vegetative communities, the loss of sensitive plants and habitats as a result of site clearing and grading, and the potential spread of invasive species and noxious weed populations (NRC, 2009).

The construction phase of the Proposed Action could potentially impact the local ecology during the Applicant's clearing vegetation and leveling the site; constructing the CPP, auxiliary structures, and surface impoundments; developing the wellfields, including drilling wells, laying pipelines, constructing header houses, and other wellfield components; constructing access roads; clearing storage, parking, and laydown areas; and installing associated infrastructures such as utility and lighting systems. The ecological impacts of these construction activities are evaluated for protected species, vegetation, and wildlife.

Terrestrial Species

Vegetation

The construction of the Ross Project facility (i.e., CPP and surface impoundments) as well as the installation of wellfields would take place within the nine vegetation communities present at the Project area (upland grassland, sagebrush shrubland, pastureland, hayland, reservoir/stock pond, wetland, disturbed land, cropland, and wooded draw) (see SEIS Section 3.2). Direct impacts of such construction would include the short-term loss of vegetation (structure modification, species composition, and areal extent of cover types). An estimated 113 ha [280 ac] of land disturbance would occur; one-half of this disturbance would occur within the upland grassland vegetation community, primarily because of wellfield-module and access-road construction.

Only 7 percent of the Ross Project area is currently hayland; however, 20 – 30 percent of the impacts would be to this vegetation community because of construction of the CPP and surface impoundments. Indirect impacts include the short-term and long-term increased potential for non-native species invasion, establishment, and expansion; exposure of soils to accelerated erosion; shifts in species composition or changes in vegetation density; reduction of wildlife habitat; and reduction in livestock foraging opportunities.

Sagebrush shrubland, the second largest vegetation type on the Ross Project area, can be difficult and time-consuming to re-establish. Consequently, preconstruction vegetation communities and sub-communities (i.e., shrub-steppe) may be different than post-construction communities (i.e., grass-dominated) for several years, or possibly decades, which could alter the composition and abundance of both plant and wildlife species in the area. Site reclamation and/or regeneration of native shrub species could be further hindered by year-long grazing pressure. Large ungulates (i.e., wild and domestic animals with hooves) are attracted to the more succulent, younger plants, and they often concentrate in newly seeded locations during the critical early-growth stage. Impacts to the sagebrush-shrubland vegetation type would be minimized by the Applicant reducing surface disturbance where possible, distributing a temporary seed mixture to prevent invasion of non-native species in disturbed areas, restoring sagebrush and other shrubs on reclaimed lands, and conducting all re-vegetation activities in accordance with an approved WDEQ/LQD reclamation plan (Strata, 2011b).

Construction activities, including the increased soil disturbance and increased traffic during construction, could stimulate the introduction and spread of undesirable and invasive, non-native species at the Ross Project area. Several species of designated and prohibited noxious weeds listed in the Wyoming *Weed and Pest Control Act* were identified on the Ross Project area. These species included field bindweed, perennial sow thistle, quack grass, Canada thistle, hounds tongue, leafy spurge, common burdock, Scotch thistle, Russian olive, and skeletonleaf bursage (Strata, 2011a). These species could be locally abundant in small areas, especially around the Oshoto Reservoir and along the Little Missouri River and Deadman Creek, but they were not common over the entire Ross Project area.

The impact from vegetation removal and surface disturbance would affect approximately 113 ha [280 ac] of land, or about 16 percent of the Proposed Action's area. Construction would be phased over time, further reducing the amount of surface area disturbed at any one time. Noxious weeds would be controlled with appropriate spraying techniques. Therefore, the impacts to terrestrial vegetation would be SMALL.

In addition, the potential impacts to vegetation during the Proposed Action's construction would be mitigated by the Applicant's ensuring that disturbed areas would be both temporarily and permanently revegetated in accordance with WDEQ/LQD regulations and its WDEQ Permit to Mine. The Applicant would seed disturbed areas to establish a vegetative cover to minimize wind and water erosion and the invasion of undesirable plant species. The impacts would be further mitigated by a phased approach to construction, and therefore surface disturbance would be phased. A temporary seed mix could be used in wellfields and other areas where the vegetation would be disturbed again prior to final decommissioning and final revegetation. The temporary seed mix typically would consist of one or more of the native wheatgrasses (e.g., western wheatgrass and thick-spike wheatgrass). Permanent seeding is accomplished with a seed mix approved by the WDEQ/LQD and with County conservation district requirements. Two permanent reclamation seed mixtures (upland and pastureland/hayland) would be used to reseed disturbed areas. Wellfield areas would be fenced as necessary to prevent livestock access, which would also enhance the establishment of temporary vegetation (Strata, 2011a). The Applicant would conduct weed control as needed to limit the spread of undesirable and invasive, non-native species on disturbed areas (Strata, 2011a).

Wildlife

As discussed in GEIS Section 4.4.5, in general, wildlife species would disperse from an area undergoing construction, although smaller, less-mobile species could perish during clearing and grading. Habitat fragmentation, temporary displacement, and direct or indirect mortalities are possible, and thus the GEIS concluded that construction impacts on wildlife could range from SMALL to MODERATE (NRC, 2009). These types of impacts could be mitigated during the Proposed Action if standard management practices suggested by the WGFD were to be followed. Moreover, impacts on raptor species from power distribution lines could be mitigated by the Applicant's following the Avian Power-Line Interaction Committee (APLIC) guidance and avoiding disturbance of areas near active nests and prior to the fledging of young (APLIC, 2006).

Mammals

The Ross Project area provides year-long range to pronghorn antelope, and winter/year-long range for mule deer, but it is considered outside of the normal range for white-tailed deer and elk (see SEIS Section 3.6.1). White-tailed deer, however, were observed during the Applicant's wildlife surveys as were pronghorn antelope. No crucial big-game habitats or migration corridors are recognized by the WGFD at the Ross Project area or the surrounding 1.6-km [1-mi] perimeter. (A crucial range or habitat is defined as any particular seasonal range or habitat component that has been documented as the determining factor in a population's ability to maintain and reproduce itself at a certain level.) Therefore, there would be no direct impact on big-game's crucial habitat, critical or key winter or summer ranges, or migration corridors. Direct impacts on white-tailed deer and elk could include direct loss and modification of habitat, increased mortality from increased traffic collisions on local and regional roads, increased competition for and reduction of available forage, increased conflicts with vehicles because of changes in wildlife movement patterns, and increased disturbance due to the presence of humans. White-tailed deer and elk could be indirectly affected during construction by displacing portions of these populations from the Ross Project area into offsite suitable regional habitat. Because the Project area provides only nonessential habitat for white-tailed deer and elk, impacts to these species would be SMALL.

The direct impacts on pronghorn antelope and mule deer could be the same as those described previously for white-tailed deer and elk. The construction phase of the Proposed Action has been estimated to last 12 months. Adequate habitat for pronghorn antelope and mule deer exists in the surrounding area, and these species could return to the areas affected by construction when the activities are complete. The staged restoration of disturbed areas that the Applicant proposes would provide grass and forage within a few years of habitat disturbance. The movement of big game through the Ross Project would not be significantly impacted by the Proposed Action. The Applicant has committed to implementing mitigation measures, such as reduced speed limits to reduce the risk of vehicular collision, fences designed to permit big game passage, and use of existing roads where possible to avoid altering wildlife movement patterns. Because pronghorn antelope and mule deer are highly mobile species, the potential impact to these species would be SMALL.

A variety of small- and medium-sized mammals are also potentially found on the Ross Project area (see SEIS Section 3.6.1) (Strata, 2011a). These include a variety of predators and furbearers, such as coyote, red fox, raccoon, bobcat, badger, beaver, and muskrat. Prey species observed during the Applicant's field surveys included rodents (e.g., mice, rats, voles, gophers, ground squirrels, and chipmunks), jackrabbits, and cottontails. These species are cyclically common and widespread throughout the region and are important food sources for raptors and other predators.

Medium-sized mammals (e.g., coyotes, foxes) could be temporarily displaced to other habitats during construction activities. Direct losses of limited-mobility, small-mammal species (e.g., voles, ground squirrels, mice) could be higher than for other wildlife because of the likelihood they would retreat into burrows if disturbed, and thus potentially be killed by topsoil scraping or staging activities. However, given the limited, noncontiguous area that would be disturbed (approximately 113 ha [280 ac]), no major changes or reductions in small- or medium-sized mammal populations would be expected. The species that occur in the area have shown an ability to adapt to human disturbance in varying degrees, and each also has a high reproductive potential and tend to re-occupy and adapt to altered or reclaimed areas quickly. Because only a few individuals would be affected, and most mammal species would likely travel to suitable habitat near the Ross Project area during its construction, the Proposed Action would have a SMALL impact on these mammals.

Birds

Potential impacts to upland game birds at the Ross Project area include nest destruction or nest desertions, reproductive failure as a result of proposed construction activities and increased presence of humans, or increased mortalities associated with traffic. Four upland game-bird species occur within or near the Ross Project area (i.e., wild turkey, sage-grouse, sharp-tailed grouse, and mourning doves) (Strata, 2011a). Suitable habitat (for nesting, brood-rearing, and foraging) for these four species exists in the Ross Project area; however, as previously discussed, there are no sage-grouse core areas or connectivity corridors within the Project area. Because of the type of disturbance (the relatively small areas of disturbance and the sequential nature of the disturbance), impacts to upland game birds as a result of the Proposed Action would be SMALL.

Potential impacts to raptors within the Ross Project area also include nest desertions or reproductive failure as a result of construction activities and increased presence of humans; temporary reductions in prey populations; and mortality associated with traffic. Six raptor species on the USFWS SMC list (i.e., bald eagle, Swainson's hawk, ferruginous hawk, golden eagle, prairie falcon, and short-eared owl) have been observed within or near the Project area (Strata, 2011a). Swainson's and ferruginous hawks are the only species known to nest in the area. One intact raptor nest (a Swainson's hawk nest, No. SH1) was located at the Ross Project area during the Applicant's field surveys. Seven intact nests and one nest no longer intact were located with 1.6 km [1 mi] of the Project area. The nest within the Ross Project area would not be directly disturbed during the Proposed Action's construction, so nesting raptors would not be directly impacted. Foraging raptors are expected to be able to avoid any areas of disturbance. Because of the type of disturbance (again, the relatively small areas of disturbance and the sequential nature of the disturbance) and the fact that no raptor nests would be directly affected, impacts to raptors during the Proposed Action would be SMALL.

Potential impacts to nongame or migratory birds within the Ross Project area include nest destruction or desertions, or reproductive failure as a result of construction activities during the Proposed Action. Increased mortality associated with the increased traffic during the construction phase could also occur. The field surveys completed by the Applicant identified 27 nongame or migratory avian species within the Ross Project area (Strata, 2011a). Because of the type and sequence of land disturbance, the Proposed Action's construction impacts to nongame or migratory birds would be SMALL.

Thus, all impacts to terrestrial wildlife would be SMALL.

Reptiles, Amphibians, and Aquatic Species

Potential impacts to reptiles, amphibians, and fish during construction of the Proposed Action would primarily be the result of the mortality of individuals and destruction of habitat. Sediment loads in surface waters and wetlands from surface-disturbing activities could also potentially impact aquatic habitat, although potential impacts would be greatly reduced through sediment-control BMPs. Up to 0.8 ha [2 ac] of wetland habitat could be disturbed as a result of construction; however, all wetland disturbance would be mitigated in accordance with USACE requirements found in the CWA permit.

Because of the type of disturbance, which would be relatively small, and the sequential nature of the disturbance as well as the fact that aquatic habitats would be avoided if at all possible during construction, impacts to reptiles, amphibians, and fish during the Proposed Action would be SMALL.

Protected Species

As discussed in SEIS Section 3.6.1.4, a protected species of bird, the Greater sage-grouse could occur on the Ross Project area. The nearest active sage-grouse lek (i.e., Cap'n Bob), a mating-strutting area for male sage-grouse, is located approximately 3.5 km [2.2 mi] southeast of the Ross Project area. There is also an inactive-status lek (for 2010) within 1.6 km [1 mi] of the Project's boundary. Wyoming policy states that surface-disturbing and/or disruptive activities are prohibited or restricted from March 15 through June 30. This restriction is typically only applied to suitable sage-grouse nesting and early brood-rearing habitat, within mapped habitat important for connectivity, or within 3 km [2 mi] of any occupied or "undetermined lek." The leks observed by the Applicant are outside of the Proposed Action area and are not located in proximity to any proposed construction or operation activities at the Proposed Action. However, if a Greater-sage-grouse lek were to be identified within the Ross Project area at any time during the Ross Project, including construction, the Applicant would follow WGFD policy regarding construction-activity restrictions. The Applicant would continue to consult with WGFD and WDEQ/LQD to determine if a sage-grouse monitoring, protection, and habitat enhancement plan would be necessary for the Ross Project, and a plan would be developed and implemented, if warranted.

During the Applicant's field surveys, the northern leopard frog was the only U.S. Bureau of Land Management (BLM)-listed reptile, amphibian, or fish sensitive species actually observed in the Ross Project area; three amphibian and five reptile Wyoming SOC were observed (Strata, 2011a). Impacts to protected avian, amphibian, and reptile species would be no different than those for other similar species because the Applicant would observe appropriate activity restrictions, attempt to avoid aquatic habitats during road construction, and implement the mitigation measures below.

The potential impacts to ecological resources associated with construction activities during the Proposed Action would be limited due to the relatively small area of surface disturbance. Nevertheless, mitigation measures to prevent or further reduce impacts to wildlife would include one or more of the following, as addressed by the various regulatory and permit-issuing agencies:

- Design of fencing to permit big-game passage as required by the WGFD.
- Use of existing roads when possible and location of newly constructed roads to access more than one well location according to BLM requirements.

- Implementation of speed limits to minimize collisions with wildlife, especially during the breeding season, according to a MOU between the Applicant and Crook County transportation authorities (Strata, 2011d).
- Adherence to temporal and spatial restrictions within specified distances of active sage-grouse leks as determined through consultation with the WGFD and the WDEQ/LQD.
- If direct impacts to raptors or migratory-bird SMC result from construction, a monitoring and mitigation plan (MMP) for those species would be prepared and approved by the USFWS, and would include one or more of the following provisions:
    - Relocation of active and inactive raptor nests that would be impacted by well drilling and other construction activities in accordance with the approved raptor MMP
    - Institution of buffer zones to protect raptor nests where necessary and restriction of uranium-recovery-related disturbances from encroaching within buffers around active raptor nests (from egg-laying until fledging) to prevent nest abandonment or injury to eggs or young
    - Restoration of the ground cover necessary to attract and sustain a suitable raptor-prey base after drilling, construction, and future uranium-recovery activities, and
    - Requirement for the use of raptor-safe construction for overhead power lines according to current guidelines and recommendations by the APLIC and/or the USFWS.
- Restoration of sagebrush and other shrubs on reclaimed lands and grading of reclaimed areas to create swales and depressions for sagebrush obligates (sagebrush obligates are those species that need sagebrush to survive, e.g., sage grouse) and their young per WDEQ/LQD requirements.
- Restoration of preconstruction, native habitats for species that nest and forage in those vegetative communities according to WDEQ/LQD and WGFD requirements.
- Restoration of diverse landforms, replacement of topsoil, and the construction of brush piles, snags, and/or rock piles to enhance habitat for wildlife per WDEQ/LQD requirements.
- Restoration of habitat provided by jurisdictional wetlands as required by both the WDEQ/LQD and the USACE.

Thus, with the measures listed above, the environmental impacts to terrestrial, aquatic, and protected species during Ross Project construction would be SMALL.

4.6.1.2 Ross Project Operation

As discussed in GEIS Section 4.4.5, alteration of wildlife habitats could result from uranium-recovery activities (e.g., fencing, traffic, and noise), and conflicts between species habitat and uranium-recovery activities could occur (NRC, 2009). The GEIS further noted the occurrence of temporary contamination of soils from spills and leaks during ISR operation. However, rapid discovery and response to spills and leaks (i.e., spill containment and cleanup of potentially impacted soil), and the eventual survey for radiation during decommissioning, would limit the magnitude of overall impacts to terrestrial ecology during the Proposed Action's operation. Leak-detection systems and spill-response plans would reduce the potential impacts to aquatic species from spills around wellheads and leaks from pipelines by preventing contamination of soils, surface waters, or wetlands. Additional mitigation measures such as perimeter fencing, surface-impoundment netting or other avian deterrents, and periodic wildlife surveys would also limit impacts during the Proposed Action's operation.

Terrestrial Species

Vegetation

During the operation phase of the Proposed Action, the wellfields and CPP would be frequently accessed by use of the existing roads. The installation and operation of the wellfields would involve the excavation of trenches for trunk lines and utilities; this surface disturbance would increase the susceptibility of the disturbed area to invasive and noxious weeds. However, surface disturbance would continue to be minimized during operation as new, additional wellfields are installed, and vehicular access would be restricted to specific roads. The potential for these impacts to occur during operations is less than that during construction, due to fewer hectares or acres of land being disturbed. There is a potential for impacts to vegetation from spills around wellheads and leaks from pipelines during the Ross Project's operation. Based upon the small amount of land that would be disturbed during operation, and the lower number of vehicles accessing the Ross Project, the impacts would be SMALL during the operation phase of the Proposed Action.

Wildlife

Wildlife use of areas adjacent to and near the Proposed Action would likely initially decline because of human presence during the Project's operation and steadily increase to near-normal levels once animals become habituated to the uranium-recovery activities. Because wildlife could be in fairly close proximity to the CPP, surface impoundments, wellfields, and roads, some impacts to wildlife would be expected from direct conflict with vehicular traffic and the presence of Strata's onsite personnel. In addition, wildlife could be exposed to contaminated soil resulting from spills and leaks. All of these impacts would be SMALL, however, because only a few individual animals would be affected, the potential for spills and leaks is low, and the continued existence of any particular species at the Ross Project area would not be affected. Potential impacts to terrestrial wildlife during the Ross Project's operation phase from process waste water and sediment in the facility's lined surface impoundments would be reduced by the fencing that would be installed around the entire facility (i.e., around the CPP and the surface impoundments) (see Figure 3.1 in SEIS Section 3.2). Therefore, during the operation of the Proposed Action, the potential impacts to wildlife would be SMALL.

Mammals

The potential impact to big game during the Proposed Action's operation phase would either be similar to or less than that described earlier for the construction phase, because limited earth-moving activities would occur. Therefore, there would be only SMALL impacts to big game species during the operation phase of the Proposed Action. The potential impacts to other mammals during operation of the Ross Project would also be similar to or less than that described earlier for the construction phase. Because only a few individual mammals would be affected, and most mammal species would likely travel to suitable habitat outside of the operating facility and wellfields, the Proposed Action would have SMALL impacts on these mammals during its operation.

Birds

The potential impacts to upland game birds, waterfowl, shorebirds, and raptors during the Proposed Action's operation would either be the same or less than that described earlier for the construction phase because earth-moving activities would be more limited during its operation phase.

For avian control at the surface impoundments, the Applicant is considering three options, including netting, "bird balls" (hollow or water-filled balls), or a radar-hazing system (Strata, 2012a). Following an extensive literature review and contact with knowledgeable individuals regarding avian deterrents for impoundments, a radar-hazing system has been identified by the Applicant as the most likely solution for its deterring avian species from the surface impoundments associated with uranium-recovery activities. This system uses radar to detect incoming waterfowl and then uses hazing techniques (primarily noise) to scare the birds away. The avian-deterrent system would require setup and routine maintenance, including calibration of the radar to site-specific conditions to avoid false activations. The potential for other wildlife to access the surface impoundments would be minimized by the installation of fencing around the CPP and surface impoundments. Additionally, BMPs would be the same as those used by the Applicant during construction; therefore, the potential impacts of the Proposed Action's operation would be SMALL for these birds.

Reptiles, Amphibians, and Aquatic Species

The potential impact to reptiles and amphibians from the Proposed Action's operation would be comparable to that described earlier for its construction. Because the potential habitat for reptiles and amphibians is limited within the Ross Project area, the potential impacts would be limited and SMALL. Because of the limited occurrence of surface water and, thus, of aquatic species at the Project area, the potential impact to aquatic species would be SMALL.

Protected Species

No impacts to Federally-listed threatened and endangered species would occur during the operation phase because these species have not been identified at the Ross Project area. Potential impacts to the protected species during the Project's operation would be the same or less than those discussed above for the construction of the Ross Project because there would be fewer humans present outdoors on the site itself and fewer vehicles being used. In general, outdoor activities would be limited. Thus, the impacts would be SMALL to all protected species. In addition, mitigation measures implemented during the Project's construction would continue to be employed to ensure that potential impacts to protected species remain SMALL.

As noted in SEIS Section 4.6.1.1, specific mitigation measures for all ecological resources would be required by several Federal and State agencies; these measures would be implemented during the Proposed Action's operation. These include the Applicant reseeding disturbed areas with WDEQ- and County-approved seed mixtures to prevent the establishment of competitive weeds and monitoring of invasive and noxious weeds. If these weeds become an issue, then the Applicant would employ other control alternatives, such as the application of herbicides, to minimize their impacts. In addition, impacts to vegetation and wildlife resulting from spills and leaks would be mitigated by the Applicant's use of BMPs. BMPs would include several leak-detection systems and spill-response plans, where released solutions would be contained and affected soils would be removed, thereby reducing the impacts of such releases.

All impacts of the Proposed Action's operation would be SMALL to the ecology of the area.

4.6.1.3 Ross Project Aquifer Restoration

In GEIS Section 4.4.5, the potential impacts to ecological resources during the aquifer-restoration phase of an ISR facility are described (NRC, 2009). These impacts were noted to include habitat disruption. As noted above, however, in the case of the Ross Project, the already in-place infrastructure from the construction and operation phases (i.e., roads) would continue to be used, and little additional ground disturbance would be expected.

Contamination of soils and surface waters could result from spills and leaks, which could impact the ecological resources of the Ross Project. The leak-detection systems and spill-response protocols described earlier, and the eventual radiation survey of all potentially impacted soils and sediments, would limit the magnitude of overall impacts to terrestrial and aquatic ecology during the aquifer restoration at the Proposed Action. In addition, continued implementation of mitigation measures, such as perimeter fencing and the avian-deterrent system would ensure that impacts to vegetation and terrestrial species would be minimized during aquifer restoration at the Ross Project. Also, because the existing infrastructure would be in place, the potential impacts to ecological resources from aquifer-restoration activities would be similar or less than that experienced during the Proposed Action's operation phase, wildlife would have already retreated or learned to tolerate the presence of humans or noise. Therefore, the potential impacts to vegetation and wildlife would be SMALL.

There would be no expected impacts to protected species during aquifer restoration beyond those which occurred during the construction and operation phases of the Proposed Action, because the existing infrastructure would be in place and no further excavation of habitat would be necessary. Additionally, to date, no threatened or endangered species have been observed at the Ross Project area. Therefore, the overall impact to threatened, endangered, or protected species during aquifer restoration would be SMALL.

4.6.1.4 Ross Project Decommissioning

As discussed in GEIS Section 4.4.1, temporary land disturbance during the decommissioning of ISR facilities would be a result of excavation and disturbance of soils; excavation and removal of buried pipelines; and the decontamination, dismantling, demolition, and removal of buildings and structures (NRC, 2009). However, any recontouring of land and its revegetation would assist in the restoration of habitats previously altered during an ISR facility's construction and operation. Wildlife would be temporarily displaced during the decommissioning phase, but species could return upon completion of this phase, when the restoration of vegetation and habitat has been accomplished. Although facility decommissioning and site restoration would result in temporary increases in sediment load in local streams, aquatic species would recover quickly as the additional sediment load decreased. For all of these reasons, the GEIS concluded the overall potential impact during the decommissioning of an ISR facility would be SMALL.

The Proposed Action's decommissioning would be phased over approximately the last five years of the Ross Project. The Applicant estimates a 12-month duration for the decommissioning of the CPP, surface impoundments, pipelines, roads, and other infrastructure (if the CPP does not continue to operate for satellite and/or other offsite uranium-loaded IX-resin processing). Stockpiled topsoil would be used to regrade the land to its pre-licensing baseline contours, as required, and be reseeded with native vegetation when the buildings and structures are removed as described earlier (see SEIS Section 2.1.1). No loss of vegetative communities beyond that disturbed during the construction phase would occur. Pipeline removal would impact vegetation that could have re-established itself, although this, too, would be temporary as the disturbed areas are reseeded. Thus, the impacts of the Proposed Action's decommissioning would not be expected to be greater than those experienced during its construction, and mitigation measures would continue to be employed. Consequently, the decommissioning impacts to vegetation would be SMALL.

The decommissioning of the Proposed Action would create increased noise and traffic as buildings and structures are decontaminated, dismantled, demolished, and transported offsite to an appropriate waste-disposal facility. During this time, wildlife could either come in conflict with heavy equipment or be disrupted by the higher-than-normal noise. As a result of these impacts, wildlife would move elsewhere either on the Ross Project area or onto other lands. Temporarily displaced wildlife could return to the Ross Project area after the Proposed Action's decommissioning and site restoration are complete. Further, as required by NRC regulations, the Applicant would be required to submit a decommissioning plan for Commission review and approval, which would address ecological impacts such as these. Thus, decommissioning impacts of the Ross Project would not be more than those experienced during the Proposed Action's construction. Thus, the impacts to terrestrial wildlife, aquatic species, and protected species during decommissioning would be SMALL.

4.6.2 Alternative 2: No Action

Under the No-Action Alternative, the Ross Project would not be licensed and the land would continue to be available for other uses. However, activities such as the plugging and proper abandonment of existing drillholes would occur as well as continued environmental monitoring, data collection, and field surveying. These activities, however, would be temporary in nature and the surface area affected would be very limited.
The Ross Project area would continue to support vegetation communities and wildlife habitat typical of the region (as described in SEIS Section 3.). Land use would continue as pastureland, and existing grazing leases would continue. Grazing of existing vegetation, particularly in the grassland communities, would continue. Existing wildlife on the Ross Project area would be affected only if continued cattle grazing destroys wildlife habitat or if species are displaced by cattle populations because of lack of forage and cover. However, in this Alternative, only a few individual species would be affected, and they would relocate to suitable nearby habitats. Therefore, vegetation and wildlife impacts would be SMALL.

4.6.3 Alternative 3: North Ross Project

Under Alternative 3, the North Ross Project would generally be the same as the Proposed Action, except that the facility (i.e., the CPP, associated buildings, and auxiliary structures as well as the surface impoundments) would be located to the north of where it would be located in the Proposed Action, as described in SEIS Section 2.1.3. The Applicant's construction of the

CPP at this location would produce a slight increase in the travel distance for vehicles accessing the Ross Project's facility and wellfields. This could slightly raise the potential for vehicular collisions with wildlife. However, the potential impacts during construction of Alternative 3 would be similar to those described for the Proposed Action. In addition, the surface impoundments would be located farther away from the Oshoto Reservoir, which would reduce the likelihood of waterfowl and other wildlife entering the surface impoundments. This would reduce the impacts to wildlife during the operation and aquifer-restoration phases of Alternative 3. All other impacts would be the same as for the Proposed Action, and the same mitigation measures would be implemented. The impacts of the North Ross Project would be of the same magnitude as during the Proposed Action, and they would be SMALL.

4.7 Air Quality

The Proposed Action could impact air quality during all phases of the Project's lifecycle. As discussed in GEIS Section 3.4.6 and in SEIS Section 3.7.1, Wyoming is generally a very windy state and ranks first in the U.S. with an annual average wind speed of 6 m/s [13 mi/hr]. During winter, wind speeds in Wyoming can reach 13 – 18 m/s [30 – 40 mi/hr] with gusts to 22 – 27 m/s [50 – 60 mi/hr] (NRC, 2009). During the 12 months of pre-licensing baseline monitoring at the Ross Project area, the onsite meteorology station recorded average annual wind speeds of 19 km/hr [12 mi/hr], with a maximum wind speed of 74 km/hr [46 mi/hr]. Southerly winds were predominantly recorded at the Ross Project area. These data suggest that combustion-engine and fugitive-dust emissions from the Ross Project would be moved by the highest wind speeds to the south-southeast, away from the Project area.

In addition to the winds, the Ross Project area and the surrounding region receive relatively little rainfall, with average annual precipitation ranging from 25 – 38 cm [10 – 15 in]. The region receives an average annual snowfall of 127 – 152 cm [50 – 60 in]; approximately one-half of the precipitation is associated with spring snows and thunderstorms. At the Ross Project meteorological station, the total precipitation measured in 2010 was 24.8 cm [9.8 in] (Strata, 2011a).

Because the Ross Project area is very dry and very windy, fugitive dust is readily generated and is a significant air pollutant (i.e., unwanted chemical vapor, gaseous, or particulate emissions found in the air, especially in disturbed land areas and areas where native vegetation has been removed). Conversely, these high winds could also more rapidly disperse air pollutants, lowering their concentrations. But the arid conditions in the Ross Project area are not as conducive to removal of suspended dust as areas receiving more rainfall. Therefore, in general, other mechanisms besides precipitation would need to be implemented within the Ross Project area to minimize fugitive dusts and other air emissions.

Air pollutants can also be affected by the regional landscape of an area. The Ross Project's topographical setting—an area consisting of rolling hills and intermittent drainages—provides some topographic breaks (see SEIS Section 2.1.1) (Strata, 2011a). In addition, the nearest mountain range is the Black Hills, whose westernmost edge is approximately 32 km [20 mi] from the eastern boundary of the Ross Project area. It has been suggested that this range may shield easterly winds and channel predominant winds into a north-south pattern (Strata, 2011a).

Finally, atmospheric-stability classification and mixing height are environmental variables that also influence the ability of the atmosphere to disperse air pollutants. The "stability class" is a measure of atmospheric turbulence and "mixing height" characterizes the vertical extent of contaminant mixing in the atmosphere. Stability-class information was collected at the Ross Project meteorological station (Strata, 2011a) and indicated that the class distributions were predominantly neutral (approximately 62 percent of the time).

This background information indicates that potential impacts to air quality could occur during all phases of the Ross Project, and the impacts could be related to both the particulate emissions (e.g., fugitive dust) as well as gaseous emissions (or effluents) (e.g., combustion-engine emissions) that would be released during the Ross Project. Consistent with the GEIS, the air quality impacts analyzed in Section 4.7 only cover nonradiological emissions. Radiological emissions and dose information are addressed in the public and occupational health and safety impacts analyses in Section 4.13.

4.7.1 Alternative 1: Proposed Action

Alternative 1, the Proposed Action, consists of four phases: construction, operation, aquifer restoration, and decommissioning of a uranium-recovery facility and wellfields. The GEIS in Section 4.2.6 as cited in GEIS Section 4.4.6 determined that uranium-recovery facilities are not, in general, major air-emission sources (NRC, 2009). Given the low levels of particulate and gaseous emissions predicted in GEIS Section 4.2.6, the GEIS determined that the overall potential air-quality impacts of an ISR facility are SMALL, if the following three conditions could be applied to a specific facility: 1) particulate and gaseous emissions are within regulatory limits and requirements; 2) air quality in the [region] is in compliance with the National Ambient Air Quality Standards (NAAQS); and, 3) the facility would not be classified as a major source under the New Source Review or operating (Title V) air-quality permit programs which were described in the GEIS (NRC, 2009). As noted in GEIS Section 4.4.6, the entire NSDWUMR is an attainment area for NAAQS (see SEIS Section 3.7.3).

These three conditions do describe the proposed Ross Project area. The Ross Project would be designed to ensure that its emissions are within regulatory limits and requirements; it would be located in the NSDWUMR which, as described in SEIS Section 3.7.3, is an attainment area for all NAAQS primary pollutants (i.e., is in compliance with NAAQS) (see Table 3.17 in SEIS Section 3.7.3); and, the Ross Project would not be classified as a major air-emissions source under New Source Review or Title V of the CAA. The Ross Project also would not impact the nearest prevention of significant deterioration (PSD) Class I areas. These conditions would apply to all phases of the Ross Project.

4.7.1.1 Ross Project Construction

Generation of fugitive dust during land-disturbing activities conducted during ISR facility construction would be the same as discussed in GEIS Section 4.3.6.1, and would be short-term. Other air-quality impacts from fugitive dust would result from road dust being suspended by moving vehicles over nearby and Ross Project roads as well as from construction equipment while it is used to clear and grade portions of the Project area where construction would occur. During the Proposed Action's construction phase, the Applicant estimated a disturbance area of 113 ha [280 ac] during construction of Ross Project buildings and auxiliary structures, surface impoundments, access roads, and other infrastructure. Traffic associated with the Ross Project would use the primary access route of New Haven Road or D Road, which are paved, such that impacts, including fugitive dust generation, would be limited to more occasional access on local dirt roads within the Ross Project area.

Fugitive dust and other particulate emissions are regulated under the *Wyoming Air Quality Standards and Regulations* (WAQSR), Chapter 3, Section 2(f), "Fugitive Dust." The WAQSR quantifies opacity and emission-specific constituent concentrations that apply exclusively to any point sources at the Ross Project (e.g., combustion engines) (WDEQ/AQD, 2011). In contrast to point sources, WDEQ/Air Quality Division (AQD) also regulates generalized fugitive-dust emissions by imposing BMPs rather than numerical limits.

In a study of air-quality impacts of road construction, Roberts et al. (2010) found that near-road pollutant concentrations decline substantially within 100 – 150 m [330 – 490 ft] of the road, and they can reach background conditions at approximately 300 – 500 m [980 – 1,600 ft] from the road (Roberts, 2010). Similarly, a study by Countess et al., undertaken to improve the modeling of windblown and mechanically re-suspended fugitive-dust emissions, found that not all particles that could be suspended are in fact transported long distances; this is due to deposition rates, vertical mixing, and transport times. Countess found that PM_{10} (less than 10 μm in diameter) particulates (i.e., dusts) deposit relatively quickly at a rate of 0.5 – 5 cm/s [0.2 – 2 in/s]; $PM_{2.5}$ particulates deposit more slowly at 0.05 – 0.2 cm/s [0.02 – 0.08 in/s], with a continuum of values between these two extremes for cropland, prairie land, and paved surfaces. In general, the fraction of the mechanically generated fugitive dust from roads and bare surfaces that is removed from the atmosphere by gravitational settling and by impacting nearby obstacles (such as vegetation) is much larger than that associated with fugitive windblown dust. This is because of the fact that the mechanically-generated particulates tend to remain closer to the ground for longer periods after suspension in the air than windblown dusts, such that there is a higher probability that these mechanically generated particles, such as those generated by vehicles, are removed from the atmosphere close to their sources.

Windblown fugitive-dust emissions can be lofted vertically to great heights above the ground by the sustained energy provided by the vertical component of the wind, especially for strong winds, and consequently, can be transported much longer distances from their sources than mechanically generated fugitive-dust emissions. A typical wind speed of 2.5 m/s [8 ft/s] results in the transport of particulates to 100 m [330 ft] in 40 seconds, 1,000 m [3,300 ft] in 400 seconds (or approximately 7 min), and 10,000 m [33,000 ft] in 4,000 seconds [1.1 hr]. In general, PM_{10} particulates are deposited at a rate that is about an order of magnitude greater than $PM_{2.5}$ because of the greater gravitational settling velocity (Countess, 2001). These data indicate that the majority of fugitive-dust impacts would not extend beyond the 80-km [50-mi] radius around the Ross Project area, although winds with large vertical components can transport dust over longer distances when they occur. This physical phenomenon is a *de facto* mitigation measure

The greatest combustion-engine gaseous emissions from diesel- and gas-powered equipment operation would occur primarily during the construction and decommissioning phases of the Ross Project because of the equipment used during those phases. To determine the potential air-quality impacts from the passenger vehicles of the commuting workforce as well as delivery and shipment trucks to and from the Ross Project area, the Applicant provided the anticipated number of passenger vehicle trips to and from the Ross Project during each of the Ross Project's phases (see Table 4.2) (Strata, 2011a; Strata, 2012a). The Applicant also estimated the number of each type of supply, product, and waste shipment during each phase. Finally, the Applicant estimated the annual operating time of these vehicles and other construction equipment (Strata, 2011a).

All of this information is important when modeling air-quality impacts, as the Applicant did for each phase of the Proposed Action. In its air-quality modeling results, the Applicant provides (primarily diesel) combustion-engine emission and fugitive-dust estimates. These modeled emissions are provided in Table 4.4 for each phase of the Ross Project (Strata, 2011c; Strata, 2011a). In the NRC's evaluation, the assumptions used by the Applicant in its air-quality modeling efforts were conservative (e.g., each worker was assumed to commute to and from the Ross Project area alone). All emission levels were estimated to be below the major-source threshold for NAAQS attainment areas.

In order to determine impacts to air quality from diesel combustion emissions, the GEIS (NRC, 2009) reported emissions for the ISR facility in Crownpoint, New Mexico, as described in the NRC's Environmental Impact Statement (EIS) for that facility (NRC, 1997). Therefore, emissions from the Crownpoint ISR facility were examined for relevance to the Ross Project. Estimated maximum production of the Ross Project and Crownpoint are both 3 million pounds per year. The estimated gaseous particulate and gaseous emissions were presented in the Crownpoint EIS and in Table 2.72 of the GEIS. The results of the Crownpoint preliminary emissions inventory were similar to the Ross Project, with the exception of particulate matter (PM). PM emissions associated with the Crownpoint facility were approximately 10 T/yr, while combustion and fugitive PM emissions for the Ross Project were estimated at 177 T/yr. In addition, estimated combustion emissions for the Ross Project were significantly higher than those presented in the Crownpoint EIS. The differences can be attributed to the source of emissions factors (AP-42 emission factors were used in the Ross Project, which are significantly more conservative than the assumptions used for the Crownpoint analysis) as well as the estimated operating hours associated with each piece of equipment. The depth to ore deposits is greater at the Ross Project site than at Crownpoint, which would require that the equipment to reach the ore at the Ross Project would be operated for longer time periods and thus create more emissions.

Table 4.4
Non-Radioactive Emissions Summary

Construction Equipment and Truck Tailpipe Emissions (t/yr [T/yr])

Phase	TOC	NO_X	CO	PM_{10}	SO_2	CO_2
Construction	13.27 [12.04]	181.77 [164.90]	39.50 [35.83]	11.89 [35.83]	10.83 [8.82]	7,014.9 [6,363.8]
Operation	3.09 [2.80]	38.78 [35.18]	8.36 [7.53]	2.75 [2.49]	2.56 [2.32]	1,438.6 [1,303.3]
Aquifer Restoration	1.8 [1.63]	22.7 [20.6]	4.9 [4.5]	1.61 [1.46]	1.50 [1.36]	842.6 [764.4]
Decommissioning	5.1 [4.63]	64.3 [58.3]	13.9 [12.6]	4.56 [4.14]	4.25 [3.86]	2,385.0 [2,163.6]

Fugitive-Dust PM_{10} Emissions (t/yr and T/yr)

Phase	Activity	PM_{10} (t/yr)	PM_{10} (T/yr)
Construction Equipment	Site preparation for facility	10.60	11.69
Construction Equipment	Wellfield and roads preparation	15.86	17.48
Construction	Vehicles on unpaved roads	129.40	142.64
Construction	Wind erosion from exposed areas	11.25	12.40
Operation	Vehicles on unpaved roads	13.23	14.29
Operation	Wind erosion from exposed areas	1.03	1.14
Operation	Year five of ISR operation	5.69	6.27
Aquifer Restoration	Vehicles on unpaved roads	8.89	9.80
Aquifer Restoration	Wind erosion from exposed areas	1.03	1.14
Decommissioning	Site preparation for CPP	2.01	2.21
Decommissioning	Wellfield and roads preparation	4.64	5.12
Decommissioning	Vehicles on unpaved roads	70.52	77.73
Decommissioning	Wind erosion from exposed areas	5.79	6.38

Storage Tank Emissions Totals (kg/yr and lb/yr)

Hydrochloric Acid	42.92	47.31
Hydrogen Peroxide	0.98	1.08
Diesel	10.80	11.90
Gasoline	1,176.99	1,297.41

Source: Strata, 2011a; Strata, 2011b.

Note: t = Tonnes, or Metric tons.
T = Short tons, or U.S. tons.

The annual average particulate concentration at Crownpoint was estimated to be less than 2 percent of the Federal $PM_{2.5}$ ambient-air standard, less than 1 percent of the previous Federal and current Wyoming PM_{10} ambient-air standards, and less than 2 percent of the Class II PSD allowable increment. However, this estimate for annual average particulate concentration did not categorize the particulates as PM_{10} or $PM_{2.5}$. The annual average SO_2 concentration was estimated in the Crownpoint analysis to be less than 1 percent of both the Federal and the Wyoming ambient-air standards and less than 1 percent of the Class II PSD allowable increment. Finally, the annual average NO_2 concentration at Crownpoint was estimated to be slightly over 2 percent of the Federal and Wyoming ambient-air standards, but less than 9 percent of the Class II PSD allowable increment. Therefore, although PM emissions at the Ross Project could exceed those at Crownpoint, the low percentages of the ambient air quality standards estimated for the Crownpoint facility emissions indicate that the Ross Project emissions would also be below NAAQS and PSD standards.

Additionally, the meteorology used at the Crownpoint site to estimate average annual air concentrations of emitted pollutants is more stable than at the proposed Ross Project site, based on review of wind stability classes. At Crownpoint, winds that fall into stability classes E and F occur over twice as frequently as winds in stability classes E and F at the Ross Project site. Good dispersion conditions (stability classes A through D) occur approximately 80 percent of the time at the Ross Project site versus approximately 55 percent of the time at the Crownpoint site. Based on the information reviewed, the dispersion conditions at the Ross Project site are more favorable than at Crownpoint and would therefore help to reduce the impacts due to PM emissions.

The Applicant proposes several onsite best available control technology (BACT) mitigation measures as well as many BMPs to control fugitive dust (e.g., fugitive dust would be minimized by the Applicant's wetting soils down during earth-disturbing activities). The Applicant's mitigation of fugitive dust from roads would also include setting appropriate speed limits for vehicle traffic, strategically placing water load-out facilities near access roads, using chemical dust suppressants (e.g., magnesium chloride), encouraging employee carpooling, and selecting road surfaces that would minimize fugitive dust. The placement of soil stockpiles on the leeward side of hills and the Applicant's prompt revegetation of disturbed areas would also reduce the potential for fugitive dust.

In addition, mitigation of all types of impacts to air quality, actual particulate- and gaseous-emission concentrations from the Ross Project area, would be required to be monitored and to comply with the conditions of the WDEQ-issued Construction Air Quality Permit No. CT-12198 (WDEQ/AQD, 2011). The gaseous-emission controls that the Applicant must employ during the Ross Project are outlined in its Air Quality Permit Application, which becomes part of the Air Quality Permit itself (Strata, 2011c). As specified, gaseous emissions would be controlled by the BACT for critical air-emission sources, such as acid-fume scrubbers on acid storage tanks (Strata, 2011c). Other BACTs are listed in the regulations implementing the CAA (40 CFR Subpart C).

The Applicant also plans to use visual observation on at least an hourly basis to monitor air quality in the Ross Project area and on a twice-daily basis at locations along the primary access route leading to the Ross Project. Further, to ensure compliance, the WDEQ/AQD would conduct regular inspections as well as unannounced inspections of permitted facilities (Strata,

2012a). Finally, the Applicant would respond aggressively to any dust-related concerns expressed by its employees, contractors, or members of the public (Strata, 2012a).
Given the predominant winds (in terms of both speed and direction) in the region, the remote location of the Ross Project area, and the BACT controls and BMPs that the Applicant is required by its Air Quality Permit to implement, many of the air emissions impacts from the Proposed Action would be fully mitigated (WDEQ/AQD, 2011). Because construction at the Ross Project would be typical of ISR facilities considered in the GEIS, anticipated gaseous-emission and fugitive-dust impacts would be limited in duration during the construction phase, and they would be mitigated. Therefore, the impacts of the Proposed Action on air quality during the construction phase would be short-term and SMALL.

4.7.1.2 Ross Project Operation

Air-quality impacts during the Ross Project's operation phase could include the same as those identified earlier for the construction phase of the Proposed Action (i.e., fugitive-dust and gaseous combustion-engine emissions), and they would be generated by many of the same sources. Estimates for these sources are provided by project phase in the Applicant's Air Quality Permit Application and are summarized here in Table 4.4 (Strata, 2011c).

Impacts from fugitive-dust and gaseous combustion-engine emissions during the operation phase would be less than the construction phase impacts, however, because fewer vehicles would be in use on or near the Ross Project area. Worker commutes would be approximately 60 workers during the operation phase (less than the 200 during construction). Construction-equipment operation (where most portions of the Ross Project area would have been cleared and graded during construction, so little earth movement would occur during operation—only the installation of wellfields would continue to generate fugitive dust) would diminish substantially, thus generating less fugitive dust and gaseous emissions.

Several point sources could release emissions while the Ross Project is in its operation phase. These point sources of gaseous emissions would be located at the CPP. These would include process-pipeline, process-vessel, and storage-tank vents; emergency generators and space heaters; and other sources such as storage vessels and tanks containing acids and bases (Strata, 2011a). Gaseous emissions from the yellowcake dryer are not expected because of the design of the proposed Ross Project's yellowcake circuit, which would include the BACT design of an indirect heat source as well as an integrated filter and condenser.

Gaseous emissions could also be released during the venting of excess vapor pressure from pipelines within the CPP, with small amounts of chemical vapor released. According to GEIS Section 4.4.6, excess vapor pressure in pipelines could be vented at various relief valves throughout the system. These emissions would be rapidly dispersed into the atmosphere, resulting in SMALL impacts (NRC, 2009). In addition, there could also be gaseous emissions during resin transfers or during resin elution (e.g., liquefied oxygen or carbon dioxide that come out of solution). The GEIS determined that a low volume of gaseous emissions would be released during resin transfer and elution at an ISR facility.

The Applicant's refilling of acid, sodium carbonate, or bicarbonate tanks would produce only small quantities of emissions; nonetheless, during the process of refilling the acid storage tanks, the BACT standard of a closed-loop system, which routes displaced vapors back to the tank truck during transfer, would be used (Strata, 2011c). The tanks would be located away from other chemical-storage tanks and away from the process vessels at the chemical-storage area (Strata, 2011b). Any emissions would be scrubbed for acid vapors prior to release to the atmosphere. Sodium carbonate and sodium bicarbonate would be delivered dry by truck and be blown into a storage silo; the vent of this silo would be filtered with a dust-vent bag to capture particulate emissions (Strata, 2011). The emissions from other storage vessels and tanks are summarized by the Applicant in its license application and additional information it has provided the NRC (Strata, 2011a; Strata, 2011b; Strata, 2012a) as well as in its Air Quality Permit Application (Strata, 2011c).

An emergency generator would be required to supply power to critical process equipment in the event of a power failure. The Applicant's Air Quality Permit restricts the generator's operation to 500 hours per year (WDEQ/AQD, 2011). Strata's Air Quality Permit Application provides a summary of generator emissions. Emissions from the vacuum dryers and space heaters in the CPP (natural-gas-burning equipment) are also listed in the emissions inventory (Strata, 2011c). Table 4.4 summarizes the Applicant's estimates of gaseous and particulate emissions, including from the point sources above, as they were modeled for the Air Quality Permit Application (Strata, 2011c).

Other types of air-quality-impact mitigation measures include gaseous-emission control systems that minimize emissions, BMPs that have demonstrated success at controlling emissions, and BACT engineering controls that reduce airborne emissions as well as minimize the potential for accidental releases. For example, powdered-form chemicals that would be delivered to the Ross Project would be delivered in covered trucks and unloaded through sealed pathways into tanks vented through dust-vent bags or fabric filters. Earth-moving and excavation activities would be governed by BMPs to minimize fugitive dust from disturbed areas, such as the Applicant watering dry soils thoroughly during such activities. To ensure that all requirements of the Air Quality Permit are being met, WDEQ/AQD would conduct regular inspections and unannounced visits of the Proposed Action (Strata, 2012a).

During operations, the Applicant will be required to monitor the effluent and selected environmental media to establish the impacts. Thus, the air-quality impacts of the Proposed Action during the operation phase would be SMALL.

4.7.1.3 Ross Project Aquifer Restoration

According to GEIS Section 4.4.6, potential nonradiological air-quality impacts during the aquifer-restoration phase of an ISR facility would include combustion-engine and fugitive-dust emissions from many of the same sources identified during the construction and operation phases. These impacts were found to be SMALL.

During the aquifer-restoration phase of the Proposed Action, the plugging and abandonment of injection and recovery wells would begin after a wellfield has undergone restoration and has met its ground-water quality goals. The emissions associated with the related equipment would be limited in duration and result in small, short-term effects. Vehicular traffic during the aquifer-restoration phase would be limited to delivery of supplies and commuting personnel; however, the workforce at the Ross Project would decrease to 20 during aquifer restoration and, consequently, the vehicular emissions of commuting traffic would substantially decrease. A significant decrease in the frequency of offsite yellowcake shipments would also occur as aquifer restoration proceeds. Thus, the emission-generating activities during the aquifer-restoration phase would be many fewer than during either the construction or operation phases. Therefore, air-quality impacts of aquifer restoration would be SMALL.

4.7.1.4 Ross Project Decommissioning

According to Section 4.4.6 of the GEIS, potential air-quality impacts during an ISR facility's decommissioning phase include fugitive dust, vehicle emissions, and the combustion-engine emissions from many of the same sources identified for the earlier phases of the facility's lifecycle (NRC, 2009). At the Ross Project, in the short term, emissions could increase, especially particulates, because decommissioning of the ISR facility would generate fugitive dust and the related construction equipment would also generate some gaseous emissions. The Applicant's dismantling and demolition of Ross Project buildings, structures, surface impoundments, and process equipment; its excavation and removal of any contaminated soils; its relocation of construction equipment to the different areas where decommissioning activities would take place; and its grading and re-contouring of the site during reclamation and restoration would produce particulate matter that would impact air quality. Combustion-engine gaseous emissions would also be generated by not only construction vehicles, but also vehicles transporting workers to and from the Ross Project (an additional 70 workers would be employed at the Ross Project during its decommissioning phase) (Strata, 2011a). Truck traffic related to the shipment of demolition and other wastes would also increase during the decommissioning phase as the wastes are shipped to various disposal facilities. However, the truck traffic would be only approximately 40 percent of that during the construction phase.

All of the respective mitigation measures identified for the other phases of the Proposed Action would continue to be implemented by the Applicant during decommissioning. Consequently, the overall decommissioning-phase impacts would be similar to or less than construction-phase impacts; therefore, decommissioning phase impacts would be SMALL.

4.7.2 Alternative 2: No Action

Under the No-Action Alternative, the Ross Project would not be licensed and the land would continue to be available for other uses. However, the Applicant could choose to continue with some preconstruction activities, such as its abandonment of exploration drillholes and its data collection and monitoring of the area. These activities would be similar to or of smaller scale as those activities currently occurring at the Ross Project area. These activities would require some equipment and vehicular access to the Ross Project area, which would result in small fugitive-dust and gaseous emissions. Other potential sources of air-quality impacts in the region (including oil-production activities) would continue as well, where emission releases from oil-recovery activities within the area could result from accidental pipe breaks or equipment and infrastructure-system failures. All of these potential emissions would be limited and short term. Thus, the air-quality impacts would be SMALL for the No-action Alternative.

4.7.3 Alternative 3: North Ross Project

Under Alternative 3, the North Ross Project would generally be the same as the Proposed Action, except that the facility (i.e., the CPP, associated buildings, and auxiliary structures as well as the surface impoundments) would be located to the north of where it would be located in the Proposed Action, as described in SEIS Section 2.1.3. At the north location, a CBW would not be required. Therefore, the incremental contribution to air quality impacts that would result from the construction and partial removal of the CBW would not occur under Alternative 3. However, additional construction activities in Alternative 3, such as greater land disturbance due to surface-impoundment construction due to the north site's topography, would be somewhat greater than those in the Proposed Action. The air quality impacts associated with these activities are not significant relative to the air quality impacts that would occur due to the activities that these two alternatives have in common. Therefore, the air-quality impacts of Alternative 3 would be expected to be similar to the air-quality impacts of the Proposed Action. Thus, the air-quality impacts of Alternative 3 would be SMALL.

4.8 Noise

The Proposed Action will generate noise during all phases of the Project's lifecycle. As noted in GEIS Section 3.3.1, most ISR facilities are proposed for undeveloped rural areas at least 16 km [10 mi] from the nearest communities. However, as described in SEIS Section 3.2, there are eleven residences within the surrounding 3 km [2 mi] radius of the proposed Ross Project. Four of these residences are located within 300 m [1,000 ft] of the Ross Project's boundary. The GEIS indicates that 300 m [1,000 ft] is the distance outside of which noise from construction activities will return to background. The nearest two residences of the four within 300 m [1,000 ft] of the Project are 210 m [690 ft] and 250 m [835 ft] from the Project's boundaries and 800 m [2,500 ft] and 1,700 m [5,600 ft] from the proposed location of the CPP and surface impoundments (i.e., the facility) (see SEIS Figure 3.3). There are no sensitive areas, such as schools, churches, synagogues, or mosques or community centers, located less than 300 m [1,000 ft] from the Ross Project's boundaries (Strata, 2011a). There are no residences within the Project area itself.

As described in SEIS Section 3.3, the primary access routes to or from the Ross Project area would be from I-90 north on either D or New Haven Roads (Strata, 2011a). As noted in SEIS Section 3.8, both of the two nearest residences to the Ross Project are located along New Haven Road. Truck traffic, in particular bentonite hauling from the Oshoto bentonite mine 5 km [3 mi] north of the Ross Project area and, less frequently, livestock hauling, are the main contributors to existing traffic noise on D and New Haven Roads. Two noise studies were conducted by the Applicant to establish the baseline noise levels in and around the Ross Project area (see SEIS Section 3.8). One study measured baseline noise with a sound-level meter at two of four nearby residences (i.e., the nearest offsite "receptors"). Pre-licensing baseline noise levels at these residences averaged between 35.4 and 37.4 dBA, depending upon simultaneous factors such as wind speed, traffic volume, vehicular speed, and the type of load being transported (Strata, 2011a). The Applicant's second noise study collected baseline noise level data at its Field Office in Oshoto, 15 m [50 ft] away from New Haven Road and adjacent to the Ross Project area (see Figure 3.1 in SEIS Section 3.2). The latter study demonstrated that the average, daily duration of noise levels above 55 dBA at the Field Office was 62 minutes per day (Strata, 2011a). This noise was attributed to traffic, because of the Office's close proximity to New Haven Road. The EPA identifies noise at or greater than 55 dBA, with a margin of safety determined to protect hearing, as causing outdoor activity interference and annoyance. The EPA identifies noise at or greater than 45 dBA, with a margin of safety determined to protect hearing, as causing indoor activity interference and annoyance (EPA, 1978).

4.8.1 Alternative 1: Proposed Action

Alternative 1, the Proposed Action, consists of four phases: construction, operation, aquifer restoration, and decommissioning of a uranium-recovery facility and wellfields. At the Ross Project, impacts from noise could be a result of vehicular traffic, such as those from commuter vehicles; deliveries of supplies, materials, and equipment; and shipments of yellowcake and wastes within and outside of the Ross Project area. In addition, equipment operation, such as trucks and other heavy pieces of construction equipment, as well as smaller equipment, such as pump jacks and compressors, and wellfield and CPP operation could be sources of noise. Both humans and wildlife are defined as potential receptors in the vicinity of the Ross Project area.

4.8.1.1 Ross Project Construction

The GEIS (Section 4.4.7.1) stated that because of the use of heavy equipment (e.g. bulldozers, graders, drill rigs, compressors), potential noise impacts would be greatest when an ISR facility is being built (NRC, 2009). This section of the GEIS concluded that the noise impacts during construction would be SMALL to MODERATE, where facility construction and wellfield installation would be expected to have only SMALL and temporary noise impacts for residences or communities that are located more than about 300 m [1,000 ft] from noise-generating activities. The MODERATE rating would be limited to temporary noise impacts to the very nearest residences traffic (NRC, 2009).

Table 4.5 indicates the noise levels that have been calculated for the different types of construction equipment planned for use at the Proposed Action, at three different distances: 15 m [50 ft], which would represent nearby workers; 210 m [690 ft], which would represent the residence nearest the Project's boundary; and 762 m [2,500 ft], which would represent the residence nearest the Ross Project's proposed CPP (Strata, 2011a).

Table 4.5
Respective Noise Levels of Construction Equipment

Equipment Type	Noise Level[a] (15 m [50 ft]) (dBA)	Noise Level[b] (210 m [690 ft]) (dBA)	Noise Level[c] (762 m [2,500 ft]) (dBA)
Heavy Truck	82-96	59-73	24-38
Bulldozer	92-109	69-86	34-51
Grader	79-93	56-70	21-35
Excavator	81-97	58-74	23-39
Crane	74-89	51-66	16-31
Concrete Mixer	75-88	52-65	17-30
Compressor	73-88	50-65	15-30
Backhoe	72-90	49-67	14-32
Front Loader	72-90	49-67	14-32
Generator	71-82	48-59	13-24
Jackhammer/Rock Drill	75-99	52-76	17-41
Pump	68-80	45-57	10-22
Drill Rig[d]	52-74	29-51	18-40

Source: NRC, 2009b; Strata, 2011a.

Notes for Table 4.5:
a = Taken from the GEIS.
b = Minimum distance between the Ross Project's boundary and nearest residence.
c = Minimum distance between the CPP and nearest residence.
d = Based upon Strata's 2010 noise study.

Heavy equipment operation within the Ross Project area would peak during the Applicant's construction of the CPP, surface impoundments, wellfields, and associated infrastructure. The majority of construction equipment would only be operated during daylight hours, and these activities would be more than 300 m [1,000 ft] from the nearest residences; thus, associated noise would not exceed the 24-hour average sound-energy guideline of 70 dBA or the daytime average of 55 dBA, the level EPA identifies as protective against interference of receptor activities and receptor annoyance, with a margin of safety determined to protect hearing (EPA, 1978). The noise impacts to nearby residents due to heavy equipment operation would thus be SMALL. Impacts to workers during the Ross Project's construction would be SMALL, because the Applicant would comply with Occupational Safety and Health Administration (OSHA) regulations concerning noise. Further, a Hearing Conservation Program would be conducted by the Applicant, which would require assessment of noise exposures, provision of hearing protection when noise levels exceed the daily permissible exposure levels, performance of periodic audiograms, and stipulation of worker training regarding noise and hearing, all consistent with 29 CFR Part 1910.95.

Impulse or impact noises from certain equipment, such as impact wrenches and pneumatic attachments on rock breakers, could be particularly annoying to residents. These types of equipment could be present during some construction activities of the Proposed Action. However, the primary locations of these noises would be at least 335 m [1,100 ft] from the nearest residence, significantly reducing their perception by residents. The average noise at residences resulting from equipment-related impact or impulse noises would not be expected to reach the 55 dBA nuisance level (Strata, 2012a). Thus, the impacts of impulse noise would be SMALL.

Indoor noise levels due to outside activities typically range from15 to 25 dBA lower than outdoor levels, depending on whether windows are open or closed. With windows open during daytime hours, indoor noise levels could be have the potential to be greater than the average 55 dBA outdoor level that the EPA defines as preventing receptor activities, interfering with their lives, and annoying them, largely because of truck traffic (EPA, 1978). However, since distances would be greater than 300 m [1,000 ft] from ongoing construction activities, potential indoor noise impacts would be SMALL.

Approximately 85 percent of the overall construction workforce would commute during the daytime (Strata, 2012a), where such commutes would occur to and from the Ross Project in single-occupant cars. Additional traffic would occur due to the relocation of construction equipment to and from the Ross Project area. Noise resulting from vehicle and truck traffic could occasionally be annoying to residents within 300 m [1,000 ft] of noise sources at the Proposed Action, particularly during nighttime hours. However, the Applicant estimates that 90 – 95 percent of all deliveries of supplies, materials, process chemicals, and equipment would occur during daytime hours. Because the county roads to and from the Ross Project area currently have very low average daily and annual traffic counts, there would be a high relative increase in vehicular traffic and, thus, noise impacts to nearby residents would be MODERATE; the more distant local communities would experience only SMALL, temporary impacts.

Elevated noise levels associated with construction activities could also affect wildlife behavior onsite. For example, continuous elevated noise levels could reduce the breeding success of the Greater sage-grouse, if the birds were located near construction equipment, making it more difficult for the female sage-grouse hens to locate and respond to the vocalizations of the males. In general, however, wildlife would likely avoid the areas where noise-generating activities are ongoing (see SEIS Section 4.6). Thus, noise impacts to wildlife would be SMALL.

To minimize noise impacts to all receptors, the Applicant proposes additional mitigation measures. For example, the USDOT reports that, for heavy trucks, speeds of 80 – 160 km/hr [50 – 100 mi/hr] result in noise levels of 80 – 97 dBA, while noise levels of 62 – 74 dBA result when passenger vehicles travel within the same speed range (USDOT, 1995). On rough roads, noise levels would be higher. Therefore, the speed limits for onsite and local county roads are a component of the Applicant's planned mitigation of noise impacts. Traffic-related noise impacts would be minimized by the Applicant's working with Crook County to implement and enforce additional speed limits on the roads as well as to develop its own speed-limit policy for employees and contractors. Regular maintenance of all road surfaces to avoid ruts, potholes, and uneven wear patterns would also minimize noise impacts from vehicle and truck traffic.

The presence of vegetation and topographic features between the noise-generating activity and the receptor would reduce noise levels even more (Countess, 2001). The large topographic features that exist in the Ross Project area (i.e., steep hills and ridges) between the noise-generating construction activities and the nearest receptors would act as barriers to noise propagation. Mitigation measures that would be implemented by the Applicant would include nighttime drilling restrictions within a specified distance of residences, daylight-hour use of construction equipment, "first move forward" driving policies to limit backup alarms from trucks, and speed limit enforcement on access roads. The Applicant would also limit the use of equipment with loud engines, unrestricted exhaust systems, and compression brakes (Strata, 2011a).

Thus, the noise impacts during the Proposed Action's construction would be SMALL to MODERATE, where only the closest residents to the Ross Project would experience MODERATE, but short term, exposures to noise.

4.8.1.2 Ross Project Operation

As noted in GEIS Section 4.4.7, the noise impacts of an ISR facility during the operation phase would be SMALL to MODERATE (NRC, 2009). Truck traffic would be present during the Proposed Action's operation phase and would be associated with yellowcake, vanadium, and waste shipments (16 trucks would be expected during operation vs. 24 during construction). Commuter-traffic noise would decrease because of the smaller workforce required during ISR operations (60 workers would commute per day during operation vs. 200 during construction). Thus, traffic noise impacts produced at the Ross Project during operation would be SMALL to MODERATE, but these would be short term and limited to the nearest receptors (i.e., residences).

During the operation phase, most of the Proposed Action's uranium-recovery activities would be conducted inside buildings (although some wellfield activities would take place outdoors) and fewer pieces of heavy machinery would be used. Therefore, the potential noise impacts from the operation of equipment during the operation phase would be less than those discussed under the construction phase and would be SMALL. Noise emanating from the CPP from a variety of mechanical equipment (e.g. generators; pumps; air compressors; and heating, ventilation, and air conditioning systems) is not expected to exceed the 55 dBA nuisance level because the doors to the CPP would be kept closed as much as possible. Since noise levels decrease significantly with distance and because the CPP would be located approximately 760 m [2,500 ft] from the Ross Project boundary, impacts due to noise emanating from the CPP are expected to be SMALL.

As during the construction phase, noise from the Ross Project's operation would have SMALL impacts to wildlife, which would likely avoid areas where noise-generating activities are ongoing. Similarly, health and safety impacts to personnel at the Ross Project would be SMALL because most of the noise associated with construction would no longer take place.

The specific mitigation measures related to noise impacts adopted by the Applicant during Ross Project construction would continue through its operation. Every plant worker would be periodically retrained to understand the hazards of excess noise and how to decrease noise impacts under the hearing conservation program the Applicant would develop.

4.8.1.3 Ross Project Aquifer Restoration

As noted in GEIS Section 4.4.7.1, the overall noise impacts during aquifer restoration would be SMALL to MODERATE (NRC, 2009). However, noise impacts during the aquifer-restoration phase at the Ross Project would be SMALL because truck traffic would subside to only approximately 12 shipments per day, because overall density of residences and receptors near the Ross Project area is sparse, and because the noise-mitigation measures that the Applicant would undertake would minimize noise. All noise impacts would also be temporary. In addition, the workforce employed during aquifer restoration would be smaller (i.e., 20 workers) than that during the construction and operation phases of the Proposed Action and, thus, there would be fewer workers, less traffic, and fewer noise-producing activities. Finally, the Applicant's continued compliance with OSHA noise standards would minimize noise impacts to workers. Wildlife would continue to avoid the areas where noise-generating activities are ongoing (e.g., the wellfields). All of these factors would ensure that the noise impacts during the aquifer-restoration phase of the Proposed Action are SMALL.

4.8.1.4 Ross Project Decommissioning

The GEIS indicated that noise impacts emanating from an ISR facility undergoing decommissioning would be SMALL to MODERATE. At the Ross Project, noise levels during the decommissioning phase of the Proposed Action would be similar to or less than those identified for the construction phase, for both onsite receptors (i.e., workers) and offsite receptors (i.e., nearest residents). Most potential impacts to nearby residences would occur as a result of the increased noise due to commuter and truck traffic to and from the Ross Project area during decommissioning (i.e., 90 workers and additional waste shipments) and would be SMALL to MODERATE.

Many decommissioning activities would be focused at the ISR facility itself (i.e., the CPP, the surface impoundments, and auxiliary structures), where activities would include decontamination, dismantling, and demolition of these structures, which would be accomplished through the use of heavy equipment. However because this area is approximately 762 m [2,500 ft] from the nearest residential receptor, noise impacts to the nearest residents would be SMALL. In the wellfields, equipment used during plugging and abandonment of recovery, injection, and monitoring wells, such as cement mixers, compressors, and pumps, would produce significant levels of short-term noise. Impacts to workers during the Proposed Action's decommissioning would be SMALL, due to the same variables indicated earlier for its construction and operation as well as for aquifer restoration (i.e., OSHA noise-standard compliance). The same is true for wildlife noise receptors, which would avoid the locations where decommissioning activities are taking place.

Despite the standard mitigation measures taken during decommissioning—the same as those identified for the other phases of the Proposed Action—the distance from the closest residences to the Ross Project would cause the noise impacts to be SMALL to MODERATE, but short-term.

4.8.2 Alternative 2: No Action

Under the No-Action Alternative, the Ross Project would not be licensed and the land would continue to be available for other uses. However, the preconstruction activities the Applicant has undertaken, such as the plugging and abandonment of wells, could continue under the No-Action Alternative. Thus, the noise levels within the Ross Project area, where the measured baseline noise levels are 36 to 40 dBA, could continue (Strata, 2011a). This noise would occasionally be elevated by the passing of heavy trucks and passenger vehicles, nearby agricultural activities, and nearby oil-production activities (Strata, 2011a). Thus, the noise impacts of Alternative 2 would be SMALL.

4.8.3 Alternative 3: North Ross Project

Under Alternative 3, the North Ross Project would generally be the same as the Proposed Action, except that the facility (i.e., the CPP, associated buildings, and auxiliary structures as well as the surface impoundments) would be located to the north of where it would be located in the Proposed Action, as described in SEIS Section 2.1.3. However, because the north site of Alternative 3 is farther away from main roads than the south site of the Proposed Action, the north site's nearest residential receptors are farther away than from the location of the south site. Therefore, the noise generated by construction equipment would be even less likely to exceed the 55 dBA nuisance level at the closest residences. Within the fenced facility area itself, the noise levels during construction of Alternative 3 would be similar to those in the Proposed Action because the same types of construction activities would take place.

The noise levels associated with vehicle and truck traffic volume under Alternative 3 would be essentially the same as described for the Proposed Action, because the uranium-recovery activities would be identical to those of the Proposed Action, including the vehicular traffic on county roads. Thus, residents nearest these roads would experience the same noise impacts as described under the Proposed Action. Workers and wildlife would experience the same impacts under this Alternative as in the Proposed Action. Mitigation measures for noise impacts under Alternative 3 would be same as well. Thus, although the impacts from noise associated with Ross Project construction, operation, aquifer restoration, and decommissioning would be slightly lower than those described above for the Proposed Action because of the slightly greater distance to receptors, the noise impacts of the North Ross Project would be SMALL to MODERATE.

4.9 Historical, Cultural, and Paleontological Resources

As discussed in GEIS Section 4.4.8, potential environmental impacts to cultural resources, which are defined in the GEIS as historical, cultural, archaeological, and traditional cultural properties (TCPs), could occur during all phases of an ISR facility's lifecycle (i.e., during construction, operation, aquifer restoration, and decommissioning) (NRC, 2009). As described in SEIS Section 1.7.3.2 and SEIS Section 3.9, the NRC staff's National Historic Preservation Act (NHPA) Section 106 consultation process for identifying and evaluating historical and cultural resources that could be adversely affected by the Proposed Action is still ongoing. Table 3.18 lists the 25 historic and cultural properties that have been identified to-date within the Ross Project area. The NRC staff's evaluations to determine whether these properties are eligible for listing on the National Register of Historic Places (NRHP) are ongoing. Additionally, the Ross Project area is located between mountains considered sacred by various Native American cultures (the Big Horn Mountains to the west and the Black Hills and Devils Tower to the east). Additional sites of Tribal religious and cultural significance, therefore, could potentially be identified during a TCP survey of the Ross Project area that would be conducted by Tribes and that is currently being coordinated by the NRC staff in consultation with the Tribes and the Applicant (see SEIS Section 1.7.3.2). Once more information becomes available regarding the historical and cultural resources that could be adversely affected by the Ross Project and any mitigation measures that would be agreed to by the Applicant to reduce the adverse effects, this SEIS will be revised accordingly.

4.9.1 Alternative 1: Proposed Action

Alternative 1, the Proposed Action, consists of four phases: construction, operation, aquifer restoration, and decommissioning of the Ross Project facility and wellfields. The impacts of the Ross Project would include the potential to disturb or destroy historical, cultural, and paleontological resources, including NRHP-eligible archaeological sites. In general, adherence to strict mitigation measures can avoid or minimize impacts. These measures could include avoidance, where practical, of NRHP-eligible sites through adjustments in the Ross Project's design, timely consultations with Wyoming State Historic Preservation Office (SHPO) and affected Tribes, and mandated protocols when inadvertent discovery(ies) of unrecorded resources are unearthed during ground-disturbing activities. Once site identification and evaluation is complete, mitigation measures to avoid, minimize, or mitigate adverse impacts to historical, cultural, and paleontological resources and to plan for inadvertent discovery of cultural materials or human remains would be developed in consultation with the Wyoming SHPO, the affected Tribes, and the Applicant.

4.9.1.1 Ross Project Construction

Construction of the Proposed Action could disturb up to 113 ha [280 ac], or 16 percent, of the total Ross Project area. As noted in GEIS Section 4.4.8, most of the potential for direct and indirect adverse impacts to NRHP-eligible properties, traditional culturally significant sites, and paleontological materials would likely occur during ground-disturbing activities during construction or decommissioning (NRC, 2009).

Ground-disturbing activities during construction with the potential to destroy the spatial integrity of archaeological sites and to damage artifacts as well as paleontological resources include, but are not limited to, grading or excavation for roads and parking lots; pipes, wells, and wellfields; buildings and structures; domestic-sewage facilities; utility transmission lines and poles; facility lighting; and surface impoundments. Buried archeological and cultural features as well as deposits of paleontological resources that are not visible on the surface during the initial cultural-resource inventories could be exposed during earth-moving activities. Other potential impacts come from compaction of the soil by heavy equipment, causing damage to subsurface site integrity by crushing or scattering artifacts or features.

Certain paleontological specimens have been located at the Ross Project area; however, they are believed not to be in situ (i.e., they had already been disturbed). Ground disturbance in excess of a few feet during construction could have a limited impact on the geological units themselves, including the Lance Formation, which have the potential to contain a variety of fossils. In addition, increased access to surface-evident archaeological sites during construction could result in vandalism. TCPs could be affected by temporary visual and aural intrusions.

The mitigation measures related to historical and cultural resources would include the standard industry practices that are described in GEIS Section 4.4.8. In addition, consultation by the NRC with the Wyoming SHPO, the Tribes, and the Applicant would result in an agreement clearly delineating the measures the Applicant would take to avoid, minimize, or mitigate adverse effects to historical, cultural, and paleontological resources and to plan for inadvertent discovery of cultural materials or human remains. The NRC staff concludes that the impacts to historical and cultural resources at the Ross Project site would range from SMALL to LARGE. This finding reflects the fact that the highest potential for adverse effects to historical and cultural resources would take place during the construction phase, as well as the fact that efforts to identify and evaluate historic and cultural properties and to determine effects and mitigation are incomplete and Section 106 consultation is ongoing.

4.9.1.2 Ross Project Operation

Direct and indirect adverse impacts on archaeological sites, NRHP-eligible historical properties, TCPs, and paleontological resources are expected to be minimal during the operation phase of the Ross Project. Impacts would be mitigated prior to facility construction and Ross Project operation is generally limited to previously disturbed areas (except continuing wellfield installation). Visual or aural impacts from uranium-recovery operation at the Ross Project to traditional cultural resources located within the Ross Project area and other cultural landscapes, which would be identified before construction, would be expected to continue during operation. Therefore, the impacts to historical and cultural resources during Ross Project operations would be SMALL.

4.9.1.3 Ross Project Aquifer Restoration

Impacts to archaeological sites, NRHP-eligible historical properties, TCPs, and paleontological resources from aquifer restoration would be similar to those expected during uranium-recovery operation. These impacts would primarily result from the surface disturbance associated with operation, maintenance, and repair of existing wellfields as part of the aquifer-restoration process as well as on-going visual or aural impacts. Therefore, the impacts to historical and cultural resources during aquifer restoration would be SMALL.

4.9.1.4 Ross Project Decommissioning

Surface-disturbing activities would temporarily increase during the Ross Project's decommissioning. As during construction, ground disturbance in excess of a few feet during facility decommissioning would have an impact on the geological units themselves, including the Lance Formation, which has the potential to contain a variety of fossils. However, most of the decommissioning activities would focus on previously disturbed areas and, therefore, most of the historic, cultural, and paleontological resources would already be known as a result of the investigations that would be conducted prior to construction. Unavoidable visual and aural impacts, however, could increase temporarily during the decommissioning of the Proposed Action. Therefore, the impacts to historical and cultural resources during decommissioning would be SMALL.

4.9.2 Alternative 2: No Action

Under the No-Action Alternative, the Ross Project would not be licensed and the land would continue to be available for other uses. Under the No-Action Alternative, no major disturbance of land and concomitant potential impacts to historic, cultural, and paleontological resources would occur, except for natural processes such as erosion, although some preconstruction activities could potentially disturb historic, cultural, and/or paleontological resources. The impacts to historical and cultural resources under Alternative 2 would be SMALL.

4.9.3 Alternative 3: North Ross Project

Under Alternative 3, the North Ross Project would generally be the same as the Proposed Action, except that the facility (i.e., the CPP, associated buildings, and auxiliary structures as well as the surface impoundments) would be located to the north of where it would be located in the Proposed Action, as described in SEIS Section 2.1.3. Any impacts to historical, cultural, or paleontological resources from the construction, operation, aquifer restoration, and decommissioning of the Ross Project under Alternative 3 could occur as described in the Proposed Action. Therefore, the impacts to historical and cultural resources due to Alternative 3 also would be SMALL to LARGE during construction and SMALL during operation, aquifer restoration, and decommissioning. However, as with the Proposed Action, mitigation measures such as avoidance would be developed prior to construction and would reduce the construction impacts.

4.10 Visual and Scenic Resources

The Proposed Action could impact visual and scenic resources during all phases of the Project's lifecycle. The visual-resources impacts analysis below is an evaluation of the landscape changes that could occur as a result of the Proposed Action. Most of the visual and scenic impacts would be associated with construction activities, which would be short term, as well as with the new buildings and roads, which would exist until all phases of the project are completed. The Ross Project would introduce new elements of form, line, color, and texture into the landscape of the Ross Project area. Because of the small surface footprint (only 113 ha [280 ac]) and low profile of the uranium-recovery facility and wellfields, no major visual or scenic impacts would be expected to occur.

The visual-resources study area for the Ross Project is currently categorized by Strata as a VRM Class III, according to the BLM scale noted in SEIS Section 3.10. Consequently, the level of change to the characteristic landscape in Class III areas can be moderate (BLM, 2010).

4.10.1 Alternative 1: Proposed Action

Alternative 1 consists of four phases: construction, operation, aquifer restoration, and decommissioning of an ISR uranium-recovery facility and wellfields. Potential visual and scenic impacts at the proposed Ross Project could result from earth moving and surface disturbance as well as the construction, operation, and decommissioning of the following: 1) wellfields (including drill rigs, wellhead covers, header houses, and roads); 2) the CPP; 3) the surface impoundments; 4) the CBW; 5) secondary and tertiary access roads; 6) power and utility lines; and 7) fencing. The visual impacts from these site components would, however, be consistent with the BLM VRM Class III designation (NRC, 2009).

4.10.1.1 Ross Project Construction

GEIS Section 4.4.9 noted that visual-resource impacts could result from heavy equipment use (drill-rig masts and cranes), dust and hydrocarbon emissions, and hillside and roadside cuts into the native topography during construction. In addition, construction activities within a rural setting could give the area a more industrial appearance, thereby decreasing the local visual appeal. However, at the proposed site the existing landscape already includes visual alterations as a result of oil recovery, existing roads, and existing utilities. Construction activities would be short term, and following completion of facility construction, many of the areas where temporary ground disturbance has occurred would be reclaimed and restored to the pre-licensing baseline conditions.

The largest visible surface features of the Proposed Action that would emerge during the construction phase would include the CPP and surface impoundments, wellhead covers and header houses; electrical and other utility distribution lines, which are mounted on 6-m [20-ft] wooden poles; and more roads. The Applicant proposes to use both existing and new roads to access each wellfield and the ISR facility itself (i.e., the CPP and surface impoundments) (see SEIS Section 3.10).

Short-term visual contrasts with the characteristic landscape of the Proposed Action would also result from actual activities associated with construction of the Ross Project. Site clearing and grading; facility and surface impoundment construction and wellfield installation; access road construction; vehicular and pedestrian traffic increases; and underground and overhead pipeline and utilities installation all would result in visual contrasts to the color of the Ross Project area. Irregularity of the natural landscape would occur during the construction phase. Construction activities would typically occur during daylight hours and would be consequently visible, with the exception of some drilling and equipment maintenance that could occur at night (Strata, 2011a).

Wellfield construction would involve the use of drill rigs, water trucks, backhoes, supply trailers, and passenger vehicles. This equipment would be temporarily concentrated at each well or wellfield. A typical truck-mounted drill rig can be about 9 – 12 m [30 – 40 ft] tall and would be the most visible piece of equipment used in wellfield construction. Once a well is completed and developed for use, the drill rig would be moved to a new location. Strata anticipates that up to 12 drill rigs could be operated at one time during wellfield construction. As with the construction activities above, drilling would primarily occur during daylight hours; however, it is possible drilling would continue into the night. For nighttime operation, the drill rigs would be lighted, increasing the potential visual impacts.

Additional construction impacts would include visible fugitive dust that would be generated during ground clearing and grading for header houses and drilling pads; access roads and parking lots; storage and laydown pads; the CPP, auxiliary structures, and surface impoundments; injection, recovery, and monitoring wells; and pipelines. In addition, the drill rigs, trucks, and other vehicles employed during the construction phase at the Ross Project could potentially emit visible emissions (see SEIS Section 3.7.3). These impacts would be temporary and short-term. In the long term (i.e., greater than one year), as major construction activities are completed, fugitive dust and vehicle emissions would decrease.

The Applicant would mitigate visual and scenic impacts related to fugitive dust by wetting the soil and using chemical dust suppressants, as necessary, when clearing and grading activities are underway as well as by establishing diminished speed limits for vehicle traffic, strategically placing water load-out facilities near access roads, encouraging personnel to carpool, and selecting road surfaces that would minimize fugitive dust. Following completion of wellfield installation, disturbed areas would be reclaimed and restored within a single construction season, if at all possible (Strata, 2011a). These mitigation measures are discussed in more detail in SEIS Section 4.7.1.1.

The viewshed analysis introduced in SEIS Section 3.10.1 demonstrates that the Ross Project would not be visible from the base of Devils Tower or from the Visitor's Center. The Proposed Action would be visible (as determined by the cross-section shown in Figure 3.21 in SEIS Section 3.10.1) to climbers scaling the Tower. During initial construction, fugitive dust, other emissions, and construction traffic could impact the viewshed for the Devils Tower climbers. As major construction activities are completed, however, fugitive dust and other emissions would decrease. The Ross Project would not be visible from Keyhole State Park, Black Hills National Forest, or Thunder Basin National Grassland during any phase of the Project due to the long distances between these recreational areas and the Ross Project as well as to the screening effects of topography (Strata, 2011a).

The Applicant would mitigate visual impacts during its construction activities by phasing construction activities; limiting the extent of land disturbance at any one time; promptly restoring and reseeding disturbed areas; using existing roads wherever possible; following existing topography during access road construction to minimize cut and fill and thus reduce contrast; minimizing secondary and tertiary access road widths; and locating access roads, pipelines, and utilities in common corridors (Strata, 2011a).

Prior to construction of the Ross Project, baseline monitoring for potential light pollution would be conducted at eight sites. Based on the results of this preconstruction baseline evaluation, a light-pollution monitoring plan would be prepared by the Applicant. This plan would finalize the locations for both continuous and intermittent light sources; in addition, it would provide a schedule for periodic checks on sky brightness during the construction and operation of the Ross Project to ensure worker safety and to measure, and to mitigate if necessary, obtrusive light emanating from the Proposed Action (Strata, 2012a).

The Applicant proposes the following mitigation measures to limit light-pollution impacts at the Ross Project:

- Designing lighting plans with an emphasis on the minimum lighting requirements for operation, safety, and security purposes;
- Using light sources of minimum intensity (as measured in lumens) necessary to accomplish the light's purpose;
- Specifying lighting fixtures that direct light only where it is needed (i.e., shine down, not out or up) in conjunction with shielding that further directs the light towards the respective work area;
- Turning lights off when not needed at proposed intermittent light locations either manually, with timers, or occupancy sensors;
- Adjusting the type of lights used so that the light waves emitted are those that are less likely to cause light-pollution problems such as those attendant with high-pressure sodium lamps;
- Fitting building windows with shutters, where appropriate, to block light emissions, including the CPP and other buildings;
- Using natural and/or in situ screens to reduce perceptible light (i.e., locating buildings and other facility components to take advantage of the natural topography and any trees; and
- Evaluating the results of the light-pollution monitoring to ensure that, as necessary, the mitigation measures suggested previously have been implemented successfully (Strata, 2012a).

Finally, the Applicant is committed to evaluating the extent of the light pollution to nearby residences following installation of the final lighting system. Additionally, the Applicant is committed to acting on any concerns of local residents as long as worker safety is not compromised (Strata, 2012a).

Because the management objective of VRM Class III is to partially retain the existing character of the landscape so that the level of change to the characteristic landscape can be moderate, the impacts from the Ross Project's construction are in fact consistent with VRM Class III. Thus, in the short-term (i.e., less than one year), construction activities at the proposed Ross Project would result in SMALL to MODERATE visual impacts to the nearest four residences, each of which has a view of the Ross Project area. For the remaining 7 of the 11 nearby residences, however, the visual impacts would be SMALL.

4.10.1.2 Ross Project Operation

SEIS Section 2.1.1 describes the Proposed Action's uranium-recovery operation. Most of the wellfield and surface infrastructure would have a low profile, and most piping and cables would be buried. The irregular layout of wellfield surface structures, such as wellhead covers and header houses, would further reduce visual contrast. Because uranium-recovery operations are generally located in sparsely-populated areas, typically in generally rolling topography, most visual impacts during facility and wellfield operation would not be visible from more than approximately 1 km [0.6 mi] away. As described in GEIS Section 4.4.9.2, the potential visual and scenic impacts from uranium-recovery operation are SMALL.

At the Ross Project, wellhead covers and header houses (wellhead covers would be typically low at approximately 1 – 2 m [3 – 6 ft] high), the CPP and auxiliary buildings, the surface impoundments, access roads, buried utilities, and unburied facility lighting and power lines would be similar to those discussed in the GEIS and, therefore, the potential impacts to the visual resources during Ross Project operation would also be SMALL. Most of the pipelines and cables associated with wellfield operation are anticipated to be buried to protect them from freezing; thus, they would not be visible during the Proposed Action's operation. Other potential impacts include the conduct of wellfield activities, such as monitoring-well sampling, module-building inspections, and mechanical-integrity testing; these impacts would also be SMALL. Because the location of the uranium ores underlying the Ross Project are typically irregular, the network of pipes, wells, and power lines (6 m [20 ft] tall) would not be regular in pattern or appearance (i.e., not a grid); this lack of a pattern would reduce visual contrast and associated potential impacts. The overall visual impact of an operating wellfield would be SMALL (NRC, 2009).

Because the uranium-recovery processing and support facilities, such as the CPP, offices, and maintenance buildings, would be located in one area, they would be more noticeable to the casual observer due to their size and density. The CPP would be the largest structure. These components would be prominent in the foreground and middle-ground views, and they would be silhouetted in the background view from public access points (i.e., the adjacent county roads). As described in SEIS Section 3.10, however, the Proposed Action would be located in gently rolling topography, where the visibility of aboveground infrastructure would vary and would be relative, depending upon the location and elevation of an observer as well as on nearby topography, total distance, and lighting characteristics.

Lighting from the Ross Project would be visible from five of the residences to the east and from various locations directly to the west, north, and southeast. Figure 3.22 in SEIS Section 3.10.2 shows where lighting emanating from the Proposed Action would be visible within the 3-km [2-mi] vicinity surrounding the Project area. Mitigation measures for local light-pollution impacts would be the same as those described above for the construction phase of the Ross Project.

In addition to the mitigation measures employed during the Proposed Action's construction phase, the Applicant identifies a number of additional mitigation measures to reduce the visual impacts during its operation. The wellhead-cover color would be selected to blend with the environment. Pipelines and electrical lines between the wells and module buildings would be buried as new wellfields come online, and disturbed areas would be immediately reclaimed, reseeded, and restored. The electrical-distribution poles would be wooden so that the natural color would tend to blend with the landscape. Another mitigation measure for screening the CPP and surface impoundments would include the Applicant's planting trees at a density that would limit views into the Project area from public roads and nearby residences. The tree species would be a conifer or another species native to the area. The approximate tree locations are depicted on Figure 4.3.

Thus, the impacts to visual and scenic resources during the operation of the Proposed Action would be SMALL.

4.10.1.3 Ross Project Aquifer Restoration

GEIS Section 4.4.9 concluded that the visual impacts during aquifer restoration would be similar to those experienced during uranium-recovery operation, and therefore the impacts would be SMALL (NRC, 2009). Much of the same equipment and infrastructure used during Ross Project operation would be employed during aquifer restoration, so that impacts to the visual landscape would be expected to be similar to or less than the impacts during the Proposed Action's operation phase. In the wellfields, the greatest source of visual contrast would be from equipment used as injection and production wells are being plugged and abandoned during the natural sequence of the installation of a new wellfield(s) and restoration of the aquifer in a spent wellfield(s). Because there is no active drilling in any wellfield undergoing aquifer restoration, potential visual impacts during this phase would be expected to be less than those during facility construction and wellfield installation, and these impacts would be of short duration.

The mitigation measures presented for both the Proposed Action's construction and operation phases would continue to be implemented during the aquifer-restoration phase, and these would continue to limit potential visual impacts. Vehicular traffic during the aquifer-restoration phase would be much more limited: worker commutes would diminish significantly (i.e., from a workforce of 200 persons to one of 20 persons during aquifer restoration) and there would be fewer deliveries of supplies. There would also be a decreasing-to-zero frequency of offsite and potential onsite yellowcake shipments as aquifer restoration proceeds. Therefore, fewer trips would occur than during the earlier phases, with concomitant lower levels of fugitive dust and combustion engine emissions as *de facto* mitigation measures.

Because aquifer-restoration activities at the Ross Project would be very similar to those described in the GEIS (NRC, 2009), the impacts of the Project during the aquifer-restoration phase would also be SMALL.

4.10.1.4 Ross Project Decommissioning

As discussed in GEIS Section 4.4.9.4, the impacts on visual and scenic resources during the decommissioning of an ISR facility would be SMALL (NRC, 2009). The Proposed Action would not cause any significant impacts to the landscape that would persist after facility decommissioning and site restoration are completed. Most visual impacts during decommissioning would be temporary and diminish as structures, equipment, and other facility components are removed; the disturbed land surface is reclaimed and restored; and the vegetation is re-established. NRC licensees are required to conduct final decommissioning and site restoration under an NRC-approved decommissioning plan, with the goal of returning the landscape to the visual conditions of the area prior to any NRC-licensed activities. While some roadside cuts and hill-slope modifications could persist beyond facility and wellfield decommissioning and site restoration (depending upon a landowner's wishes), the re-contouring, re-vegetating, and restoring of the Ross Project area would consist of the same activities described in the GEIS and, hence, the visual and scenic impacts from the Proposed Action's decommissioning would be SMALL.

When the Ross Project's decommissioning efforts have been accepted by the NRC, all buildings and equipment would have been decontaminated, dismantled, decommissioned, and either disposed of or relocated to another facility. Site reclamation efforts would be designed to return the visual landscape of the Ross Project to its baseline contours. Re-contouring of disturbed areas on the Ross Project (including access roads) and the reseeding of those areas with native vegetation or an approved seed mix would both be accomplished during site restoration. All of these activities would minimize any permanent impacts on visual and scenic resources.

The Applicant would mitigate the fugitive-dust impacts that could result from decommissioning activities by its use of water spray during dismantling and demolition activities and on unimproved roads to reduce dust emissions (Strata, 2011a). Areas of disturbance would be restored and reseeded to the pre-construction condition. All facility-decommissioning and site-restoration activities would be done in accordance with NRC and WDEQ/LQD guidelines. Once these activities are complete, the visual landscape would have been returned to its pre-construction, pre-operational condition.

4.10.2 Alternative 2: No Action

Under the No-Action Alternative, the Ross Project would not be licensed and the land would continue to be available for other uses. Therefore, there would be no change to the existing visual and scenic resources at the Ross Project area. In general, the existing site conditions and land uses would persist. All existing roads, fences, utilities, landscape formations, and vegetation would remain. No additional structures or land uses associated with the Ross Project would be introduced to affect the existing viewscapes, and the existing scenic quality would be unchanged. The visual resource classification would remain BLM Class III, as described in SEIS Section 3.10. Thus, visual and scenic impacts would be SMALL.

4.10.3 Alternative 3: North Ross Project

Under Alternative 3, the North Ross Project would generally be the same as the Proposed Action, except that the facility (i.e., the CPP, associated buildings, and auxiliary structures as well as the surface impoundments) would be located to the north of where it would be located in the Proposed Action, as described in SEIS Section 2.1.3. The Alternative 3 facility would remain within the Ross Project area, albeit in a location that is more shielded by topographical features than where it would be located in the Proposed Action. Thus, some of the Ross Project views from neighboring properties would be diminished, and the nearby residences would be more shielded from light pollution than they would be under the Proposed Action. As a result, the visual- and scenic-resource impacts would, at the least, not differ from those of the Proposed Action and, most likely, they would be reduced from those of the Proposed Action. Therefore, the visual-resource impacts would be SMALL to MODERATE in the short-term and SMALL in the long-term.

4.11 Socioeconomics

The Proposed Action could impact local socioeconomics during all phases of the Project's lifecycle. During socioeconomic impact analyses, several areas are examined; these include employment, demographics, income, housing, finance, education, and social and health services.

4.11.1 Alternative 1: Proposed Action

Alternative 1, the Proposed Action, consists of four phases: construction, operation, aquifer restoration, and decommissioning of a uranium-recovery facility and wellfields.

4.11.1.1 Ross Project Construction

The Ross Project would employ approximately 200 people during construction (Strata, 2012a). The peak construction workforce of 200 workers is within the range of the construction workforce estimates provided in the GEIS (i.e., also 200 workers) (NRC, 2009). The GEIS assumed that the majority of the construction personnel positions would be filled by skilled workers from outside the NSDWUMR and that this influx of workers would be expected to result in SMALL to MODERATE socioeconomic impacts, with impacts the greatest for communities with small populations (NRC, 2009). However, due to the short duration of construction, the GEIS also noted that these workers would have only a limited effect on public services and community infrastructure. Further, construction workers would be less likely to relocate their families to another region, and if the majority of the construction workforce would be filled from within the region of the facility, socioeconomic impacts would be SMALL (NRC, 2009).

Because the size of the Ross Project's construction workforce is of similar size to that presented in the GEIS, and the Applicant is committed to hiring locally—it projects that 90 percent of the construction workforce would be local hires (Strata, 2012a)—the employment, demographic, income, housing, education, and health and social services impacts during the construction phase of the Ross Project would be SMALL: Employment increases would represent only 1.2 percent of all jobs in the Region of Influence (ROI) (i.e., Crook and Campbell Counties). The population increases, and consequent increases in public and private services, would represent only a 0.1 percent increase over pre-licensing baseline levels. MODERATE impacts are projected for the finance sector as a result of the additional property-tax revenues generated by the Project (see Table 4.6).

Table 4.6
Estimated Major Tax Revenues

Revenue Source	Tax Revenues: Average Per Year	Tax Revenues: Over 10 Years
Severance Taxes	$855,000	$8,550,000
State Royalties	$243,000	$2,430,000
Gross Production Taxes	$1,337,000	$13,370,000
Property Taxes	$350,000	$3,500,000
TOTAL	$2,785,000	$27,850,000

Source: Strata, 2012a.

The following sections provide impact estimates for each of the specific resource areas within socioeconomics during all phases of the Ross Project.

Employment

The 200 construction workers that would be employed at the proposed Ross Project could generate an additional 140 indirect jobs in the ROI (NRC, 2009), for a peak employment impact of 340 workers as a result of the Project's construction phase. With an employment base in the ROI of 28,842 workers (see SEIS Section 3.11.4), impacts on the Region's employment would be SMALL, representing approximately 1.2 percent of all jobs in the two Counties.

Demographics

It is estimated that less than 10 percent of the construction workforce would come from outside the immediate Ross Project vicinity, or approximately 20 workers (Strata, 2012a). As workers could potentially travel from anywhere in the U.S., based upon the average household size of 2.58 for the U.S. (USCB, 2012), this would translate into 52 additional residents in the ROI. It is likely that most new construction workers for the Ross Project would not relocate their families, however for the purposes of this SEIS, it is assumed that they would move their families. This number is less than 0.1 percent of the combined population base of 53,216 persons in Crook and Campbell Counties as of 2010 (see SEIS Section 3.11.1). This would be a SMALL demographic impact.

Income

It is expected that workers would be paid the regional rates typical of Crook and Campbell Counties, where a higher percentage of jobs are in the relatively higher-paying energy industry. Based upon a weighted-average annual earnings per job of $61,400 (see SEIS Section 3.11.2), the 200 workers would generate approximately $12.3 million in annual earnings. With an estimated $2.6 billion in total personal income in both Crook and Campbell Counties, the impacts of the construction of the Ross Project on local income would represent less than 1 percent of total income in the two Counties and would be a SMALL impact.

Housing

According to GEIS Section 4.4.10, the impacts to housing from ISR-facility construction would be expected to be SMALL (and short term), even if the workforce were to be primarily filled from outside the region (NRC, 2009). It is likely, however, that the majority of workers would use temporary housing such as apartments, hotels, or trailer camps (NRC, 2009). At the maximum, if the additional 20 new workers to the Ross Project vicinity represent a demand for 20 housing units in the ROI (see above), this additional demand for housing would represent less than 0.1 percent of the total housing stock of 22,550 units in the region (see SEIS Section 3.11.3), and this would be a SMALL impact.

Finance

As noted in GEIS Section 4.4.10, the construction of an ISR facility could have a MODERATE impact on finances within a ROI (NRC, 2009). Local-government finances would be affected by ISR-facility construction by the additional taxes collected and the purchase of goods and services in support of construction activities. Although Wyoming does not have an income tax, it does have a state sales tax, a lodging tax, and a use tax. Construction workers would contribute to these as they purchase goods and services within the Ross Project ROI, while they work on the construction of the Proposal Action. Based on a valuation of $50 million for the Ross Project facility and wellfields, as well as the related and real property, multiplied by an 11.5 percent assessment ratio and the Crook County mill levy of 0.062545, local property taxes that would accrue to Crook County would be estimated to be approximately $350,000 per year, reflecting approximately 13 percent of Crook County property-tax collections (Strata, 2012a). These benefits would be offset, however, by the cost of additional public services required by the new residents in the vicinity. This additional demand would be associated with just the estimated 52 additional residents in the ROI, representing less than 0.1 percent of the population in the two Counties; the additional cost for public services also would represent less than a 0.1 percent increase in local-government expenditures. Because the size and scale of the Ross Project is similar to that described in the GEIS, and given the foregoing information, the impacts to local finance would be MODERATE.

Education

As discussed above, it is likely that most new construction workers for the Ross Project would not move their families. However, at a maximum, if all 20 workers were to bring their families, and based upon a school-age population representing 20.4 percent of the population nationwide (USCB, 2012), the 52 additional residents in the Ross Project vicinity would generate 11 additional elementary and secondary students in the ROI schools. This would represent less than 0.1 percent of the total enrollment in area schools and would represent a SMALL impact on education.

Health and Social Services

Increased demand for health and social services is a function of the additional population in the ROI. As discussed above, the population increase in the ROI due to construction activities would represent less than a 0.1 percent increase in the local population because most workers would already reside within a commuting radius of the Project. Thus, only a 0.1 percent increase in the demand for health and social services would occur, and this increased demand for such services would represent a SMALL impact.

In addition, as noted in the GEIS, accidents resulting from construction of the Proposed Action would not be expected to be different than those from other types of similar industrial facilities (NRC, 2009). In the case of an industrial accident, the Applicant would commit to maintaining emergency-response personnel on staff and would train local emergency responders in preparing and responding to potential environmental, safety, and health emergencies resulting from Ross Project construction (Strata, 2011a), thereby minimizing any potential decrease in or impact to the availability of local emergency health services.

4.11.1.2 Ross Project Operation

The Ross Project would employ approximately 60 people during its operation (Strata, 2012a). This number is within the range of the operation-workforce estimates provided in the GEIS (50 – 80 workers) (NRC, 2009). According to the GEIS, if the majority of the operation workforce is filled by personnel from outside the area, potential population and public services impacts would range from SMALL to MODERATE, depending upon the proximity of the ISR facility to population centers (NRC, 2009). However, because an outside workforce would be more likely to settle in more populated areas, with increased access to housing, schools, services, and other amenities, these impacts could be reduced (NRC, 2009). If the majority of the workforce during ISR-facility operation is of local origin, the potential impacts to population and public services would be expected to be SMALL (NRC, 2009).

Because the size of the Ross Project's proposed workforce during the operation of the Ross Project would be within the range evaluated in the GEIS, and because the Applicant would commit to hiring locally—80 percent of the operation workforce would be expected to be local hires (Strata, 2012a)—the employment, demographic, income, housing, education, and health and social services impacts during the Ross Project's operation phase would be SMALL. Employment and population increases, and consequent increases in public and private services, would represent less than 1 percent over pre-licensing baseline levels. MODERATE impacts, however, would be projected for finance as a result of the additional tax revenues that would accrue to Crook County (see Table 4.6).

4.11.1.3 Ross Project Aquifer Restoration

The GEIS assumed that the workforce during aquifer-restoration activities at an ISR facility would be the same as the operation phase (i.e., 50 – 80 workers) and, thus, the impacts would be similar and would be SMALL (NRC, 2009). The Applicant indicates that at the Ross Project there would be a workforce of 20 – 30 workers during the aquifer-restoration phase , without concurrent operations (Strata, 2012a), a smaller workforce than that projected in the GEIS.

The need for regulatory, management, and health and safety personnel would continue throughout aquifer restoration, but this need would be met by personnel transitioning from operation-phase work to aquifer restoration, and no new personnel would necessarily be required (Strata, 2012a). Thus, the impacts of the Proposed Action's aquifer-restoration phase would likely be at most the same, or, would more likely be less than those noted for the Ross Project's operation phase. Because the aquifer-restoration workforce at the Project would be less than that estimated in the GEIS, and with an employment base in Crook and Campbell Counties of 28,842 workers (see Section 3.2.10.4), the socioeconomic impacts of the Ross Project on area employment would be SMALL, representing less than 1 percent of all jobs in the two Counties. Severance tax revenues accruing to local jurisdictions would decrease as uranium production ceases during this phase of the Ross Project.

4.11.1.4 Ross Project Decommissioning

In GEIS Section 4.4.10, the workforce examined for an ISR facility's decommissioning was estimated to be similar to that of the construction phase (i.e., up to 200 persons) and, thus, the impacts would be similar and would be SMALL to MODERATE, with MODERATE impacts for areas with small populations (NRC, 2009). The Applicant indicates, however, that about only 90 workers would be required during decommissioning of the Ross Project (Strata, 2011a). Only 12 of these workers would be non-local hires (Strata, 2012a). These personnel generally represent the regulatory, management, and health and safety personnel that would have been present at the Ross Project during the earlier Project phases. Because the size of the workforce for the Ross Project's decommissioning phase is less than that estimated in the GEIS, and only 12 workers would be expected to be non-local hires, the overall socioeconomic impacts of the Proposed Action's decommissioning phase would be SMALL. Tax revenues accruing to local jurisdictions would decrease to zero as uranium production is concluded during decommissioning of the Ross Project.

4.11.2 Alternative 2: No Action

Under the No-Action Alternative, the Ross Project would not be licensed and the land would continue to be available for other uses. There would be no new jobs created; no changes in income levels in the ROI; no changes in population; no increased demand for education, health, or social services; and no changes in local finances. Other forms of energy development in the ROI would continue to impact regional socioeconomic resources. The economic benefits and socioeconomic impacts described for the Proposed Action would not accrue to Crook and Campbell Counties, nor to the State of Wyoming. Thus, the socioeconomic impacts of the No-Action Alternative would be SMALL.

4.11.3 Alternative 3: North Ross Project

Under Alternative 3, the North Ross Project would generally be the same as the Proposed Action, except that the facility (i.e., the CPP, associated buildings, and auxiliary structures as well as the surface impoundments) would be located to the north of where it would be located in the Proposed Action, as described in SEIS Section 2.1.3. The construction of the CPP and surface impoundments at the north site would not change workforce levels, and therefore the impacts would be the same as those described under the Proposed Action. Because changes in employment are the principal driver of socioeconomic impacts, the socioeconomic impacts of Alternative 3 would be the same as for the Proposed Action, SMALL to MODERATE during Alternative 3's construction and operation, and SMALL during aquifer restoration and its decommissioning.

4.12 Environmental Justice

On February 11, 1994, President Clinton signed Executive Order (EO) No. 12898, entitled *Federal Actions to Address Environmental Justice in Minority Populations and Low-Income Populations*, which directs each Federal agency to "... make achieving environmental justice part of its mission by identifying and addressing, as appropriate, disproportionately high and adverse human health or environmental effects of its programs, policies, and activities on minority populations and low income populations" (EOP, 1994).

On December 10, 1997, the Council on Environmental Quality (CEQ) issued its *Environmental Justice Guidance Under the National Environmental Policy Act*. The CEQ developed this guidance to "... further assist Federal agencies with their National Environmental Policy Act (NEPA) procedures." As an independent agency, the CEQ's guidance is not binding on the NRC. However, the NRC considered the CEQ's guidance on environmental justice in developing its own environmental justice analytical procedures (NRC, 2003).

The CEQ provided the definitions listed in the text box to the left in its Guidance for consistent use during environmental-justice analyses (CEQ, 1997).

The NRC has required an environmental-justice analyses be included in its environmental impact statements (EISs) (NRC, 2004; NRC, 2003, Appendix C). NRC environmental-justice guidance discusses the procedures to evaluate potential disproportionately high and adverse impacts associated with physical, environmental,

What is the terminology used during an environmental-justice analysis ?

- ***Low-Income Populations***

These populations are identified by annual statistical poverty thresholds from the U.S. Census Bureau (USCB). In identifying low-income populations, agencies may consider a community as either a group of individuals living in geographic proximity to one another *or* a set of individuals (such a migrant workers or Native Americans), where either type of group experiences common conditions of environmental exposures or impacts.

- ***Minority Individuals***

Minority individuals are those who identify themselves as members of the following population groups: Hispanic or Latino, American Indian or Alaska Native, Asian, Black or African American, Native Hawaiian, or Other Pacific Islander *or* are two or more races, meaning individuals who identified themselves on a Census form as being a member of two or more races, for example, Hispanic and Asian.

- ***Minority Populations***

Minority populations must be identified when the minority population of an affected area exceeds 50 percent *or* the minority-population percentage of the affected area is meaningfully greater than the minority-population percentage in the general population or other appropriate unit of geographic analysis.

- ***Disproportionately High and Adverse Human Health Effects***

Adverse health effects are measured in risks and rates that could result in latent cancer fatalities as well as other fatal or nonfatal adverse impacts on human health. Adverse health effects may include bodily impairment, infirmity, illness, or death. Disproportionately high and adverse human health effects occur when the risk or rate of exposure to an environmental hazard for a minority or low-income population is significant (as determined during NEPA analysis) and appreciably exceeds the risk or exposure rate for the general population or for another appropriate comparison group.

- ***Disproportionately High and Adverse Environmental Effects***

A disproportionately high environmental impact that is significant (as defined by NEPA) refers to an impact or risk of an impact on the natural or physical environment in a low-income or minority community that appreciably exceeds the environmental impact on the larger community. Such effects may include ecological, cultural, human health, economic, or social impacts. An adverse environmental impact is an impact that is determined to be both harmful and significant (as employed by NEPA). In the assessment of cultural and aesthetic environmental impacts, impacts that uniquely affect geographically dislocated or dispersed minority or low-income populations or American Indian tribes are considered.

socioeconomic, health, and cultural resources to minority and low-income populations (NRC, 2004).

4.12.1 Minority and Low-Income Population Analysis for the Ross Project

Demographic and socioeconomic data for the Ross Project area and surrounding communities was assembled to identify minority or low-income populations within a 6-km [4- mi] radius of the area and is shown in Tables 4.7 and 4.8.

Table 4.7 compares race and ethnicity characteristics by census block group to Crook County and Wyoming. The percentage of the population in Wyoming and Crook County that is nonwhite is 9.3 percent and 2.9 percent, respectively (100 percent minus percent white alone equals percent nonwhite). The percentage of nonwhite population that lives in the block groups within a 6-km [4-mi] radius of the Ross Project area ranges from 0.4 – 2.9 percent. In addition, the percentage of the population in Wyoming and Crook County who are Hispanic or Latino is 8.9 percent and 2.0 percent, respectively. The percentage of Hispanic or Latino populations that lives in the block groups within a 6-km [4-mi] radius of the Ross Project area ranges from 1.3 – 4.7 percent. When these numbers are compared to the State and Crook County proportions, they do not exceed the 20-percent level that is commonly considered of environmental-justice significance.

Table 4.8 compares poverty and income characteristics by census tract to Crook County and Wyoming. The percentage of the population living below poverty for Wyoming and Crook County as well as Census Tracts 9502 and 9503 are 9.8 percent, 7.8 percent, 7.2 percent, and 9.0 percent, respectively. When these numbers are compared to the State and Crook County proportions, they also do not exceed the 20-percent level that is considered of environmental-justice significance.
Because no minority or low-income populations, as defined by EO 12898, have been identified in the Ross Project area, no further environmental-justice analysis (Steps 3 – 5) was conducted.

4.12.2 Alternative 1: Proposed Action

Under the Proposed Action, there are no minority or low-income populations identified that are greater than 20 percent within a 6-km [4-mi] radius of the Proposed Action. Therefore, there are no disproportionately high and adverse impacts to minority and low-income populations under the Proposed Action.

4.12.3 Alternative 2: No Action

Under the No-Action Alternative, the Ross Project would not be licensed and the land would continue to be available for other uses. The conditions affecting minority and low-income populations in the vicinity of the Ross Project area would remain unchanged. Therefore, there would be no disproportionately high and adverse impacts to minority and low-income populations under the No-Action Alternative.

Table 4.7
Ross Project Area Race and Ethnicity Characteristics

Area of Comparison	Total	White Alone	% White Alone	Black or African American Alone	% Black or African American	American Indian and Alaska Native Alone	% American Indian and Alaska Native Alone	Asian Alone	% Asian Alone	Hispanic or Latino	% Hispanic or Latino
Wyoming	563,626	511,279	90.7	4,748	0.8	13,336	2.4	4,426	8.2	50,231	8.9
Crook County	7,083	6,884	97.1	14	0.2	48	0.7	11	0.2	141	2.0
Block Group 1 Census Tract 9502	1,211	1,176	97.1	0	0	7	0.6	1	0.1	20	1.6
Block Group 2 Census Tract 9502	1,880	1,843	98	2	0.1	11	0.6	2	0.1	22	1.2
Block Group 3 Census Tract 9502	1,390	1,333	95.9	6	0.4	9	0.6	2	0.1	65	4.7
Block Group 1 Census Tract 9503	1,280	1,171	96.9	4	0.3	8	0.7	5	0.4	16	1.3
Block Group 2 Census Tract 9503	1,394	1,361	97.6	2	0.1	13	0.9	1	0.1	18	1.3

Source: USCB, 2012b (P1 and QT-P4).

Table 4.8 Ross Project Area Poverty and Income Characteristics		
Area of Comparison[a]	Percent Living Below Poverty	Median Household Income
Wyoming	9.8	$53,802
Crook County	7.8	$49,890
Census Tract 9502	7.2	$52,106
Census Tract 9503	9.0	$46,848

Source: USCB, 2012b.
Notes:
a = Income data is not available at the Census-Block-Group level for 2010.
b = Source: USCB, 2012b (S1701).
c = Source: USCB, 2012b (B19013)

4.12.4 Alternative 3: North Ross Project

Under Alternative 3, the North Ross Project would generally be the same as the Proposed Action, except that the facility (i.e., the CPP, associated buildings, and auxiliary structures as well as the surface impoundments) would be located to the north of where it would be located in the Proposed Action, as described in SEIS Section 2.1.3. As there are no minority or low-income populations identified that are greater than 20 percent within a 6-km [4-mi] radius of the Ross Project area under this Alternative, there are no disproportionately high and adverse impacts to minority and low-income populations.

4.13 Public and Occupational Health and Safety

All phases of the proposed Ross Project could result in potential nonradiological and radiological impacts to public and occupational health and safety. Impacts to occupational health and safety could result from both routine exposures to hazardous chemicals and radiation emitted from radionuclides present during uranium-recovery activities, as well as from exposures following an accident. Public nonradiological impacts are unlikely, except under accident conditions. Radiological impacts to the public could occur during both routine Ross Project activities as well as during accidents.

4.13.1 Alternative 1: Proposed Action

Alternative 1, the Proposed Action, consists of four phases: construction, operation, aquifer restoration, and decommissioning of a uranium-recovery facility and wellfields.

4.13.1.1 Ross Project Construction

Proposed construction activities at the Ross Project are very similar to those described in GEIS Sections 4.4.1 and 4.4.11, where the greatest risk to a worker is the inhalation of radionuclides (e.g., radon) during well drilling and installation and inhalation of fugitive dust containing uranium or its progeny during construction activities. The 10 CFR Part 20 public dose limit is 1 mSv/yr [100 mrem/yr] and the 40 CFR Part 190 annual limit is 0.25 mSv [25 mrem]. The corresponding occupational dose limit is 50 mSv [5 rem] for total effective dose equivalent (TEDE) exposures. The GEIS states that an internal exposure to radiation via ingestion is unlikely without substantial intake of the soils and that radiological impacts to both the public and site workers from inhalation of fugitive dust during construction would be SMALL because the radionuclide concentrations would be low (NRC, 2009). The GEIS concluded that the radiological impacts to both the general public as well as construction workers during ISR facility construction would be SMALL.

As described in SEIS Section 2 and consistent with the GEIS, construction activities associated with the Ross Project would include site preparation and the construction of buildings, storage ponds, access roads, wellfields, and other structures and systems. The important radiation exposure pathway during the construction phase would be through direct exposure and inhalation or ingestion of radionuclides during well construction, construction activities that disturb surface soil, and fugitive dust from vehicular traffic during construction. However, the concentrations of these naturally occurring radionuclides are low; for example, the total concentration of uranium in the native surface soils at the Ross Project area is only 0 to 2.80 mg/kg [2.80 ppm] (on the order of 1 – 2 pCi/g). The low concentrations of radionuclides and the atmospheric dispersion of radionuclides in fugitive dust would minimize impacts from exposures to workers. For direct (i.e., gamma) radiation, the public's potential exposure would be equivalent to approximately 5.3 to 25.3 microRoentgens (μR) per hour (μR/hr), which is much lower than the radiation exposure from naturally occurring radionuclides that the public has during day-to-day activities. Thus, the sparse population near the Ross Project area and its vicinity, the lack of public access, the low concentrations of radionuclides, and the atmospheric dispersion of radionuclides in fugitive dust would be sufficient to minimize impacts from any such exposures to the public.

During the Applicant's proposed use of mud-rotary drilling techniques during wellfield installation, some drilling fluids and muds (i.e., cuttings), originating from the ore zone into which the wells would be drilled, would be brought to the surface. This type of well drilling technique involves the use of a drilling fluid that is introduced through the drill's stem, out the drill bit (i.e., end), and then back up to the surface through the drillhole and the drill stem. These fluids and muds would be collected in pre-dug pits near the well being installed. After drying out, the pits would be covered with native topsoil and then re-vegetated (see SEIS Section 2.1.1.5) (Strata, 2011a). However, because these fluids have been passed through the ore-bearing zone, they have the potential to have higher concentrations of naturally occurring radionuclides than do surficial soils. As the discussion of the radiological baseline conditions in SEIS Section 3.12.1 establishes, however, the relative concentration of radionuclides would still be small. Thus, the radiological impacts to the occupational health and safety of workers, including the well-drillers, would also be SMALL.

Construction equipment would likely be diesel powered and would emit diesel exhaust, which includes small particles (<PM_{10}). The impacts and potential human exposures from these emissions would be small because the releases are usually short and are readily dispersed into the atmosphere. SEIS Section 4.7 describes in greater detail the potential impacts to air quality from proposed diesel emissions including comparisons with health-based standards. Therefore, the NRC staff concludes that the impact and potential human exposure from these particulate emissions would be SMALL, consistent with the GEIS conclusions in Section 4.4.11.1 (NRC, 2009).

Thus, the potential impacts to public and occupational health and safety during construction of the Proposed Action are SMALL.

4.13.1.2 Ross Project Operation

Radiological

Normal Conditions

As discussed in GEIS Section 4.4.11.2.1, some amount of radioactive materials will be released to the environment during normal ISR operations. The potential impact from these releases can be evaluated by the MILDOS-AREA computer code (MILDOS), which Argonne National Laboratory developed for calculating offsite facility radiation doses to individuals and populations. MILDOS uses a multi-pathway analysis for determining external dose; inhalation dose; and dose from ingestion of soil, plants, meat, milk, aquatic foods, and water. The primary radionuclide of interest at an ISR facility is radon-222. MILDOS uses a sector-average Gaussian plume dispersion model to estimate downwind concentrations. This model typically assumes minimal dilution and provides conservative estimates of downwind air concentrations and doses to human receptors.

GEIS Section 4.4.11.2.1 presented historical data for ISR operations, providing a range of estimated offsite doses associated with six current or former ISR facilities. For these operations, doses to potential offsite exposure (human receptor) locations range between 0.004 mSv [0.4 mrem] per year for the Crow Butte facility in Nebraska and 0.32 mSv [32 mrem] per year for the Irigaray facility in Johnson County, Wyoming. In each case, the estimated dose is well below the 10 CFR Part 20 annual radiation public dose limit of 1 mSv/yr [100 mrem/yr] (NRC, 2009).

How is radiation measured?

Radiation dose is measured in units of either Sievert or rem and is often referred to in either milliSv/mSv or millirem/mrem where 1,000 mSv = 1 Sv and 1,000 mrem = 1 rem. The conversion for Sieverts to rem is Sv=100 rem. These units are used in radiation protection to measure the amount of damage to human tissue from a dose of ionizing radiation. Total effective dose equivalent, or TEDE, refers to the sum of the deep-dose equivalent (for external exposures) and the committed effective dose equivalent (for internal exposures).

GEIS Section 4.4.11.2.1 also provided a summary of doses to occupationally exposed workers at ISR facilities. As stated, estimated doses at an ISR facility are not dependent on a facility's location and are well within the 10 CFR Part 20 annual occupational dose limit of 0.05 Sv [5 rem] per year. The largest annual average dose to a worker at a uranium recovery facility over a 10-year period [1994–2006] was 0.007 Sv [0.7 rem]. More recently, the maximum total dose equivalents reported for 2005 and 2006 were 0.00675 and 0.00713 Sv [0.675 and 0.713 rem]. Similarly, the average and maximum worker exposure to radon and radon daughter products ranged from 2.5 to 16 percent of the occupational exposure limit of 4 working-level months. NRC staff concluded in the GEIS that the radiological impacts to workers during normal operations at ISR facilities will be SMALL.

For occupational doses at the proposed Ross Project, the planned ISR facility design and operations are consistent with those analyzed in the GEIS. To mitigate radiological exposure to workers, the applicant will (i) install ventilation designed to limit worker exposure to radon; (ii) install gamma exposure rate monitors, air particulate monitors, and radon daughter product monitors to verify that expected radiation levels are met; and (iii) conduct work area radiation and contamination surveys to help prevent and limit the spread of contamination (Strata, 2011a). The applicant's airborne radiation monitoring program is further described in SEIS Section 6.

For estimated maximum dose to members of the public, GEIS Section 4.4.11.1.2 noted that radon gas is emitted from ISR wellfields and processing facilities during operations and is the only radiological airborne effluent during normal operations for facilities using vacuum dryer technology (NRC, 2009). The Applicant plans to dry yellowcake using a rotary vacuum dryer (Strata, 2011a). Therefore, during normal operations, emissions other than radon are not expected.

The Applicant evaluated the potential consequences of radiological emissions at the proposed Ross Project (Strata, 2011a). Sources of radon emanation the Applicant identified and modeled consisted of point sources (i.e., those operations that have their exhaust confined in a stack, duct, pipe, etc., prior to atmospheric release, such as process tank vents) and area sources (i.e., ore pads and wellfields). The Applicant described its implementation of the computer code MILDOS that was used to model radiological impacts on human and environmental receptors (e.g., air and soil) using site-specific data that included Rn-222 release estimates, meteorological and population data, and other parameters. The estimated radiological impacts from routine site activities were compared to applicable public dose limits in 10 CFR Part 20 {1 mSv/yr [100 mrem/yr]}, as well as to baseline radiological conditions (see SEIS Section 3).

The NRC review of the Applicant's radiological impact modeling independently verified that appropriate exposure pathways were modeled and reasonable input parameters were used. The Applicant also listed the origin of the input parameters and provided justification for their use. The Applicant described the source terms, and the NRC staff review concluded that the source terms represented operations at full capacity and consisted of ISR operations at the wellfields and releases from the CPP and deep disposal wells. The Applicant calculated the TEDE across the projected area on a grid system centered about the CPP and extending beyond the site boundary for a total of 287 locations, 14 members of the public including children that could be living at the four nearest residences and the Oshoto Field Office, 5 ranchers, 2 oil-field workers, and 2 vendors/couriers working both within and outside of the project area.

Results of the Applicant's modeling indicate that the maximum TEDE of 0.016 mSv/yr [1.6 mrem/yr] is located near the Ross Project boundary in the vicinity of the CPP area. The Applicant's calculations also demonstrate that inhalation accounted for 98 percent of the TEDE at this location (Strata. 2011a). Thus, the 10 CFR Part 20 public dose limit is not expected to be exceeded at any property boundary. The annual background dose to the population within 80 km of the Ross Project is estimated at 10,500 person-rem based on a background radiation dose of 2.57 mSv/yr for Wyoming. For comparison, the TEDE from the Ross Project to the population based on the Applicant's modeling is estimated to be 0.361 person-rem. This TEDE represents 1.6 percent of the 10 CFR Part 20 public dose limit of 1 mSv/yr [100 mrem/yr].

Because Rn-222 is the only radionuclide emitted during normal operations, the public dose requirements in 40 CFR Part 190 and the 0.1 mSv/yr [10 mrem/yr] constraint rule in 10 CFR Part 20.1101 do not apply. The Applicant calculated that radon emissions from the wellfields accounted for 75 percent of the total emissions. In its calculations, the Applicant assumed that 100 percent of the radon in the liquids was released to the atmosphere. The estimated radon release from the facility is listed in Table 7.3-4 of the application (Strata, 2011a). The dose to the public is below the 10 CFR Part 20 public dose limit, thus, radiological dose impacts to the public from normal operations will be SMALL.

In summary, potential radiation doses to occupationally exposed workers and members of the public during normal operations would be SMALL. Calculated radiation doses from the releases of radioactive materials to the environment are small fractions of the limits in 10 CFR Part 20 that have been established for the protection of public health and safety. In addition, the applicant is required to implement an NRC-approved radiation protection program (RPP) to protect occupational workers and ensure that radiological doses are as low as reasonably achievable (ALARA). The applicant's RPP includes commitments for implementing management controls, engineering controls, radiation safety training, radon monitoring and sampling, and audit programs (Strata, 2011a).

Accident Conditions

The GEIS identified, discussed, and assessed the consequences for bounding abnormal and accident conditions that might occur with an ISR operation. The GEIS information was based on previous radiological hazard assessments (Mackin, et al., 2001) that considered the various stages of an ISR facility. The GEIS considered three separate accidents, which represent events resulting in higher levels of radioactivity being released: thickener failure and spill, pregnant lixiviant and loaded resin spills (radon release), and yellowcake dryer accident release. The GEIS concluded that potential impacts to workers could be MODERATE based on the estimated consequences of an unmitigated dryer release, but doses to the general public would be SMALL.

An overview of these three accident scenarios, as evaluated in the GEIS along with a specific application to the Ross Project, is presented in the following paragraphs.

Thickener Failure and Spill

Thickeners are used to concentrate yellowcake slurry before it is transferred to a dryer or packaged for offsite shipment. Radionuclides could be inadvertently released to the atmosphere through thickener failure or spill. This accident scenario, as evaluated in the GEIS, assumed a tank or pipe leak that releases 20 percent of the thickener inside and outside of the processing building. The analyses included a variety of wind speeds, stability classes, release durations, and receptor distances. A minimum receptor distance of 500 m [1,640 ft] was selected because it was found to be the shortest distance between a processing facility and an urban development for currently operating ISR facilities. Offsite, unrestricted doses from such a spill could result in a dose of 0.25 mSv [25 mrem], or 25 percent of the annual public dose limit of 1 mSv [100 mrem] per year with negligible external doses based on sufficient distance between facility and receptor (NRC, 2009). The nearest two residences to the Ross Project facility are located at a distance of 800 m [2,500 ft] and 1,700 m [5,600 ft], which are further than the minimum distance analyzed in the GEIS. Therefore, the potential public dose from a thickener spill at the Ross Project would be less than the dose estimated in the GEIS.

As stated in the GEIS, doses to unprotected workers inside the facility have the potential to exceed the annual dose limit of 0.05 Sv [5 rem] if timely corrective measures are not taken to remediate the spill. Typical protection measures such as monitoring, respiratory protection, and radioactive material control, which would be a part of the applicant's radiation protection program, would reduce worker exposures and resulting doses to a small fraction of those evaluated (NRC, 2009). The Applicant has proposed a radiation protection program and a spill response program that would include similar commitments to those described in the GEIS, such as requiring the use of personal protective equipment (PPE) (Strata, 2011a). Therefore, the potential dose to workers at the Ross Project from a thickener spill is expected to be consistent with the dose estimate provided in the GEIS but this dose would be reduced significantly, as described in the GEIS, by the Applicant's implementation of radiation protection and spill response programs.

Pregnant Lixiviant and Loaded Resin Spills

Process equipment (e.g., ion-exchange columns) at the Ross Project would be located on curbed concrete pads to prevent any liquids from spills or leaks from exiting the building and contaminating the outside environment of the facility. In the event of a process tank failure, released fluids would be captured in concrete berms in the process building, which would be designed to contain a volume of 110 percent of the largest tank in the building (Strata, 2011b). Collected fluids would be pumped via a sump to other process vessels, a lined surface impoundment, or a deep disposal well and the contaminated area would be washed down. Additionally, personnel would follow spill response procedures, which would require the use of PPE (Strata, 2011a). Therefore, except for wellfield leaks, the NRC staff does not consider an accidental liquid release with liquid pathways of exposure to be realistic. The primary radiation source for liquid releases within the Ross Project facility would be the resulting airborne radon-222 released from a liquid or resin tank spill.

In the case of a wellfield leak at the Ross Project, pregnant lixiviant could be released from the pipes containing the fluid onto the soil below. The Applicant would be able to identify such a leak by monitoring the pipelines to detect changes in pressure or flow. If a significant change in pressure or flow is detected, an alarm would sound at the CPP, which would prompt the Applicant's personnel to investigate the cause and identify any leaks. If the pressure or flow change is outside of acceptable operating parameters, the pumping system would automatically shut down. Additionally, wellfield operators would visually inspect all piping and equipment within the module buildings, wellheads, and valve vaults at least weekly (Strata, 2011a). Potentially contaminated soil will be sampled and contaminated soil would be removed and disposed of in accordance with NRC and State requirements. In the event of a spill that meets NRC criteria for reporting, the Applicant will notify the NRC within 24 hours and submit a report within 30 days that describes the conditions leading to the spill, the corrective actions taken, and the results achieved.

The GEIS assumed a radon accident release scenario in which a pipe or valve of the ion-exchange system, containing pregnant lixiviant, develops a leak and releases (almost instantaneously) all the radon-222 at a high activity level (2.96×10^7 Bq/m^3 [8×10^5 pCi/L]). For a 30-minute exposure, the dose to a worker located inside the building performing light activities without respiratory protection was estimated as 10 mSv [1,300 mrem], which is below the 10 CFR Part 20 occupational dose limits (NRC, 2009). The Ross Project would include a piping system containing pregnant lixiviant consistent with the system evaluated in the GEIS and, therefore, the potential dose estimated in the GEIS is consistent with the dose expected during this type of accident scenario at the Ross Project. Ventilation systems and alarms at the Ross Project that would alert workers to immediately evacuate the building would further reduce the potential exposure and resulting dose to workers. Considering that atmospheric transport offsite would reduce the airborne levels by several orders of magnitude, any dose to a member of the public would be less than the 1 mSv [100 mrem] public dose limit of 10 CFR Part 20.

Yellowcake Dryer Accident Release

In GEIS Section 4.4.11.2.2, the consequences of an explosion involving a multiple-hearth yellowcake dryer at an ISR facility were evaluated. The analysis assumes that about 4,409 kg [9,500 lb] of uranium yellowcake is released within the building housing the dryer and that, due to the nature of the material, most of the yellowcake would rapidly fall out of airborne suspension. Therefore, only 1 kg [2.2 lb] of the yellowcake is assumed to be subsequently released as an airborne effluent to the outside atmosphere as a 100 percent respirable powder. The calculated maximum dose to workers in this scenario would be 0.088 Sv [8.8 rem], which exceeds the annual occupational dose limit of 0.05 Sv [5 rem] established in 10 CFR Part 20. The atmospheric dispersion of the fraction of the yellowcake that is assumed to be released as an airborne effluent would significantly reduce the exposure to members of the public to about 6.5 x 10-4 Sv [65 mrem], which is less than the 10 CFR Part 20 public dose limit of 1 mSv [100 mrem] (NRC, 1980).

The Applicant proposes to use a vacuum dryer for both yellowcake and vanadium, which is the current industry standard for ISR facilities. In a vacuum dryer, the heater combustion source is separated from the dryer itself. This configuration mostly eliminates the possibility of an explosion, which is the initiating event for the accident scenario considered in the GEIS. Therefore, the vacuum dryer accident release that could occur at the Ross Project is expected to have less significant consequences than the multiple-hearth yellowcake dryer accident release scenario considered in the GEIS. The Applicant analyzed the potential for a release of yellowcake from a vacuum dryer into the dryer room due to a seal rupture. Operating procedures proposed by the Applicant such as conducting regular inspections of the seals and monitoring for pressure changes and other indicators of problems with the seal during dryer operations would reduce the likelihood of an unnoticed seal rupture. However, in the event of a yellowcake release due to a seal rupture, dose to workers would be minimized because they would be required to wear respiratory protection when the dryer is in operation and would immediately evacuate the area. Public exposure would be significantly reduced, as described in the GEIS, due to atmospheric dispersion of any fraction of the yellowcake that is released from the dryer building.

Accident Analysis Conclusions

The NRC staff reviewed and evaluated site-specific and project-specific information related to potential accidents and determined that the types of accidents analyzed in the GEIS and their potential consequences bound those that could occur for the proposed Ross Project. There would be no significant radiological impacts from potential accidents to the public or occupationally exposed workers beyond those described in the GEIS. Based on this finding, the potential doses may result in a MODERATE impact to occupational health and safety, in the case of an unmitigated accident, and a SMALL impact to public health and safety. Occupational health and safety impacts from accidents would be reduced by the Applicant implementing protection measures such as routine monitoring, spill response and cleanup procedures, and respiratory protection. Therefore, the overall radiological impacts to public and occupational health and safety from accidents during operations would be SMALL.

Nonradiological

Normal Conditions

GEIS Section 4.4.11.2.4 identified the various chemicals, hazardous and nonhazardous, that are typically used at ISR facilities. The GEIS also identifies the typical quantities of these chemicals that are used. The following hazardous chemicals would be used in the largest quantities at the CPP during the Ross Project's operation:

- Anhydrous ammonia
- Sodium hydroxide
- Sulfuric acid and/or hydrochloric acid
- Oxygen
- Hydrogen peroxide
- Carbon dioxide
- Sodium carbonate
- Sodium chloride
- Ammonium sulfate

Each of these chemicals would be purchased in bulk, would be transported to the Project area by motorized vehicles, and would be stored within the controlled area of the Ross Project (i.e., in the fenced facility itself). Typical onsite quantities for some of these chemicals exceed the regulated, minimum reporting quantities and trigger an increased level of regulatory oversight regarding possession (type and quantities), storage, use, and disposal practices. The use of hazardous chemicals at ISR facilities is controlled under several regulations that are designed to provide adequate protection to workers and the public. The primary regulations applicable to use and storage include the following:

- **40 CFR Part 68:** ***Chemical Accident Prevention Provisions***. This regulation lists regulated toxic substances and threshold quantities for accidental-release prevention.
- **29 CFR Part 1910.119:** ***Occupational Safety and Health Administration Standards/Process Safety Management of Highly Hazardous Chemicals***. This regulation lists highly hazardous chemicals as well as toxic and reactive substances (i.e., chemicals that can potentially cause a catastrophic event at or above the threshold quantity).
- **29 CFR Part 1910.120:** ***Hazardous Waste Operations and Emergency Response***. This regulation instructs employers to develop and implement a written health and safety program for their employees involved in hazardous-waste operations. The program should

be designed to identify, evaluate, and control health and safety hazards and provide for emergency response during hazardous-waste operations.

- **40 CFR Part 355: *Emergency Planning and Notification*.** This regulation lists extremely hazardous substances and their threshold planning quantities so that emergency response plans can be developed and implemented. There are approximately 360 extremely hazardous substances listed. Over a third of these are defined by the *Comprehensive Environmental Response, Compensation, and Liability Act* (CERCLA), the "Superfund" law. The regulations associated with this statute also list so-called "reportable quantity" values for these substances.

- **40 CFR Part 302.4: *Designation, Reportable Quantities, and Notification/Designation of Hazardous Substances*.** This regulation identifies the reportable quantities for the CERCLA hazardous substances on the promulgated list. There are approximately 800 of these substances, and they are compiled from the 1) CWA, Sections 311 and 307(a); 2) CAA, Section 112; 3) *Resource Conservation and Recovery Act* [RCRA], Section 3001; and 4) *Toxic Substance Control Act*, Section 7.

The Applicant's compliance with applicable regulations would reduce the likelihood of continuing or significant releases, which may result in injury or illness to an exposed worker. The risk of offsite impacts to the public due to a chemical spill is not significant because chemicals would be stored and used in or near the facility and wellfields. Therefore, impacts to the public would be SMALL.

To promote occupational health and safety, the Applicant would issue a formal Safety Policy Statement to define its overall health- and safety-protection policy and the requirements that must be met by all employees and contractors at all times while at the Ross Project (Strata, 2012a). In addition, the Applicant proposes the development of several plans, SOPs, and other management tools to further decrease and mitigate occupational health and safety impacts (Strata, 2011a). All workers and contractors would receive required health and safety training. This training would include indoctrination to plans such as the Project's HASP, as well as all pertinent SOPs and BMPs. The Ross Project would operate under a comprehensive Project HASP, which would include specific industrial-hygiene SOPs and other health and safety plans. These SOPs would govern a worker entering a confined space, trenching and excavation of utility and pipeline corridors, referring to appropriate Material Safety Data Sheets (MSDSs), decanting a hazardous chemical, and donning appropriate levels of PPE. Other health and safety plans could include a respiratory protection plan, a hearing conservation plan, and a health and safety training plan. These latter plans would be developed and instituted by the Applicant only when it is not practical to use process or other engineering controls [Strata, 2012a]). The Applicant's HASP would also include specific training requirements and hazard identification and mitigation policies and procedures. The HASP would define the protocols, methods, and procedures the Applicant would use to ensure compliance with the OSHA requirements found at 29 CFR Part 1910.

The types and quantities of chemicals (hazardous and nonhazardous) identified for use at the proposed Ross Project are consistent with those evaluated in the GEIS. Additionally, the Applicant proposes to implement the occupational health and safety protection plans evaluated for typical ISR facilities in the GEIS and to comply with the requirements of regulations governing the use and storage of chemicals. Therefore, the NRC staff concludes that the nonradiological impacts to public and occupational health and safety during normal operations of the Proposed Action would be SMALL.

Accident Conditions

Potential nonradiological accidents are consistent with the typical accidents at other industrial facilities, including high consequence chemical release events. In GEIS Section 4.4.11.2.2, the likelihood of such a release is determined to be low based on historical operating experience at ISR facilities, primarily due to operators following commonly applied chemical safety and handling protocols. Past history at current and former ISR facilities demonstrates that these facilities can be designed and operated with measures that adequately reduce the risks to worker and public health and safety. The GEIS concluded that the nonradiological impacts due to accidents at an ISR facility would be SMALL offsite and potentially MODERATE for workers involved in accident response and cleanup.

If a large quantity of one or more of the chemicals that would be present in significant quantities at the Ross Project were to be released during the Ross Project's operation, the nonradiological impacts to public health and safety would depend on the proximity of potentially impacted populations. Potential receptors are sparse in the area around the Ross Project (the nearest residents to the Ross Project are identified in Figure 3.1 in SEIS Section 3.2). In addition, the Ross Project area is large and affords distance that would allow released hazardous chemicals to be either deposited or dispersed before reaching the Project boundaries, thereby diminishing individual impacts. Workers involved in a response and cleanup of an accident could experience MODERATE impacts, but training requirements and the establishment of and adherence to applicable procedures would reduce the impact to SMALL. Thus, consistent with the GEIS, impacts to public and occupational health and safety due to an onsite accident during Ross Project operations would be SMALL.

4.13.1.3 Ross Project Aquifer Restoration

GEIS Section 4.4.11 indicated that the activities that would take place during aquifer restoration are similar to ISR facility operation (i.e., wellfield operation, uranium extraction, waste-water treatment, and waste disposal), except that each would begin to diminish as less and less uranium is recovered from the production aquifer. The gradual cessation of many of these processes as the Ross Project, such as loaded-IX-resin elution, yellowcake drying and packaging, vanadium recovery and packaging, further limits the relative magnitude of potential public and occupational health and safety hazards. There would be fewer opportunities for accidents with the decreasing number of operations and the decreasing workforce as well as fewer chemicals used onsite and smaller volumes of chemicals stored onsite. The same mitigation measures and management controls, such as the RPP and the Project's HASP, as discussed earlier for the Ross Project's construction and operation would be observed during its aquifer-restoration phase. Thus, the nonradiological and radiological impacts to public and occupational health and safety during aquifer restoration would be SMALL.

4.13.1.4 Ross Project Decommissioning

The GEIS found in Section 4.4.11 (NRC, 2009) that the radiological impacts to the public and occupational health and safety from the decommissioning of an ISR project would be SMALL. Consistent with the description in the GEIS, the magnitudes of potential impacts from the decommissioning of the Ross Project facility and its wellfields would be less significant than impacts during operations because hazards would be reduced and eliminated; and soils, structures, and equipment would be decontaminated.

In addition to the mitigation measures described in SEIS Section 4.13.1.1, the NRC would require that the Applicant submit a decommissioning plan for the Ross Project for its review and approval. Protection of workers and the public is ensured through NRC approval of the decommissioning plan and verification that doses from exposures during decommissioning would comply with 10 CFR Part 20 limits. Following decommissioning, the Ross Project site could be released for unrestricted use in conformance with the conditions of the NRC license and the dose criteria for site release in 10 CFR Part 40, Appendix A. The criteria in 10 CFR Part 40, Appendix A limit the dose from radiological contamination that may exist at the site after decommissioning is complete to levels that are sufficiently low to protect public health and safety. Therefore, the impacts to public and occupational health and safety from the decommissioning of the Ross Project would be SMALL.

4.13.2 Alternative 2: No Action

Under the No-Action Alternative, the Ross Project would not be licensed and the land would continue to be available for other uses. However, until the NRC has made its decision regarding the licensing of the Project, the Applicant could continue with some of preconstruction activities (e.g., monitoring well installation). In addition, if the NRC license is not issued, there would need to be some additional work to properly abandon the wells that would have been installed by the Applicant. However, the public and occupational impacts to health and safety of this No-Action Alternative would be less than those impacts associated with the construction of the Proposed Action (i.e., Alternative 1). Thus, the public and occupational impacts of the No-Action Alternative would be SMALL.

4.13.3 Alternative 3: North Ross Project

Under Alternative 3, the North Ross Project would generally be the same as the Proposed Action, except that the facility (i.e., the CPP, associated buildings, and auxiliary structures as well as the surface impoundments) would be located to the north of where it would be located in the Proposed Action and the construction of a CBW would not be necessary, as described in SEIS Section 2.1.3.

Under Alternative 3, the length of the wellfield pipelines may be increased and, thus, there would be more pipeline subject to failure. However, the Applicant would implement the same procedures described under the Proposed Action to reduce the risk and severity of pipeline failures (e.g. monitoring the pipelines to detect changes in pressure or flow, allowing for automatic shut down of the pumping system, visually inspecting piping at least weekly, and removing contaminated soil).

Alternative 3 would be located, constructed, and operated farther away from the primary roads to the Ross Project area, which would require the construction of additional road extensions. This road construction would generate additional fugitive dust. However, the nearest residential receptors would be farther away from the CPP under the North Ross Project than they would be from the location of the CPP under the Proposed Action; thus, they would be less affected overall by fugitive dust and/or the impacts of accidents. Construction activities and chemical use would be similar to the Proposed Action because the construction footprint of the facility would be consistent with the Proposed Action. Construction activities associated with constructing and decommissioning the CBW with the Proposed Action and the associated incremental contribution to public and occupational health and safety would not be present under Alternative 3. All other potential public and occupational health impacts would be the same as described for the Ross Project in this SEIS Section 4.13.1. Consequently, as with the Proposed Action, workers involved in a response and cleanup of an accident could experience MODERATE impacts, but training requirements and the establishment of and adherence to applicable procedures would reduce the impact to SMALL. Thus, the impacts to public and occupational health and safety of Alternative 3 would be SMALL.

4.14 Waste Management

The Proposed Action could have potential waste-management impacts during all phases of its lifecycle. Waste volumes, disposal practices, and associated mitigation measures for the four phases of the Proposed Action are evaluated and compared to the impacts identified in the GEIS (NRC, 2009). The waste management practices, waste types, and estimated waste volumes that the Applicant proposes are generally consistent with the typical ISR facility described in the GEIS. The impacts of the Applicant's management of liquid and solid waste streams for each phase of the Proposed Action as well as the two Alternatives are evaluated in this section. All of the three Alternatives are described in SEIS Section 2.1; impacts from the transportation of solid wastes offsite for disposal are evaluated in SEIS Section 4.3.1; impacts to the geology, soils, and water resources as a result of spills, leaks, and other accidental releases of liquid wastes as well as onsite disposal of liquid wastes are assessed in SEIS Sections 4.3.1 and 4.4.1, respectively.

4.14.1 Alternative 1: Proposed Action

Alternative 1, the Proposed Action, consists of four phases: construction, operation, aquifer restoration, and decommissioning of a uranium-recovery facility and wellfields.

The volumes of each type of waste the Applicant expects to be generated by the Ross Project and the Applicant's proposed management approach and disposal activities are fully described in SEIS Section 2.1.1 and are shown in Table 4.9. As described, the specific permits that the Applicant would need to obtain for its UIC Class I deep-injection wells would mitigate many of the impacts of liquid-waste disposal at the Project. The pre-operational agreements with solid-waste and radioactive-waste disposal facilities that are required to be in place prior to the NRC's issuing a license to the Applicant would mitigate impacts from solid-waste management (NRC, 2009). As part of these agreements, the Applicant would need to ensure that sufficient capacity for solid byproduct wastes (liquid byproduct wastes would be disposed of onsite in the deep-injection Class I UIC wells) would be available throughout the lifecycle of the Ross Project (NRC, 2009). NRC license conditions and inspections would ensure that proper practices are used by the Applicant to comply with safety requirements to protect workers and the public during waste management (NRC, 2009). The Applicant would implement waste-minimization and volume-reduction BMPs, as possible, to further mitigate the impacts of waste management (Strata, 2011a).

Each of the disposal facilities noted in Table 4.9 has indicated to the Applicant that it has sufficient disposal capacity to accept the volumes of wastes shown in Table 4.9 (see Table ER RAI Waste-1-1 in Strata, 2012a).

Table 4.9
Ross Project Waste Streams

Waste Stream	Source	Disposal Method	Estimated Typical Quality
NRC-Regulated Wastes			
Excess Permeate	Uranium Production Aquifer Restoration RO Circuits	Reinjection into Wellfield CPP Make-Up Water Deep-Well Injection	C: 0 m^3/min [0 gal/min] O: 0.2 m^3/min [57 gall/min] R: 0 m^3/min [0 gal/min] D: 0 m^3/min [0 gal/min]
Brine and Other Liquid Byproduct Wastes	Uranium Production Aquifer Restoration RO Circuits Spent Eluate Process Drains Contaminated Reagents Filter Backwash Wash-Down Water Decontamination Showers	Deep-Well Injection Evaporation from Surface Impoundments	C: 0 m^3/min [0 gal/min] O: 0.2 m^3/min [62 gal/min] R: 0.9 m^3/min [227 gal/min] D: 0.04 m^3/min [<10 gal/min]
Solid Byproduct Wastes	Filtrate and Spent Filters Scale and Sludges from Equipment Maintenance Contaminated Soils Damaged IX Resins Contaminated Solids from Wells Contaminated PPE Contaminated Materials and Equipment	Shipment to NRC- or Agreement State-Licensed Disposal Facility	C: 0 m^3 [0 yd^3] O: 76 m^3/yr [100 yd^3/yr] R: 76 m^3/yr [100 yd^3/yr] D: 3,058 m^3 [4,000 yd^3]
Non-NRC-Regulated Wastes			
TENORM	Drilling Fluids and Muds	Mud Pits	C: Per Well = Drilling Fluids 23 m^3 [6,000 gal] Drilling Muds 0.1 m^3 [15 yd^3] O: 0 m^3 [0 gal] R: 0 m^3 [0 gal] D: 0 m^3 [0 gal]
Industrial or Municipal Solid Waste	General Office Trash	Shipment to Municipal Landfill	C: 11 m^3/wk [15 yd^3/wk] O: 11 m^3/wk [15 yd^3/wk] R: 11 m^3/wk [15 yd^3/wk] D: 11 m^3/wk [15 yd^3/wk]
Recyclable Solid Waste	Plastic, Glass, Paper, Aluminum, and Cardboard	Shipment to Municipal Recycling Facility Recyclable Waste-Collection Facility	C: 4 m^3/wk [5 yd^3/wk] O: 4 m^3/wk [5 yd^3/wk] R: 4 m^3/wk [5 yd^3/wk] D: 4 m^3/wk [5 yd^3/wk]

Table 4.9
Ross Project Waste Streams
(Continued)

Waste Stream	Source	Disposal Method	Estimated Typical Quality
Construction and Demolition Debris	Construction Debris Decontaminated Materials and Equipment	Shipment to Demolition-Debris Landfill	C: 4 m^3/wk [5 yd^3/wk] O: 4 m^3/wk [5 yd^3/wk] R: 4 m^3/wk [5 yd^3/wk] D: 1,529 m^3 [2,000 15 yd^3]
Petroleum-Contaminated Soil	Equipment Spills and Leaks	Shipment to WDEQ/SHWD-Permitted Disposal Facility	C: < 0.8 m^3/mo [< 1 yd^3/wk] O: < 0.8 m^3/mo [< 1 yd^3/wk] R: < 0.8 m^3/mo [< 1 yd^3/wk] D: < 0.8 m^3/mo [< 1 yd^3/wk]
Hazardous Waste	Used Batteries Expired Laboratory Reagents Fluorescent Bulbs Solvents, Cleaners, and Degreasers	Shipment to WDEQ/SHWD-Permitted Recycling or Disposal Facility	C, O, R, D: < 100 kg/mo [< 220 lb/mo]
Used Oil	Vehicle Maintenance	Shipment to Used-Oil Recycling Facility	C: 0.02 m^3/mo [5 gal/mo] O: 0.02 m^3/mo [5 gal/mo] R: 0.02 m^3/mo [5 gal/mo] D: 0.02 m^3/mo [5 gal/mo]
Used Oil Filters and Oily Rags	Vehicle and Equipment Maintenance	Shipment to Used-Oil Recycling Facility	C: < 9 m^3 [< 20 lb/mo] O: < 9 m^3 [< 20 lb/mo] R: < 9 m^3 [< 20 lb/mo] D: < 9 m^3 [< 20 lb/mo]
Domestic Sewage	Restrooms	Onsite Waste-Water Disposal or Treatment System Holding Tanks/Portable Toilets during Construction and Decommissioning	C: 9.8 m^3/d [2,600 gal/d] O: 3 m^3/d [800 gal/d] R: 1.1 m^3/d [300 gal/d] D: 4.5 m^3/d [1,200 gal/d]

Source: Strata, 2012a.

Notes:
C = Construction
O = Operation
R = Aquifer Restoration
D = Decommissioning

4.14.1.1 Ross Project Construction

As described in GEIS Section 4.4.12, construction activities would be expected to generate low volumes of wastes. No radioactive wastes that are regulated by the NRC would be generated during the Proposed Action's construction phase. The GEIS found that the waste management impacts from the construction of an ISR facility would be SMALL due to the limited volumes of wastes (NRC, 2009).

Liquid Waste

Non-byproduct liquid waste would be generated during construction of the Ross Project from the Applicant's drilling and development of injection, recovery, and monitoring wells. Construction of the Class I deep-injection wells would produce drilling fluids and muds. The Applicant estimates that a volume of 22,000 L [6,000 gal] of water and 12 m^3 [15 yd^3] of drilling muds would be produced per well (Strata, 2012a). These fluids would be stored onsite in mud pits which would be constructed adjacent to the respective drilling pad(s) and evaporated. The GEIS found that the liquid waste management impacts from the construction of an ISR facility would be SMALL due to the limited volumes of wastes (NRC, 2009).

Construction releases from the mud pits would be mitigated by the implementation of sediment-control BMPs (Strata, 2011a). The dried pits would ultimately be backfilled, graded, covered with topsoil, and reseeded to achieve the reclamation standards required by WDEQ/LQD (Strata, 2011a). The Applicant would attempt to complete reclamation of the mud pits within one construction season to minimize wind and water erosion. The reclaimed mud pits would be included in the radiation surveys that would be accomplished during the Proposed Action's decommissioning so that no potential long-term impacts from radioactivity are present (Strata, 2011a).

The Applicant estimates that 19 L/mo [5 gal/mo] of used oil would be generated and shipped to a local commercial recycler (Strata, 2012a). The Applicant also estimates that 9,842 L/d [2,600 gal/d] of domestic sewage would be generated during construction; this waste would be managed in an onsite domestic waste-water system designed according to WDEQ/WQD standards (Strata, 2011a).

The potential impacts of the management of liquid wastes during construction, therefore, would be SMALL.

Solid Waste

Solid wastes generated during the construction of the Proposed Action would be of limited quantity and volume. The estimated volume of each type of waste and the respective disposal practices that would be used by the Applicant to manage the wastes are described in SEIS Section 2.1.1 and are summarized as follows:

- Less than 9 kg/mo [20 lb/mo] of used oil filters and oily rags would be produced and shipped to a local commercial recycler.
- 19 m^3/wk [25 yd^3/wk] of solid waste not regulated by the NRC nor the EPA would be generated and disposed or recycled at an offsite local landfill.
- Less than 1 m^3/wk [1 yd^3/wk] of petroleum-contaminated soil would be transported by a waste-disposal contractor to a permitted facility in northeast Wyoming, such as the Campbell County Landfill.
- Less than 100 kg/mo [220 lb/mo] of hazardous waste would be securely and appropriately accumulated at the Ross Project and transported by a hazardous-waste contractor to an

appropriately permitted, commercial treatment, storage and disposal (TSD) facility outside of Wyoming (Strata, 2012a).

The Applicant proposes to minimize the volume of used oil and hazardous waste by servicing its vehicles and equipment offsite and by limiting its chemical-reagent orders to quantities that can be consumed within the regents' shelf lives (Strata, 2011a).

Waste volumes are similar to those described in Section 4.4.12 of the GEIS. Thus, the potential impacts of the management of solid wastes during the construction of the Proposed Action would be SMALL.

4.14.1.2 Ross Project Operation

As described in GEIS Section 4.4.12, waste-management impacts during the operation of an ISR facility would be SMALL, based upon the required preoperational disposal agreement(s) for solid radioactive wastes in addition to regulatory controls such as the applicable permit and license conditions with which an Applicant must comply as well as the inspections the NRC and other regulatory agencies would perform (NRC, 2009). At the Ross Project, the UIC Permit for the Class I injection wells that has already been obtained by Strata for deep-well injection of liquid byproduct (i.e., radioactive) waste specifies operating conditions and reporting requirements with which the Applicant must comply (WDEQ/WQD, 2011). Design specifications related to radioactive waste that would need to be approved by the NRC include waste treatment and volume reduction techniques, surface-impoundment leak detection systems, and other routine monitoring activities that would further minimize the potential for impacts to the environment (NRC, 2009).

Liquid Waste

As described in SEIS Section 2.1.1, liquid byproduct waste generated during ISR operations would include process bleed (an average of 1.5 percent of injection volume) and other process waste waters. The process bleed would be treated by a two-stage RO circuit during the Proposed Action, producing a minimized volume of brine and permeate. Permeate from the RO process would be re-used as plant make-up water or lixiviant. Excess permeate requiring disposal would be only generated during the first two and one-half years of ISR operations before aquifer restoration begins (Strata, 2011a). The Applicant proposes that excess permeate, up to 190 L/min [50 gal/min] would be discharged to the surface impoundments. The double-liner, leak-detection system the Applicant proposes for its surface impoundments, in addition to the monitoring and reserve-capacity requirements mandated by NRC regulations and NRC license conditions, would allow detection of any surface-impoundment spills or leaks before any significant release of material occurs (NRC, 2009). These requirements were also anticipated by the GEIS, when it concluded that similar waste-management techniques would result in SMALL impacts. Thus, the potential impacts of the Proposed Action's use of surface impoundments for the management of liquid byproduct waste would be SMALL.

The Applicant estimates that approximately 240 L/min [62 gal/min] of brine and other process waters would be disposed of into the UIC-permitted Class I deep-injection wells that the Applicant has already obtained from the WDEQ/WQD (WDEQ/WQD, 2011). The lined surface impoundments and a storage tank with secondary containment would be used to manage the brine before its disposal in the deep-injection wells (Strata, 2012b). The use of the surface impoundments for waste management and the disposal by deep-well injection that the Applicant proposes are consistent with the waste-management practices described in the GEIS.

The Applicant expects that ground water generated during the construction and development of recovery and injection wells would be disposed of in mud pits similarly to the disposal of drilling fluids generated during the construction phase. However, drilling fluids generated during development of wells completed in an aquifer affected by uranium-recovery operations would be disposed of in the lined retention ponds or via the deep disposal wells (Strata, 2012b).

The volume of used oil that would be produced during the Proposed Action's operation and its management would be the same as during its construction (Strata, 2012a). The volume of domestic sewage, which would be managed in an onsite system, would be approximately 3,000 L/d [800 gal/d] (Strata, 2012a).

The potential impacts of the management of liquid wastes during operation would therefore be SMALL.

Solid Waste

As described in SEIS Section 2.1.1, the Applicant estimates that approximately 80 m^3/yr [100 yd^3/yr] of solid byproduct (i.e., radioactive) waste would be generated during the operation phase of the Proposed Action (Strata, 2012a). The Applicant proposes to minimize the quantity of byproduct solid waste by selecting high-efficiency filter media for uranium-recovery and aquifer-restoration circuits (Strata, 2011a). Getting more use out of filter media would minimize the quantity used as well as the waste generated during operation. This byproduct waste would be accumulated inside 208-L [55-gal], lined drums and stored in a restricted area of the CPP (Strata, 2011a). Full drums would later be sealed and then moved into a 15-m^3 [20-yd^3] roll-off container. Roll-off containers would be stored in a restricted area outside of the CPP where access is secured and restricted. Sealed roll-off containers would be transported to a radioactive-waste disposal facility licensed by the NRC or an Agreement State. This disposal would only be allowed by the NRC after preoperational agreements between the Applicant and the licensed facility(ies) have been executed. The Applicant has identified four facilities currently licensed to receive such byproduct waste and that can ensure adequate capacity for the solid byproduct waste generated by the Ross Project (Strata, 2012a).

Solid non-byproduct waste and hazardous-waste volumes generated during the Proposed Action's operation would be similar to or less than that generated during its construction (Strata, 2011a). Therefore, the potential impacts of the management of all solid wastes during Ross Project operation would be SMALL.

4.14.1.3 Ross Project Aquifer Restoration

In GEIS Section 4.4.12.3, the impacts associated with waste management during an ISR facility's aquifer-restoration phase were evaluated. These were determined to be generally the same as those during its operation. Thus, the GEIS found that waste-management impacts would be SMALL.

Liquid Waste

Liquid byproduct (radioactive) wastes generated during the Proposed Action's aquifer restoration would amount to approximately 740 L/min [190 gal/min] of brine. The Applicant proposes to minimize the volume of liquid byproduct waste that would be generated while the Ross Project is in the aquifer-restoration phase by its limiting the ground-water sweep to the perimeter of a wellfield module, rather than throughout the entire module. As during operation, the two-stage RO circuit would reduce the volume of brine requiring disposal. Evaporation of stored brine from the surface impoundments would further reduce the volume of brine needing disposal by an estimated 36 L/min [9.3 gal/min]. All permeate from the RO process would be used for process water and aquifer restoration.

The volume of used oil that would be produced during the Proposed Action's aquifer-restoration phase would be the same as that produced during its construction and operation (Strata, 2012a). The volume of domestic sewage managed with the Ross Project's onsite treatment system would decrease to approximately 1,100 L/d [300 gal/d] (Strata, 2012a) due to the smaller number of workers at the Ross Project during aquifer restoration. Thus, the potential impacts of the management of all types of liquid wastes during aquifer restoration at the Proposed Action would be SMALL.

Solid Waste

The management of solid wastes, including byproduct, radioactive and hazardous wastes, generated during the aquifer-restoration phase of the Proposed Action would be similar to its construction and operation phases (Strata, 2011a). The volume of office and municipal solid wastes would decrease due to the smaller workforce during aquifer restoration (i.e., 200 and 60 vs. 20 workers), while the volume of byproduct and other radioactive wastes would also diminish, producing less and less waste contaminated by byproduct materials, as the aquifer is restored. Thus, the potential impacts of the management of solid wastes during aquifer restoration would be SMALL.

4.14.1.4 Ross Project Decommissioning

As described in GEIS Section 4.4.12, the impacts associated with liquid-waste management during decommissioning at an ISR facility would be SMALL and would be similar to the respective construction and operational impacts. However, the volume of solid byproduct waste and all other types of solid wastes generated during decommissioning would be substantially greater than during the other phases due to the decontamination, dismantling, demolishing, and disposal of the Ross Project components (Strata, 2012a).

Liquid Waste

The Applicant estimates that less than 38 L/min [10 gal/min] of brine would be generated and disposed of by deep-well injection during the Proposed Action's decommissioning (Strata, 2012a). This volume would be a significant reduction from that generated during the other phases of the Proposed Action. The volume of used oil that would be generated during decommissioning and its management would be the same as that generated during operation (Strata, 2012a). The volume of domestic sewage that would be treated in the onsite system would be approximately 4,500 L/d [1,200 gal/d] (Strata, 2012a). Thus, the potential impacts of the management of liquid wastes during the decommissioning phase of the Proposed Action would be SMALL.

Solid Waste

The Applicant estimates that decommissioning would generate 3,000 m^3 [4,000 yd^3] of solid byproduct waste (Strata, 2012a). The nature of this waste is described in SEIS Section 2.1.1. A typical ISR Project generates approximately 4,593 m^3 [6,008 yd^3] of byproduct waste, and Strata would generate less, thus the analysis in the GEIS is bounding (NRC, 2009).

The onsite collection, minimization, and storage of this solid byproduct waste would follow the same techniques and SOPs as those described for the Proposed Action's operation. The pre-operational agreements with one or more appropriately licensed waste disposal facilities would govern the disposal of this waste the same as during the Ross Project's operation. The Applicant proposes to reduce the quantity of solid byproduct waste by decontaminating as many surfaces as technically possible while using decontamination techniques such as high pressure washing, sand blasting, and acid rinsing that allow waste volumes to be reduced (Strata, 2011a). Where possible, the Applicant intends to decontaminate equipment and building surfaces so that the mobile equipment, dismantled process equipment, and demolished building components could be reclassified for unrestricted use by demonstrating that radioactivity levels are below regulatory concern.

The Applicant estimates that decommissioning would generate 1,500 m^3 [2,000 yd^3] of solid non-byproduct waste. Such waste would consist of construction debris and decontaminated equipment and materials (Strata, 2012a). As described in Section 2.1.1 of this SEIS, the Applicant proposes this waste would be disposed of in local solid-waste landfills. The estimated volume of solid waste would be about twice the amount generated by the typical ISR facility described in the GEIS (NRC, 2009), however the capacity of the local landfills are shown in the Applicant's responses to the NRC's Requests for Additional Information and the Applicant's corresponding table indicates there would be sufficient local capacity for disposal of this volume (Strata, 2012a).

The volumes of other typical solid and hazardous wastes including industrial or municipal waste, recyclable, demolition, and petroleum contaminated soil generated during the Proposed Action's decommissioning would be similar to those generated during construction and operation (Strata, 2012a). The potential impacts of the management of solid wastes during decommissioning, therefore, would be SMALL.

4.14.2 Alternative 2: No Action

Under the No-Action Alternative, the Ross Project would not be licensed and the land would continue to be available for other uses. However, the Applicant could continue preconstruction activities until that decision has been made. Thus, drilling fluids and muds from drillholes and wells installed to delineate the ore zone and to characterize the ground-water and the geotechnical, subsurface conditions at the Ross Project area would continue to generate wastes under the No-Action Alternative. These wastes would continue to be contained in mud pits constructed at the well sites (as described in SEIS Section 2.1.1) and then evaporated to dryness. The dried pits would be backfilled, graded, covered with topsoil, and reseeded to achieve reclamation standards required by WDEQ/LQD (Strata, 2011a). No additional, distinct waste management impacts would result from the No-Action Alternative; thus, the potential impacts of waste management in the No-Action Alternative would be SMALL.

4.14.3 Alternative 3: North Ross Project

Under Alternative 3, the North Ross Project would generally be the same as the Proposed Action, except that the facility (i.e., the CPP, associated buildings, and auxiliary structures as well as the surface impoundments) would be located to the north of where it would be located in the Proposed Action, as described in SEIS Section 2.1.3. The wastes generated during this Alternative would be essentially the same as those generated during the Proposed Action during each of its phases: Alternative 3 would be constructed and operated the same as the Proposed Action, and its aquifer restoration and decommissioning would also be the same. Thus, the waste-management techniques and disposal strategies employed for the Proposed Action would be employed for Alternative 3.

However, as described in SEIS Section 2.1.1, the lined surface impoundments would not require the construction of the CBW included in the design of the Proposed Action because of the south site's higher water table. Consequently, the volume of liquid wastes generated at north site would be reduced by the volume of any leaks and/or ground water that would need to be dewatered from inside the CBW during facility operation, aquifer restoration, and decommissioning of Alternative 1. In addition, the volume of solid waste ultimately requiring disposal would be reduced by the small amount of material generated during the breach of the CBW during decommissioning. Therefore, potential impacts of waste management for Alternative 3 would be SMALL.

4.15 References

10 CFR Part 20. Title 10, "Energy," *Code of Federal Regulations*, Part 20 "Standards for Protection Against Radiation." Washington, DC: U.S. Government Printing Office.

10 CFR Part 40. Title 10, "Energy," *Code of Federal Regulations*, Part 40, "Domestic Licensing of Source Material." Washington, DC: Government Printing Office. 1985, as amended.

10 CFR Part 40, Appendix A. Title 10, "Energy," *Code of Federal Regulations*, Part 40, "Domestic Licensing of Source Material," Appendix A, "Criteria Relating to the Operation of Uranium Mills and the Disposition of Tailings or Wastes Produced by the Extraction or Concentration of Source Material from Ores Processed Primarily for their Source Material Content." Washington, DC: Government Printing Office. 1985, as amended.

29 CFR Part 1910. Title 29, "Title," *Code of Federal Regulations*, Part 1910, "Occupational Safety and Health Standards." Washington, DC: Government Printing Office. 1974, as amended.

40 CFR Part 192. Title 40, "Protection of the Environment," *Code of Federal Regulations*, Part 192, "Health and Environmental Protection Standards for Uranium and Thorium Mill Tailings." Washington, DC: Government Printing Office. 1983, as amended.

40 CFR Subpart C. Title 40, "Protection of the Environment," *Code of Federal Regulations*, Subchapter C. Washington, DC: Government Printing Office. 1971, as amended.

49 CFR Part 171. Title 49, "Transportation," *Code of Federal Regulations*, Part 171, "General Information, Regulations, and Definitions," Subpart B, "Incident Reporting, Notification, BOE Approvals and Authorization." Washington, DC: Government Printing Office. 2003, as amended.

49 CFR Part 172. Title 49, "Transportation," *Code of Federal Regulations*, Part 185, "Hazardous Material Table, Special Provisions, Hazardous Materials Communication, Emergency Response Information, Training Requirements, and Security Plans." Washington, DC: Government Printing Office. 1972, as amended.

APLIC (Avian Power Line Interaction Committee). *Suggested Practices for Avian Protection on Power Lines: The State of the Art in 2006*. CEC-50-2006-022. Washington, DC: Edison Electric Institute; Sacramento, CA: Avian Power Line Interaction Committee; Sacramento, CA: California Energy Commission. 2006.

BLM (U.S. Bureau of Land Management). *Visual Resource Management*. Manual 8400. Washington, DC: Government Printing Office. 2007.

BLM. *Visual Resource Inventory*. Manual H–8410–1. Washington, DC: USBLM. 2010. Agencywide Documents Access and Management System (ADAMS) Accession No. ML13014A644.

(US)CB (U.S. Census Bureau, U.S. Department of Commerce [DOC]). "State and County QuickFacts, Wyoming, Crook County, Campbell County." At http://quickfacts.census.gov/qfd/states/56000.html (as of June 25, 2012). 2012a.

(US)CB. "2006 – 2010 American Community Survey." At http://factfinder2.census.gov (as of April 26 and May 2, 2012). 2012b.

CEQ (Council of Environmental Quality). *Environmental Justice, Guidance under the National Environmental Policy Act*. Prepared for The Executive Office of the President. Washington, DC: Government Printing Office. December 1997. ADAMS Accession No. ML13022A298.

Countess, R., W. Barnard, C. Claiborn, D. Gillette, D. Latimer, T. Pace, and J. Watson. "Methodology for Estimating Fugitive Windblown and Mechanically Resuspended Road Dust Emissions Applicable for Regional Air Quality Modeling" in *10th International Emission Inventory Conference Proceedings*. Westlake Village, CA: Countess Environmental. May 2001. ADAMS Accession No. ML13022A448.

(US)DOT (U.S. Department of Transportation). *Highway Traffic Noise Analysis and Abatement Policy and Guidance*. Washington, DC: Noise and Air Quality Branch, Office of Environment and Planning, Federal Highway Administration, USDOT. June 1995. Accession No. ML13023A373.

EOP (Executive Office of the President). *Federal Actions to Address Environmental Justice in Minority Populations and Low-Income Populations.* EO No. 12898. Washington, DC: Government Printing Office. February 1994. ADAMS Accession No. ML13023A255.

(US)EPA (U.S. Environmental Protection Agency). *Protective Noise Levels.* Condensed Version of EPA Levels Document. EPA 550/9-79-100. Washington, DC: USEPA. November 1978. ADAMS Accession No. ML13015A552.

Ferguson, D. *A Class III Cultural Resource Inventory of Strata Energy's Proposed Ross ISR Uranium Project, Crook County, Wyoming* (Redacted Version). Prepared for Strata Energy, Inc., Gillette, Wyoming. Butte, MT: GCM Services, Inc. 2010.

Mackin, P.C., D. Daruwalla, J. Winterle, M. Smith, and D.A. Pickett. NUREG/CR–6733, "A Baseline Risk-Informed Performance-Based Approach for In Situ Leach Uranium Extraction Licensees." Washington, DC: NRC. September 2001.

(US)NPS (U.S. National Park Service, US Department of Interior). Jeffery R. Hanson and Sally Chirino. *Ethnographic Overview and Assessment of Devils Tower National Monument, Wyoming.* Paper 4. Washington, DC: USNPS. 1997.

(US)NRC (U.S. Nuclear Regulatory Commission). NUREG–0706. "Final Generic Environmental Impact Statement on Uranium Milling Project M-25." Washington, DC: NRC. September 1980. ADAMS Accession Nos. ML032751663, ML0732751667, and ML032751669.

(US)NRC. NUREG–1508. "Final Environmental Impact Statement To Construct and Operate the Crownpoint Uranium Solution Mining Project, Crownpoint, New Mexico." Washington, DC: NRC. February 1997. ADAMS Accession No. ML082170248.

(US)NRC. "Environmental Assessment for Renewal of Source Material License No. SUA-1534, Crow Butte Resources Incorporated, Crow Butte Uranium Project, Dawes County, Nebraska." Washington, DC: NRC. 1998. Accession No. ML071520242.

(US)NRC. NUREG–1748. "Environmental Review Guidance for Licensing Actions Associated with NMSS Programs." Washington, DC: NRC. August 2003. ADAMS Accession No. ML032450279.

(US)NRC. "Policy Statement on the Treatment of Environmental Justice Matters in NRC Regulatory and Licensing Actions." No. 7590-01-P. Washington, DC: NRC. 2004. ADAMS Accession No. ML033380930.

(US)NRC. NUREG–1910. "Generic Environmental Impact Statement for In-Situ Leach Uranium Milling Facilities." Volumes 1 and 2. Washington, DC: NRC. 2009. ADAMS Accession Nos. ML082030185 and ML091480188.

(US)NRC. NUREG–1910, Supplement 1. "Environmental Impact Statement for the Moore Ranch ISR Project in Campbell County, Wyoming, Supplement to Generic Environmental Impact Statement for In-Situ Leach Uranium Milling Facilities, Final Report." Washington, DC: USNRC. August 2010. ADAMS Accession No. ML102290470.

(US)NRC. NUREG–1910, Supplement 2. "Environmental Impact Statement for the Nichols Ranch ISR Project in Campbell and Johnson Counties, Wyoming. Supplement to the Generic Environmental Impact Statement for In-Situ Uranium Leach Milling Facilities." Washington, DC: USNRC. January 2011a. ADAMS Accession No. ML103440120.

(US)NRC. NUREG–1910, Supplement 3. "Environmental Impact Statement for the Lost Creek ISR Project in Sweetwater County, Wyoming. Supplement to the Generic Environmental Impact Statement for In-Situ Uranium Leach Milling Facilities." Washington, DC: USNRC. June 2011b. ADAMS Accession No. ML11125A006.

(US)OSHA (U.S. Occupational Safety and Health Administration). "Wyoming Plan." 2011. At http://www.osha.gov/dcsp/osp/stateprogs/wyoming.html (as of April 24, 2012).

Stout, Dennis E. and R. Mark Stover. "The Smith Ranch Uranium Project" from *The Uranium Institute Twenty-Second Annual Symposium*, September 3–5, 1997, London, United Kingdom." London: The Uranium Institute. December 1997.

Strata (Strata Energy, Inc.). *Ross ISR Project USNRC License Application, Crook County, Wyoming, Environmental Report, Volumes 1, 2 and 3 with Appendices.* Docket No. 40-09091. Gillette, WY: Strata. 2011a. ADAMS Accession Nos. ML110130342, ML110130344, and ML110130348.

Strata. *Ross ISR Project USNRC License Application, Crook County, Wyoming, Technical Report, Volumes 1 through 6 with Appendices.* Docket No. 40-09091. Gillette, WY: Strata. 2011b. ADAMS Accession Nos. ML110130333, ML110130335, ML110130314, ML110130316, ML110130320, and ML110130327.

Strata. *Air Quality Permit Application for Ross In-Situ Uranium Recovery Project*. Prepared for Strata Energy, Inc. Sheridan, WY: Inter-Mountain Laboratories, IML Air Science. 2011c. ADAMS Accession No. ML11222A060.

Strata and Crook County. *Memorandum of Understanding for Improvement and Maintenance of Crook County Roads Providing Access to the Ross ISR Project.* Sundance, WY: Crook County. April 6, 2011d. ADAMS Accession No. ML111170303.

Strata. *Ross ISR Project USNRC License Application, Crook County, Wyoming, RAI Question and Answer Responses, Environmental Report, Volume 1 with Appendices.* Docket No. 40-09091. Gillette, WY: Strata. 2012a. ADAMS Accession No. ML121030465.

Strata. *Ross ISR Project USNRC License Application, Crook County, Wyoming, RAI Question and Answer Responses, Technical Report, Volumes 1 and 2 with Appendices.* Docket No. 40-09091. Gillette, WY: Strata. 2012b. ADAMS Accession Nos. ML121020357 and ML121020361.

WDEQ/AQD (Wyoming Department of Environmental Quality/Air Quality Division). "Air Quality." *Rules and Regulations*, Chapter 3. Cheyenne, WY: WDEQ/AQD. 2010.

WDEQ/AQD. *Permit to Construct, Air Quality Permit #12198.* Cheyenne, WY: WDEQ/AQD. 2011. September 13, 2011. ADAMS Accession No. ML112770430.

WDEQ/LQD (Wyoming Department of Environmental Quality/Land Quality Division). "Noncoal In Situ Mining," *Rules and Regulations*, Chapter 11. Cheyenne, WY: WDEQ/LQD. 2005.

WDEQ/LQD. *Cumulative Hydrologic Impact Assessment of Coal Mining in the Middle Powder River Basin, Wyoming.* WDEQ-CHIA-27. Cheyenne, WY: WDEQ/LQD. March 2011.

WDEQ/WQD. "Implementation Policy for Radium Effluent Limits in WYPDES Permits." At http://deq.state.wy.us/wqd/wypdes_permitting/index.asp (as of June 25, 2012). 2005. ADAMS Accession Nos. ML13015A683.

WDEQ/WQD (Wyoming Department of Environmental Quality/Water Quality Division). "Water Quality," *Rules and Regulations*, Chapter 1. Cheyenne, WY: WDEQ/WQD. 2007.

WDEQ/WQD. *Strata Energy, Inc. – Ross Disposal Injection Wellfield, Final Permit 10-263, Class I Non-hazardous, Crook County, Wyoming.* Cheyenne, WY: WDEQ/WQD. April 2011.

WDWS (Wyoming Department of Workforce Services). "Labor and Employment," *Rules and Regulations*, Chapter 11. Cheyenne, WY: WDWS. 2012.

5 CUMULATIVE IMPACTS

5.1 Introduction

The Council on Environmental Quality's (CEQ's) *National Environmental Policy Act* (NEPA) regulations, as amended (Title 40 *Code of Federal Regulations* [CFR] Parts 1500 – 1508) (40 CFR Parts 1500 – 1508), define cumulative effects as "the impact on the environment that results from the incremental impact of the action when added to other past, present, and reasonably foreseeable future actions regardless of what agency (Federal or non-Federal) or person undertakes such other actions" (40 CFR Parts 1500 – 1508). Cumulative impacts can result from individually minor, but collectively significant, actions that take place over a period of time. (For the purposes of this analysis, the phrase "cumulative impacts" is synonymous with the phrase "cumulative effects.") A proposed project could contribute to incremental cumulative impacts when its environmental impacts overlap with those of other past, present, or reasonably foreseeable future actions in a given area. For this Supplemental Environmental Impact Statement (SEIS), other past, present, and future actions near the Ross Project include (but are not limited to) cattle and sheep grazing, agricultural production, other uranium-recovery production, coal mining, oil and gas production, and wind-farm operation.

This analysis of the cumulative impacts of the Proposed Action is based upon publicly available information on existing and proposed projects, information in the Generic Environmental Impact Statement (GEIS) (NRC, 2009), and general knowledge of the conditions in Wyoming and in the nearby communities. The primary activities currently taking place in the area of the Ross Project are mineral mining and uranium recovery as well as oil and gas development. The Power River Basin contains the largest deposits of coal in the United States as well as significant reserves of other natural resources including uranium, oil, and gas (NRC, 2010). There has been a resurgence in interest in these mining and recovery activities.

This section evaluates the potential for cumulative impacts associated with the Ross Project and other past, present, and reasonably foreseeable future actions as described below in Section 5.2. The GEIS provides an example methodology for conducting a cumulative-impacts assessment (NRC, 2009). This methodology, which has been used by U.S. Nuclear Regulatory Commission (NRC) staff in its cumulative-impact analysis in this SEIS, is discussed in Section 5.3.

5.2 Other Past, Present, and Reasonably Foreseeable Future Actions

The Ross Project area, where the Proposed Action would be sited, is located just within the Nebraska-South Dakota-Wyoming Uranium Milling Region (NSDWUMR) as defined in the GEIS (NRC, 2009). The Ross Project encompasses approximately 697 ha [1,721 ac] of land, all of which is located in Crook County. It is located within the Lance District (see Figure 2.1 in Section 2), so-called due to its location above the uranium-rich Late Cretaceous Lance Formation as discussed earlier in Section 3.4. The surface owners of the Ross Project area include private parties (553 ha [1,367 ac]), the State of Wyoming (127 ha [314 ac]), and the U.S. Bureau of Land Management (BLM) (16ha [40 ac]). The subsurface-mineral owners include the same parties, except that of 553 ha [1,367 ac] of privately owned land, 65 ha [161 ac] of subsurface mineral rights are administered by BLM. Somewhat unusually, the surface water at the Ross Project predominantly flows in a northeasterly direction to the Little Missouri River, while the ground water, which is part of the Powder River Basin regime, flows mostly westerly. This bifurcation is important to note as cumulative impacts are identified and evaluated. The Ross Project area, at approximately 7 km^2 [somewhat less than 3 mi^2] in size, represents approximately 0.03 percent of the 25,900 km^2 [10,000 mi^2] of the entire Powder River Basin.

5.2.1 Actions

The historical and current actions (land uses) on and near the Ross Project area include livestock grazing, crop cultivation and agriculture, wildlife habitats, oil recovery, and, to the northeast, bentonite mining (Strata, 2011a). The historical Nubeth Joint Venture (Nubeth) also was operated on the lands, which comprise the proposed Ross Project. SEIS Section 3.2 discusses these historical and present land uses in more detail; these land uses are expected to continue into the future, albeit to a lesser extent, while the Ross Project is operating in the area. It should be noted that no long-term, permanent changes to the environment are anticipated as a result of the Ross Project within about 8 km [5 mi] of the Ross Project area, except for the potential installation of additional roads. The extensive aquifer restoration and site reclamation activities the Applicant would perform during the Ross Project's decommissioning would ensure that no permanent land-use changes occur on the Ross Project area itself.

Several industries presently conduct activities in and near Crook County, activities which could have environmental impacts that, when combined with those of the Ross Project, could be greater than the individual impacts of the Ross Project. In addition, some of these activities, such as uranium recovery as well as oil and gas recovery, could be actively expanded within Crook County and into its neighboring counties. These activities are described below.

5.2.1.1 Uranium Recovery

Uranium was first mined in Wyoming in 1920. Uranium discovered in the Powder River and Wind River Basins during the 1950s, and continued exploration for uranium resulted in discovery of additional sedimentary uranium deposits in the major basins of central and southern Wyoming, including the Powder River Basin. Continued uranium exploration resulted in discovery of additional sedimentary uranium deposits in the major basins of central and southern Wyoming. Uranium production in Wyoming declined in the mid-1960s, but increased again in the late 1960s and 1970s. Conventional mine production peaked in 1980 and then decreased in the early 1980s through the early 1990s when in situ recovery (ISR) facilities were developed. The total uranium mine production in the United States in 2007 was 2.1 million kg [4.5 million lb], almost half of which occurred in the southernmost Powder River Basin. ISR replaced conventional mining and milling as the preferred means for extracting uranium in the U.S. Currently, only ISR facilities are extracting uranium in Wyoming.

Interest in uranium-recovery has translated into several ISR projects in Wyoming. The Ross Project is one. In addition, the Applicant indicates that it could develop at least four additional satellite uranium-recovery areas within the larger Lance District over the next few years. Several other ISR projects are currently licensed in Wyoming as well, with two facilities operating and two ready for construction in the Powder River Basin (see Figure 5.1). None of these operating and/or licensed ISR projects are located in Crook County (the location of the proposed Ross Project) nor have any other Crook County ISR facilities be officially proposed to the U.S. Nuclear Regulatory Commission (NRC). However, four ISR projects are reportedly in the very early stages of development in Crook County (Strata, 2012a). In addition, two licensed ISR facilities are located in adjacent Campbell County (satellite areas of the Smith Ranch ISR Project, which is currently operating, and the Moore Ranch, which is still to be constructed). Two other ISR facilities overlap both Campbell and Johnson Counties (Willow Creek, which is currently operating, and Nichols Ranch, which is licensed and under construction).

The Applicant describes in its license application the types and sequence of its planned development of the Lance District. The Applicant has identified significant uranium resources within the District, and it intends for the Ross Project to be the first of several ISR areas. These potential satellite areas could consist of those shown in Figure 2.2 in SEIS Section 2.1.1, including, to the north, Ross Amendment Area 1 and, to the south end of the Lance District, the Kendrick, Richards, and Barber satellite areas (Strata, 2012a). If additional wellfields were to be developed by the Applicant and licensed by the NRC, the Ross Project's Central Processing Plant (CPP) would be used to process pregnant solutions from these satellite areas into yellowcake. In addition, the Applicant also proposes that ion-exchange (IX) resins loaded with uranium would be accepted at the Ross Project's CPP from other offsite ISR facilities (referred to as "toll milling") or companies and/or from water-treatment plants (Strata, 2011a). This additional potential use of the CPP at the Ross Project is the reason that the Plant is designed for four times the capacity needed for only the Ross Project.

Lance District

The four satellite areas within the Lance District that the NRC staff identifies as reasonably foreseeable are as follows:

Ross Amendment Area 1

This area would be an extension of the proposed Ross Project to the north and west. This area would not increase the overall production rate of yellowcake, but rather it would increase the operating life of the Ross Project. As uranium production from early wellfields within the Ross Project area begins to diminish and the wellfields begin to enter the aquifer-restoration phase of the proposed Project, additional wellfields in the Ross Amendment Area 1 could be brought into production by the Applicant. The Ross Amendment Area 1 could extend the lifetime of the Ross Project by several years as shown in Figure 5.4 (Strata, 2012a).

Kendrick Satellite Area

The Kendrick satellite area would be contiguous to the Ross Project area as shown in Figure 2.2 in SEIS Section 2.1.1. However, unlike the Ross Amendment Area 1, the Kendrick satellite area would allow the Applicant to increase its production of yellowcake to approximately 680,000 kg/yr [1.5 million lb/yr] (Strata, 2012a).

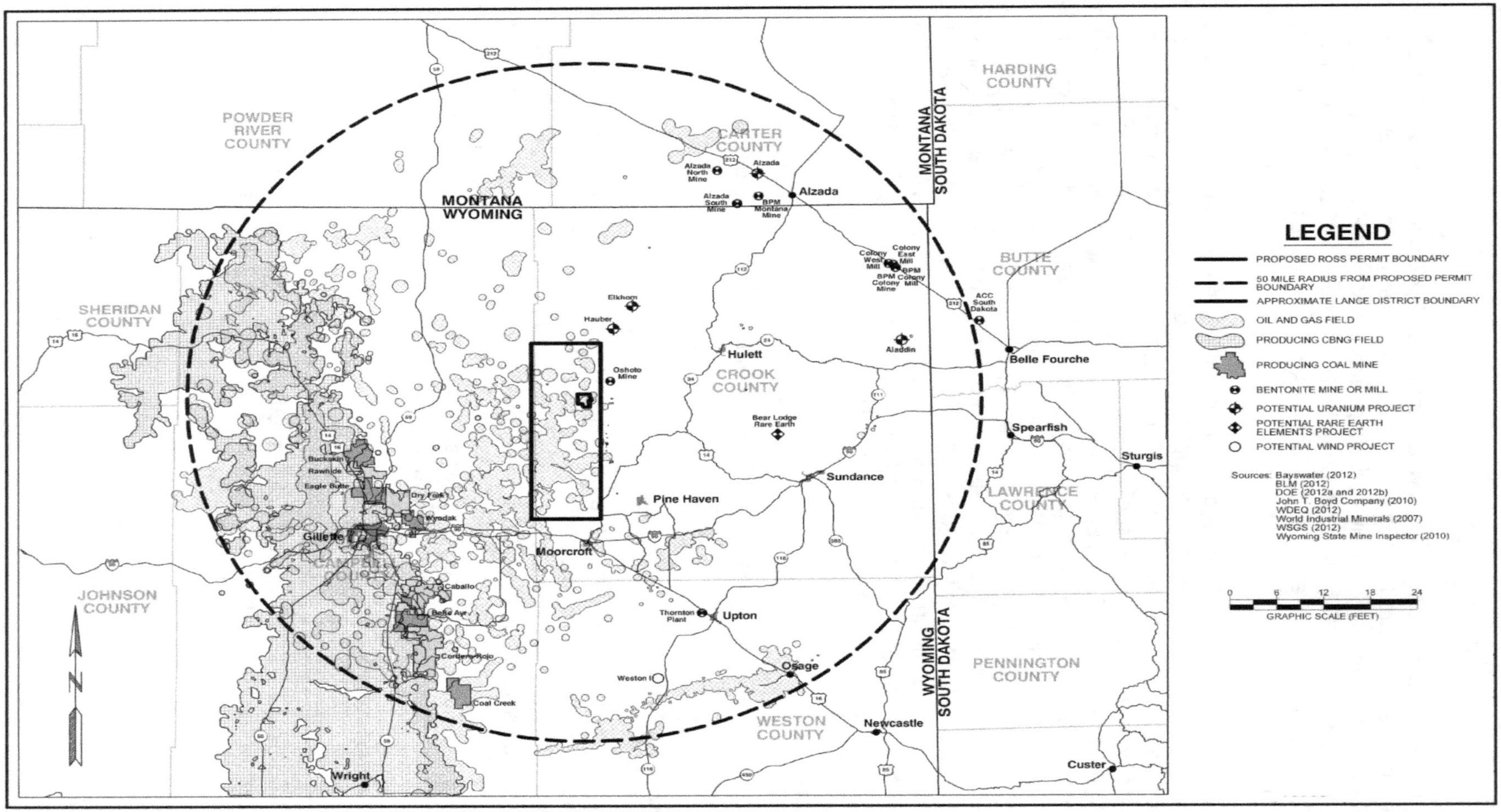

Source: Strata, 2012a.

Figure 5.1

Eighty-Kilometer- (Fifty-Mile-) Radius Area around Ross Project Area

Richards Satellite Area

The Richards satellite area would be contiguous to the Kendrick satellite area. The uranium-rich solutions extracted from this satellite area would be piped to the Ross Project's CPP for uranium recovery or, potentially, piped to the Barber satellite area as described below (Strata, 2012a).

Barber Satellite Area

Although the Applicant's plans for development of the Lance District are not yet complete, Strata anticipates that a remote IX-only plant could be constructed at the Barber satellite area. This would mean that the pregnant, uranium-rich solutions brought to the surface at the Barber satellite area would be treated by IX to yield uranium-loaded resins, which would then be trucked to the Ross Project's CPP for further processing (e.g., resin elution) (Strata, 2012a). This additional uranium would increase the CPP's output to approximately 993,000 kg/yr [2.19 million lb/yr]. In addition, the Applicant would investigate the possibility of transferring pregnant solutions from wellfields in the Richards satellite area to the remote IX facility at the Barber satellite area before transfer to the CPP at the Ross Project area.

Other Potential ISR Facilities within 80 Kilometers [50 Miles] of the Ross Project

There are no uranium recovery or nuclear-fuel-cycle projects currently located within 80 km [50 mi] of the Ross Project area nor have any Letters of Intent or license applications been filed with the NRC for any ISR projects within 80 km [50 mi] (Strata, 2011a; NRC, 2013). An 80-km [50-mi]-radius area from the Ross Project is shown in Figure 5.1. There are, however, four other uranium-recovery operations in various very early planning stages located within 80 km [50 mi] of the Ross Project, including the following:

Potential Aladdin Project

This potential ISR Project would be located in Crook County, approximately 66 km [41 mi] east-northeast of the Ross Project, although the driving distance to this project would be approximately 113 km [70 mi]. The Aladdin Project is being considered by Powertech Inc. and comprises approximately 7,099.8 ha [17,554 ac].

Potential Elkhorn Project

This potential ISR Project is currently being evaluated by NCA Nuclear, Inc. (a wholly owned subsidiary of Bayswater Uranium Corporation). This Project would also be located in Crook County, approximately 26 km [16 mi] from the Ross Project (driving distance would be approximately 32 km [20 mi]). It is currently estimated that this Project's area of 2,110 ha [5,215 ac] may ultimately yield approximately 544,000 kg [1.2 million lb] of uranium. The Project is located near the former, and decommissioned, Homestake Hauber Uranium Mine (see below).

Potential Hauber Project

The potential Hauber ISR Project would also be owned by NCA Nuclear, Inc., in a joint venture with Ur-Energy Inc. This Project would be located approximately 23km [14 mi] from the Ross

Project area, or 32 km [20 mi] if driven, and would comprise approximately 469 ha [1,160 ac]. The total uranium production from this Project is estimated at 680,000 kg [1.5 million lb] (Strata, 2012a). This Project would be located near the now-closed Hauber Uranium Mine, which was operated between 1958 and 1966 (Strata, 2011a), which is discussed below.

Potential Alzada Project

This Project would be owned and operated by NCA Nuclear, Inc. and would comprise approximately 10,000 ha [25,000 ac]. It would be located approximately 62 km [39 mi] north-northeast of the Ross Project area (driving distance would be approximately 129 km [80 mi]) (Strata, 2012a).

Other ISR Facilities within the Powder River Basin

There are four ISR projects in various stages of NRC's licensing process and/or currently operating or being constructed within the Powder River Basin, all of which are located in Wyoming. The 80-km [50-mi] cumulative-impacts area does not include the entire Powder River Basin. Two of these facilities are currently operating; two have been licensed, one of which has begun construction. The owner of a fifth ISR project has conveyed a Letter of Intent to submit a license application to the NRC, but the application has not yet been submitted. These ISR projects include the following:

Smith Ranch ISR Project

This is a uranium-recovery project currently being operated by Power Resources Inc. (dba Cameco Resources Inc. [Cameco]). The Smith Ranch ISR Project is primarily located in Converse County, Wyoming, but the operation includes several remote satellite areas in other Wyoming counties that are not located in the Powder River Basin. A license application to renew and to expand Source Materials License SUA-1548 for the Smith Ranch Project was received by the NRC in February 2012 (see Docket No.40-8964). If the NRC grants a license as proposed, the renewed license would allow Cameco to continue conducting ISR activities at its Smith Ranch Project as well as to initiate and/or expand ISR activities at its associated and remote ISR satellite areas: 1) the Highlands Uranium Project and the Reynolds Ranch ISR satellite areas, both also located in Converse County; 2) the Gas Hills ISR satellite area in Fremont and Natrona Counties, Wyoming; 3) the North Butte ISR satellite area in Campbell County, Wyoming; and 4) the Ruth ISR satellite area in Johnson County, Wyoming (NRC, 2013).

Willow Creek ISR Project

The Willow Creek ISR Project is located in Johnson County in Wyoming. This Project is owned by Uranium One (see Docket No. 40-8502). Currently, its NRC license is in timely renewal as of May 2008 (i.e., a renewal license application has been submitted and the NRC is currently engaged in technical review of that application).

A license application for the Ludeman ISR Project was originally submitted to the NRC in January 2010, but it was subsequently withdrawn in May 2010. A license application was resubmitted by the owner of the Project, Uranium One, in December 2011, where three specific subdivisions of the Ludeman area, which is located in Converse County, would be satellites of the Willow Creek ISR Project, which is located in Johnson County (NRC, 2013). Both of these Projects are situated in the Powder River Basin. The Ludeman ISR Project consists of approximately 8,000 ha [20,000 ac]; the Willow Creek ISR Project is approximately 5,500 ha [13,600 ac].

Nichols Ranch ISR Project

The Nichols Ranch ISR Project is located in Johnson and Campbell Counties of Wyoming. It is owned by the Uranerz Corporation (Uranerz) and is comprised of 1,251 ha [3,091 ac]. Its NRC license has been granted, and the facility is currently under construction (see Docket No. 40-9067) (NRC, 2013). Uranerz currently has an Underground Injection Control (UIC) Permit Application pending at Wyoming Department of Environmental Quality (WDEQ). Uranerz has signed a toll-milling agreement with the owner of the Smith Ranch ISR Project, Cameco, to transfer uranium-loaded IX resins from the Nichols Ranch ISR Project to the Smith Ranch Project for final processing to yellowcake.

Moore Ranch ISR Project

The Moore Ranch ISR Project is located in Campbell County, Wyoming; it is owned by Energy Metals Corporation, a wholly owned subsidiary of Uranium One. It is comprised of approximately 2,879 ha [7,110 ac]. It is currently licensed by the NRC to operate through September 2020 (see Docket No. 40-9073) (NRC, 2013); construction on this ISR facility has not yet begun.

Reno Creek ISR Project

AUC LLC, submitted a Letter of Intent to the NRC on November 3, 2010, indicating AUC LLC's intention to site, design, license, construct, and operate an ISR facility in Campbell County, Wyoming. According to publically available information, the NRC currently anticipates receiving AUC LLC's license application in April 2012 (NRC, 2012c).

Table 5.1 presents these Projects and indicates the respective linear distances from the Ross Project; Figure 5.1 shows these Projects' locations.

Table 5.1
Uranium-Recovery Projects within 80 Kilometers [50 Miles] of Ross Project Area

Project	Owner	County	Direction and Distance[a] (km [mi])	Status
Smith Ranch License SUA-1548 North Butte Ruby Ranch	Cameco Resources Inc./ Power Resources Inc.	Converse Campbell Campbell	SSW 180 km [110 mi]	Operating. Renewal and expansion (additional satellite areas) license application in technical review. Construction activities are occurring at the North Butte site. Ruby Ranch expansion license application not yet submitted.
Willow Creek (Formerly Irigaray/ Christianson Ranch) License SUA-1341 Ludeman Allemand-Ross	Uranium One	Johnson and Campbell Converse Converse	WSW 120 km [75 mi]	Operating. Renewal license application in technical review. Expansion to include Ludeman (license application has been submitted) and, later, Allemand-Ross (license application has not been submitted) satellite areas.
Nichols Ranch License SUA-1597	Uranerz Energy Corporation	Johnson and Campbell	SW 120 km [75 mi]	Licensed and under construction.
Moore Ranch License SUA-1596	Energy Metals Corporation/ Uranium One	Campbell	SW 150 km [90 mi]	Licensed, but not yet under construction.
Reno Creek	AUC LLC	Campbell		Letter of Intent filed, license application is not yet submitted.

Source: Strata, 2012a.

Note:

[a] Approximate distance from the Ross Project area to the respective ISR project in "as the crow flies" (i.e., straight line) in kilometers [miles].

Past ISR Facilities within 80 Kilometers [50 Miles] of the Ross Project

In addition to the present and reasonably foreseeable uranium-recovery facilities described above, it should be noted that, historically, two uranium-recovery facilities were located in the 80-km [50-mi] area surrounding the Ross Project area. The first was a historic uranium mine near Hulett, and the second, Nubeth, was identified above and has been included in this SEIS's analysis of pre-licensing baseline data as well as cumulative impacts in this section.

The historic Homestake Hauber Uranium Mine was operated by the Homestake Mining Company between 1958 – 1966; the mine closed in 1966. It is also located in Crook County, approximately 19 km [12 mi] to the northeast of the Ross Project. This mine is no longer a contributor to cumulative impacts in the area because it is not operating and, thus, no longer producing impacts related to traffic, water resources, air quality, noise, and so forth. However, it is now a part of the area currently being explored for additional potential uranium recovery by NCA Nuclear, Inc., in a joint venture with Ur-Energy Inc. The potential Hauber ISR Project is described above; the Project is currently in the planning stages. This Project would be the nearest ISR uranium-recovery project to the proposed Ross Project

Nubeth was described in SEIS Sections 2.1.1 and 3.5.3. This research and development ISR uranium-recovery operation operated between 1978 – 1986. Nubeth was decommissioned according to NRC and WDEQ requirements, and final approval for its decommissioning was issued between 1983 – 1986. Additional information regarding potential impacts from this historical operation is included in this SEIS Section assessing cumulative impacts.

5.2.1.2 Mining

Both coal as well as other natural resources are mined in and around Crook, Weston, and Campbell counties. Indeed, Powder River Basin coal mines supplies over 96 percent of the coal produced in Wyoming each year (BLM 2005a; BLM 2005b; BLM2005c), and Wyoming produces the greatest amount of coal in the U.S. Thus, substantial mining activities occur throughout the Basin, and coal mining continues to be the most prolific mining activity in the region.

Coal Mining

Coal mining in the Powder River Basin began during 1883, and underground coal mines began operation during 1894. The Powder River Basin emerged as a major coal-production area during the 1970s and early 1980s. The largest area, the Gillette coalfield, is approximately 24 km [15 mi] wide and extends from approximately 35 km [22 mi] north of Gillette, Wyoming, to approximately 40 km [25 mi] south of Wright, Wyoming. A second coal area is approximately 32 km [20 mi] wide, extending from Sheridan, Wyoming, north to the Wyoming-Montana state line. In 2007, this region accounted for approximately 97 percent of Wyoming's production and hosted the 10 largest coal mines in the U.S. Coal production in the Wyoming portion of the Basin is expected to grow at an annual rate of 2 – 3 percent per year. Additional coal leases and associated lands may be required to keep up with the world's demand (BLM, 2009e).

The Powder River Federal Coal Region was decertified as a federal coal production region by the Powder River Regional Coal Team in 1990, which allowed leasing to occur in the region on an application basis. Because of decertification, United States coal production increased 11 percent, from 900,000 t [1 million T] in 1990 to 1.1 million [1.2 million T] in 2007 (BLM 2009a). From 1990 to 2008, the BLM Wyoming State Office held 25 competitive lease sales and issued 19 new federal coal leases containing more than 5.7 billion tons of coal using the "lease by application" process (BLM 2005a; BLM 2005b; BLM 2005c). In 2003, the cumulative disturbed land area attributable to coal mines within the Powder River Basin totaled nearly 28,000 ha [70,000 ac]. Reasonably foreseeable future development projects contributing to the estimate of the cumulative acreage disturbed range from 47,400 – 50,600 ha [117,000 – 125,000 ac] in 2015. Other development related to coal includes railroads, coal-fired power plants, major (230 kV) transmission lines, and coal technology projects. The total land area of other coal-related disturbance in the Powder River Basin in 2003 was nearly 2,000 ha [5,000 ac].

Within 80 km [50 mi] of the Ross Project there are nine active coal mines (Strata, 2012a). Table 5.2 lists surface coal mines within 80 km [50 mi]; the respective locations are shown in Figure 5.1.

Table 5.2
Active Coal Mines within 80 Kilometers [50 Miles] of Ross Project Area

Mine Name	Owner	Straight-Line Distance km [mi]	Driving Distance km [mi]
Belle Ayr Mine	Alpha Coal West, Inc.	64 [40]	103 [64]
Buckskin Mine	Buckskin Mining Company	47 [29]	108 [67]
Caballo Mine	Peabody Caballo Coal L.L.C.	63 [39]	109 [68]
Coal Creek Mine	Thunder Basin Coal Co. L.L.C.	72 [45]	137 [85]
Cordero Rojo Mine	Cloud Peak Energy/ Cordero Rojo Mine	68 [42]	119 [74]
Dry Fork Mine	Western Fuels Wyoming Inc.	45 [28]	85 [53]
Eagle Butte Mine	Alpha Coal West Inc.	48 [30]	93 [58]
Rawhide Mine	Peabody Energy Rawhide Mine	47 [29]	100 [62]
Wyodak Mine	Wyodak Resources Development	45 [28]	71 [44]

Source: Wyoming State Mine Inspector, 2010; BLM, 2012, as included in Strata, 21012a.

Bentonite Mining

Bentonite is weathered volcanic ash that is used in a variety of products, including drilling muds and cat litters, because of its absorbent properties. There are 10 bentonite-producing mines in in the 80-km [50-mi] area surrounding the proposed Ross Project area. One, the Oshoto Mine, is 8 km [5 mi] (driving distance) from the Ross Project area. The next two closest bentonite mines are approximately 56 – 69 km [35 – 43 mi] from the Ross Project area.

Other Mining

Sand, gravel, and clinker (or "scoria") have been and continue to be mined in the Powder River Basin. Aggregate, which is sand, gravel, and stone, is used for construction purposes. The largest aggregate operation is located in the Powder River Basin in northern Converse County, and it has an associated total disturbance area of approximately 27 ha [67 ac], of which 1.62 ha [4 ac] have been reclaimed. Scoria is used as aggregate where alluvial terrace gravel or in-palace granite/igneous rock is not available. Scoria generally is mined in Converse and Campbell Counties, in the western portion of the Powder River Basin (BLM, 2005a; BLM, 2005b; BLM, 2005c). None of these are within 80 km [50 mi] of the Ross Project area.

Table 5.3
Active Bentonite Mines within 80 Kilometers [50 Miles] of Ross Project Area

Mine Name	Owner	Straight-Line Distance km [mi]	Driving Distance km [mi]
ACC South Dakota	American Colloid Company	74 – 80 [46 – 55]	129 [80]
Alzada North	American Colloid Company	56 – 65 [35 – 40]	89 [55]
Alzada South	American Colloid Company	56 – 65 [35 – 40]	72 [45]
BPM Colony Mill	Bentonite Performance Minerals L.L.C.	71 [44]	151 [94]
BPM Colony Mine	Bentonite Performance Minerals L.L.C.	71 [44]	151 [94]
BPM Montana	Bentonite Performance Minerals L.L.C.	56 – 64 [35 – 40]	72 [45]
Colony East Mill	American Colloid Company	71 [44]	151 [94]
Colony West Mill	American Colloid Company	69 [43]	151 [94]
Oshoto Mine	Black Hills Bentonite	5 [3]	8 [5]
Thornton Plant	Black Hills Bentonite	56 [35]	69 [43]

Source: Wyoming State Mine Inspector, 2010; WDEQ, 2012; BLM, 2008; BLM, 2011 as cited in Strata, 2012a.

Oil and Gas Production

Regional oil and gas development activities (e.g., exploration, production, and pipeline development) could have the potential to generate cumulative impacts (BLM, 2005b) when evaluated in conjunction with the Ross Project. There are approximately 472 oil and gas production units evenly dispersed throughout the Powder River Basin in various stages of production. The Wyoming Oil and Gas Conservation Commission reported that in 2003, oil and gas wells in the Powder River Basin produced approximately 113 million barrels of oil and 1.1 billion m^3 [40 billion ft^3] of conventional gas (BLM, 2005a; BLM, 2005b; BLM, 2005c).

Most of Wyoming current oil production is from old oil fields with declining production and the level of exploration drilling to discover new fields has been low (BLM, 2005a). From 1992 to 2002, oil production from conventional oil and gas wells in Campbell and Converse Counties within the Powder River Basin decreased approximately 60 percent. Oil- and gas-related development includes major transportation pipelines and refineries. In 2003, the cumulative disturbed land area in the Powder River Basin from oil and gas, coal-bed methane (CBM), and related development was nearly 76,081 ha [188,000 ac]. The corresponding projection for the year 2015 is 123,429 ha [305,000 ac] (BLM, 2005a; BLM 2005b; BLM, 2005c). The depth to producing gas and oil-bearing horizons generally ranges from 1,219 – 4,115 m [4,000 – 13,500 ft], but some wells are as shallow as 76 m [250 ft] (BLM, 2005a; BLM, 2005b; BLM, 2005c).

There are three oil-producing wells on the Ross Project area itself in addition to three oil-field water-supply wells and two injection wells. These are used for enhanced oil recovery (EOR) and were discussed during this SEIS's evaluation of ground-water impacts in Section 4.5.1. Figure 3.2 indicates the locations of all of the oil- and gas-producing wells in a 3-km [2-mi] radius of the Ross Project area.

Coal-Bed Methane Development

Natural gas production has been increasing in Wyoming. CBM is located where there are abundant coal resources. For this reason, the majority of CBM production in Wyoming occurs in the Powder River Basin. Annual CBM production in the Powder River Basin increased rapidly between 1999 and 2003, with nearly 15,000 producing CBM wells in the Powder River Basin in 2003 and a total production volume of 10.3 billion m^3 [364 billion ft^3] (BLM, 2005a; BLM, 2005b; BLM, 2005c). However, there are no CBM-producing wells in the 80-km [50-mi] radius vicinity of the Ross Project area. This is because the local stratigraphy at the Ross Project area falls below the Wasatch and Fort Union Formations where CBM production occurs (Strata, 2012a).

Wind Power Development

While there is potential in the Powder River Basin for wind-power generation to contribute to meeting forecasted electric power demands, they are dependent on 1) the location of sage-grouse core breeding areas and 2) available transmission capacity to send power to users. Both the location of Greater sage-grouse core breeding areas and transmission capability may be constraining factors (BLM, 2008; WOG, 2010). There are currently no wind power projects within the 80-km [50-mi] vicinity of the Ross Project area, and only one is proposed (see Figure 5.1) (Strata, 2012a).

This wind-power project, as proposed, would have a 250 MW capacity with 166 turbines generating approximately 600 million kWh annually (Strata, 2012a). It would be constructed and operated by Wind Energy America. This wind-power project would be located approximately 42 miles south-southeast of the Ross Project area, while it would be approximately 97 km [60 mi] to drive. It is south of I-90, where the Ross Project area is north of I-90.

5.3 Cumulative Impacts Analysis

5.3.1 EISs as Indicators of Past, Present, and Reasonably Foreseeable Future Actions

One indicator of present and reasonably foreseeable future actions (RFFAs) in a particular region of interest is the number of recent draft and final environmental-impact-statement (EIS) documents prepared by Federal agencies. The NRC used information in the GEIS, Section 5.1.1, as well as publicly available information, several site-specific EISs and SEISs for projects in the Powder River Basin, and draft and final programmatic EISs for large-scale actions related to several states including Wyoming to accomplish its cumulative-impacts analyses (NRC, 2009).

5.3.2 Methodology

For the determination of potential cumulative impacts, the NRC staff reviewed Appendix F of the GEIS and determined that a Level 2 cumulative effects analysis was appropriate for this SEIS due to the fact that concerns were identified during the site-specific analysis (SEIS Section 4) with respect to the sustainability or quality of some of the resource areas within the uranium milling region (NRC, 2009). Therefore, the following methodology was developed, based on CEQ guidance (CEQ, 1997) for a Level 2 cumulative effects analysis as described in the GEIS (NRC, 2009):

- Identify for each resource area the potential environmental impacts that would be of concern from a cumulative-impacts perspective. The impacts of the Proposed Action and the two Alternatives are described and analyzed by resource area in SEIS Section 4, Environmental Impacts and Mitigation Measures.
- Identify the geographic scope for the analysis of each resource area. This scope is expected to vary from resource area to resource area, depending on the geographic extent of the potential impacts.
- Identify the timeframe over which cumulative impacts would be assessed. The timeframe selected for this SEIS begins in approximately 2013, when the Applicant would receive a source material license from the NRC for the Ross Project, and includes any contemporary effects of past activities that persist at the Ross Project area. After receiving a license, the Applicant could begin facility construction and wellfield installation. After the NRC approves the Applicant's definition of its target background values (for excursion detection and aquifer restoration), the Applicant could begin operation. In general, the cumulative-impact analyses timeframes terminate in 2027, which represents the projected license termination data at the end of the decommissioning period (see Figure 2.6 in SEIS Section 2.1.1). In some resource areas, however, the NRC's analysis considers impacts beyond 2027 to the extent that some resources, such as ground-water resources, could require additional time to equilibrate after the complete decommissioning of the Ross Project.
- Identify past, existing, and anticipated future projects and activities in and surrounding the project area. These projects and activities are identified in this section.
- Assess the cumulative impacts for each resource area from the Proposed Action and reasonable alternatives and other past, present, and reasonably foreseeable future actions.

This analysis would take into account the environmental impacts of concern identified in Step 1 and the resource area-specific geographic scope identified in Step 2.

The following terminology was used to define the level of cumulative impact:

SMALL: The environmental effects are not detectable or are so minor that they would neither destabilize nor noticeably alter any important attribute of the resource considered.

MODERATE: The environmental effects are sufficient to alter noticeably, but not destabilize important attributes of the resource considered.

LARGE: The environmental effects are clearly noticeable and are sufficient to destabilize important attributes of the resource considered.

In conducting this assessment, NRC staff recognized that for many aspects of the activities associated with the proposed Ross Project, there would be SMALL impacts on affected resources. It is possible, however, that an impact that may be SMALL by itself, but could result in a MODERATE or LARGE cumulative impact when considered in combination with the impacts of other actions on the affected resource. Likewise, if a resource is regionally declining or imperiled, even a small individual impact could be important if it contributes to or accelerates the overall resource decline. The NRC staff determined an appropriate level of analysis that was merited for each resource area potentially affected by the Proposed Action and alternatives. The level of detailed analysis was determined by considering the impact level to that resource, as described in SEIS Section 4, as well as the likelihood that the quality, quantity, or stability of the given resource could be affected.

The subsequent sections document the NRC's cumulative impact analyses in the following areas:

- Land Use
- Transportation
- Geology and Soils
- Water Resources
- Ecology
- Air Quality
- Global Climate Change and Greenhouse-Gas Emissions
- Noise
- Historical, Cultural, and Paleontological Resources
- Visual and Scenic Resources
- Socioeconomics
- Environmental Justice
- Public and Occupational Health and Safety
- Waste Management

5.4 Land Use

The geographic area within which cumulative impacts to land use were evaluated were Crook and Weston counties, which are within the BLM's Newcastle Field Office planning area, and Campbell County, which is within the planning area administered by the BLM Buffalo Field Office (see Figure 2.1 in SEIS Section 2). These three counties include over 26,000 km^2 [10,000 mi^2] and incorporate the approximately 42 km^2 [25 mi^2] of the Ross Project area. These three counties serve as the geographic boundary area where socioeconomic factors that could relate to land use (i.e., within commuting distance, within shopping distance, and/or within lodging or new home construction distance) would occur. This area is referred to in this section as the "land-use cumulative-impacts study area." Thus, the Ross Project would be approximately 1/4 of 1 percent of the entire land-use cumulative-impacts study area. The timeframe for this cumulative-effects analysis is from 2013, when the Applicant could be issued a license by the NRC, through 2027, when the Ross Project would be completely decommissioned and the aquifers would have been restored.

Land use within the Powder River Basin is diversified and cooperative, with CBM as well as oil and gas extraction activities sharing the land with livestock. Although Federal grasslands and forests cover approximately 21 percent of the Powder River Basin area, most rangeland is privately owned (68 percent) and is used primarily for grazing cattle and sheep. In Crook County, the land ownership is also primarily private. Within Campbell County, however, land ownership is primarily Federal and is allocated by BLM for use as pasture (see Figure 5-5).

As noted in SEIS Section 3.2, the land-use impacts of the Ross Project would result primarily in the interruption, reduction, or impedance of livestock grazing and wildlife habitat; there is not public access to the area generally (e.g., for hunting or fishing) nor is there significant agriculture occurring currently at the Ross Project area (see Table 2.1 in SEIS Section 2.1.1). There are no longer any impacts from historical operations at the Ross Project area (i.e., Nubeth.) In addition, the area that would be disturbed by the Ross Project encompasses a total of 113 ha [280 ac] of land, which represents 16 percent of the Ross Project area. The permanent impacts of the Ross Project would be limited, because the Applicant would be required to return the land to the post-licensing, pre-operational conditions described in SEIS Section 2.1.1.2, unless the respective landowners wish to have certain roads, for example, remain. Thus, the potential land-use impacts from the Ross Project would be temporary and SMALL through all of its phases, as discussed in SEIS Section 4.2.

Mining in the form of coal, mineral, oil, and gas production are all important land uses of the cumulative-impacts study area. As noted Section 5.2, both conventional and CBM oil and gas production are expected to continue in upcoming years. As of 2010, there were over 2,600 conventional oil- and gas-well permits in the land-use cumulative-impacts study area (USGS, 2011), with 889 producing wells (or less than 1 producing well per 26 km^2 [10 mi^2]. A typical drilling location, including the access road, disturbs approximately 1.11 ha [2.75 ac] of land; at a density of 1 well per 26 km^2 [10 mi^2], this would represent up to 0.04 percent of the land affected by these wells. In addition, over 1,570 of the permitted wells have been abandoned and are no longer being used. Through 2008, 547 CBM wells had been drilled within the three-county study area (or approximately one producing well per 52 km^2 [20 mi^2], affecting approximately 0.02 percent of the total land area) (USGS, 2011). Because of the small area of impact for each well and the moderate number of wells currently being operated, the cumulative impacts of the use of land for oil and gas production is SMALL.

As noted in Section 5.2, coal production in the Wyoming portion of the Powder River Basin is expected to grow at an annual rate of 2 – 3 percent per year. It is predicted that from 2010 to 2020, the land area impacted by coal development in the Powder River Basin will increase from 39,927 ha – 55,621 ha [98,662 ac – 137,443 ac]. By 2020, these impacts would represent 1.3 percent of the land in the Powder River Basin. However, most of this coal-mining growth would be in the central area of Campbell County and in an area where the nearest coal mine is over 45 km [28 mi] from the Ross Project area. In the 80-km [50-mi] area shown in Figure 5.1, there are 9 operating coal mines (Strata, 2012a). This coal-mining land use has and would continue to have a MODERATE impact in the land-use cumulative-impacts study area.

There are no operating nor licensed ISR facilities within 83 km [50 mi] of the Ross Project, although there are four uranium-recovery projects in the very early stages of development as described in SEIS Section 5.2 (i.e., Aladdin, Elkhorn, Hauber and Altzada). There is also a potential for development of other uranium facilities to the south of the Ross Project as part of the entire Lance District as described earlier. Thus, some land-use changes as a result of these reasonably foreseeable future developments could occur. To assess the projected land area that would be affected by the development of these present and foreseeable future actions, the NRC staff assumed that approximately the same area affected by the Ross Project and its disturbance of 113 ha [280 ac] would also be approximately the same as by these other ISR projects. Using this assumption, the NRC estimated that the four other non-Strata projects and the four other Strata Lance District projects would impact an additional 904 ha [2,240 ac], for a total area disturbed by potential ISR projects in the land-use cumulative-impacts study area of 1,017 ha [2,520 ac]. This acreage accounts for only approximately 0.04 percent of the total study area. Therefore, these ISR projects would have a SMALL impact on land use.

The NRC staff has concluded that the cumulative impacts on land use in the study area resulting from past, present, and reasonably foreseeable future actions is MODERATE. The Ross Project would have a SMALL incremental effect on land use when added to the MODERATE cumulative land-use impacts.

5.5 Transportation

An area with an 80-km [50-mi] radius was used as the geographic boundary in the evaluation of the cumulative impacts of transportation for this SEIS (referred to in this section as the "transportation cumulative-impacts study area"). This study area was selected because it incorporates the area that would likely be used by the majority of the workers at the Ross Project and includes the distance to the nearest Interstate highway (i.e., Interstate-90). The analysis of transportation-related cumulative impacts is the timeframe of 2013 – 2027, which would be the entire lifecycle of Ross Project from licensing to final decommissioning. The analysis assumes that within this timeframe the four potential satellite areas within the Lance District would be developed sufficiently by the Applicant to begin construction and operation.

The environmental impacts identified in SEIS Section 4.3.1 for the Ross Project would result from the transport of chemical supplies, building materials, yellowcake product, vanadium product, solid byproduct wastes, other hazardous and nonhazardous wastes, and the commuting workforce, all of which increase traffic volumes to and from the Ross Project area. During the phases of the Ross Project examined in SEIS Section 4.3, traffic volume was estimated to increase up to 200 percent. This traffic would predominantly be present on the local Crook County roads. As a result, the wear and tear of the county roads would be significantly increased, and the potential for wildlife mortality and vehicular accidents would increase as well. Therefore, the transportation impacts were found to be SMALL TO LARGE, as discussed in Section 4.3. With the mitigation measures discussed in Section 4.3, the transportation impacts would be reduced to SMALL to MODERATE. Once the Ross Project is decommissioned, most wellfield roads constructed as part of the Ross Project would be removed, and the traffic volume would subside to a little more than the 2010 volume.

Direct impacts to the roads and highways within the transportation cumulative-impacts study area include increased vehicular-traffic volumes and increased risk of vehicular accidents during daily commutes by workers and the trips their families take, especially on roads such as New Haven and D Roads. Ross Project workers would use these roads as would workers from the Lance District satellite areas and two of the five potential ISR projects currently being planned. If the same workforce is assumed for the two other potential ISR projects; if they are assumed to be under construction at the same time; and if it is assumed that the workers at both the Elkhorn and Hauber projects were to use D or New Haven Roads to commute to and from work, this would increase D and New Haven Roads traffic to approximately and conservatively 920 additional automobiles on this road alone per day (it was assumed here that the Ross Project would be already in its operation phase and its workforce would have been reduced to 60 workers). In addition, all of the supply and materials deliveries during their construction phase and uranium-product shipments would need to be added to this traffic volume. The volume that results, assuming the same number of deliveries and shipments by the other ISR projects would rise to almost 1,000 vehicles per day. (Also, D Road is already being used by the Oshoto bentonite mine northeast of the Ross Site area, although there are only a reported eight workers currently commuting to that facility; consequently, this traffic was already considered under the Ross Project's transportation impacts in SEIS Section 4.3.) This would be a LARGE cumulative impact for D and New Haven Roads. Traffic on I-90 is expected to be similarly increased during this period. However, the Interstate highway has been designed to provide sufficient capacity for this increase (as discussed in SEIS Section 4.3). Thus, the transportation impacts on the Interstate-highway system of the U.S. would be SMALL.

All of indirect impacts identified for the proposed Ross Project, including increased wear and tear on existing roads, air emissions, fugitive dusts, noise, and risk of vehicle collisions with livestock, wildlife, and other vehicles, would occur as a result of this increased traffic volume on the county roads. This would be a MODERATE to LARGE impact.

The NRC staff has concluded that the cumulative impacts within the study area resulting from past, present, and reasonably foreseeable future actions is MODERATE to LARGE. The proposed Ross Project would have a SMALL to MODERATE incremental effect on transportation when added to the MODERATE to LARGE cumulative transportation impacts.

5.6 Geology and Soils

The geographic area for the evaluation of geology and soils cumulative impacts ("geology and soils cumulative-impacts study area") is defined as the approximately 9,000-ha [22,200-ac] Lance District shown on Figure 2.2 in SEIS Section 2.1.1. Limiting the cumulative impacts assessment for soils to this area is appropriate since geology and soil impacts are limited to the area in which they occur. The Ross Project itself would result in the disturbance of 113 ha [280 ac] of surface soil, a very small fraction of the total study area (i.e., approximately 0.013 percent).

Previous ISR activities at the Ross Project site include research and development activities conducted by Nubeth in the late 1970's. These activities included construction and operation of a small 5-spot wellfield for one year that likely resulted in some soil disturbance to a small area of land (Strata, 2011a). Regulatory approval of Nubeth's decommissioning was granted by 1986. The Nubeth area was restored and these past activities are consequently no longer relevant for the geology and soils cumulative impacts analysis.

As noted in Section 5.3.2, the proposed schedule for construction, operation, and decommissioning as well as the restoration of the aquifer(s) at the Ross Project show activities taking place over an approximate nine-year period from the time the Project would be licensed by the NRC (Strata, 2012a). The other Lance District wellfield-development activities (i.e., satellite areas) could extend the processing of loaded IX resins at the Ross Project's CPP by another five years or more (Strata, 2012a) to 2027 (see Figure 2.6 in Section 2.1.1). However, the geology and soils impacts within the Ross Project area where the soils would have been disturbed would need additional time to recover. These impacts would dissipate quickly once site restoration is complete, within five years or less; therefore, the time period for this geology and soils cumulative-impacts evaluation is 19 years from the licensing of the Ross Project, or the year 2032.

During the lifecycle of the Ross Project, as discussed in SEIS Section 4.4, potential impacts to Ross Project area geology would be predominantly associated with drillholes, wells, and wellfields. At the conclusion of the Ross Project, an average density of approximately 4.3 wells/ha [1.7 wells/ac], each properly plugged and abandoned, would remain. The Applicant's proper plugging and abandoning of these holes would mitigate their impact vis-à-vis the local geology. Also, the records required by the Applicant's permits for well plugging and abandonment would allow a final assessment of geology impacts after the Ross Project has been decommissioned, if necessary.

The most significant impacts for soils would be soil loss and compaction, soil-productivity loss, and potential soil contamination. There would also be soil disturbance associated with the construction of the CPP, surface impoundments, and access roads as well as pipeline and wellfield installation. Accidental releases of drilling fluids and muds, process solutions, and other liquids could cause soil contamination throughout the Project's lifecycle. As noted in SEIS Section 4.4, facility- and wellfield-design features, best management practices (BMP), and permit requirements, such as the Applicant's Permit to Mine, UIC, and Wyoming Pollution Discharge Elimination System (WYPDES) Permits would minimize these potential impacts during the Ross Project's construction, operation, aquifer restoration, and decommissioning. The Project's decommissioning would include reclamation of soils and the restoration of the site to baseline conditions. Baseline conditions have been documented by soils and vegetation surveys of the Ross Project area. The surveys have established a baseline conditions against which soils impacts at the Ross Project can be measured (see Figure 3.10).

Thus, the geology and soil impacts of the Ross Project would be SMALL in the geology and soils cumulative-impact study area.

To assess cumulative impacts to soils, the area of soil disturbances need to be quantified. The Applicant has identified four potential ISR satellite areas within the Lance District (see Figure 2.2 in SEIS Section 2.1.1) (Strata, 2012a). The NRC assumed that each of these satellite areas would require the same area of soil disturbance as the Ross Project; thus, their development would result in 450 ha [1,120 ac] of soil disturbance. The density of wells at the satellite facilities would be similar to the density at the Ross Project. The impacts to geology and soils would be mitigated as those at the Ross Project would, including complete site reclamation at the end of the Project's lifecycle. If the density of drillholes and wells at these areas would be the same as the Ross Project, and the requirements for plugging and abandonment of the holes would be the same, the potential impacts to geology and soils at each satellite facility would be generally equivalent to those of the Ross Project, which were determined to be SMALL.

As shown on Figure 5.1, there are numerous oil and gas fields that are located within the Lance District. There are no publicly announced plans for further oil and gas development in the area. The impacts to local geology would be the depletion of the oil and gas resources and the remaining, plugged wells after production. For soils, the current wells and any future wells would cause soil impacts due to the drilling of recovery wells, constructing new roads, and conducting other operating activities. These soil impacts would also be required to be mitigated with site-specific BMPs and site-restoration requirements.

The NRC staff has determined that the cumulative impacts to geology and soils in the geology and soils cumulative-impacts study area would be SMALL. The soil disturbance associated with the Ross Project area and the other satellite projects in the Lance District would be limited to approximately 5 percent of the approximately 9,000-ha [22,200-ac] Lance District with 95 percent of the area remaining undisturbed. This disturbance to geology and soils would be dispersed throughout the Lance District and site restoration would be required. The proposed Ross Project would have a SMALL incremental impact on the SMALL cumulative impacts to geology and soils in the geology and soils cumulative-impacts study area.

5.7 Water Resources

The analysis of the cumulative impacts to both surface and ground waters are described below.

5.7.1 Surface Water

The geographic area for the evaluation of surface-water cumulative impacts has been defined as Little Missouri River Basin, from the Ross Project downstream to the Wyoming/Montana border (see Figure 3.10 in SEIS Section 3.4.2). Within this stretch of the Little Missouri River, which begins in within the Ross Project area, the mean flow increases from an average of less than 0.05 m^3/s [1.7 ft^3/s] at SW-1, near the downstream Ross Project boundary, to an average of 2 m^3/s [77 ft^3/s] just downstream of the Wyoming/Montana border. The 45-fold increase in flow within 80 km [50 mi] indicates that cumulative impacts associated with the Ross Project could only be measured in the upper reaches of the Little Missouri River Basin, which is why this geographic area was selected for cumulative-impacts analysis. As the River's flow substantially increases downstream of the Ross Project, any cumulative impacts would be lessened by the additional volume of water.

As discussed in Section 5.3.2, the timeframe defined for the cumulative-impact analysis is 14 years after license issuance. The schedule shown in Figure 2.6 in SEIS Section 2.1.1 indicates that the construction, operation, aquifer restoration, and decommissioning of the Ross Project facility and wellfields would take place during this time period. Since the impacts of the Ross Project on surface-water flows and surface-water quality would dissipate quickly upon completion of the decommissioning phase, this cumulative-impact analysis for surface water ends at 2027 after final Ross Project decommissioning is complete.

The Ross Project would use surface water from the Little Missouri River for dust control and construction-related activities. The Applicant would need to obtain a WYPDES Permit for storm-water management and for the discharge of ground water from wells outside the exempted ore-zone aquifer during the Ross Project's lifecycle. As described in SEIS Section 4.5.1, the impacts to surface-water quantity would be minimal, and the potential water-quality impacts would be mitigated by BMPs, management plans, and permit requirements. The potential impacts of erosion in the small area of temporary land disturbance as well as from accidental process-solution and other liquid spills and leaks would be localized and short-term because of the management plans and standard operating procedures (SOPs) the Applicant would adopt. The potential impacts to the surface-water quantity and quality from the Ross Project would be SMALL.

With respect to wetlands, the Ross Project's construction would have the potential to impact up to 0.8 ha [2 ac] of wetlands. A USACE-required permit would oblige the Applicant to provide a site-specific mitigation plan for all Project-related disturbance of jurisdictional wetlands. This plan would ensure that appropriate mitigation measures would be in place so that there is no net loss of wetlands. As described in SEIS Section 4.5.1, the Ross Project's potential impacts to wetlands would be SMALL.

Measurements of pre-licensing baseline surface-water flows and baseline water-quality parameters provide the baseline characteristics for assessment of cumulative impacts to surface-water quantity and quality (Strata, 2011a). The monitoring program that the Applicant would implement during all phases of the Ross Project would ensure that the Applicant meets NRC license conditions and WDEQ/Land Quality Division's (LQD's) Permit to Mine requirements. This monitoring program is discussed in SEIS Section 6.

The cumulative impacts for surface water would be related to water quantity and water quality. All streams within the upper reaches of the Little Missouri River and for 67 km [40 mi] downstream of the Ross Project are classified by WDEQ/Water Quality Division (WQD) as 3B streams (i.e., intermittent or ephemeral stream incapable of supporting fish populations or providing drinking water). At the confluence with Government Canyon Creek (approximately 67 km [40 mi] downstream of the Ross Project area), the River's flow increases to the point that the stream classification changes to 2ABWW (i.e., it is protected as a drinking-water source and can support warm-water fisheries). Surface-water quality in the upper reaches of the Little Missouri River currently meet Wyoming's surface-water criteria for a Class 3B stream (Strata, 2011a). Current surface-water flows would define the baseline conditions against which impacts can be measured over time. Data on surface-water flows are available from three monitoring stations within the Ross Project area for 2010 and 2011 (Strata, 2012a). These data, combined with flow data from the Wyoming/Montana border would provide a dataset against which changes in surface-water flow can be evaluated.

Surface-Water Quantity

Strata's potential uranium-recovery satellite areas in the Lance District, as described in SEIS Section 5.2, could impact the Little Missouri River (Strata, 2012a). Of the four identified potential satellite areas, only the Ross Amendment Area 1 lies within the Little Missouri River Basin. The others are located within the drainage basin of the Belle Fourche River. However, because the uranium-recovery and water-treatment from the satellite areas would continue to occur at the Ross Project's CPP, these areas are considered in this evaluation of surface-water-quality cumulative impacts, later.

Crop irrigation and stock watering are the primary uses of surface water in the Wyoming portion of the Little Missouri River Basin (WWDC, 2002a). Irrigation use is estimated to range from 1,200 ha-m [9,700 ac-ft] to 1,400 ha-m/yr [11,600 ac-ft/yr] and evaporative loss from stock reservoirs is less than approximately 120 ha-m/yr [1,000 ac-ft/yr] (WWDC, 2002a). There are no other significant uses of surface water in the Wyoming portion of the Little Missouri River. The high estimate of current surface-water use is approximately 22 percent of the mean annual flow in the Little Missouri River at the Wyoming/Montana border (6,900 ha-m/yr [55,800 ac-ft/yr]). Agricultural uses of surface water in the northeastern portion of Wyoming are estimated to grow between 0 and 9 percent, or an increase up to 140 ha-m/yr [1,130 ac-ft/yr], over the next 30 years (WWDC, 2002a).

During the lifecycle of the Ross Project, the annual surface-water use for construction and dust control is estimated to range from 0.71 ha-m/yr [5.8 ac-ft/yr] to 4.6 ha-m/yr [37 ac-ft/yr]. If the Ross Amendment Area 1 were to be permitted and developed concurrently with the Ross Project, and if it were to use a similar quantity of water for construction and dust control, surface-water use would double. However, the potential for increasing water-quantity impacts would continue to be mitigated by BMPs, management plans, and permit requirements. The remaining Lance District potential uranium-recovery areas are expected to rely upon surface water from outside the Little Missouri River Basin.

Other projects that could potentially affect surface-water use within the surface-water cumulative-impacts study area (i.e., the Little Missouri Basin within Wyoming) are described as follows.

- ***Oshoto Mine***: Bentonite mining typically does not use surface water. Water quality could be impacted by sediments, due to erosion and runoff (see **Water Quality** below) (BLM, 2011).

The two uranium-recovery projects that have been identified for potential development within the Little Missouri River Basin are the Hauber and Elkhorn projects. Because there are no existing plans for these projects, the amount of surface water usage is unknown. However, the quantity of uranium targeted by each project has been used to scale and calculate the approximate water use by each, based upon the quantity of uranium reported to occur at each site.

- ***Hauber Uranium Project***: This project targets approximately 1.5 million pounds of U_3O_8, approximately 12 – 25 percent of the 3 – 6 million pounds targeted by the Ross Project. Thus, this project could use between 12 – 25 percent of the surface water the Ross Project would use.
- ***Elkhorn Uranium Project***: This project targets approximately 1.2 million pounds of U_3O_8, approximately 10 – 20 percent of the 3 – 6 million pounds targeted by the Ross Project.

This project would use between 10 – 20 percent of the surface water as the Ross Project would use.

The numerous oil- and gas-recovery projects identified in Figure 5.1 have been assumed to rely upon ground water for water supply and are not expected to impact surface-water quantity. In addition, the projected changes in agricultural and industrial uses of surface water over the next 14 years are predicted to increase surface-water use of the Little Missouri River from 22 percent to approximately 24 percent of the total flow in the Little Missouri River. Agriculture would account for about 1.8 percent increase. The two areas that the Applicant could develop (i.e., the Ross Project and the Ross Amendment Area 1) and the two other planned uranium-recovery projects, the Hauber and Elkhorn projects, all in the Little Missouri Basin, would account for a 0.2 percent increase over the current use. Thus, the cumulative impact, a two-percent decline in the flow of the Little Missouri at the Wyoming/Montana border, due primarily to an increase of agricultural withdrawals over the next 14 years, is small. In addition, the reduction in flow due to uranium-recovery projects would be short-term and minor compared to agricultural use. Thus, surface-water cumulative-impacts related to water quantity would be SMALL.

Surface-Water Quality

The water quality at the Ross Amendment Area 1 and the two uranium-recovery projects described above would also be protected by BMPs, management plans, and permit requirements. Increases in sediment and other water-quality parameters from uranium-recovery projects and other mining (bentonite) activities would be mitigated by the owner/operator implementing BMPs and management plans as well as complying with WYPDES Permits, WDEQ/LQD Permits to Mine, and NRC's license conditions that would be included if a license amendment for this satellite were to be issued to the Applicant. Increases in impacts to water quality from agriculture would be mitigated through compliance with Wyoming's Watershed Protection Program. Thus, the cumulative impacts to surface-water quality in the Little Missouri River Basin would be SMALL.

The cumulative impacts to water quantity and quality in the upper reaches of the Little Missouri River would be SMALL. The proposed Ross Project would contribute SMALL incremental impacts to the SMALL cumulative impact.

5.7.2 Ground Water

The geographic area for the cumulative-impact analysis of ground-water impacts was based upon the hydrogeology of the Lance and Fox Hills Formations within the Powder River Basin, the practical maximum depth for water-supply wells, and the availability of ground-water sources as alternatives to the Lance and Fox Hills Formations. As described in SEIS Section 3.5.3, the ore zone at the Ross Project area is within the lower interval of the Lance Formation and upper interval of the Fox Hills Formation, which are separated from the aquifers above and below by confining units. NRC's evaluation of cumulative effects is therefore limited to only the stratigraphic horizon targeted by the Ross Project, because the ore-zone aquifer is not in contact with aquifers above and below it.

The Black Hills Monocline east of the Ross Project area brings the Lance and Fox Hills Formations to outcrop. Recharge occurs primarily in the area of outcrop and where the Formations are directly below alluvium-filled drainages. The geographic extent for the "ground-water cumulative-impacts analysis study area" is therefore delimited by the extent of the outcrop of the Fox Hills Formation to the east and by the 0 m [0 ft] elevation contour of the top of the Fox Hills Formation to the west. Along the other Ross Project boundaries, the geographic extent is defined by the 80-km [50-mi] radius from the Ross Project.

The schedule for construction, operation, aquifer restoration, and decommissioning at the Ross Project indicates a period of 14 years, from the licensing of the Ross Project to its complete decommissioning (see Figure 2.6 in SEIS Section 2.1.1) (Strata, 2012a). Ground-water modeling demonstrates that 10 years after restoration is complete, ground-water levels would have nearly recovered to a pre-uranium-recovery state (Strata, 2011b). Thus, the time period of 24 years from the start of the Ross Project was defined for this cumulative-impacts evaluation (i.e., the year 2037). The Applicant estimates that recharge to the Lance Formation would be between 0.03 to 0.09 cm/yr [0.07 and 0.22 in/yr] (Strata, 2011b). Because of the limited Lance and Fox Hills Formations recharge area and their low recharge rates, small residual drawdowns in the vicinity of the Lance District would likely be present for tens of years after cessation of uranium-recovery activities.

The primary cumulative impacts for ground water would be related to both water quantity and water quality. During uranium-recovery at the Ross Project, there would be a net withdrawal of water from the ore-zone aquifer. This withdrawal rate would produce decreases in ground-water levels in Ross Project wellfields. Other ground-water users that operate wells completed in the same hydrostratigraphic unit would also affect water levels in the vicinity of their wells. Extraction of ground water in excess of the rate of recharge to the aquifer in the same hydrostratigraphic unit would result in the decline in ground-water levels with time. Upon termination of water extraction, however, recharge of the aquifer would then increase ground-water levels. As described in SEIS Section 4.5.1, the potential impacts to the ground-water quantity from the Ross Project would be SMALL as its consumptive use would be mitigated by alternative water supplies as necessary.

Data on ground-water levels and water-quality data are available for a number of wells within the Ross Project area from early 2010 (Strata, 2011a; Strata, 2011b; Strata, 2012a). These data, together with individual wellfield post-licensing, pre-operational baseline data that would be required as part of the NRC license, would provide a dataset against which changes in ground-water quality can be evaluated. Long-term observations of ground-water levels and ground-water monitoring within the hydrostratigraphic unit would provide a metric for assessing the cumulative ground-water quantity impacts. The monitoring program proposed by the Applicant to meet NRC and WDEQ/LQD Permit to Mine requirements are discussed in SEIS Section 6.

At the Ross Project area, ground-water flow is to the northwest, into the Powder River Basin. The top of the Fox Hills Formation is at approximately an elevation of 1,100 m [3,600 ft] in the area of the Ross Project. A review of ground-water resources in the Powder River Basin notes that ground-water quality and drilling economics generally limit the maximum depth of wells to less than 300 m [1,000 ft] (WWDC, 2002b). However, the City of Gillette does have wells approximately 1,050 – 1,350 m [3,500 – 4,500 ft] deep, tapping the Fox Hills Formation where the top of the Fox Hills Formation is at an elevation 150 m [500 feet] (WSGS, 2012). At this location, the high total dissolved solids (TDS) in the ground water requires it be mixed with waters from deep wells, which are located near Moorcroft; they are drilled into the Madison Formation, where fewer TDSs are present. Because both the depth to the Fox Hills Formation and the fact that TDS concentrations increase farther into the Powder River Basin, the municipal water-supply wells for Gillette mark the westernmost practical limit for extraction of potable water from the Ross Project's ore-zone aquifer. Therefore, the western edge of the ground-water area defined for cumulative-impact analysis is the 0 m [0 ft] structural contour, on the top of the Fox Hills Formation, which is located about 60 km [37 mi] west of the Ross Project area. At this point, the Fox Hills aquifer is approximately 1,200 – 1,500 m [4,000 – 5,000 ft] deep.

During the operation and aquifer-restoration phases of the Ross Project, the weighted average ground-water consumption has been estimated to be 462 L/min [122 gal/min] over a period of 6 years (Strata, 2011a). The Ross Project area has a predicted U_3O_8 production of 340,000 kg/yr [750,000 lb/yr] over 4 – 8 years, and the Ross Amendment Area 1 would extend this rate of production for several years (Strata, 2012a). Production would rise to 993,000 kg/yr [2.19 million lb/yr] U_3O_8 (i.e., yellowcake) with the Kendrick, Richards and Barber satellite areas. If consumptive water use is assumed to be proportional to U_3O_8 production, then ground-water consumption would increase to an average of 1,347 L/min [356 gal/min] over the period of maximum production within the Lance District. It is likely that ground-water drawdowns at the uranium-recovery wellfields in the Lance District would overlap both spatially and temporally.

As noted earlier, the Wyoming State Engineer's Office (SEO) maintains a database of ground-water rights, including water use, well yield, well location, and well depth; however, the geologic interval from which the ground water is extracted is not recorded. Furthermore, data on the yield may not be representative of the actual volumes pumped. Thus, the current rate of ground-water withdrawal from the Lance and Fox Hills formations, and in particular the ore-zone interval, cannot be estimated. The Applicant reviewed the Wyoming SEO's database and concluded that most of the permitted stock and domestic wells within the region of the Ross Project were completed within the Lance Formation sandstones above the ore zone and were not in hydrologic communication with the ore zone. The depth of the ore zone, typically greater than 120 m [400 feet], and the fact that there are other aquifers above the ore zone would make the ore-zone (OZ) aquifer unattractive as a ground-water source (Strata, 2011b). In addition, any future ground-water development of the Lance and Fox Hills aquifer system would be localized and limited, due to poor water quality (WWDC, 2002a).

There are a number of existing or potential resource-extraction projects within the ground-water cumulative-impacts study area that have water demands. These are:

- ***Uranium Recovery*:** Other existing or planned uranium-recovery projects are outside the specific geographic area selected for ground-water-related cumulative-impact analysis, and are in a different stratigraphic horizon than is the Ross Project (Strata, 2012a). The planned Aladdin, Elkhorn, Hauber, and Alzada uranium-recovery projects, if they come to fruition, would target uranium in the Fall River and Lakota Formations. These Formations are of lower Cretaceous age, located several thousand feet below the Lance and Fox Hills Formations, and are separated by the thick Pierre Shale. Thus, uranium-recovery activities in those Formations would not impact the same ground water at the Ross Project.

- ***Coal Mining and CBM Extraction*****:** The mining of coal and extraction of CBM occur along the western margin of the geographic area (see Figure 5.1). The principal coal seams are in Tongue River Member of the Fort Union Formation, which is separated from the Lance and Fox Hills Formations by several thousand feet of the Upper Hell Creek and Lebo confining units (Hinaman, 2005). Ground-water pumping associated with CBM production, coal mining and processing, and mine-mouth power generation would therefore not impact ground water within the Lance and Fox Hills Formations.
- ***Bentonite Mining*****:** Bentonite-mining operations take place in the shale intervals stratigraphically below the Lance and Fox Hills Formations and are, therefore, outside the geographic area for the analysis of ground-water cumulative impacts.
- ***Other Mining*****:** Other potential mining projects, for example, the Bear Lodge Rare Earth project, are also outside the geographic area defined for ground-water cumulative impacts.
- ***Oil Recovery*****:** In the mature oil fields of northeast Wyoming, water is used for EOR and is described as "water flooding" (De Bruin, 2007). At the Ross Project area, the Lance and Fox Hills aquifers show approximately 46 m [150 ft] of drawdown due to withdrawals by the three water-supply wells that have been used since 1980 for oil production (see SEIS Section 4.5.1) (Strata, 2011b). The oil-field water-supply wells within the cumulative-impacts study area would continue to be used during the period of active uranium recovery at the Ross Project. Only a portion of the water requirements, however, would be provided by the Lance and Fox Hills Formations, as stratigraphically higher aquifers are available in the western portion of this area.

<u>Ground-Water Quantity</u>

The NRC staff has determined that the cumulative impacts to ground-water quantity in the ground water cumulative-impacts study area would be SMALL. There would be no increases in water consumption for oil recovery, agriculture, or domestic uses in the Lance and Fox Hills Formations. The drawdown from the pumping of water for EOR is expected to be greater than any of the other uses in areas where the Lance and Fox Hills aquifers supply water for oil-production activities. The effects on ground-water quantity from uranium recovery in the Lance District would also be essentially restored within 24 years after the issuance of the NRC license to the Applicant. Cumulative impacts to ground-water quantity in the Lance and Fox Hills Formation, therefore, would be SMALL. The proposed Ross Project would have a SMALL incremental impact on the SMALL cumulative impacts to ground-water quantity in the ground water cumulative-impacts study area.

<u>Ground-Water Quality</u>

Impacts from previous uranium recovery at Nubeth are part of cumulative impacts to the area. Past impacts can be evaluated by comparing Nubeth's pre-operational baseline water-quality data to Nubeth's post-restoration data as summarized in Table 5.4 (Nuclear Dynamics, 1980; ND Resources, 1982) and to Strata's pre-licensing baseline data as described in Section 3.5. The data in Table 5.4 show that aquifer restoration at Nubeth returned the TDS to levels below pre-licensing conditions except for the injection well No. 20X, which also contained levels of radiological parameters above pre-operational baseline values at the close of restoration. Of the seven non-injection wells in the ore zone, three were restored to pre-operational values for both gross alpha and radium-226. Uranium concentrations after restoration exceeded pre-operational baseline in all of the ore-zone wells except for No. 5X. The well monitoring in the shallow-monitoring (SM) zone (No. 7X) did not show excursions of TDS and uranium. The pre-operational baseline and post-restoration radium-226 values in No. 7X were equivalent within the analytical error of the measurement. The gross-alpha measurement of 180 pCi/L [6.7 Bq/L] in well No. 7X for 4/1980 in Table 5.4 shows excursion of radioactivity into the aquifer above the ore zone. However, gross-alpha measurements in well No. 7X during the 1979 restoration period were much lower than 180 pCi/L [6.7 Bq/L], ranging from 1.4 – 4.7 pCi/L [0.1 – 0.2 Bq/L] (Nuclear Dynamics, 1980).

Table 5.4
Comparison of Baseline and Post-Restoration Water Quality at Nubeth

Well in Zone	Well Use	Sample Date	TDS (mg/L)	Gross Alpha (pCi/L)	Radium-226 (pCi/L)	Uranium (mg/L)
3X in OZ	Buffer	Baseline 4/1978	1680	290	73	0.071
		Restoration 10/1981	1500	130	22	0.24
4X in OZ	Buffer	Baseline 4/1978	1670	180	16	0.08
		Restoration 10/1981	1510	180	26	0.22
5X in OZ	Monitoring	Baseline 4/1978	1600	157	0.3	0.1
		Restoration 4/1980	1550	37	0.5	0.035
6X in OZ	Monitoring	Baseline 4/1978	1740	128	0.6	0.075
		Restoration 4/1980	1650	66	0.1	0.095
7X in SM	Observation	Baseline 4/1978	1530	<3*	0.5	0.008
		Restoration 4/1980	1400	180	0.6	<0.001
11X in OZ	Monitoring	Baseline 4/1978	1750	112	1.4	0.079
		Restoration 4/1980	1730	116	1	0.082
12X in OZ	Monitoring	Baseline 4/1978	1620	72	2.3	0.073
		Restoration 4/1980	1520	111	1.6	0.076
19X in OZ	Recovery	Baseline 4/1978	1680	310	97	0.3
		Restoration 10/1981	1510	300	31	0.48
20X in OZ	Injection	Baseline 4/1978	1270	7.7	0.6	0.006
		Restoration 10/1981	1520	85	20	0.068

Source: Nuclear Dynamics, 1980; ND Resources, 1982.
Note:
* "<" = "Less than," where the value following the "<" is the detection limit.

Evaluation of the restoration conditions in Nubeth's wells provides a short-term assessment of past impacts. The longer-term impacts from Nubeth are determined by a comparison of Nubeth's pre-operational baseline water-quality data with Strata's pre-licensing baseline data as described in SEIS Section 3.5.3. The data presented in Table 3.7 in SEIS Section 3.5.3 suggest that the current water quality in the ore zone and the SM zone are the same as Nubeth's pre-operational baseline values. Thus, the aquifers are not currently impacted by past uranium-recovery activities by Nubeth.

As described in SEIS Section 4.5.1, water quality at the Ross Project could be impacted by excursions from the ore zone into surrounding aquifers. The lixiviant injected into the ore zone causes metals such as uranium, vanadium, arsenic, selenium, and molybdenum as well as other parameters such as radium to dissolve into the ground water. Despite the design of the wellfields and the pumping methods, which are to contain the uranium-recovery process within the exempted aquifer, short-term impacts from excursions do occur. As described in SEIS Sections 2.1.1 and 4.5.1, a network of monitoring wells around the perimeter of each wellfield would provide the capability for early detection, control, and reversal of such excursions; ground-water restoration would return the exempted aquifer to levels that would be established in the NRC license. As described in SEIS Section 4.5.1, therefore, the potential impacts to the ground-water quality from the Ross Project would be SMALL.

The TDS of ground water in the Lance and Fox Hills aquifers generally increases with greater well depth and distance into the Powder River Basin (i.e. down-gradient of the Lance District) (Langford, 1964). Also, NRC license conditions would require the Applicant to recover any excursions into aquifers surrounding the ore zone. Thus, in the unlikely event that increased concentrations of metals mobilized by the lixiviant at the Ross Project migrate down-gradient, the geochemical conditions of the ore-zone aquifer outside the exempted zone would promote lower dissolved metal concentrations (i.e., would cause the dissolved metals to precipitate out). As the dissolved metals enter portions of the aquifer that had not been subjected to the oxidizing lixiviant, the naturally occurring oxygen-deficient conditions would cause chemical reactions that would precipitate the dissolved metals as minerals into the rock of the impacted aquifer. Thus, cumulative impacts to ground-water quality would be SMALL.

Thus, the incremental impacts of the proposed Ross Project in terms of both ground-water quantity and quality would be SMALL when added to the SMALL ground-water quantity and quality cumulative impacts in the ground-water cumulative-impacts study area.

5.8 Ecology

The geographic area considered in the analysis of cumulative impacts is the entire Powder River Basin (the "ecology cumulative-impacts study area") because grassland and sagebrush shrubland habitats are important features of the Basin's entire landscape, and these habitats occur on the Ross Project area as well. The Powder River Basin includes approximately 1,801,401 ha [4,451,360 ac] of land (BLM, 2009e). Approximately 222,568 acres, or 5%, of the Powder River Basin land area has been disturbed by past development activities. Of this amount, one-half of the disturbed area has been reclaimed (BLM, 2009e).

The timeframe for the ecological-resource cumulative-impacts analysis is from 2013 to 2032. This time frame was chosen to allow impacts to ecology of the Ross Project area and its vicinity to mature. It would take some time (the NRC has assumed five years) for the flora and fauna to fully recover after site restoration.

5.8.1 Terrestrial Ecology

Activities occurring in the vicinity of the Ross Project include livestock and wildlife grazing, agricultural production, and mineral exploration. These activities take place over a larger area of the Powder River Basin as well. As discussed in SEIS Section 4.6, potential impacts to ecological resources, both flora and fauna, include reduction in wildlife habitat and forage productivity; modification of existing vegetative communities; and potential spread of invasive species and noxious weed populations. Impacts to wildlife could involve loss, alteration, and incremental habitat fragmentation; displacement of and stresses on wildlife; and direct and indirect mortalities.

5.8.1.1 Vegetation

Vegetation at the Ross Project area is primarily sagebrush shrubland and upland grasslands, which are typical of the Powder River Basin. As discussed in Section 4.6, the impacts to vegetation at the Ross Project area would be SMALL.

There are no operating or licensed ISR facilities within 83 km [50 mi] of the Ross Project area, although there is a potential for development of satellite areas as part of the Applicant's development of the entire Lance District. There are also four potential ISR uranium-recovery projects in the very early stages of development as described earlier (i.e., Aladdin, Elkhorn, Hauber and Altzada). To assess the projected extent of vegetation that would be affected by the development of these prospects, the NRC staff assumed that approximately the same area affected by the Ross Project (113 ha [280 ac]) would also be affected by these other ISR projects. With this assumption, the four Lance District areas and the four other independent ISR projects would impact approximately 904 ha [2,240 ac], for a total potential vegetation impact from ISR projects in the study area of 1,017 ha [2,520 ac]. This accounts for approximately 0.05 percent of the total ecology cumulative-impacts study area. Therefore, these ISR projects would have a SMALL impact on vegetation.

Other mineral development activities described in Section 5.2, including coal-, oil-, and gas-recovery developments, occur within the Powder River Basin. Currently, 53, 680 ha [132,645 ac] of land is disturbed by these activities (BLM, 2009e). Reclamation would be required for these activities within the Powder River Basin in their respective permits. It is estimated that all but approximately 0.8 percent of the disturbed vegetation would be reclaimed (BLM, 2009e). The remaining areas would be associated with permanent infrastructure components. Therefore, the impact to vegetation within the Powder River Basin due to the identified activities would also be SMALL.

5.8.1.2 Wildlife

Loss and degradation of native sagebrush shrubland habitats has affected much of this ecosystem type as well as sagebrush-obligate species including the Greater sage-grouse. Most of the sagebrush shrublands in the Powder River Basin have already been significantly changed by land uses such as livestock grazing, agriculture, or resource extraction. These uses can influence habitats either directly or indirectly; for example, an indirect effect would be the alteration of the natural regime, which could change the frequency of land-clearing fires (Naugle, et al., 2009). For example, the long-term viability of the Greater sage-grouse continues to be at risk because of population declines related to habitat loss and degradation. Because of its spatial extent, oil- and gas-resource development is regarded as playing a major role in the decline of this species in the eastern portion of its range (Becker, et al., 2009). Therefore, there are currently MODERATE cumulative impacts to the Greater sage-grouse. As of NRC's cumulative-impacts analysis, the USFWS has designated the Greater sage-grouse as a "candidate species" under the *Endangered Species Act* (ESA) and would consider the bird on an annual basis for listing as a threatened or endangered species.

However, the impact to sagebrush shrubland communities at the proposed Ross Project would be SMALL because only 113 ha [280 ac], 16 percent of the total Project area, would be disturbed. Additionally, only 21 percent of the Ross Project area consists of sagebrush shrubland habitat. Most of the habitat disturbance would consist of scattered drilling sites for wells; these would not result in large expanses of habitat being dramatically transformed from its original character as in other surface-mining operations; no substantial long-term impact would be expected. No leks or wintering areas have been identified on the Ross Project area, and the area is not located within a designated core area for the Greater sage-grouse.

In addition, potential impacts (e.g., habitat loss, habitat fragmentation, and noise disturbance) would also likely occur at mines and oil and gas facilities throughout the geographic ecological-resource cumulative-impacts area, and would potentially impact other localized wildlife populations. The impacts to other species would be similar; therefore, impacts from the other Lance District and other ISR projects would be SMALL. Other past, present, and reasonably foreseeable future actions discussed in the Powder River Basin could result in the disturbance of tens of thousands of acres. However, site-reclamation permit requirements and BMPs would mitigate these impacts, and it is expected that the cumulative impacts on terrestrial ecological resources would be SMALL in the Powder River Basin. Cumulative impacts to the Greater sage-grouse would continue to be MODERATE.

5.8.2 Aquatic Ecology

Three amphibians and five reptiles designated as Wyoming Species of Concern (WSOC) have been observed on the Ross Project area. However, because aquatic areas would be avoided during construction and operation, the proposed Ross Project would have a SMALL impact on aquatic resources. Similarly, due to the amount of surface disturbance in the Powder River Basin (5 percent), and the mitigation requirements associated with the regulatory permits and licenses, the cumulative impacts on aquatic ecology anticipated from the other past, present, and reasonably foreseeable future actions within the Powder River Basin would be SMALL (BLM, 2009e).

5.8.3 Protected Species

No Federal- or State-listed protected plant species or designated critical habitats occur within the proposed Ross Project area. With regard to protected species, the Ross Project has the potential to impact 12 avian species known to be present on the Ross Project area (see SEIS Section 4.6). Impacts would be SMALL, however, due to the limited footprint of the actual buildings and other structures across the entire Ross Project area.

There are Federally listed protected species within the Powder River Basin, including the Ute ladies'-tresses orchid, the Preble's Meadow Jumping Mouse, the Boreal Toad and the Mountain Plover (BLM, 2003). Additionally, the Bald Eagle is located throughout the Powder River Basin. On the lists of sensitive species maintained by the BLM, WGFD, and the USFS, there are 3 plants, 3 amphibians, 1 snake, 10 fish, 25 birds, and 8 mammals that are known to occur within the Powder River Basin. For the majority of these species, the BLM determined that there may be an affect due to development (BLM, 2003); however, considering the location of development activities compared with the occurrence of many of these species, and with the permit requirements that are in place, the impacts to all but one species would be SMALL.

Potential impacts to the greater sage grouse were identified by the BLM to be of particular concern.

The USFWS has designated the Greater sage grouse as a "candidate species" under the ESA, and will consider the bird on an annual basis for listing as a threatened or endangered species. Within the Power River Basin, potential impacts were identified due to loss of habitat and connectivity, construction of disposal ponds for produced waters generated during oil and gas activities, and disturbance related to increased vehicular traffic (BLM, 2003). Because of these factors, the BLM concluded that the cumulative impacts would likely result in a downward trend for the sage grouse population, and may lead to its federal listing.

Therefore, the NRC staff determined that the cumulative impact on protected species within the ecological resources study area resulting from all past, present, and reasonably foreseeable future actions is SMALL to MODERATE.

Thus, the proposed Ross Project would have a SMALL incremental impact when added to the SMALL to MODERATE cumulative impacts on the ecology of the Powder River Basin.

5.9 Air Quality

The geographic area for the cumulative impacts analysis was based on the NRC staff's consideration of other regional air-modeling studies that address larger-scale emissions sources applicable to oil and gas activities as well as a general understanding of the effect of source-emission strength on the spatial extent and magnitude of downwind air impacts (i.e., larger plumes transport air emissions longer distances downwind before diminishing to insignificant levels). The "air-quality cumulative impacts study area" was therefore defined for air-quality emissions as a circular area formed by an 80-km [50-mi] radius around the Ross Project area. However, significant air-pollution contributors and prevention of significant deterioration (PSD) sensitive areas up to approximately160 km [100 mi] were included, as appropriate, in this analysis. As shown on Figure 5.1, an 80-km [50-mile] radius area encompasses the northeast corner of Wyoming, including the city of Gillette and numerous small towns, and extends into South Dakota and Montana.

Any immediate air-quality impacts of the Ross Project would dissipate quickly once wellfield closure and facility decommissioning is complete and as vegetation is re-established in the areas where there was soil disturbance. The generally windy conditions present at the Ross Project readily disperse airborne pollutants and suspended particulates under the influence of gravity fall out of suspension. As described in Section 5.3.2, the timeframe considered in this assessment of air-quality cumulative impacts begins in 2013, when the NRC could issue a license for the Ross Project, and ends in 2027 when the license would be terminated at the end of the Ross Project's decommissioning phase. After license termination, there would be no impacts on air quality by the Ross Project.

As noted in SEIS Section 4.7, the potential impacts to air quality from the Ross Project would be SMALL during each phase of the Project. Air-quality impacts primarily involve combustion-engine emissions from both the equipment that would be used predominantly during the construction and decommissioning phases of the Ross Project as well as the combustion-engine emissions associated with the commute of Project workers and Project deliveries and shipments. In addition, there would be measurable fugitive-dust emissions from roads traveled by vehicles used for commuting, deliveries, and shipments to and from the Ross Project facility, as well as from the land-disturbing activities during, especially, the construction and decommissioning phases.

Very small emissions are possible from processes at the CPP and/or the storage of waste liquids in the surface impoundments at the Ross Project facility. These could include minor chemical emissions during tank and vessel refilling, chemical delivery, or waste shipments. Windblown emissions from the surface impoundments are also possible. However, BMPs, SOPs, and other air-quality-related management plans, such as monitoring plans, that the Applicant would adhere to, would help mitigate air emissions and air quality impacts. Other facility-design attributes, such as exhaust-point filters, would help to reduce these potential air-quality impacts.

The Ross Project could contribute to air-quality cumulative impacts when its environmental impacts overlap with those of other present or reasonably foreseeable future actions. As described in SEIS Section 5.2, other past, present, and future actions in the air-quality cumulative-impacts study area could include additional ISR uranium-recovery projects, both those by the Applicant in the Lance District and four other planned ISR projects in the study area; coal, bentonite, and rare-earth element mining; oil and gas production; electricity generation by a wind farm; and the current uses of cattle and sheep grazing. However, air-quality impacts from past operations in the study area have been resolved as demonstrated by the discussion in SEIS Section 3.7.

Three of the most important metrics in the estimate of the cumulative impacts of combustion-engine and fugitive-dust emissions is the amount of soil that is disturbed during a project's construction, road installation, and wellfield drilling as well as the types of roads used to access the project (e.g., gravel roads), their maintenance, and the number of vehicles on the roads (see SEIS Section 4.7). In general, undisturbed surfaces produce much less dust than disturbed surfaces, because the undisturbed surfaces usually require considerably higher wind speeds to pick up and suspend particles that then become a significant emission source (Countess, 2001). Also in general, fugitive dusts are usually generated by ground-level activities.

The Ross Project would ultimately disturb 113 ha [280 ac] of soil; there are, however, no other existing ISR projects within the air-quality cumulative-impacts study area that could, at the present time, generate impacts to air quality because of the disturbance of native soils. Studies have been performed to better understand the characteristics of the windblown fugitive dust and mechanically re-suspended road dust that contribute to regional haze (i.e., visible air pollutants such as fugitive dust). These studies are summarized in SEIS Section 4.7.1.1 and indicate that the majority of fugitive-dust-related air-quality impacts caused by the Ross Project would not be expected to extend beyond the 80-km [50-mi] radius around the proposed Ross Project area during its entire lifecycle.

However, as described in SEIS Section 5.2, four satellite areas within the Lance District could be developed for uranium recovery by the Applicant (Strata, 2012a). The NRC staff has made the assumption that each of these satellite areas would involve the same amount of soil disturbance as the Ross Project. (This is a conservative approach, as the satellite areas would not include a CPP and surface impoundments.) Thus, the satellite areas would result in approximately 450 ha [1,120 ac] of soil disturbance. It was further assumed that any air-quality impacts of these satellite areas would be mitigated with the same measures identified in SEIS Section 4.7 for the Ross Project itself. These dust-control measures would include the Applicant minimizing the area of soil that would be disturbed at any one time, spraying water to suppress dust, and promptly revegetating disturbed areas. Further, the Applicant's enforcement of speed limits, treatment roads to minimize dust, and restriction of equipment-operation hours would further mitigate fugitive-dust impacts.

Although no other nuclear-fuel-cycle or ISR projects are currently operated within 80 km [50 mi] of the Ross Project, within the 80-km [50-mi] radius of the Ross Project area there are four other, potential uranium-recovery projects in the early planning stages as noted in Section 5.2. These include the Aladdin Project (7,100 ha [17,550 ac]), the Elkhorn Project (2,110 ha [5,215 ac]), the Hauber Project (469 ha [1,160 ac]), and the Alzada Project (10,000 ha [25,000 ac]).

It has been assumed that these projects would be developed similarly to the Ross Project and that 16 percent of the total area of each would be disturbed during these projects' lifecycles. This would result in approximately 3,150 ha (7,840 ac) of soil disturbance. Because ISR uranium-recovery commonly employs a phased approach to well drilling and wellfield construction, and because the four facilities would not begin construction simultaneously (as each must go through an average two-year licensing process), the degree of overlap for activities associated with these four ISR projects would likely occur predominantly during the wellfield-drilling phase, not the plant construction phases. Thus, the surface disturbances likely would not occur simultaneously and would not be additive. Once fugitive dust was suspended in the air, the dust would settle out within the distances described earlier (not exceeding 80 km [50 mi]. In this assessment of air-quality cumulative impacts, it has been further assumed that combustion-engine and fugitive-dust emissions as well as any processing plant emissions would be managed and mitigated in a manner similar to the Ross Project. Therefore, the relative contribution of reasonably foreseeable future ISR projects to any regional air-quality impacts would be SMALL.

As shown on Figure 5.1, 9 coal mines are located within 80 km [50 mi] of the Ross Project area, southwest of the Project (Strata, 2012a). The straight-line distances to the nine active coal mines within 80 km [50 mi] range from 45 – 72 km [28 – 45 mi]. Five surface coal mines are within approximately 48 km [30 mi] of the proposed Ross Project. This distance is sufficient to ensure that any fugitive dusts that would be generated at either the Ross Project or the coal mines would not be additive and that the particulates, whether mechanically suspended or windblown, would settle out prior to traveling those distances.

As noted in SEIS Section 3.7.3, no violations of the ozone standard have occurred in the area. The levels reported by the nearby air-quality monitoring stations described earlier, however, are close to the respective ozone standard (see Table 3.17 in SEIS Section 4.7.1). Reasonably foreseeable future actions, if conducted concurrently with the Ross Project, could result in occasional exceedances of the ozone standard because of the cumulative number of vehicles associated with all of the activities. However, because of the distance to these mines and the pollutant mixing afforded by the winds in Wyoming, air-quality impacts related to ozone would also be SMALL.

This conclusion is consistent with a previous evaluation by BLM of potential air-quality impacts from future coal and CBM mining, and oil and gas production in the Powder River Basin (BLM, 2003; BLM, 2006; BLM, 2009b; BLM, 2009e; BLM, 2010; ENSR, 2006; BLM, 2009e). This recent BLM cumulative-impacts analysis of air quality in the Powder River Basin was conducted to support the development of increased coal production (BLM, 2009e). Emissions data were acquired for the base year of 2004 for NO_2, SO_2, $PM_{2.5}$, and PM_{10}; these were then modeled to 2020. The estimated impacts of the modeled emissions indicated that air-pollutant concentrations were compliant with (i.e., below) the National Ambient Air Quality Standards (NAAQS), except for the 2020 estimates where short-term and annual $PM_{2.5}$ and PM_{10} standards were exceeded in localized areas. Therefore, although future coal-mine expansion and development of other projects could result in some increase in emissions in the Powder River Basin and downwind areas during the cumulative-impacts study's general timeframe, such impacts would be SMALL.

Ten current bentonite mining operations are within 80 km [50 mi] of the Ross Project area. The straight-line distances to the ten active bentonite mines from the Ross Project range from 5 – 88 km [3 – 55 mi]. The Oshoto bentonite mine is approximately 5 km [3 mi] from the Ross Project area; the next closest bentonite mine is approximately 56 km [35 mi] distant (Strata, 2012a). Surface mining of bentonite can result in significant removal and disturbance of soils during operation, resulting in both combustion-engine and fugitive-dust emissions. However, bentonite mines must apply the same BMPs and other air-quality-management tools as would the Ross Project, including spraying exposed soils to ensure that fugitive particulates are not generated. Currently, bentonite mining has a SMALL impact on air quality.

Finally, numerous oil fields are located within 80 km [50 mi] of the Ross Project area. In general, future development of these resources would include well installation and operation activities which would cause combustion-engine emissions and some soil disturbance, generating fugitive dust. However, it has been assumed that combustion-engine and fugitive-dust emissions would be managed and mitigated in a manner similar to the Ross Project. Both the potential rare-earth metals extraction and wind-power projects have also been assumed to be required to manage and minimize each project's respective soil disturbance and combustion emissions during construction and operation. Thus, the air-quality cumulative impacts related to these present or reasonably foreseeable future projects would be SMALL.

Because nonradiological emissions associated with uranium recovery would be very low, as would those from existing and reasonably foreseeable future actions in the region, the NRC staff has concluded the incremental air-quality impacts of the Ross Project would be a SMALL contribution to the SMALL cumulative impacts to air quality resulting from past, present, and future actions.

5.10 Global Climate Change and Greenhouse-Gas Emissions

5.10.1 Global Climate Change

While there is general agreement in the scientific community that some change in climate is occurring, considerable uncertainty remains in the magnitude and direction of some of these changes, especially during the prediction of trends in a specific geographic location. To predict the effect on climate change of the proposed Ross Project (and vice-versa), temperature and precipitation data for Wyoming were evaluated. Data have been collected over the period of 1895 – 2010. On average, the temperature in Wyoming has increased approximately 0.09 °C [0.16°F] per decade during this time period (NCDC, 2011a). In its report, the U.S. Global Change Research Team (USGCRT) indicated that the temperatures in the past 15 years have risen faster (i.e., 0.83°C [1.5 °F] for the Great Plains), most of which is attributed to warmer winters (GCRP, 2009). The projected change in temperature over the period from 2000 – 2020, which encompasses the period that the Ross Project would be licensed and operated, ranges from a decrease of approximately 0.28°C [0.5 °F] to an increase of approximately 1.1 °C [2 °F] (GCRP, 2009).

For the same period (i.e., 1895 – 2010), a slight downward trend in precipitation (0.30 cm [0.12 in] per decade) has been measured (NCDC, 2011b). Nevertheless, the USGCRT has predicted that the Great Plains region would receive increased precipitation in future decades. Most of the precipitation is expected to fall in the colder months (i.e., winter and spring), and the summer and fall are predicted to become drier. In addition, with the colder months expected to warm over the next several decades, more precipitation would fall in liquid form, resulting in less snow pack in the higher elevations (GCRP, 2009).

The small predicted increases in temperatures and precipitation over the next decade would have no effect on any of the phases of the Ross Project. Because one of the most significant activities at the Ross Project would be below ground, the effects of the surficial and atmospheric environments are not expected to impact significantly uranium recovery. There could be an increase in recharge to aquifers underlying the Ross Project area in future years, which would result from the predicted increased precipitation (i.e., higher precipitation would consequently increase infiltration into the ground water regime). This could affect the Ross Project by increasing the volume of ground water in the ore-zone and improving the effectiveness of the aquifer-restoration process. Similarly, while potential changes to the Ross Project area environment and its resources, such as ecology, are plausible, the small magnitude of the predicted climate change during the period when the uranium recovery would be conducted is not sufficient to alter the environmental conditions at the Ross Project area in a manner that would significantly change the environmental impacts from those that have been evaluated in this SEIS. Based on the above analysis, the proposed Ross Project's incremental impact to predicted climate change is SMALL.

5.10.2 Greenhouse-Gas Emissions

The evaluation of cumulative impacts of greenhouse-gas (GHG) emissions requires the use of a global-climate model. A comparison of annual carbon dioxide emissions by source is included as Table 5.5. A U.S. Global Change Research Program (GCRP) report provided a synthesis of the results of numerous climate-modeling studies (GCRP, 2009). NRC staff has concluded that the cumulative impacts of GHG emissions around the world, as presented in the GCRP report, are an appropriate basis for its evaluation of cumulative impacts. Based upon the impacts identified in the GCRP report, the national and worldwide cumulative impacts of GHG emissions are noticeable, but they are not destabilizing (refer to SEIS Section 5.3 which defines the impact magnitudes that the NRC uses). Consequently, a meaningful approach to address the cumulative impacts of GHG emissions, including carbon dioxide, is to recognize that such emissions contribute to climate change and that the carbon footprint is a relevant factor in the evaluation of potential impacts of alternatives.

Table 5.5 Comparison of Annual Mass of Carbon-Dioxide Emissions by Source			
Source	**Annual CO_2 Emissions (tonnes [T])**	**Percent of World Emissions**	**Percent of U.S. Emissions**
Global Emissions (EPA, 2009)	28,000,000,000 [30,884,000,000]	100%	500%
United States (EPA, 2009)	6,000,000,000 [6,618,000,000]	21%	100%
Current/Proposed ISR Facilities	7,380 [8,140]	0.000026%	0.00012%
Average U.S. Passenger Vehicles (FHWA, 2006)	4.5 [5]	Negligible	Negligible
Estimated Proposed Ross Project (Strata, 2011c)	11,872 [13,087]	0.000042%	0.0002%

Note: t = Tonnes, or Metric tons.
T = Short tons, or U.S. tons.

The Center for Climate Strategies (CCS) prepared a report for the WDEQ that provides an inventory and forecast of Wyoming's GHG emissions (CCS, 2007). These emissions data were based on projections from electricity generation, fuel use, and other GHG-emitting activities. Emissions are reported as carbon-dioxide equivalents (CO_2e); this conversion renders all of the various gases emitted (i.e., methane or nitrous oxides) during an operation or activity into an equivalent GHG effect compared to carbon dioxide. Gross CO_2e emissions in 2005 for Wyoming were 56 million t [62 million T]; these account for less than 1 percent (i.e., 0.8 percent) of the total U.S. gross GHG emissions. This total is reduced to 36 million t [40 million T] CO_2e as a result of annual sequestration (i.e., removal) due to forestry and other land uses (CCS, 2007).

Wyoming has a higher per-capita emission rate than the national average (i.e., greater than 4 times the national average), due primarily to the State's fossil-fuel-production industry, industries that consume great amounts of fossil fuels, a large agricultural industry, great distances between Wyoming cities, and a small population (EPA, 2008). The report shows that the Wyoming GHG emissions would continue to grow as demand for electricity is projected to increase, followed by emissions associated with transportation. It is estimated that Wyoming gross GHG emissions will be 69 million t [76 million T] by 2020 (EPA, 2008).

According to the Wyoming Oil and Gas Conservation Commission (WOGCC), the State of Wyoming contains over 33,000 active gas and oil wells, 45 operational gas-processing plants, 5 oil refineries, and over 14,484 km [9,000 mi] of gas pipelines (CCS, 2007). Because there is no regulatory requirement to track carbon dioxide or methane emissions, there is a high degree of uncertainty associated with the estimated Wyoming GHG emissions from the oil and gas industry. However, the CCS estimated that approximately 13.5 million t [14.9 million T] of CO_2e was emitted by fossil-fuel industries (CCS, 2007). Of this amount, 80 percent was due to the natural-gas industry. This amount is expected to grow an additional 8 to 10 percent in the next decade (CCS, 2007). No data currently exists for the non-fossil fuel industries, including the uranium-recovery industry.

In response to current concerns related to GHG emissions, the Applicant evaluated carbon-dioxide emissions for the lifecycle of the Ross Project and then compared them with other forms of resource extraction. Annual and cumulative carbon-dioxide emissions from the Ross Project during the construction and decommissioning phases were estimated by the Applicant during the air-permitting process for the WDEQ (Strata, 2011c). Combustion-engine exhaust calculations performed for the Ross Project were based upon a combination of Project-specific and representative information appropriate to support a conservative emissions screening analysis. The primary source of carbon-dioxide emissions at the Ross Project would result from combustion-engine emissions from construction equipment, including drill rigs (see Table 5.6). The GHG inventory was calculated for the maximum yellowcake production rate of 1,360 t/yr [1,500 T/yr]. Construction equipment is used most frequently during initial facility construction and wellfield installation, but also later during the decommissioning phase to demolish buildings, dismantle equipment, and reclaim the land.

Table 5.6
Maximum Annual Greenhouse-Gas Emissions
(CO_2 in t [T])

Activity	Carbon Dioxide (t [T])
Uranyl Tricarbonate Breakdown	640 [705]
Sodium Bicarbonate in Eluate	776 [855]
Product Drying	871 [960]
Space Heaters	1,049 [1,156]
Diesel Powered Equipment	8,433 [9,296]
Diesel Generators	104 [115]
TOTAL	11,872 [13,087]

Source: Strata, 2011c.

The Applicant found that minor amounts of methane and nitrous oxides, both of which are considered GHG, would be emitted during natural-gas combustion. The GHG potential or $C0_2e$ of these emissions is a fraction of one percent of the carbon-dioxide emissions, and they were therefore omitted from the calculations. The maximum GHG emissions per year coincide with the year where some wellfield installation, facility and wellfield operation, and aquifer restoration would occur concurrently (i.e., Year 4).

As described above, the total gross amount of GHGs produced in Wyoming in 2005 was 56 million t [61.7 million T], without the reducing effects of sequestration (EPA, 2008). If the 36 million t [39.7 million T] of GHGs sequestration is taken into account (EPA, 2008), the net total of GHGs produced annually in Wyoming is 20 million t [22 million T]. The Ross Project would conservatively produce a maximum annual GHG total of 11,872 t [13,087 T] (as carbon dioxide). This figure equates to approximately 0.06 percent of the net total GHGs produced in Wyoming in 2005. If there has been an increase in GHG emissions, or a decrease in sequestration since 2005, the effect of the Ross Project would be even less.

The Applicant's use of BMPs and other mitigation measures could minimize the emission of GHGs at the Ross Project. These mitigation measures could include, but are not limited to, the Applicant:

- Using fossil-fuel vehicles that meet latest emission standards.
- Ensuring that diesel-powered construction equipment and drill rigs are properly tuned and maintained.
- Using low-sulfur diesel fuel.
- Using newer, cleaner-running equipment.
- Avoiding equipment idling or equipment running unnecessarily.
- Minimizing the number of trips to drilling pads and wells.

Therefore, the potential impact of GHGs from the Ross Project would be SMALL and the cumulative impacts of GHG within the cumulative impacts study area would be SMALL.

5.11 Noise

Cumulative noise impacts were assessed within a rectangular area, a 300-m [1,000-ft] distance from all points of the Lance District, so as to include the potential development by the Applicant of satellite areas within the Lance District (the "noise cumulative impacts study area") (see SEIS Section 5.2). Although some noises would be detectable beyond the Lance District, this distance was considered appropriate because noise dissipates a short distance from the source.

As described in SEIS Section 5.3.2, the timeframe considered in the assessment of potential noise cumulative impacts begins in 2013 and ends in 2027. All Ross Project-related noise of any type would cease at the end of the decommissioning phase. There would be no more activities taking place at the Project area to generate noise, nor would there be any further worker commutes to and from the Project area, supply deliveries to the area, and yellowcake shipments from the area.

As discussed in SEIS Section 4.8, the potential impacts because of noise at the Ross Project result from both activities taking place at the Project itself as well as automobiles and trucks coming and going from the Project area. The noise generated at the Ross Project area would be the greatest during its construction phase and second greatest during the decommissioning phase. Vehicular noise would be generated during all phases, however, as workers commute; as supplies, materials, and uranium-loaded resins are delivered to the Project; and as yellowcake and wastes are taken away from the Project. All of these sources of noise would generate SMALL to MODERATE impacts during the lifecycle of the Ross Project.

As shown in Figure 2.6 in SEIS Section 2.1.1, the potential development of the Lance District would occur in significantly overlapping phases. Each of the phases (i.e., construction, operation, aquifer restoration, and decommissioning) at each of the satellite areas would produce the same noise as discussed in SEIS Section 4.8.1 for each phase of the Ross Project. At the Ross Project itself, the sources of noise are primarily associated with the operation of construction and drilling equipment during facility construction and wellfield installation as well as vehicular noise. In general, the noise generated during construction would occur during only the Ross Project's construction, not at any of the satellite areas because the satellite areas would be predominantly only additional wellfields. However, the Applicant has indicated it expects to construct an IX facility at the Barber satellite area to treat pregnant lixiviant by IX. Thus, some construction noise can be expected there while that smaller facility is built.

As Figure 2.6 in SEIS Section 2.1.1 shows, wellfield installation would begin at the very start of the Ross Project and continue through at least 2021. During this time, other wellfields would begin to enter the aquifer-restoration phase and even decommissioning. Nonetheless, this cumulative-impacts analysis has assumed that the noise generated within the Lance District would be the same as the construction phase throughout the Project's lifecycle, including all satellite areas. Thus, this noise—the maximum of which would occur during the CPP's and surface impoundments' construction during the same time the first wellfields are being installed—would be SMALL.

Based upon a construction phase where 400 passenger vehicles and 24 heavy truck trips per day would be the single highest traffic volume anticipated for the four phases of the Ross Project, the maximum estimated impacts of vehicular noise would not exceed the noise evaluated in SEIS Section 4.8, and thus these impacts would be short-term and SMALL to MODERATE (for the nearest residences). The transportation of process chemicals and supplies to the Ross Project, and yellowcake and waste shipments from the Ross Project, were predicated on the maximum yellowcake production rate of 1.4 million kg/yr [3 million lb/yr], which would include the truck delivery of uranium-loaded IX resins from the Barber satellite area to the Ross Project's CPP. With respect to noise generated by vehicular traffic as a result of the Lance District's development, there would be some increase in noise because of the additional uranium-loaded resins produced at the Barber satellite area being trucked to the Ross Project CPP for further treatment and production of yellowcake. As well, the anticipated maximum workforce of 60 Project-operation workers at the Ross Project was predicated on this maximum yellowcake production rate. That is, the workforce would not increase because of the additional Lance District satellite areas, were they to be developed (Strata, 2012). Thus, vehicular noise would not increase with the additional Lance District satellite areas, because the number of vehicles has already been considered in SEIS Section 4.8.

There are no past, present, or additional reasonably foreseeable future actions within the noise cumulative-impacts study area than those in the Lance District itself; all of the ISR facilities in the preplanning stage near the Ross Project are over ten miles away from the Project area. Similarly, all other past, present, and future actions are greater than 300 m [1,000 ft] away, and no cumulative noise impacts would occur. This cumulative-impacts analysis included a search for any planned oil- and gas-extraction projects that would take place in the Lance District; however, none were identified. Because the Applicant is also unaware of any such plans, this analysis did not include any noise related to future oil- and gas-recovery wells in the Lance District. Thus, construction noise cumulative impacts would be SMALL.

Some of the present and reasonably foreseeable future actions that could be constructed near the Ross Project area as described in SEIS Section 5.2, however, could produce noise cumulative impacts related to vehicular traffic. For example, the primary access route to and from the Ross Project and the Lance District would be along D Road (County Road [CR] 68) for 18.3 miles, then the New Haven Road (CR 164) for 3.0 miles to the appropriate access roads onto the Ross Project area itself (see Figure 2.1 in SEIS Section 2) (Strata, 2011a). Virtually all traffic associated with the Ross Project would use this access route (Strata, 2012a). Of the present and potential projects identified during the noise cumulative-impacts analysis, the only potential projects that would share the route on D and New Haven Roads would be the Elkhorn and Hauber ISR projects. Because of the uncertainty of uranium recovery and processing methods that would be proposed, no estimate of the number of employees or truck traffic is possible at this time (Strata, 2012). However, if it is assumed that the same workforce would be required for those two developments (as was assumed in the transportation cumulative-impact analysis in SEIS Section 5.5), then there would be SMALL to MODERATE cumulative impacts with regard to noise along D and New Haven Roads.

In addition, the existing bentonite mine just northeast of the Ross Project area would contribute to noise along some of the routes potentially taken by the Applicant's personnel at the Ross Project. Highway-legal trucks (as opposed to heavy mine-haul trucks) transport bentonite from the Oshoto Mine to a processing and packaging plant in Upton (see Figure 5.1). The transportation route between the Oshoto Mine and Upton includes portions of D and New Haven Roads, which are adjacent to the Ross Project area and the Lance District. The bentonite truck routes also include roads north and east of the Ross Project that would not be used by Ross Project-related traffic. The degree to which the increased traffic would contribute to potential cumulative noise impacts would depend on hiring and production at Oshoto. The daily Oshoto Mine traffic is estimated at eight commuter trips and ten truck trips. This traffic was already included in the analysis of both transportation and noise impacts in SEIS Sections 4.3 and 4.8 (see also Table 3.2 in SEIS Section 3.8). Thus, the noise associated with the present operation of the nearby bentonite mine has already been considered in the noise impacts found to be SMALL to MODERATE during the Ross Project's lifecycle.

All of the sources of noise described above would be short-term and dissipate quickly with distance. For noise levels typical of drilling and construction, including multiple simultaneous noise sources in close proximity, calculations show that at the residences nearest to the Ross Project, the average noise from equipment would be significantly less than 55 dBA based on the noise data collected by the Applicant (EPA, 1978; Strata, 2011a). Given the distance between potential and existing projects, the Ross Project and Lance District areas would only contribute SMALL incremental impacts. However, given the potential noise from increased traffic on local roads as a result of present and reasonably foreseeable future projects, there would be MODERATE noise cumulative impacts to the residents living nearest the roads traversed by traffic associated with these projects. These MODERATE impacts would continue insofar as the two potential ISR projects (Elkhorn and Hauber) use the primary access roads to the Ross Project.

5.12 Historical, Cultural, and Paleontological Resources

The assessment of cumulative impacts on historical, cultural, and paleontological resources has been geographically defined as the area of Area of Potential Effect (APE) that has been established through the Section 106 consultation process. The APE is discussed in Section 3.9. It includes the Ross Project area, the access roads to and from the area, and a buffer outside the proposed Ross Project's boundaries as well as the area established for potential effects to traditional cultural properties (TCPs). In relationship to other proposed undertakings with the potential to affect these resources, the regional cultural sub-area constituted by the headwaters of the Little Missouri River and the Cretaceous-era Lance Formation provide vectors for analysis of cumulative effects to the archaeological and paleontological record.

The cumulative-impacts analysis timeframe begins in 2013, when the Applicant would be issued a license by the NRC, and concludes in 2027, the estimated year the license would be terminated after the decommissioning and site restoration of the Ross Project.

The Class I and Class III cultural resource survey conducted for the Applicant at the Ross Project area in 2010 resulted in the identification of 24 new sites and 21 isolated resources. A previously recorded site, 48CK1603, which was not found during the survey in 2010, was identified and included in an updated report provided by Strata in 2011. 15 of the sites are recommended by the Applicant as eligible for the National Register of Historic Places (NRHP). The remaining sites were determined to be ineligible for listing (Ferguson, 2010). However, the NRC staff's evaluations to determine whether these properties are eligible for NRHP listing are ongoing. Also, sites of Tribal religious and cultural significance could potentially be identified during a TCP survey of the Ross Project area (see SEIS Section 1.7.3.2). At present, there is already some disturbance from past livestock grazing and agricultural activities as well as some encroachment due to road construction, but other effects from human activities are minimal. Erosion is currently causing some site damage as archeological and paleontological materials erode out of cut banks. In some portions of the APE where alluvium is present, some sites as yet unidentified likely remain protected by intact terraces, and they may be deeply buried.

Archaeological investigations for the Ross Project and other undertakings in the vicinity show that humans have occupied the area for at least 12,000 years (Ferguson, 2010). The Ross Project area is situated in a known culturally-sensitive area at the headwaters of the Little Missouri River, where there is potential for deeply buried archaeological materials that could provide information on earlier periods of regional culture. Ground disturbance during construction activities would be the greatest threat to archaeological sites. This includes the impacts of excavation as well as from construction of access roads. There is a risk of damaging Native American archaeological sites that may be eligible for the NRHP, depending on the depth and location of such ground disturbances.

Ground disturbances could also have an adverse impact on TCPs by damaging landforms or other organic relationships that create or enhance a TCP's setting. A TCP could also be damaged by compromising of the very qualities that make it significant to a community and help it to maintain and perpetuate cultural identity and values. Significant qualities could include integrity of visual setting, a sense of privacy, silence, and other factors that support the general ambiance of a natural setting.

The Project could also damage paleontological resources, as the APE is situated within the Late Cretaceous-age Lance Formation, which is known for its potential to contain a variety of fossil types. Paleontological remains in two of the prehistoric sites recorded during the Class I survey were brought to the site from elsewhere, but, as in the case of the potential for buried sites, paleontological materials of varying ages could be encountered wherever the Lance Formation is penetrated or otherwise disturbed.

To determine cumulative effects, other proposed projects in the nearby Powder River Basin were reviewed for activities that have the potential to impact historical, cultural, and paleontological resources. Other ongoing developments include activities related to energy development, including other potential ISR uranium-recovery projects, coal mines, and oil- and gas-recovery operations. The potential projects related to changing population demographics and public-service needs throughout the general vicinity include wind-power facilities; utility transmission and distribution lines; transportation infrastructure; reservoir development; agricultural activities; livestock grazing; and other economic endeavors. Activities related to all of these pursuits—in addition to natural effects, particularly erosion—have the potential to amplify the impacts of the Ross Project. These impacts taken cumulatively can lead to incremental damage to the archaeological and paleontological record by the elimination of potential data points from the cumulative record of the entire vicinity.

The Applicant expects to develop subsequent areas of the Lance District for uranium-recovery satellite operations (see Figure 2.2 in SEIS Section 2.1.1). No information on identified cultural resources is available for the Lance District; however, similarity in landscape and existing conditions make it likely that the impacts to historical, cultural, and paleontological resources would be similar to those resulting from Ross Project.

Cumulative-impacts analysis for the Moore Ranch project, which is the nearest operating ISR facility to the Ross Project, indicated that the potential impacts of its construction and operation would be small, because the Moore Ranch project is not expected to directly impact eligible archaeological sites when added to the moderate cumulative impacts to the resources from other past, present, and reasonably foreseeable future actions (NRC, 2010). The Nichols Ranch ISR facility, approximately the same distance from the Ross Project area as the Moore Ranch project, identified numerous "pre-contact" sites (i.e., the period of time prior to the arrival of Euroamericans) and deemed the impacts from that project to be small to moderate, and cumulative effects to be moderate.

The BLM has identified proposed coal-mining operations in the Powder Basin as well as continuing development trends. Impacts arising from development of mines, access roads, and related transportation infrastructure, such as extensions of railways, could have a varying effect on historical, cultural, and paleontological resources, depending on where they are sited, but such development is projected to increase at least over the next few years in the Powder River Basin. The same is true of quarries for sand, gravel, and scoria, all of which are used in road construction and maintenance.

CBM and oil and gas exploration and delivery are also expected to continue increasing with population growth and its attendant energy demands. These increases, however, are tempered by economic and regulatory factors. Development of these projects would also be similar to uranium-recovery projects, potentially involving the construction of access roads, pipelines, utility transmission lines, and support facilities of various types as well as ground-water-well installation and facility decommissioning activities.

Mitigation measures can reduce or minimize some impacts to historical, cultural, and paleontological resources. Sites could be deliberately avoided during construction, by flagging them or protecting them with a barrier. Careful monitoring during construction and the implementation of an inadvertent discovery plan can also provide a measure of avoidance or minimize impacts to sites as well as to paleontological discoveries. When impacts are unavoidable, data recovery is often proposed as mitigation measure. A Memorandum of Agreement (MOA) between the NRC, BLM, Wyoming State Historic Preservation Office (SHPO), the Applicant, and the respective Tribes would stipulate the management and treatment of discovered sites and would support ongoing consultation with the Tribes designed to avoid adverse impacts to archaeological sites, TCPs, and other cultural resources. Activities which are on Federally-managed lands or are subject to Federal licenses and permits would be expected to generate fewer impacts, as each is required to undertake the consultation process stipulated in Section 106 of the NHPA. Impacts can be greater on lands, including private or even State, that are not Federally administered. These would include impacts to physical remains as well as the integrity of their settings.

The *National Historic Preservation Act* (NHPA) provides regulatory thresholds for the assessment of impacts to historic properties, which would include the identification of the loss of characteristics that make the properties eligible for the NRHP as well as loss of integrity. For archaeological sites, these impacts could entail an incremental loss of data. For TCPs, these impacts could entail a gradual decline of the very qualities that make a property a functioning element, important for its role in maintaining a living culture.

While data recovery is a mitigation option that is often included in a treatment plan, archaeological sites are nonrenewable resources, and loss of any data contributes to the net loss of information on local and regional cultural history. Whether sites are removed by inadvertent destruction or intended data collection, this loss of these properties precludes any additional investigation in the future, when advances in the field could change interpretations or allow new methodologies to be applied. Paleontological resources are also non-renewable, and they are subject to the same cumulative risks.

Due to urbanization, population growth, and its attendant development, Tribal peoples are experiencing an ongoing loss of TCPs, places that play a vital role in maintaining and perpetuating cultural identity and values. Along with other threats to their life ways, the loss of any culturally empowering resource has a cumulative impact on a group's ability to maintain its cultural identity.

The NRC staff has concluded that the cumulative impacts on historical, cultural, and paleontological resources in the study area resulting from past, present, and reasonably foreseeable future actions is MODERATE to LARGE. The Ross Project would have a SMALL to LARGE incremental effect on historical, cultural, and paleontological resources when added to the MODERATE to LARGE cumulative impacts of the facilities and operations described above. However, the NRC staff's Section 106 consultation for the Ross Project is ongoing as are efforts to identify properties, determine effects, and develop mitigation measures to reduce impacts. The Ross Project is located within an archaeologically rich area; the activities described above could result in a cumulative loss of historical, cultural, and paleontological resources. The impacts to TCPs cannot be determined at this time as TCP identification is still ongoing, as described in Section 1.7.3.2 and Section 4.9. However, any past, present, and reasonably foreseeable future actions that occur on Federal land or require a Federal license or permit would require Section 106 consultations, which would be expected to ensure that historical, cultural, and paleontological resources are adequately protected.

5.13 Visual and Scenic Resources

The geographic area used in this analysis of visual and scenic cumulative impacts (the "visual-resources cumulative-impacts study area") is a circular area with a 32-km [20-mi] radius around the Ross Project area. This area was established as the geographic boundary because it includes the recreational destinations in the immediate vicinity of the Ross Project (described in SEIS Section 3.10), and it addresses the highest (i.e., most sensitive) visual-classification areas in the vicinity of the Ross Project as well. Devils Tower, Thunder Basin National Grassland, Keyhole Reservoir State Park, and the Black Hills National Forest all fall within this visual-resources cumulative-impacts study area. As discussed in SEIS Section 5.3.2, the time frame evaluated for the cumulative-impacts analysis is 14 years, to the year 2027.

As described in SEIS Section 4.10, the potential impacts on visual and scenic resources from the Ross Project include the contrast of surface infrastructure (e.g., drilling rigs, the CPP, access roads, and utility lines) with the existing visual inventory. These types of visual impacts are consistent with the management objectives of the VRM Class IV area in which the Ross Project area is located. Thus, the potential impacts to visual and scenic resources from the surface structures and equipment of the Ross Project would be SMALL during all phases, except during construction phase. The short-term impacts to visual and scenic resources from construction activities would be MODERATE.

Many of the construction and operation activities (e.g., drilling, pipeline and wellfield installation, and surface infrastructure assembly, such as access-road, utility-corridor, and lighting-system construction) at the present and reasonably foreseeable future projects identified in SEIS Section 5.2, both uranium recovery as well as oil production, are very similar to those described in SEIS Section 4.10. In addition, the bentonite mine has already become a fixture of the landscape in the cumulative-impacts area. There are no coal mines within the 32-km [20-mi] radius of the visual and scenic resources cumulative-impacts area. Thus, the same types of impacts to visual and scenic resources described in SEIS Section 4.10 would be associated with these other mineral-extraction and energy-production activities that occur or could occur within the 32-km [20-mi] radius of the Ross Project.

All of these developments, however, would take place in the existing classifications of VRM Class III or IV, where change to an environment can be moderate or even undergo significant modification. In addition, many of the mitigation measures that would be used to reduce the contrast of the Ross Project structures with the existing visual inventory would also be required of new areas and projects. The lower profile and smaller footprint associated with the Ross Project, and presumably with the other satellite areas and planned ISR projects, would diminish visual impacts as well.

Thus, the NRC staff concluded that the cumulative impacts to the viewshed within the 32.2 km [20 mi] visual-resources cumulative-impacts study area as a result past, present, and reasonably foreseeable future actions would be MODERATE. The Ross Project would contribute a SMALL incremental impact and a MODERATE short-term incremental impact to the

MODERATE potential cumulative impacts to the viewshed within the 32.2 km [20 mi] visual-resources cumulative-impacts study area.

5.14 Socioeconomics

The geographic scope for this cumulative socioeconomics analysis are the six counties of Crook, Campbell, Weston, Sheridan, Johnson, and Converse, consistent with the geographic scope of the BLM's *Report for the Powder River Basin Coal Review Cumulative Social and Economic Effects* (BLM, 2005b), the "socioeconomics cumulative-impacts study area" for coal-related impacts. The timeframe for this analysis is 2013 through 2027.

The potential socioeconomic impacts of the Ross Project range from SMALL to MODERATE, with the MODERATE impacts associated with the benefits of the additional tax revenue projected to accrue to Crook County. Because the size and scope of the Ross Project relative to existing employment levels in a two-county ROI are small (see SEIS Section 4.11), and the Applicant is committed to hiring locally, the population impacts and the associated increase in demand for public and private services are expected to be SMALL.

There have been, however, a number of energy-related developments recently completed in the region as well as the proposed projects in the ROI that have the potential to cause additional impacts to socioeconomics areas of study. The projects considered in the BLM report cited above include two additional coal mines over the 2003 – 2010 period; 9,519 additional conventional oil and gas wells, with over one-half of these in place over the 2003 to 2010 period; 62,868 additional CBM wells, with about 40 percent of these in place over the 2003 – 2010 period; and 3 – 4 new coal-fired power plants, with three in place over the 2003 – 2010 period and 1 additional plant planned in the 2016 to 2020 period.

Socioeconomic impacts have been projected over both a low-production scenario and a high-production scenario. Under the low production scenario, the 2020 population in the six-county area is projected to increase by 24,100 persons over 2003 levels, reflecting an increase of 25.1 percent, with 55.8 percent of the increase attributed to projects already in place by 2010 (BLM, 2005b). Under the high-production scenario, the 2020 population in the six-county area is projected to increase by 28,625 persons over 2003 levels, reflecting an increase of 29.8 percent, with 54.0 percent of the increase attributed to projects already in place by 2010. Under both scenarios the large majority (over 70 percent) of the increase is projected in Campbell County, the regional commercial and services center for the region.

The population increases through 2010 already have shown up in the U.S. Census Bureau data for 2010. Population over the 2000 – 2010 period in Campbell County increased 36.9 percent and increased 20.3 percent in Crook County (see Section 3.11). In contrast, population growth in Wyoming was 14.1 percent per year over the same period.

Population increases associated with other current and proposed ISR projects in the ROI would be in addition to those discussed above. Some of the additional potential projects would involve only wellfield construction at satellite areas, including those associated with the Applicant's development of satellite areas in the Lance District. However, in this cumulative-impacts analysis, the NRC staff has assumed that the other planned ISR projects in the 80-km [50-mi] vicinity have the same construction and operating characteristics as the Ross Project, meaning that, at peak construction employment, including the employment associated with the Ross Project, all ISR projects within 80 km [50 mi] would create approximately 2,080 jobs. If these additional projects are online and operating through 2027, operation-phase employment levels would total approximately 540 jobs. If these other ISR projects follow the Applicant's local hiring and purchasing patterns, peak construction population increases would amount to an additional 436 residents in the two-county ROI while the operation-phase population increases by 2027 would total an additional 248 residents. The additional operation-phase population would increase the projected six-county population in 2027 to 24,348 residents, or a 25.4 percent increase over 2003 levels under the low-production scenario, and to 28,873 residents under the high-production scenario, or a 30.1 percent increase over 2003 levels.

Campbell County and local jurisdictions throughout the Powder River Basin have shown their ability to respond these periods of rapid growth. As an example, in response to Campbell County population increases of 36.9 percent over the 2000 – 2010 period, new housing construction increased 42.5 percent over the same period (USCB, 2002; USCB, 2012). Similarly, new housing construction in Crook County increased 22.5 percent compared to population growth of 20.3 percent over the same period.

Periods of rapid growth can stress other public and private service delivery systems. Over the 2010 – 2027 period, population in the six-county area, including the additional residents associated with operation-phase activities of the additional planned ISR projects, is projected to increase by another 10,900 persons, a 10.0 percent increase, under the low-production scenario, and another 13,419 persons, a 12.2 percent increase, under the high-production scenario. Under the low-production scenario, BLM (2005b) also projects enrollment in Campbell County School District No. 1 to increase by 1,587 additional students by 2020, reflecting a 22 percent increase over recent levels; this could cause short-term capacity shortfall. Under the high-production scenario, enrollments could increase another 10 percent. Water and waste-water systems in all communities in the six-county area would have the capacity to accommodate the projected increases in demand through 2020. However, if ongoing and planned improvements are completed (BLM, 2005b), short-term peak demands might result in the need for temporary rationing. This would be a MODERTAE impact.

While local county jurisdictions are expected to benefit from the increased tax revenues from these various projects, some directly from increased property taxes and others indirectly from worker spending and local purchases of goods and services from project proponents, this benefit would be offset by additional demands for public services. Additional street and highway improvements would likely be required in response to the increasing population as well (see SEIS Section 5.5) (BLM, 2009a). Increased traffic levels would also result in increased demand for law-enforcement services and emergency-response services, and similar increases in the demand for health services are expected.

Although the incremental socioeconomic impacts of the Ross Project are SMALL with MODERATE impacts to finance, as the cumulative population increases and their consequent impact on the demand for other public and private community services rises as well, there would be MODERATE socioeconomic cumulative impacts.

5.15 Environmental Justice

Because no minority or low-income populations, as defined by Executive Order 12898 have been identified in the Ross Project area, no disproportionate human-health and environmental impacts were determined. Therefore, there are no cumulative impacts expected in minority and low income populations near the Ross Project.

5.16 Public and Occupational Health and Safety

Cumulative impacts to public and occupational health and safety were assessed along the roads of the circular area defined by an 80-km [50-mi] radius around the Ross Project area (the "public and occupational health and safety cumulative-impacts study area"). This area includes the potential development of satellite areas within the Lance District by the Applicant, four other potential ISR projects, and the other past, present, and other reasonably foreseeable future projects described in SEIS Section 5.2. As described in SEIS Section 5.3, the timeframe for this cumulative-impacts analysis is 2013 to 2027, the expected lifecycle of the Ross Project, including potential uranium-recovery activities in the Lance District. There would be no potential impacts on public or occupational health and safety from the Ross Project following its license termination.

The public and occupational health and safety impacts from the proposed Ross Project would be SMALL and are discussed in Section 4.13. During normal activities associated with all phases of the project lifecycle, radiological and nonradiological worker and public health and safety impacts would be SMALL. Annual radiological doses to the population within 80 km [50 mi] of the proposed project would be far below applicable NRC regulations. For accidents, radiological and nonradiological impacts to workers could be MODERATE if the appropriate mitigation measures and other procedures to ensure worker safety are not followed. Typical protection measures, such as radiation and occupational monitoring, respiratory protection, standard operating procedures for spill response and cleanup, and worker training in radiological health and emergency response, would be required as part of the Applicant's NRC-approved Radiation Protection Program (RPP) (Strata, 2011a). These procedures and plans would reduce the overall radiological and nonradiological impacts to workers from accidents to SMALL.

As shown in Figure 5.1 and discussed in SEIS Section 5.2, in addition to the Ross Project, four satellite areas could be developed by the Applicant and four other ISR projects could be brought to construction and operation during the timeframe of this cumulative-impacts analysis. If constructed and operated, all of these facilities would have similar radiological and nonradiological impacts on the public and occupational health and safety to those at the Ross Project site. Potential radiological cumulative impacts from these facilities would result from incremental increases in annual radiological doses to the population when combined with the impacts of the proposed Ross Project. As stated in Section 4.13, for normal operations, Rn-222 and its progeny would be the most prevalent radionuclides, by dose contribution, anticipated to be released during normal operations at the proposed Ross Project. As further described in SEIS Section 4.13, the maximum expected exposure to a member of the public is estimated to be 0.008 mSv/yr [0.799 mrem/yr] and is consistent with estimates of exposure levels at other operating ISR facilities in the United States (NRC, 2009). This exposure, combined with exposures from other potential ISR facilities in the study area, would remain far below the 10

CFR Part 20 public dose limit of 1.0 mSv/yr [100 mrem/yr] and have a negligible contribution to the 6.2 mSv [620 mrem] average yearly dose received by a member of the public from all sources.

As described in SEIS section 4.13, both worker and public radiological exposures are addressed in NRC regulations at 10 CFR Part 20. Licensees are required to implement an NRC-approved RPP to protect workers and ensure that radiological doses are "as low as reasonably achievable" (ALARA). The Applicant's RPP includes commitments for implementing management controls, engineering controls, radiation safety training, radon monitoring and sampling, and audit programs (Strata, 2011a). Measured and calculated doses for workers and the public are often only a fraction of regulatory limits. Analyses of various radiological accident scenarios, described in section 4.13, also estimate that the dose to the public would be a fraction of the applicable regulatory limits.

Other developments in the 80-km [50-mi] area include existing and potential coal, oil, gas, and bentonite projects. The concomitant major nonradiological occupational hazards of all of these existing or future facilities would be similar to those at the Ross Project; that is, they would include slips, trips, and falls, which could then result in musculoskeletal injuries; potential exposures to excessive noise; potential inhalation of particulates, gasses, or vapors; and skin contact with corrosive materials. These impacts would only be present at the actual facilities where occupational risks are located; the distance between the facilities and operations in the public and occupational health and safety cumulative-impacts analysis study area suggests that, if an occupational hazard were to be experienced, such as a chemical release into the air, the distance itself would mitigate the resulting impacts and would limit impacts to the onsite workers.

All of the facilities and operations identified above would be required to implement the same or similar mitigation measures as at the Ross Project. For example, all such facilities would be required to have spill-response plans, Occupational Safety and Health Administration (OSHA)-compliant SOPs, and health and safety plans as a matter of course because all such facilities are subject to State and Federal occupational health and safety requirements. Thus, nonradiological cumulative impacts to occupational workers would be SMALL, since there would be no cumulative-effects between facilities or projects. However, in the unlikely event that an accident or spill is not mitigated, the impacts to workers could be MODERATE.

The cumulative impacts to the public from nonradiological normal operations would be SMALL, because the public would not have access to the facilities included in this cumulative-impacts analysis. Concurrent generation of fugitive dusts at various operations could occur, if they were closely located to each other, but these facilities would implement the same or similar BMPs for fugitive-dust and combustion-emissions control as described in SEIS Section 4.7. (See also SEIS Section 5.9 regarding air-quality cumulative impacts.) The very distance from the Ross Project to the other potential ISR, coal, gas, oil, and bentonite facilities preclude fugitive-dust cumulative impacts due to not only similar BMPs, SOPs, and other mitigation measures, but also due the significant winds in the study area which would disperse the fugitive dust rapidly.

Potential accidents and chemical releases could affect the public, depending upon the location of the release and the nearest receptors, the closest of which is 0.21 km [690 ft] from the Ross Project's boundary. Accidents could include bulk chemical spills during transport, during operations or maintenance, or during product or waste shipment. Spill prevention and response mitigation measures would include training of all personnel as well as standard spill-response plans. Coordination between both present and future ISR projects, especially the two that would use the same county roads as are adjacent to the Ross Project area (the Hauber and Elkhorn uranium-recovery projects), could optimize emergency-response activities and efficient response. Thus public impacts could range from SMALL to MODERATE, if accidents are not appropriately managed.

Because Strata will implement preventative and mitigation measures, the incremental impacts on public and occupational health and safety of the proposed Ross Project would be SMALL when added to the SMALL cumulative impacts of other past, present, and reasonably foreseeable future actions.

5.17 Waste Management

The cumulative impacts of waste management at the Ross Site were evaluated for both liquid and solid waste streams.

5.17.1 Liquid Wastes

There are two types of potential liquid waste disposal techniques that would be used at the Ross Project: those that employ deep-well injection and those that do not.

The Applicant estimates the completion of Ross Project's (i.e., CPP's) decommissioning and that of the Lance District satellite areas to be approximately 14 years after the NRC license would be issued. Since the impacts from deep-well injection would take some time to dissipate, 20 years is used as the timeframe for evaluation of these cumulative impacts (i.e., the year 2032). Except for the domestic sewage and the used oil, which would be managed only for the lifecycle of the Lance District satellite areas, the generation of other liquid wastes, such as excess permeate as well as fluids and ground water from monitoring wells, would cease during Ross Project operation and aquifer restoration, respectively.

5.17.1.1 Disposal by Deep-Well Injection

The geographic area selected for cumulative-impacts analysis for the management of liquid wastes into the UIC Class I deep-injection wells is similar to the area defined as the ground-water cumulative-impacts study area in SEIS Section 5.7. This area extends westward into the Powder River Basin, to the stratigraphic dip approximately 60 km [37 mi] west of the Ross Project, where the Cambrian aquifers targeted for waste injection at the Ross Project are over 3,700 m [12,000 ft] below the ground surface at that location. This depth to the aquifers make drilling Class I wells impractical; thus, the aquifers accessed at the Ross Project would not be penetrable at that western location. Also, at this location within the Basin, injection wells make use of the Upper Cretaceous aquifers at depths of 1,200 – 2,900 m [4,000 – 9,500 ft]. The aquifers in the Upper Cretaceous are: Tecla, Teapot, and Parkman members of the sandy intervals of the Pierre Shale; Lance and Fox Hills Formations; and the Tullock member of the Fort Union Formation above the Lance Formation. These aquifers are used for UIC Class I and Class V injection wells at existing uranium-recovery operations in Campbell, Johnson, and Converse Counties (NRC, 2010; NRC, 2011; WDEQ/WQD, 1999; WDEQ/WQD, 2010).

The other boundaries of the "waste-management cumulative-impacts study area" for deep-well injection would be the 80-m [50-mi] radius shown in Figure 5.1. This area includes the three ISR projects that may be located in Crook County (in addition to the Ross Project and the four Lance District satellite areas potentially operated by the Applicant) and another one just over the state line, in Montana. These potential projects were described earlier, in SEIS Section 5.2.

As described in SEIS Section 2.1.1.1, the liquid wastes generated by the Ross Project would include byproduct wastes, predominantly brine from the RO process and other process waters. These wastes would be stored in lined surface impoundments and then disposed of into the UIC Class I deep-injection wells, into the Deadwood and Flathead Formations (WDEQ/WQD, 2011). As noted earlier in SEIS Section 4.14, impacts of the management and disposal of liquid byproduct wastes into the UIC-permitted deep-injection wells at the Ross Project would be mitigated by the Applicant's adherence to permit requirements and would be SMALL.

5.17.1.2 Disposal by Other Methods

The geographic area for cumulative impacts from soil disturbances, such as the mud pits at the drilling pads and the lined surface impoundments, is the circular area with an 80-km- [50-mi]-radius around the Ross Project area as shown in Figure 5.1.

Liquid non-byproduct wastes would include drilling fluids and muds from the installation of injection, recovery, and monitoring wells; small amounts of used oil; and domestic sewage. BMPs, management plans, and WDEQ permit requirements would be implemented to mitigate such waste-management and disposal techniques. For drilling fluids and muds, the respective management technique would be their evaporation and disposal in mud pits near each drillhole, and the pits would subsequently be reclaimed when the Ross Project area is restored to pre-licensing, baseline conditions. All used oils would be taken offsite to a properly permitted oil recycler. Finally, the domestic-sewage system installed onsite would follow the required standards and practices as well as all permitting requirements. Thus, as described in SEIS Section 4.14, the impacts of the management and disposal of liquid non-byproduct wastes at the Ross Project would also be SMALL.

Four potential uranium-recovery projects outside of the Lance District, but within 80 km [50 mi] of the Ross Project have been identified. These projects are located east and northeast of the Ross Project and would recover uranium from the lower Cretaceous Fall River and Lakota sandstones. They range from 11 – 70 km [7 – 44 mi] from the Ross Project. Uranium production at each of these potential ISR uranium-recovery projects is expected to be less than the Ross Project (Strata, 2012a). The area encompassing the Ross Project and future potential projects is approximately 0.5 million ha [1.3 million ac].

The use of UIC Class I deep-injection wells for the disposal of liquid byproduct wastes would be expected at these projects, if these projects were to become licensed. It appears likely, given the stratigraphy, that the same aquifers targeted by the deep-injection wells at the Ross Project would be used for disposal at these future projects. For example, the Dewey-Burdock uranium-recovery project in the eastern portion of the NSDWUMR, is stratigraphically similar to the future projects near the Ross Project. The Dewey-Burdock project, located in the Edgemont uranium district in South Dakota, would recover uranium from the Fall River and Lakota sandstones and has proposed deep-injection wells in the Minnelusa and Deadwood Formations, the same that would be used for the Ross Project (NRC, 2009; Powertech, 2010).

The U.S. Environmental Protection Agency (EPA) has determined that the area of potential impacts from deep-well injection is generally less than 0.4 km [0.25 mi] (EPA, 2001). Thus, EPA has defined an "area of review" as the zone of endangering influence around the well, or the radius at which pressure due to injection may cause the migration of the injected wastes and/or poor-quality water in the target formation into an underground drinking water source.

In addition, earthquakes induced by underground waste disposal have been rare, because typically large, porous aquifers are targeted and injection pressures are sufficiently low so that seismic activity is avoided (Nicholson and Wesson, 1990). Nicholson and Wesson documented only two instances in which waste disposal triggered significant adjacent seismicity. If earthquakes were to be induced by fluid-injection activities, they would be located within a few miles from the point of injection.

The WDEQ/WQD's UIC Class I Permit prescribes well design, injection rates, permitted wastes, and injection pressures. Careful monitoring is required to characterize post-licensing, pre-operational baseline water quality of the targeted aquifer and pressures of the lowermost drinking-water aquifer for a new well. Operational monitoring is required to record continuously the rate, volume, and pressure of injection. Every two years, wells must be tested to determine the radius of influence and to compare the results with historical and expected future responses. These required data would provide the information necessary for an assessment of cumulative impacts.

During this analysis, the NRC assumed that all five UIC Class I wells that are already permitted for the Ross Project would be installed and that an average of three UIC Class I wells would be installed at each of the four potential future projects near the Ross Project; thus, there would be 17 deep-injection wells within the approximately 0.5 million-ha [1.3 million-ac] area. The overall density of injection wells would consequently be very low. Given that the potential impacts from deep-well injection are localized, generally 0.4 km [0.25 mi], the cumulative impacts of disposal of liquid byproduct wastes would be SMALL, to which the Ross Project would contribute only a SMALL incremental impact.

5.17.2 Solid Wastes

The geographic area selected for solid waste-management cumulative-impacts analysis is the Ross Project area itself and, though disconnected, the areas that would be impacted by the actual disposal of each type of solid-phase waste that would be generated at the Ross Project (the "solid-waste-management cumulative-impacts study area). Because most of the waste-disposal facilities that would accept the Ross Project's wastes would be open through 2027, the NRC's waste-management cumulative-impacts analysis assumed that the cumulative impacts of waste management would occur through 2027.

The waste-management impacts of the Ross Project were determined to be SMALL in SEIS Section 4.14 through all of the Project phases. This impacts magnitude is primarily a result of the relatively small solid-waste volumes that would be generated at the Ross Project. Even during the decommissioning of the Ross Project, the volumes of the different types of solid wastes, including radioactive waste, would be relatively small due to the decontamination efforts anticipated by the Applicant as well as the fact that the Ross Project would not generate substantial quantities of waste when dismantled and/or demolished (uncontaminated equipment would be re-used).

For the waste-management cumulative-impacts analysis, the NRC assumed that all of the waste-disposal facilities that would accept and dispose of Ross Project wastes would have been properly licensed or permitted. (And that all Ross Project waste shipments would be managed as required in the pre-operational agreements the Applicant must set up with the respective waste-disposal facilities prior to uranium-recovery.) Every waste-disposal facility must undergo significant pre-operational planning and design. This is especially true for the radioactive-waste disposal facilities which could accept the Ross Project's radioactive waste. These facilities would have been licensed by the NRC or by an Agreement State; the other, non-radioactive facilities would have been permitted on the county- or State-level. Also, licensed or permitted facilities that generate solid byproduct material would be required to demonstrate that they have a valid agreement with a solid byproduct material disposal facility in order to continue to operate. This requirement would help to ensure that the byproduct disposal facilities have sufficient capacity to accept incoming material.

Consequently, the incremental impact of the Ross Project's waste management would be SMALL when considered with the SMALL cumulative impacts of waste management over the solid waste management cumulative-impacts study area.

5.18 References

40 CFR Part 1500 to 40 CFR Part 1508. Title 40, "Protection of the Environment," *Code of Federal Regulations,* Part 1500, "Purpose Policy and Mandate," through Part 1508, "Terminology and Index." Washington, DC: Government Printing Office. 1978, as amended.

Becker, J.M., C.A. Duberstein, J.D. Tagestad, and J.L. Downs. "Sage-Grouse and Wind Energy: Biology, Habits, and Potential Effects of Developmen.t" Prepared for the U.S. Department of Energy, Office of Energy Efficiency and Renewable Energy, Wind and Hydropower Technologies Program. Contract DE–AC05–76RL018302009. At http://www.pnl.gov/main/publications/external/technical_reports/pnnl-18567.pdf (as of December 8, 2012). . Agencywide Documents Access and Management System (ADAMS) Accession No. ML13011A248.

BLM (U.S. Bureau of Land Management). *Final Environmental Impact Statement and Proposed Plan Amendment for the Powder River Basin Oil and Gas Project.* Volumes 1 – 4. WY-072-02-065. Buffalo, WY: USBLM. January 2003.

BLM. *Task 2 Report for the Powder River Basin Coal Review, Past and Present and Reasonably Foreseeable Development Activities.* Prepared for the BLM Wyoming State Office, BLM Casper Field Office, and BLM Montana Miles City Field Office. Fort Collins, CO: ENSR Corporation. July 2005a (Revised 2009e). ADAMS Accession No. ML13014A657.

BLM. *Task 1A Report for the Powder River Basin Coal Review, Current Air Quality Conditions.* Prepared for the BLM Wyoming State Office, BLM Casper Field Office, and BLM Montana Miles City Field Office. Fort Collins, CO: ENSR Corporation. September 2005b. ADAMS Accession No. ML13014A730.

BLM. *Task 1D Report for the Powder River Basin Coal Review, Current Environmental Conditions.* Prepared for the BLM Wyoming State Office, BLM Casper Field Office, and BLM Montana Miles City Field Office. Fort Collins, CO: ENSR Corporation. June 2005c.

BLM. *Task 3D Report for the Powder River Basin Coal Review, Cumulative Environmental Effects.* Prepared for the BLM Wyoming State Office, BLM Casper Field Office, and BLM Montana Miles City Field Office. Fort Collins, CO: ENSR Corporation. December 2005d. ADAMS Accession No. ML13016A456.

BLM. *Task 3C Report for the Powder River Basin Coal Review Cumulative Social and Economic Effects.* Prepared for the Wyoming State Office and BLM Casper Field Office. Fort Collins, CO: ENSR Corporation. December 2005e. ADAMS Accession No. ML13014A730.

BLM. *Task 3A Report for the Powder River Basin Coal Review – Cumulative Air Quality Effects.* Prepared for the BLM Wyoming State Office, BLM Casper Field Office, and BLM Montana Miles City Field Office. Fort Collins, CO: ENSR Corporation. February 2006.

BLM. *Visual Resource Management.* Manual 8400. Washington, DC: USBLM. 2007.

BLM. *Update of Task 3A Report for the Powder River Basin Coal Review, Cumulative Air Quality Effects for 2015.* Prepared for the BLM Wyoming State Office, BLM Casper Field Office, and BLM Montana Miles City Field Office. Fort Collins, CO: ENSR Corporation. October 2008a. ADAMS Accession No. ML13016A474.

BLM. *Final Environmental Impact Statement for the West Antelope II Coal Lease Application WYW163340.* Volumes 1 and 2. Washington, DC: USBLM. December 2008b, 2008c. ADAMS Accession No. ML13016A478.

BLM. *Update of Task 3A Report for the Powder River Basin Coal Review, Cumulative Air Quality Effects for 2020.* Prepared for the BLM Wyoming State Office, BLM Casper Field Office, and BLM Montana Miles City Field Office. Fort Collins, CO: ENSR Corporation. December 2009a.

BLM. *Summary of the Analysis of the Management Situation Buffalo Resource Management Plan Revision.* Buffalo, WY: USBLM. March 2009b.

BLM. *Final Environmental Impact Statement for the South Gillette Area Coal Lease Applications WYW172585, WYW173360, WYW172657, WYW161248.* Volumes 1 and 2. At http://www.blm.gov/pgdata/etc/medialib/blm/wy/information/NEPA/hpdo/south_gillette/feis.Par.57426.File.dat/vol1.pdf and http://www.blm.gov/pgdata/etc/medialib/blm/wy/information/NEPA/hpdo/south_gillette/feis.Par.41361.File.dat/vol2.pdf. BLM. 2009c, 2009d.

BLM. *Update of the Task 2 Report for the Powder River Basin Coal Review Past and Present and Reasonably Foreseeable Development Activities.* Prepared for BLM High Plains District Office and Wyoming State Office. Fort Collins, CO: AECOM. December 2009e. ADAMS Accession No. ML13014A657.

BLM. "Available [GIS] Data, Wyoming." 2010. At http://www.blm.gov/wy/st/en/resources/publicroom/gis/datagis.html (as of April 24, 2012). ADAMS Accession No. ML13029A247.

BLM *Environmental Assessment Bentonite Mine Plan of Operation Update to Wyoming Mining Permit 321C WYW-165212.* DOI-BLM-WY-010-EA11-24, BLM Worland Field Office/Wind River/Bighorn Basin District. Washington, DC: USBLM. June 2011. ADAMS Accession No. ML13014A670.

(US)CB (U.S. Census Bureau). "Profile of General Demographic Characteristics: 2000, Summary File 1." 2002. At http://factfinder2.census.gov/faces/tableservices/jsf/pages/productview.xhtml (as of December 8, 2012). ADAMS Accession No. ML13014A719.

(US)CB. "State and County QuickFacts, Wyoming, Crook County, Campbell County." 2012. At http://quickfacts.census.gov/qfd/states/56000.html (as of December 8, 2012).

CCS (Center for Climate Strategies). *Wyoming Greenhouse Gas Inventory and Reference Case Projections 1990 – 2020.* ML13011A261. Washington DC: Center for Climate Strategies. Spring 2007.

CEQ (Council of Environmental Quality). *Environmental Justice, Guidance under the National Environmental Policy Act.* Prepared for The Executive Office of the President. Washington, DC: Government Printing Office. December 1997. ADAMS Accession No. ML13022A298.

Countess, R., W. Barnard, C. Claiborn, D. Gillette, D. Latimer, T. Pace, and J. Watson. "Methodology for Estimating Fugitive Windblown and Mechanically Resuspended Road Dust Emissions Applicable for Regional Air Quality Modeling" in *10th International Emission Inventory Conference Proceedings.*" Westlake Village, CA: Countess Environmental. May 2001. ADAMS Accession No. ML13022A448.

De Bruin, R. H. *Oil and Gas Map of the Powder River Basin, Wyoming.* MS-51. Laramie, WY: Wyoming State Geological Survey. 2007. Accession No. ML13024A167.

(US)EMC (U.S. Energy Metals Corporation). *Application for USNRC Source Material License, Moore Ranch Uranium Project, Campbell County, Wyoming, Environmental Report.* Casper, WY: Uranium One Americas Corporation. 2007a. ADAMS Accession Nos. ML072851222, ML072851229, ML072851239, ML07285249, ML07285253, and ML07285255.

(US)EMC. *Application for USNRC Source Material License, Moore Ranch Uranium Project, Campbell County, Wyoming, Technical Report.* Casper, WY: Uranium 1 Americas Corporation.

2007b. ADAMS Accession Nos. ML072851222, ML072851258, ML072851259, ML072851260, ML072851268, ML072851350, and ML072900446.

(US)EPA (U.S. Environmental Protection Agency). *U.S. EPA Region 8 Climate Change Strategic Plan*, Appendix B, "Summary of State Greenhouse Gas Emissions Inventories in EPA Region 8, Draft." Washington, DC: EPA. 2008. ADAMS Accession No. ML13015A528.

(US)EPA. *Class I Underground Injection Control Program: Study of the Risks Associated with Class I Underground Injection Wells.* EPA 816-R-01-007. Washington, DC: EPA, Office of Water. March 2001. ADAMS Accession No. ML13015A557.

(US)EPA. *Protective Noise Levels.* Condensed Version of EPA Levels Document. EPA 550/9-79-100. Washington, DC: USEPA. November 1978. ADAMS Accession No. ML13015A552.

Ferguson, D. *A Class III Cultural Resource Inventory of Strata Energy's Proposed Ross ISR Uranium Project, Crook County, Wyoming* (Redacted Version). Prepared for Strata Energy, Inc., Gillette, Wyoming. Butte, MT: GCM Services, Inc. 2010.

GCRP (U.S. Global Change Research Program). *Global Climate Change Impacts in the United States.* Cambridge, MA: Cambridge University Press. 2009.

(US)GS (U.S. Geological Survey). *Geodatabase of Wyoming Statewide Oil and Gas Activity to 2010*, Version 1, No. DS-625. Denver, CO: USGS. 2011.

Hinaman, Kurt. *Hydrogeologic Framework and Estimates of Ground-Water Volumes in Tertiary and Upper Cretaceous Hydrogeologic Units in the Powder River Basin*, Wyoming. U S. Geological Survey (USGS) Scientific Investigations Report 2005-5008. Prepared in Cooperation with the Bureau of Land Management. Reston, VA: USGS. 2005. ADAMS Accession No. ML13011A278.

Langford, Russell H. "The Chemical Quality of the Ground Water" in Harold A. Whitcomb and Donald A. Morris, *Ground-Water Resources and Geology of Northern and Western Crook County, Wyoming.* USGS Water-Supply Paper 1698. Prepared in Cooperation with the State Engineer of Wyoming. Washington, DC: Government Printing Office. 1964. ADAMS Accession No. ML13024A154.

Naugle, D.E, K.E. Doherty, B. Walker, M.J. Holloran, and H.E. Copeland. "Energy Development and Greater Sage-Grouse." Chapter 21, under "Ecology and Conservation of Greater Sage-Grouse: A Landscape Species and Its Habitats." Washington, DC: Studies in Avian Biology. 2009. ADAMS Accession No. ML13024A344.

National Climate Data Center (NCDC). "Wyoming Climate Summary, Temperature." 2011a. At http://www.ncdc.noaa.gov/oa/climate/research/cag3/wy.html (as of November 9, 2011). ADAMS Accession No. ML13029A254.

NCDC. "Wyoming Climate Summary, Precipitation." 2011b. At http://www.ncdc.noaa.gov/oa/climate/research/cag3/wy.html (as of December 8, 2012). ADAMS Accession No. ML13029A270.

ND Resources. *Assessment Restoration Activities, Sundance Project.* 1982. ADAMS Accession No. ML13037A140.

Nicholson, C. and Robert L. Wesson. *Earthquake Hazard Associated with Deep Well Injection – A Report to the U.S. Environmental Protection Agency.* USGS Bulletin 1951. Denver, CO: USGS. 1990. ADAMS Accession No. ML13011A329.

(US)NRC (U.S. Nuclear Regulatory Commission). NUREG–1748. *Environmental Review Guidance for Licensing Actions Associated with NMSS Programs.* Washington, DC: USNRC. 2003. ADAMS Accession No. ML032450279.

(US)NRC. NUREG–1910. *Generic Environmental Impact Statement for In-Situ Leach Uranium Milling Facilities.* Volumes 1 and 2. Washington, DC: USNRC. 2009. ADAMS Accession Nos. ML091480244 and ML091480188.

(US)NRC. NUREG–1910, Supplement 1. *Environmental Impact Statement for the Moore Ranch ISR Project in Campbell County, Wyoming, Supplement to Generic Environmental Impact Statement for* In-Situ *Leach Uranium Milling Facilities, Final Report.* Washington, DC: USNRC. August 2010. ADAMS Accession No. ML102290470.

(US)NRC. NUREG–1910, Supplement 2. *Environmental Impact Statement for the Nichols Ranch ISR Project in Campbell and Johnson Counties, Wyoming, Supplement to Generic Environmental Impact Statement for* In-Situ *Leach Uranium Milling Facilities, Final Report.* Washington, DC: USNRC. January 2011. ADAMS Accession No. ML103440120.

(US)NRC. "Expected New Uranium Recovery Facility Applications/Restarts/Expansions: Updated February 1, 2013." 2013. http://www.nrc.gov/materials/uranium-recovery/license-apps/ur-projects-list-public.pdf (February 25, 2013).

Nuclear Dynamics. *1980 Activity and Restoration Report on In Situ Solution Mining Test Site, Sundance Project Crook County, Wyoming.* Source Material License No. SUA-1331, Docket No. 40-8663. Casper, WY: Nuclear Dynamics. 1980. ADAMS Accession No. ML13038A113.

Powertech, Inc. *UIC Permit Application Draft, Class V Injection Wells, Dewey-Burdock Disposal Wells, Custer and Fall River Counties, South Dakota.* Greenwood Village, CO: Powertech, Inc. March 2010. ADAMS Accession No. ML102380514.

Strata (Strata Energy, Inc.). *Ross ISR Project USNRC License Application, Crook County, Wyoming, Environmental Report, Volumes 1, 2 and 3 with Appendices.* Docket No. 40-09091. Gillette, WY: Strata. 2011a. ADAMS Accession Nos. ML110130342, ML110130344, and ML110130348.

Strata. *Ross ISR Project USNRC License Application, Crook County, Wyoming, Technical Report, Volumes 1 through 6 with Appendices.* Docket No. 40-09091. Gillette, WY: Strata. 2011b. ADAMS Accession Nos. ML110130333, ML110130335, ML110130314, ML110130316, ML110130320, and ML110130327.

Strata. *Air Quality Permit Application for Ross In-Situ Uranium Recovery Project*. Prepared for Strata Energy, Inc. Sheridan, WY: Inter-Mountain Laboratories, IML Air Science. 2011c. ADAMS Accession No. ML11222A060.

Strata. *Ross ISR Project USNRC License Application, Crook County, Wyoming, RAI Question and Answer Responses, Environmental Report, Volume 1 with Appendices*. Docket No. 40-09091. Gillette, WY: Strata. 2012. ADAMS Accession No. ML121030465.

Wyoming (State of Wyoming, Office of the Governor). *Greater Sage-Grouse Core Area Protection*. EO No. 2011-5. Cheyenne, WY: State of Wyoming Executive Department. 2011. ADAMS Accession No. ML13015A702.

WSGS (Wyoming State Geological Survey). Powder River Basin Interactive Map. 2012. At http://ims.wsgs.uwyo.edu/PRB/ (as of December 8, 2012). ADAMS Accession No. ML13024A347.

WDEQ/LQD (Wyoming Department of Environmental Quality/Land Quality Division). "Noncoal Mine Environmental Protection Performance Standards." *Rules and Regulations*, Chapter 3. Cheyenne, WY: WDEQ/LQD. April 2006.

WDEQ/LQD. *Cumulative Hydrological Impact Assessment of Coal Mining in the Northern Powder River Basin, Wyoming*. WDEQ-CHIA-25. Cheyenne, WY: WDEQ/LQD. April 2010. ADAMS Accession No. ML13015A591.

WDEQ/LQD. *Cumulative Hydrological Impact Assessment of Coal Mining in the Middle Powder River Basin, Wyoming*. WDEQ-CHIA-27. ML13015A660. Cheyenne, WY: WDEQ/LQD. 2011a.

WDEQ/LQD. "Coal Permit Application Requirements." *Rules and Regulations* , Chapter 2. Cheyenne, WY: WDEQ/LQD. 2011b.

WDEQ/LQD. 2011. "Environmental Protection Performance Standards for Surface Coal Mining Operations," *Rules and Regulations*, Chapter 4. Cheyenne, WY: WDEQ/LQD. 2011c.

WDEQ/WQD (Wyoming Department of Environmental Quality/Land Quality Division). *Rio Algom Mining Company. Amendment to Wyoming Underground Injection Control Permit UIC 99-347 for Class I Non-Hazardous Injection Wells Smith Ranch Facility*. Cheyenne, WY: WDEQ/WQD. October 1999. ADAMS Accession No. ML102380514.

WDEQ/WQD. Uranium One USA, Inc. *Christensen Ranch UIC Class I Injection Well Permit Number: 10-219.* Cheyenne, WY: WDEQ/WQD. November 2010. ADAMS Accession No. ML103140475.

WDEQ/WQD. *Strata Energy, Inc. – Ross Disposal Injection Wellfield, Final Permit 10-263, Class I Non-hazardous, Crook County, Wyoming*. Cheyenne, WY: WDEQ/WQD. April 2011.

WWDC (Wyoming Water Development Commission). *Northeast Wyoming River Basin Plan, Final Report*. Prepared for WWDC. ML13024A156. Butte, MT: HKM Engineering Inc., Lord Consulting, and Watts and Associates. 2002a.

WWDC. *Powder/Tongue River Basin Plan, Final Report.* Prepared for WWDC. Butte, MT: HKM Engineering, Lord Consulting, and Watts and Associates. February 2002b.

6 ENVIRONMENTAL MEASUREMENTS AND MONITORING

6.1 Introduction

As described in the Generic Environmental Impact Statement (GEIS), monitoring programs are developed for in situ recovery (ISR) facilities to verify compliance with the applicable standards and requirements for the protection of worker health and safety in active uranium-recovery areas (i.e., both the facility and the wellfields) and for protection of the public and the environment beyond the licensed facility's boundary (NRC, 2009). Monitoring programs provide data on operating and environmental conditions so that prompt corrective actions can be implemented when adverse conditions are detected. It is important to note that the management of spills and leaks is not considered part of a routine environmental monitoring program (NRC, 2009). Potential spills and leaks are described in this Supplemental Environmental Impact Statement's (SEIS) Section 2.1.1, including the design components and management techniques that are intended to detect and to minimize the impacts of spills and leaks.

This section discusses the types of environmental monitoring activities that the Applicant would undertake throughout the Ross Project. These include radiological, physiochemical, meteorological, and ecological monitoring activities.

6.2 Radiological Monitoring

Radiological effluent and environmental monitoring programs are required for an U.S. Nuclear Regulatory Commission (NRC)-licensed facility. The purpose of the monitoring programs is to (i) characterize existing levels of radiological materials in the environmental media, (ii) provide data on measurable levels of radiation and radioactivity in the effluent and environmental media during the operational life of the facility, and (iii) evaluate principal pathways of radiological exposure to the public. This section describes Strata's proposed radiological monitoring programs for the Ross ISR Project as described in its license application and supporting documents and subsequent responses to NRC requests for additional information.

In accordance with 10 CFR Part 40, Appendix A, Criterion 7, an Applicant is required to establish a pre-operational monitoring program to establish facility baseline conditions prior to construction. Results of Strata's baseline radiological monitoring program are presented in SEIS Section 3.12.1. After establishing baseline conditions, an ISR facility operator must conduct an operational monitoring program to measure or evaluate compliance with standards and environmental impacts of an ISR facility under operational conditions. In accordance with 10 CFR Part 40.65, the license must submit to NRC a semiannual effluent and environmental monitoring report which would specify the quantity of each of the principal radionuclides released as effluent or their levels within various environmental media in all unrestricted areas during the previous 6 months of operation. This report would also provide other NRC required information to estimate the maximum potential annual radiation doses to the public resulting from effluent releases.

The following sections briefly describe the Applicant's proposed operational monitoring program. NRC Regulatory Guide 4.14 (NRC, 1980) provides guidance for establishing radioactive effluent and environmental monitoring programs for uranium mills, which includes ISR facilities. A summary of the effluent and environmental monitoring program is presented in Table 6.1.

Table 6.1 Summary of the Major Elements of the Ross Project Operational Environmental Monitoring Program

Table 6.1. Summary of the Major Elements of the Ross Operational Environmental Monitoring Program

Program Element	Location	Radionuclides Analyzed	Sampling Frequency	Number of Sampling Locations
Ground water – Monitor Wells	Up-gradient and down-gradient from CPP	Dissolved uranium, Ra-226, Th-230, Pb-210, Po-210, gross alpha, gross beta	Monthly first year, quarterly thereafter	3 or more down-gradient;at least up-gradient control sample
Ground water - Water Supply Wells	Private wells within 3.3 km (2mi) of project area similar to the pre-operational baseline monitoring	Dissolved and suspended uranium, Ra-226,Th-230, Pb-210, Po-210, gross alpha, gross beta	Quarterly	29
Surface Water (1)	Surface waters passing through project area and reservoirs subject to runoff similar to pre-operational baseline monitoring	Dissolved and suspended uranium, Ra-226,Th-230, Pb-210, Po-210, gross alpha, gross beta	Quarterly (as available)	3 surface water monitoring stations and11 reservoirs within project area
Air Particulates	Locations with the highest predicted concentrations, nearest residences and control location similar to pre-operational baseline monitoring	Total uranium, Th-230, Ra-226,Pb-210	Continuous - Composites of weekly filters analyzed quarterly	5 or more
Radon in Air	Particulate in air locations and other areas of interest similar to pre-operational baseline monitoring	Rn-222	Continuous viaTrack-Etch units – quarterly exchange and analysis of units	5 or more
Soil	Particulate in air locations and other locations with the highest predicted concentrations similar to pre-operational baseline monitoring	Total uranium,Ra-226, Pb -210,gross alpha	Annually	5 or more
Sediment	Surface waters passingthrough project area and reservoirs subject to runoff similar to pre-operational baseline monitoring	Total uranium,Ra-226, Pb -210,gross alpha	Annually (as available)	3 surface water monitoring stations and 11 reservoirs within project area
Direct Radiation	Particulate in air locations and other areas of interest similar to pre-operational baseline monitoring	Continuous via TLD	Quarterly	5 or more
Vegetation (2)	Animal grazing areas and other locations with the highest predicted concentrations similar to pre-operational baseline monitoring	Ra-226 and Pb-210	Three timesduring grazing season	Grazing vegetation representing 3 different sectors that have the highest predicted concentrations of radionuclides
Animal Tissue	Livestock (cattle) raised within 3 km of the site and fish from Oshoto Reservoir similar to pre-operational baseline monitoring	Ra-226 and Pb-210	Once during site decommissioning and prior to license termination	3 samples of beef; 1 fish sample (composite to meet laboratory MDL)

(1) Location of air particulate samplers used during the preoperational baseline monitoring will be re-evaluated for operational monitoring based on results of the pre-operational meteorological monitoring program and the results of the MILDOS-AREA analysis to insure at least 3 locations are selected representing 3 different sectors that have the highest predicted concentrations of radionuclides

(2) In accordance with the provisions of NRC Regulatory Guide 4.14, Footnote (o) to Table 2: " *vegetation and forage sampling need be carried out only if dose calculations indicate that the ingestion pathway from grazing animals is a potentially pathway...*" defined as a pathway which would expose an individual to a dose in excess of 5% of the applicable radiation protection standard. This pathway was evaluated by MILDOS-AREA.

Source: Table 5.7-1 (Strata, 2011a)

6.2.1 Airborne Radiation Monitoring

The Applicant proposes to conduct continuous air particulate sampling at five locations identified in Figure 6.1. The filters from air samplers will be analyzed on a weekly basis, or more frequently if required due to dust loading, for natural uranium, Th-230, Ra-226, and Pb-210 in accordance with Regulatory Guide 4.14 (Strata, 2011a; NRC, 1980). The air samplers will be calibrated per manufacturer recommendations or at least semiannually with a mass flow meter or other primary calibration standard (Strata, 2011a).

In addition to the air particulate sampling, passive track-etch detectors and thermoluminescent dosimeters (TLDs) will be deployed at each air particulate monitoring station (Strata, 2011a). The passive track-etch detectors will provide continuous monitoring of Rn-222 and the detectors will be exchanged and analyzed on a monthly basis. The TLDs will be used to assess gamma exposure rates continuously at each air particulate monitoring station. The TLDs will be exchanged and analyzed on a quarterly basis.

During operations, Strata will monitor radon gas and passive gamma radiation using Landauer radon Trak-Etch detectors and environmental low level TLDs at locations shown in Figure 6.1. In total, radon will be monitored at 17 sampling locations, of which five locations are co-located with the air particulate samplers, as recommended in Regulatory Guide 4.14 (NRC, 1980).

6.2.2 Soils and Sediment Monitoring

The Applicant proposes to collect representative soil samples to a depth of 152 cm (60 in) annually at each of the five air particulate monitoring stations shown in Figure 6.1. The soil samples will be collected similar to the baseline collection procedure (i.e., two surficial samples (to a depth of 15 cm) and two subsurface samples. The samples will be analyzed for natural uranium, Ra-226, Pb-210 and gross alpha (Strata, 2011a).

The Applicant proposes to collect sediment samples annually at the three surface water gaging stations on Little Missouri River and Deadman Creek and from the Oshoto reservoir. The sediment sampling at the stream gaging stations will occur during a runoff event between April and October. The sediment samples will be analyzed for natural uranium, Th-230, Ra-226, and Pb-210 and gross alpha (Strata, 2011a).

The proposed sampling and analyses are consistent with recommendations of Regulatory Guide 4.14 (NRC, 1980). Similarly, the analytical limits of detection for the soil and sediment sampling program are consistent with the recommendations of Regulatory Guide 4.14 (NRC, 1980) unless matrix interferences prohibit attainment of these low detection limit goals.

6.2.3 Vegetation, Food, and Fish Monitoring

Where a significant pathway to man is identified, Regulatory Guide 4.14 suggests analyzing three of each type of crop, livestock, etc., raised within 3 km of the ISR site (NRC, 1980). Vegetation samples should be collected three times during the grazing season, and food and fish samples should be collected at the time of harvest or slaughter.

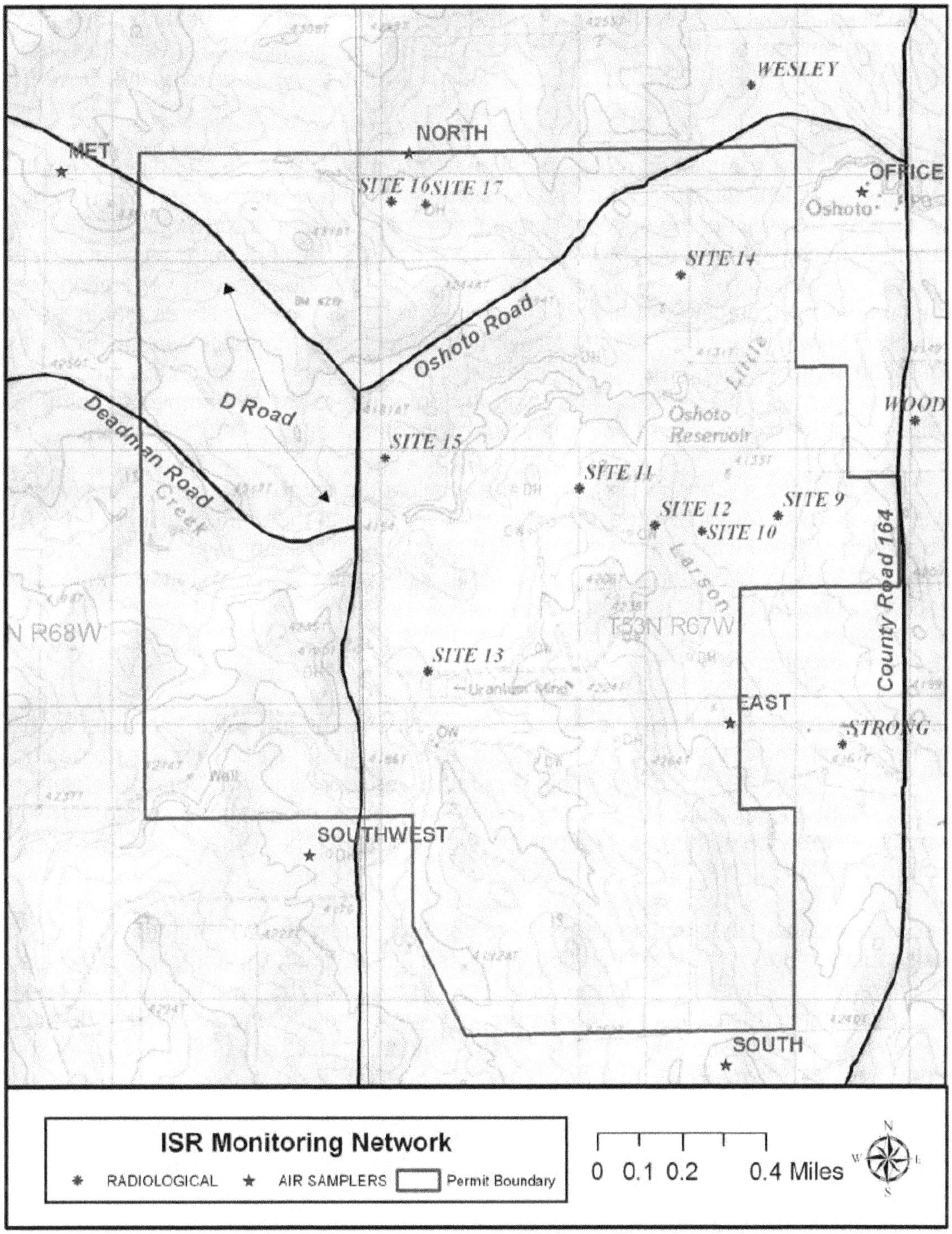

Source: Figure 3 of Addendum 3.6-A to the environmental report (Strata, 2011a)

Figure 6.1
Ross Project Meteorological and Baseline Radiological Monitoring Locations

All should be analyzed for Ra-226 and Pb-210. Note (o) in Regulatory Guide 4.14 (NRC, 1980), Table 2 clarifies that an exposure pathway should be considered important if the predicted dose to an individual would exceed 5 percent of the applicable radiation protection standard. Individual members of the public are subject to the dose limits in 10 CFR Part 20.1301. Pursuant to 10 CFR Part 20.1301, the dose limit is 100 mrem/yr Total Effective Dose Equivalent (TEDE).

The Applicant has established a pre-operational baseline. Based on modeling (i.e., MILDOS-Area), the Applicant calculates that maximum impacts to the public through all pathways would be less than 1 percent of the applicable radiation protection standard (Strata, 2011a). Therefore, because the Applicant has determined that a significant pathway to man does not exist from these sources, the Applicant does not propose to perform any vegetation, food or fish sampling during operations (Strata, 2011a). However, the Applicant states that in the event that monitoring is required, it proposes to follow the protocol used in baseline sampling for three vegetation samples during the grazing season at three locations at which the model-predicted concentrations were the highest. The Applicant proposes to collect samples of animal tissue and fish from the Oshoto Reservoir during site decommissioning.

NRC staff includes a license condition for the Applicant to establish a plan for verifying the input values used in the MILDOS-Area calculations by monitoring the effluent discharges. Should the effluent discharges invalidate the model calculations, the Applicant will be required to recalculate the model and/or verify the radiological impacts to the vegetation and food sources through routine sampling.

6.2.4 Surface Water Monitoring

During the construction phase, the Applicant proposes to conduct a surface water monitoring program consisting of sampling at the Oshoto reservoir and three on-site stream gaging stations (SW-1, SW -2 and SW-3) located within Deadman Creek or Little Missouri River (Strata, 2011a). The Applicant anticipates that, based on the preoperational monitoring program, flows in the streams will likely be ephemeral primarily during April to October (Strata, 2011a). Surface water is found year-long in the Oshoto reservoir.

During operations, the Applicant proposes to conduct a surface water monitoring program which was conducted during the pre-operational monitoring, i.e., quarterly sampling at three on-site stream gaging stations and 11 on-site or nearby reservoirs. The parameters to be analyzed for the operational surface water monitoring program are dissolved and suspended uranium, Th-230, Ra-226, Po-210 and Pb-210, and, gross alpha and gross beta unless sufficient cause can be demonstrated to measure a parameter less frequently.

The Applicant also commits to monitoring surface water should monitoring be required for a Wyoming storm water discharge permit through the WYPDES program (Strata, 2011a).

6.2.5 Groundwater Monitoring

The Applicant proposes to monitor groundwater quality at the domestic, livestock, and industrial water supply wells located within a 2 km [1.2 mi] radius of the Ross Project boundary during both construction and operation phases. The Applicant states that monitoring of the nearby water supply wells will be conducted quarterly and results provided to NRC on an annual basis.

The monitoring at a specific water supply well will be contingent upon landowner's (well owner's) consent and, for a variety of reasons (e.g., abandoned, non-functioning pump, winterized), may not be available every quarter (Strata, 2011a). The parameters to be analyzed consist of dissolved and suspended uranium, radium-226, thorium 230, lead-210 and polonium-210, and gross alpha and gross beta.

The Applicant estimates that 29 wells exist within 2 km [1.2 mi] of the Ross Project (Strata, 2011a). Based on information in the application, the water supply wells consist of 2 industrial water supply wells, 15 livestock water supply wells and 12 domestic water supply wells of which four livestock water supply wells and three industrial wells are located within the Ross Project area. The proposed monitoring program is a continuation of the pre-operational monitoring program though the parameters analyzed will be reduced from those analyzed in the pre-operational monitoring program.

By license condition, NRC staff will require that nearby water supply wells within 2 km [1.2 mi] of an active wellfield be sampled in lieu of 2 km [1.2 mi] of the project area. In addition, other license conditions will require an annual update on the nearby ground water use and require monitoring of the onsite industrial wells on a monthly basis for the effluent monitoring program if operations at the industrial wells have not been terminated.

6.3 Physiochemical Monitoring

This section describes the monitoring program proposed by the Applicant that would be initiated in compliance with applicable environmental regulations and the NRC license. This monitoring program would allow an evaluation of changes in the chemical and physical environment as a result of the proposed Ross Project. The physiochemical monitoring program would include surface water and ground water as well as flow and pressure monitoring of wellfields and pipelines as described in this section.

Pre-licensing, baseline monitoring of surface water and ground water was completed by the Applicant in 2009 and 2010, and the acquired data were used to characterize the Ross Project site according to the requirements in 10 CFR Part 40, Appendix A, Criterion 7 (Strata, 2011a). Sample collection and analysis were performed according to the recommendations found in NRC's Regulatory Guide 4.14 (NRC, 1980) as well as the specifications in ASTM D449-85a (now superseded by ASTM D4448-01), *Standard Guide for Sampling Groundwater Monitoring Wells*. In addition, the Applicant also provided supplemental environmental monitoring data in 2012 (Strata, 2012).

The surface-water monitoring stations and ground-water monitoring wells established for pre-licensing baseline monitoring would be incorporated into the post-licensing, pre-operational data-collection effort and into the active operation-phase environmental-monitoring network.

6.3.1 Surface-Water-Quality Monitoring

The Applicant proposes to continue quarterly sampling of the surface-water stations that were established for pre-licensing baseline water-quality data (Strata, 2011b). The existing surface-water monitoring stations include the Oshoto Reservoir and three surface-water monitoring stations; these surface-water stations are located on the Little Missouri River (SW-1 and SW-2) and on Deadman Creek (SW-3) (see Figure 3.12). The Applicant would add additional stations as necessary to meet additional NRC license conditions. Each station is already equipped with a pressure transducer, a data-logging system, and a runoff-event-activated sampling mechanism.

6.3.2 Ground-Water-Quality Monitoring

The Applicant proposes a ground-water monitoring program to acquire post-licensing, pre-operational data in order to establish the parameters (i.e., constituent concentrations) necessary to detect excursions outside the ore zone during active uranium-recovery operation and to observe aquifer-restoration performance as it proceeds (Strata, 2011b). The post-licensing, pre-operational baseline data would be collected from each individual wellfield as it is completed, but prior to the Applicant's initiating uranium recovery. Each wellfield's monitoring data would be used to establish NRC-approved upper control limits (UCLs) in accordance with 10 *Code of Federal Regulations* (CFR) Part 40 Appendix A Criterion 5B(5) (i.e., constituent concentration-based values for excursion detection and aquifer-restoration performance assessment). Thus, the excursion indicators (or "excursion parameters") and the aquifer-restoration target values would be wellfield specific.

Monitoring wells would be installed in the ore zone to establish post-licensing, pre-operational baseline water quality for each "mine unit" (i.e., wellfield) (see Figure 2.4 in SEIS Section 2.1.1.). In addition, monitoring wells would be installed around each wellfield as well as into the overlying and underlying aquifers. Impending potential excursions to adjacent geologic units and progress toward meeting aquifer-restoration targets would be monitored by the Applicant's sampling designated wells within the wellfields during operation and during aquifer-restoration. These samples would be analyzed by a laboratory and would yield constituent-concentration data.

6.3.2.1 Post-Licensing, Pre-Operational Ground-Water Sampling and Water-Quality Analysis

The baseline ground-water monitoring program, which has been used for the last three years at the Ross Project area, would be expanded from the pre-licensing monitoring wells installed for site characterization, to a program designed to generate data specific to a mine unit, as needed. This program would be codified in the NRC license. The post-licensing, pre-operational monitoring program would provide data to establish UCL constituent concentrations that would be used by the Applicant to identify potential horizontal excursions of lixiviant outside of a wellfield and potential vertical excursions into the overlying or underlying aquifers (Strata, 2011b). The spacing, distribution, and the number of monitoring wells would be site specific and would be codified in the NRC license (NRC, 2009).

The Applicant proposes the installation of one well cluster for every four wellfield acres for their post-licensing, pre-operational data-collection program, which is consistent with the range the of one well per 0.4 ha [1 ac] to one well per 1.62 ha [4 ac] in the GEIS and the SRP (NRC, 2009, 2003), and historically used at existing ISR facilities. At the time of preparation of this manuscript, NRC staff has developed a draft license condition to require a minimum density of one well per 0.8 ha [2 ac] for the Commission-approved background based on staff's evaluation of site-specific geologic and hydrogeologic conditions, and the applicant's proposed sequencing and area of individual wellfield modules.

The Ross Project would include approximately 45 wells completed in the ore aquifer (30 – 55 m [100 – 180 ft]-thick sand interval) in the lower Lance/Upper Fox Hills Formations (designated as the ore-zone [OZ] unit) to establish baseline data. At approximately half of those location (24 locations), an additional well will be completed in the underlying aquifer (3 – 9 m [10 – 30 ft] thick sandy interval in the Fox Hills Formation (designated as the deep-monitoring [DM] unit below the ore zone) and the overlying lying aquifer in the first water-bearing unit above all mineralized zones in the Lance Formation (designated as the SM unit) forming a three-well cluster at those locations. The wells completed in the SM and DM units would use a fully penetrating completion while the ore-zone wells would target specific roll fronts (see Figures 2.8 – 2.10). Beyond the six existing well clusters used for pre-licensing baseline monitoring and site characterization, the Applicant proposes no additional surficial-aquifer (SA) wells for the wellfield areas; however, by license condition, the Applicant would be required to monitor the uppermost SA aquifer for wellfields which overly the SA aquifer that is found at shallow depths and is comprised of alluvial deposits associated with the recent stream channels.

For post-licensing, pre-operational water-quality characterization of the wellfields, the Applicant proposes to obtain at least four samples, with a minimum of two weeks between sampling events, for all perimeter, SM, OZ, and DM baseline wells. In addition, the SA-well network would continue to be sampled on a quarterly basis through the wellfield data-acquisition phase before final licensing for uranium recovery. The first and second sampling events would include laboratory analyses for constituents listed in GEIS Table 8.2-1 (NRC, 2009). The Applicant also proposes a reduced list of constituents for the third and fourth sampling events, which would be informed by the results of the previous two sampling events. Results from the sample analyses would be averaged arithmetically to obtain an average value as well as a maximum value for use in the NRC's determination of UCLs for excursion detection. The Applicant's proposed monitoring program would be modified as required by the NRC license.

6.3.2.2 Operational Ground-Water Sampling and Water-Quality Analysis

As described in GEIS Section 8.3.1.2, the placement of monitoring wells would occur around the perimeter of wellfields, in the aquifers both overlying and underlying the ore zone, and within the ore-zone aquifer for the early detection of potential horizontal and vertical excursions of lixiviant (NRC, 2009). The spacing, placement, and number of monitoring wells would be site-specific and would be established by the NRC in its license to the Applicant (NRC, 2009).

Three configurations of monitoring wells would be constructed to ensure detection of horizontal and vertical excursions: wells through the entire targeted ore zone (i.e., the ore body) at the perimeters of the wellfields; wells completed in the aquifer underlying the ore zone; and wells completed in the aquifer overlying the ore zone (Strata, 2011b). The design of a typical monitoring well is described in SEIS Section 2.1.1 (see also Figures 2.8 – 2.10). To detect whether an excursion of lixiviant has occurred, the monitoring results would be compared against the NRC-approved UCLs.

The Applicant proposes well spacing that meets the minimum requirement described in the GEIS as necessary to detect excursions (NRC, 2009). However, NRC staff has developed a draft license condition to require a minimum density of one well per 0.8 ha [2 ac] for the Commission-approved background based on NRC staff's evaluation of site-specific geologic and hydrogeologic conditions, and the applicant's proposed sequencing and area of individual wellfield modules. Wells completed in the aquifer underlying the ore zone and in the aquifer overlying the ore zone would be installed at a density of one well per 1 – 2 ha [3 – 4 ac] of wellfield to detect vertical migration. The Applicant proposes a spacing of the perimeter monitoring wells of 120 – 180 m [400 – 600 ft] apart and at a distance of approximate 120 – 180 m [400 – 600 ft] from the edge of the wellfield to detect potential horizontal excursions. Simulations by the Applicant demonstrate that the proposed well spacing successfully detects hydraulic anomalies in the form of water-level increases well before lixiviant has actually moved beyond the active uranium-recovery areas (Strata, 2011b).

The Applicant proposes that samples from these monitoring wells would be collected every two weeks to be analyzed for the excursion parameters (i.e., constituents) (Strata, 2011b). In addition, dedicated pressure transducers and/or in situ water-quality instruments would be used in the perimeter monitoring wells to provide early detection of potential excursions or hydraulic anomalies. Water levels would be routinely measured during well sampling in the perimeter, overlying, and underlying monitoring wells in order to provide an early warning for impending wellfield problems. An increasing water level in a perimeter monitoring well has been shown to be an indication of a local-flow imbalance within the wellfield, which could result in a lixiviant excursion. An increasing water level in an overlying or underlying monitoring well could similarly be caused by the migration of lixiviant from the ore-zone aquifer, or it could indicate an injection well-casing failure. This monitoring would allow immediate corrective actions, thus reducing the likelihood of excursions.

6.3.3 Flow and Pressure Monitoring of Wellfields and Pipelines

In GEIS Section 8.3.2, the monitoring of flow rates and pressures of lixiviant pumped to injection wells and from recovery wells is described. These monitoring data would be used by the Applicant to manage the water balance for the entire wellfield and to maintain an inward gradient to reduce the likelihood of excursions (NRC, 2009). To manage the water balance at the Ross Project, the Applicant proposes flow meters and pressure transmitters on each of the pipelines between the module building and injection and recovery wells. All instrumentation would be monitored at the module building and at the central processing plant (CPP). The wellfield flows would be balanced based on the module injection and recovery feeder-line meters. An individual well's flow targets would be determined on a per-well-pattern basis to ensure that local wellfield areas are balanced on at least a weekly basis. The maximum injection pressure would be less than the formation's fracture pressure.

Each module building would have the capability of being isolated from the pipelines by manually operated butterfly valves contained in the manholes exposing the pipelines. The manholes would have leak-detection devices that would activate an audible and visible alarm at the CPP in the event of a leak. Pressure transmitters on each end of the trunk lines and feeder lines would relay pressure readings back to the CPP's control room. In the event of a pressure reading that is outside of acceptable operating parameters, an audible and visible alarm would occur at the CPP. Automatic sequential shutdown of the trunk-line pumps and/or module-building booster pumps and recovery-well pumps would then occur if operating parameters do not return to normal ranges within a specified amount of time.

6.4 Meteorological Monitoring

The Applicant proposes to continue operating the meteorological monitoring station installed in January 2010 as part of its site-characterization baseline monitoring program (Strata, 2011a).

The data collected at this station would include continuous measurements of wind speed, wind direction, temperature, relative humidity, precipitation, and evaporation.

6.5 Ecological Monitoring

Ecological monitoring would include both vegetation and wildlife surveys.

6.5.1 Vegetation Monitoring

The Applicant proposes to monitor all disturbed areas on the Ross Project area for the presence of undesirable (i.e., noxious or invasive) species and to use control measures to prevent their spreading. Vegetation monitoring in reclaimed areas would be conducted according to U.S. Bureau of Land Management (BLM) and Wyoming Department of Environmental Quality (WDEQ) requirements and would be in accordance with the decommissioning requirements that would be included in the Applicant's NRC license (Strata, 2011a). Revegetation success would be monitored by the "extended reference area" concept, as defined in WDEQ/Land Quality Division (LQD), Guideline No. 2 (Strata, 2011a). The extended reference area would include all of the undisturbed portions of any vegetation type which has experienced disturbance in any phase of the Ross Project. At the end of decommissioning, quantitative vegetation data for extended reference areas representing each disturbed vegetation type would be directly compared by statistical analysis to quantitative vegetative data from reclaimed vegetation types. The duration of vegetation monitoring, and the target goals, would be defined in the final decommissioning plan required by the NRC license.

6.5.2 Wildlife Monitoring

The Applicant proposes annual wildlife surveys in and near the Ross Project area throughout the lifecycle of uranium-recovery activities in order to document key wildlife species, population trends, and habitats (Strata, 2011a).

6.5.2.1 Annual Reporting and Meetings

The Applicant would coordinate its wildlife-monitoring program with the BLM's Newcastle Field Office and the Wyoming Game and Fish Department (WGFD). Consultation with the U.S. Fish and Wildlife Service (USFWS), BLM, and WGFD would be conducted prior to the Applicant's initiating a survey and would be documented in a work plan, with BLM and WGFD concurrence. The Applicant would prepare an annual monitoring report and submit it to the BLM, WGFD, and other interested parties by November 15 of each year. The monitoring report would include:

- Survey methods and results as well as observations of any trends and assessments of wildlife-protection measures implemented during the past year;
- Recommendations for changes in wildlife-protection measures for the coming year;
- Recommendations for modifications to wildlife monitoring or surveying; and
- Recommendations for additional species to be monitored (e.g., a newly Federal- or State-listed species).

Data and mapping would be formatted to meet BLM requirements (i.e., geographic information systems data and maps).

6.5.2.2 Annual Inventory and Monitoring

Wildlife surveying and monitoring would be performed by BLM or WGFD biologists or a qualified scientist under contract to the Applicant. All aspects of a regular and/or periodic monitoring program would be developed according to current regulatory and permitting guidelines and requirements. These would include field-survey and survey-equipment requirements; data collection, analysis, reporting, and storage procedures; agency consultations and collaborations; and any other relevant survey- and monitoring-program components.

6.5.2.3 Wildlife Species

Mammals and certain birds as well as all wildlife on the BLM Sensitive Species (BLMSS), WSOC, and USFWS's SMC lists at the Ross Project area would be monitored in the Applicant's wildlife monitoring program.

Mammals

Opportunistic observations of all wildlife species would be conducted in late spring and summer, during the Applicant's completion of the surveys discussed below for sensitive species. No big-game crucial ranges, habitats, or migration corridors are recognized by the WGFD at the Ross Project area or the surrounding 1.6-km [1-mi] perimeter. A "crucial" range or habitat is defined as any particular seasonal range or habitat component that has been documented as the determining factor in a population's ability to maintain and reproduce itself at a certain level. Due to the lack of crucial big-game habitats, the WGFD did not require big-game surveys during pre-licensing baseline monitoring the Applicant performed in 2009 and 2010 (Strata, 2011a). Long-term monitoring for big game is not anticipated and has not been proposed by the Applicant.

Protected Species and Other Birds

The Applicant proposes to monitor protected species, using the following strategy (Strata, 2011a):

- Early spring surveys for and monitoring of sage-grouse leks within 2 km [1.2 mi] of the Ross Project area. All threatened and endangered species as well as those on the BLMSS, Wyoming Species of Concern (WSOC), and USFWS's "Migratory Bird Species of Management Concern in Wyoming" (SMC) lists would be surveyed and monitored on the Ross Project area as well.
- Late spring and summer opportunistic observations of all wildlife species, including threatened, endangered, BLMSS, WSOC, SMC, and any other species of concern would occur and noted.
- Any other surveys as required by regulatory agencies.

Raptors

Only one raptor's nest was previously identified on the Ross Project area, and the opportunity for nesting is limited in the area due to a lack of suitable habitat (i.e., trees and cliffs). However, the Applicant has committed to completing the following:

- Early-spring surveys for new and/or occupied raptor territories and/or nests, and
- Late-spring and summer surveys for raptor reproduction at occupied nests.

The nearest human disturbance to active and inactive raptor nests, any visual barriers in the line of sight of raptor nests, and the prey abundance (e.g., jackrabbits and cottontails) would be reported in each annual report to allow an assessment of whether any raptor disturbance is related to uranium-recovery activities.

Migratory Birds

The Applicant would conduct nesting-bird surveys for nongame species during early summer, following recommended WDEQ techniques. All birds, observed or heard, and the vegetation and habitat type where they might be found would be recorded. These surveys would document all high-interest bird species identified by the BLM, WGFD, and USFWS.

6.6 References

10 CFR Part 20. Title 10, "Energy," *Code of Federal Regulations*, Part 20 "Standards for Protection Against Radiation." Washington, DC: U.S. Government Printing Office.

(US)NRC (U.S. Nuclear Regulatory Commission). *Radiological Effluent and Environmental Monitoring at Uranium Mills*, Revision 1. Regulatory Guide 4.14. Washington, DC: NRC. 1980. ADAMS Accession No. ML003739941.

(US)NRC. *Generic Environmental Impact Statement for In-Situ Leach Uranium Milling Facilities*, Volumes 1 and 2. NUREG–1910. Washington, DC: NRC. 2009. ADAMS Accession Nos. ML091480244 and ML091480188.

Strata (Strata Energy, Inc.). *Ross ISR Project USNRC License Application, Crook County, Wyoming, Environmental Report, Volumes 1, 2 and 3 with Appendices.* Docket No. 40-09091. Gillette, WY: Strata. 2011a. Agencywide Documents Access and Management System (ADAMS) Accession Nos. ML110130342, ML110130344, and ML110130348.

Strata. *Ross ISR Project USNRC License Application, Crook County, Wyoming, Technical Report, Volumes 1 through 6 with Appendices.* Docket No. 40-09091. Gillette, WY: Strata. 2011b. ADAMS Accession Nos. ML110130333, ML110130335, ML110130314, ML110130316, ML110130320, and ML110130327.

Strata. *Ross ISR Project USNRC License Application, Crook County, Wyoming, RAI Question and Answer Responses, Environmental Report, Volume 1 with Appendices.* Docket No. 40-09091. Gillette, WY: Strata. 2012. ADAMS Accession No. ML121030465.

7 COST-BENEFIT ANALYSIS

This section summarizes the benefits and costs associated with the Proposed Action and the two Alternatives. The discussion of costs and benefits follow the U.S. Nuclear Regulatory Commission (NRC) guidance presented in NUREG–1748 (NRC, 2003). The discussion of the costs and benefits include both the costs of each Alternative and a qualitative discussion of environmental impacts, as applicable.

7.1 Proposed Action

Benefits of the Proposed Action include the additional employment opportunities available to area residents, increased incomes to area residents, and additional tax revenues accruing to local jurisdictions and the State of Wyoming. Potential costs include both the internal costs of the Ross Project borne by the Applicant and the potential external monetary costs that may be required by local public-service providers in response to Project activities as well as non-monetary costs associated with the potential environmental impacts.

7.1.1 Ross Project Benefits

The economic benefits of the Ross Project would be positive for Crook County and generally positive for residents directly or indirectly affected by the Project. The Applicant is committed to hire local personnel and to make equipment purchases at local suppliers whenever possible (Strata, 2012), maximizing the economic benefits to Crook County and neighboring counties.

7.1.1.1 Employment and Income

The Ross Project is expected to require a peak workforce of approximately 200 workers during its construction phase; 60 workers during operation; 20 – 30 workers during the aquifer-restoration phase; and 90 workers for decommissioning activities (see Supplemental Environmental Impact Statement [SEIS] Section 4.11). This employment would be beneficial because it would reduce the local unemployment rate for the duration of construction, and some workers would likely stay through the operation phase of the Ross Project. It is expected that workers would be paid the regional rates typical of Crook and Campbell Counties, where a higher percentage of jobs are in the relatively higher-paying energy industry. Based upon weighted average annual earnings per job of $61,400 (see SEIS Section 3.11), earnings accruing to area residents would range from $1.2 million to $1.8 million during the aquifer-restoration phase to approximately $12.3 million during the Ross Project's construction phase. In addition, existing private-property landowners at the Ross Project area would be compensated for the loss of use of their land; however, the specific terms of this compensation is unknown.

7.1.1.2 Tax Revenues

Average annual tax revenues are estimated to be $2,785,000 per year during the Ross Project's operation (see Section 4.11.1) and would total $27,850,000 over the lifecycle of the Project. The State of Wyoming would benefit, in part, from the severance and royalty payments, estimated to be $10.9 million over the lifecycle of the Ross Project, whereas Crook County would benefit from the gross production and property taxes, totaling $16.9 million over the lifecycle of the Project. In addition, some portion of the State severance and royalty payments would be distributed among all Wyoming cities and counties and, thus, all jurisdictions within the State are expected to benefit from increased State tax revenues (WLSO, 2010).

7.1.2 Ross Project Costs

Potential costs include both the internal costs of the project borne by the Applicant and potential external costs that may be required by local public service providers in response to project activities, as well as non-monetary costs associated with the potential environmental impacts.

7.1.2.1 Internal Costs

All internal costs would be borne by the Applicant—that is, the direct financial costs of the construction, operation, aquifer restoration, and decommissioning of the proposed Ross Project. The primary internal costs would include:

- Capital costs associated with the Applicant's obtaining land and mineral rights as well as securing regulatory approvals including permits, licenses, and related environmental studies
- Capital costs of facility and wellfield construction
- Costs of facility and wellfield operation and maintenance
- Costs of aquifer restoration
- Costs of facility and wellfield decontamination, dismantling, and decommissioning
- Costs of site reclamation and restoration

The Applicant estimates that these internal costs would be approximately $136.7 million (Strata, 2011a). The actual, estimated decommissioning costs for the Ross Project would be determined prior to Project operation, and a surety arrangement equal to the estimated decommissioning costs would be made a condition of the NRC license. Each year, the decommissioning cost estimate would be reviewed by the NRC and Wyoming Department of Environmental Quality (WDEQ), and adjustments would be made as necessary.

7.1.2.2 External Costs

<u>Land Use</u>

During the Proposed Action, impacts to local land use would occur. Impacts would result from land disturbances during construction and decommissioning, grazing and access restrictions, and competition for access to mineral rights. Land use impacts during all phases of the Project would be SMALL. Access restrictions at the Ross Project area, however, would preclude the economic benefits from existing agricultural and grazing activities. If site access is assumed to be restricted across the entire Ross Project area—696 ha [1,721 ac]—and based upon a market value of products sold from crop and livestock sales in Crook County averaging $28 per acre in 2007 dollars (USDA, 2009), $48,188 in annual lost-agriculture sales would be estimated as the upper end of this potential loss, or $481,880 over the lifecycle of the Project. These losses would be offset by the compensation paid to the landowners, where the exact terms of the respective compensation is confidential.

Transportation

During the Proposed Action, the highest traffic volume would occur during the construction phase because of the relatively large workforce as well as the increased demand for materials and equipment at the Project area. The increased traffic is expected to be 400 passenger cars and 24 trucks per day, which, compared to 2010 levels, represents a significant traffic volume increase of approximately 400 percent on New Haven Road. Thus, construction-phase transportation impacts would be MODERATE to LARGE with respect to the traffic levels on local roads and the road surfaces, and SMALL with respect to traffic levels on I-90. All other phases would have less traffic related to commuting workers and, thus, the impacts would range from SMALL to LARGE. This traffic could result in more traffic accidents as well as wear and tear on road surfaces. Mitigation measures would be in place and would reduce the range of these impacts to SMALL to MODERATE.

Geology and Soils

Under the Proposed Action, potential impacts to geology and soils would occur due to the disturbance of 113 ha [280 ac] of the Ross Project area, or about 16 percent (Strata, 2011b). Other soil impacts would include the Applicant's clearing of vegetation; stripping of topsoil; excavating, backfilling, and compacting soil; grading of the land; and trenching for utilities and pipelines. There is limited potential impact to geology because of the minor depth of disturbance associated with construction of the Ross Project. The potential impacts from soil loss would be minimized by proper design and operation of surface-runoff features and implementation of best management practices (BMPs). Impacts to geology and soils would be SMALL.

Water Resources

The Ross Project has the potential to impact surface water and ground water during each phase of its lifecycle.

Surface Water

Under the Proposed Action, surface-water-related impacts would include potentially increased sediment concentrations. Depending upon discharge rates and locations, impacts from the discharge of water generated during aquifer testing, during well installation and pipeline integrity testing, and during the dewatering of the facility areas inside of the containment barrier wall (CBW), surface-water impacts would be SMALL. Stream-channel disturbance, surface-water contamination, and surface-water consumptive use impacts would be SMALL. Impacts to surface water would also include the potential contamination of surface water by a spill or unintended release of process solutions, which could result in SMALL impacts with mitigation. Finally, reduced flows, in particular, the Little Missouri River would be a SMALL impact.

Ground Water

Under the Proposed Action, potential impacts to ground water are primarily from the consumptive use of ground water (i.e., removing more than is injected in), disposal of drilling fluids and cuttings during well drilling, and spills and leaks of fuels and lubricants from construction equipment. Impacts to shallow (i.e., near-surface) aquifers would be SMALL. The impacts to the ore-zone and surrounding aquifers regarding the quantity of water available would also be SMALL to MODERATE, while the potential impact of improperly abandoned drillholes, over-penetration of holes, or well integrity could result SMALL to MODERATE water-quality impacts in the event of an excursion in a Ross Project wellfield and SMALL elsewhere.

Ecology

Under the Proposed Action, potential environmental impacts to ecological resources, both flora and fauna, could occur during all phases of the Project; all impacts would be SMALL. The impacts to local vegetation would include:

- Removal of vegetation from the Ross Project area
- Modification of existing vegetative communities
- Loss of sensitive plants and habitats
- Potential spread of invasive species and noxious weed populations
- Reduction in wildlife habitat and forage productivity
- Increased risk of soil erosion and weed invasion

Impacts to terrestrial wildlife could include:

- Loss, alteration, or incremental fragmentation of habitat
- Displacement of and stresses on wildlife
- Direct and/or indirect mortalities

Aquatic species could be affected by:

- Disturbances of stream channels
- increases in suspended sediments
- Pollution from spills and leaks
- Reduction of habitat

These impacts would be mitigated by, for example, implementing the standard management practices required or suggested by the Wyoming Game and Fish Department (WGFD). All ecological resource impacts would be SMALL.

Air Quality

Under the Proposed Action, impacts from nonradiological particulate emissions would primarily result from fugitive road dust created by moving vehicles and mobile equipment throughout the Ross Project area and, to a far lesser extent, the processes and circuits implemented in the Central Processing Plant (CPP). Combustion-engine emissions from diesel-equipment operation would occur primarily during the construction, operation, and decommissioning phases. In general, however, uranium-recovery activities are not major air-emission sources. Air-quality impacts during all phases of the Ross Project would be SMALL.

Noise

Under the Proposed Action, there would be very temporary, but MODERATE, noise impacts for residences very near the Ross Project area; for residences, communities, or sensitive areas that are located more than approximately 300 m [1,000 ft] from specific noise-generating activities the impacts would be SMALL because noise levels quickly decrease with distance. These impacts would be the result of uranium-recovery activities and the associated traffic that would be associated with the Ross Project. During high truck-traffic events on New Haven Road during all phases of the Ross Project, residents living on those routes could occasionally be annoyed by the noise. There are no churches, schools, or community centers located less than 300 m [1,000 ft] from the Ross Project's boundary. Impacts to workers at the Project also would be SMALL because of the Applicant's compliance with OSHA noise regulations.

Historical, Cultural, and Paleontological Resources

The costs and benefits of the Ross Project related to historic, cultural, and paleontological resources will be determined once a complete inventory of these resources within the Ross Project area has been completed.

Visual and Scenic Resources

Under the Proposed Action, MODERATE, short-term impacts to the visual and scenic resources of the area during construction would occur, and SMALL longer-term impacts for the remainder of the Ross Project (see SEIS Section 4.10). Potential visual and scenic impacts would result from the surface disturbance and construction of the following: 1) wellfields (including drill rigs, header houses, wellhead covers, and roads; 2) the CPP; 3) surface impoundments; 4) the CBW; 5) secondary and tertiary access roads; 6) power lines; and 7) fencing. The nearest protected visual resource to the Ross Project is the Devils Tower National Monument, which is approximately 16 km [10 mi] east of the Ross Project. Although the Project itself would not be visible at the lower park portion of the Tower, climbers ascending to the top of the Tower may be able to see some of the Project's largest attributes as well as, in the night sky, the lights of the Project. The visual impacts from the Ross Project would be consistent with the U.S. Bureau of Land Management's (BLM's) VRM Class III designation (NRC, 2009).

The degradation of views of the nighttime sky in the surrounding vicinity of the Project area has been evaluated using the contingent valuation method (CVM) at four national parks (i.e., Yellowstone, Great Basin, Mesa Verde, and Chaco Canyon) during summer surveys in 2007 (Mitchell et al., 2008). These surveys were designed to quantify the willingness-to-pay (WTP) to reduce light pollution in these areas. Over 50 percent of respondents were willing to pay a positive amount to address light pollution. The average amount individuals would be willing to increase their Federal tax to reduce light pollution was estimated to be $39.37 per year per person. When the self-reported survey characteristics are reviewed, there is a positive correlation between the extent individuals are exposed to light pollution and their willingness to pay to reduce it. Hence, people in rural areas are generally less willing to pay to reduce light pollution. There are 11 residences within Ross Project area where visual-resource impacts were evaluated (see SEIS Section 4.10). Based on an average household size of 2.41 persons per household in Crook County (USCB, 2012), an estimated 27 persons could be affected by light from the Project, and the external costs associated with light pollution would be $1,063 per year or $10,630 over the lifecycle of the Ross Project.

Socioeconomics

Under the Proposed Action, the impacts of the Ross Project on the demand for community services are projected to be small (see SEIS Section 4.11.1.1). The Applicant is committed to hiring locally and, during peak construction-phase activities, it is projected only 52 additional residents are expected in the ROI (i.e., Crook and Campbell Counties). Lower demographic impacts occur in subsequent Project phases. Ross Project-related population increases would represent less than 0.1 percent of the 2010 population in the two-county ROI and, in general, existing community-service providers, such as local schools, health-service agencies, and police and fire-protection agencies, are not expected to be adversely affected by this level of increased demand for public services.

There would be an increased need, however, for emergency-response services. The Applicant has entered into a Memorandum of Understanding (MOU) with Crook County (Strata and Crook County, 2011) that states the Applicant would coordinate emergency-management, hazardous-materials management, and fire-suppression planning with Crook County's Homeland Security Director and Crook County's Fire Warden and Fire Zone Warden. The Applicant commits to maintaining the onsite personnel and equipment necessary to provide emergency services when environmental, safety, or health emergencies arise at the Ross Project. As such, these services would not represent a cost to local governments (Strata and Crook County, 2011).

The MOU also states the Applicant would:

- Provide electronic warning signs that would close county roads into the Ross Project area in the case of an emergency.
- Provide dust control for the existing and increased traffic as a result of the Ross Project, as necessary, and, as required by the WDEQ. This would include dust control over each one-quarter mile of county roads fronting the residences along any road designated by the County as an access road to the Ross Project, in order to minimize dust impacts on area residents beyond the Ross Project area.
- Maintain and repair damage caused by Applicant's trucks or contracted trucks as a result of their use, as dictated and regulated by Crook County (Strata and Crook County, 2011).

These measures would minimize any costs that would be borne by local jurisdictions and area residents.

Environmental Justice

Under the Proposed Action, no minority or low-income populations have been identified in the Ross Project area. Therefore, there are no disproportionately high and adverse impacts to minority and low-income populations by the Ross Project.

Public and Occupational Health and Safety

Under the Proposed Action, potential nonradiological and radiological impacts to the public's and workers' health and safety over the course of the Ross Project could include accidental chemical or radiological releases, chemical or byproduct liquid spills, particulate and gaseous emissions, vehicular and equipment accidents, worker injuries and illnesses, or fires. The Applicant proposes to minimize these potential impacts through rigorous worker training, facility and wellfield design, operational controls, and a series of emergency-response protocols.

An important factor in the assessment of risks to public health and safety is the proximity of potentially impacted populations. The nearest incorporated community to the Ross Site is Moorcroft, Wyoming, with an estimated population of less than 1,000; Moorcroft is located approximately 35 km [22 mi] south of the Ross Project area. Unincorporated Oshoto is adjacent to the Ross Project area, but it has a population of fewer than 50 persons. In addition, the quantities of materials that could be released, even through the air pathway, would be small and, as discussed in SEIS Section 4.7, would be dispersed and diluted. Workers involved in the response and cleanup of spills and leaks could receive MODERATE impacts; these would be mitigated by establishing standard operating procedures (SOPs) and training requirements. Thus, little to no risk would be borne to the offsite public, and these impacts would be considered SMALL.

Waste Management

Under the Proposed Action, both liquid and solid wastes would be generated during all phases of the Ross Project's lifecycle. Several major waste streams are identified in SEIS Section 4.14. At least four of these waste streams have to the potential to impact the local communities.

The disposal of liquid byproduct wastes would be accomplished by injection of these wastes into a confined aquifer. The regulatory-permitting process for this type of waste disposal would ensure that all mitigation measures to minimize related potential impacts would be taken. Ordinary solid wastes would include trash, spent materials, and broken equipment. Hazardous waste would represent a very small volume of spent reagents and other items such as batteries. Radioactive solid waste would consist of Ross Project equipment, process vessels, building components, and other items that could not be decontaminated and released as nonradioactive. Although all of these wastes would be disposed of at offsite waste-disposal facilities, the relatively small volume of such wastes would have little impact on the respective disposal facilities' ultimate capacity. Waste management impacts during all phases of the Ross Project would be SMALL.

7.1.3 Findings and Conclusions

Implementation of the Proposed Action would have a SMALL to MODERATE socioeconomic impact on the ROI, with MODERATE impacts associated with the benefits of the additional tax revenue projected to accrue to Crook County. Regional benefits would include increased employment, economic activity, and tax revenues in the region and the State of Wyoming. Because the Applicant is committed to hiring locally, population increases and the subsequent need for additional public services is projected to be negligible. Access restrictions to the Ross Project area would result in the loss of some economic activities, but this loss is expected to be offset to a degree by the Applicant's compensation to the affected landowners. A limited number of residents would also be affected by light pollution from the Ross Project. However, overall, the economic benefits of the Proposed Action would be greater than the associated costs.

7.2 Alternative 2: No Action

Under the No-Action Alternative, the NRC would not issue the Applicant a license to construct, operate, restore the aquifer, and decommission the proposed Ross Project. Area residents would benefit from some limited preconstruction activities, but no longer-term economic benefits would accrue to area residents, local jurisdictions, or the State. Similarly, there would be no potential costs borne by nearby jurisdictions and residents.

7.3 Alternative 3: North Ross Project

Construction, operation, aquifer restoration, and decommissioning of the North Ross Project are not expected to result in any significant differences in this cost-benefit analysis. Overall land use impacts would be generally the same as for the Proposed Action, although impacts to dry-land crop agriculture would be lower, while impacts to grazing activities would be greater. Small changes in traffic patterns on roads to and in the Ross Project Area would result in reduced traffic volumes on New Haven Road that would be offset by increased traffic on other roads. These changing traffic patterns would slightly increase noise and air quality impacts, but the impacts would be offset by fewer affected residents. Impacts to other resources areas also are generally the same as for the Proposed Action. Thus, the major benefits and costs described for the Proposed Action would accrue similarly were the facility to be constructed and operated at the north site.

7.4 References

(US)CB (U.S. Census Bureau, USDOC). "State and County QuickFacts, Wyoming, Crook County, Campbell County." 2012. At http://quickfacts.census.gov/qfd/states/56000.html (as of June 25, 2012).

(US)DA (U.S. Department of Agriculture). *2007 Census of Agriculture, County Profile, Crook County, Wyoming.* Washington, DC: USDA. 2009. ADAMS Accession No. ML13014A730.

Mitchell, David, Julie Gallaway, and Reed Olsen. "Estimating the Willingness to Pay for Dark Skies Draft)" In *83rd Annual Meetings of the Western Economic Association International.* Springfield, MO: Missouri State University. July 2008. ADAMS Accession No. ML13024A313.

(US)NRC (U.S. Nuclear Regulatory Commission). *Environmental Review Guidance for Licensing Actions Associated with NMSS Programs.* NUREG–1748. Washington, DC: USNRC. August 2003. ADAMS Accession No. ML032450279.

(US)NRC. *Generic Environmental Impact Statement for* In-Situ *Leach Uranium Milling Facilities.* Volumes 1 and 2. NUREG–1910. ML09148024 and ML0911480188. Washington, D.C.: NRC. 2009.

Strata (Strata Energy, Inc.). *Ross ISR Project USNRC License Application, Crook County, Wyoming, Technical Report, Volumes 1 through 6 with Appendices.* Docket No. 40-09091. Gillette, WY: Strata. 2011a. Agencywide Documents Access and Management System (ADAMS) Accession Nos. ML110130333, ML110130335, ML110130314, ML110130316, ML110130320, and ML110130327.

Strata. *Ross ISR Project USNRC License Application, Crook County, Wyoming, Environmental Report, Volumes 1, 2 and 3 with Appendices.* Docket No. 40-09091. Gillette, WY: Strata. 2011b. ADAMS Accession Nos. ML110130342, ML110130344, and ML110130348.

Strata. *Ross ISR Project USNRC License Application, Crook County, Wyoming, RAI Question and Answer Responses, Environmental Report, Volume 1 with Appendices.* Docket No. 40-09091. Gillette, WY: Strata. 2012. ADAMS Accession No. ML121030465.

Strata and Crook County. *Memorandum of Understanding for Improvement and Maintenance of Crook County Roads Providing Access to the Ross ISR Project.* Sundance, WY: Strata and Crook County. April 6, 2011.

WLSO (Wyoming Legislative Service Office). *Wyoming Severance Taxes and Federal Mineral Royalties.* ML13015A700. Cheyenne, WY: WLSO. 2010.

8 SUMMARY OF ENVIRONMENTAL CONSEQUENCES

The environmental consequences of the Proposed Action, the Ross Project, and Alternative 3, the North Ross Project, are summarized next in Table 8.1.

Table 8.1
Summary of Environmental Consequences of the Proposed Action and Alternatives

	Unavoidable Adverse Environmental Impacts	Irreversible and Irretrievable Commitment of Resources	Short-Term Impacts and Uses of the Environment	Long-Term Impacts and the Maintenance and Enhancement of Productivity
LAND USE				
Alternative 1: Proposed Action Ross Project (See SEIS Section 4.2.1.)	There would be SMALL unavoidable adverse impacts on land use over the lifecycle of the Ross Project because certain areas of the Project area would be fenced during the Project's construction, operation, aquifer restoration, and decommissioning. The area disturbed, however, is small—113 hectares (ha) [280 acres (ac)]. At the end of the Project, after facility decommissioning and site restoration have been fully accomplished, the former land uses of the Ross Project area would be restored.	There would be no irreversible or irretrievable impacts to land use in the area as a result of the Proposed Action. All land would be restored to its baseline uses post-Ross Project. Access roads would be removed, or they would be left as desired by the respective landowner(s).	There would be short-term land-use impacts during the Proposed Action, predominantly due to a decrease in the total area available for livestock grazing.	There would be no long-term impacts on land use within the Ross Project area. The land would be restored to its pre-licensing baseline and former land uses would be poss ble after decommissioning of the Project and the site's restoration.
Alternative 3: North Ross Project (See SEIS Section 4.2.3.)	Approximately the same number of acres would be taken out from service during Alternative 3. The area that would be disturbed would still be small. However, some livestock grazing could be diminished during the lifecycle of Alternative 3. However, as above, after complete facility decommissioning and site restoration have been accomplished, the former land use would be restored.	There would be no irreversible or irretrievable impacts to land use in the area as a result of Alternative 3. All land would be restored to its baseline uses.	There would be short-term impacts to land use in Alternative 3. The total area available for livestock grazing would be temporarily reduced.	There would be no long-term impacts on the land use of the area of Alternative 3. As above, the land would be restored and the former pre-licensing land use would be re-established after the decommissioning of the Alternative 3.

Table 8.1
Summary of Environmental Consequences of the Proposed Action and Alternatives (Cont.)

	Unavoidable Adverse Environmental Impacts	Irreversible and Irretrievable Commitment of Resources	Short-Term Impacts and Uses of the Environment	Long-Term Impacts and the Maintenance and Enhancement of Productivity
TRANSPORTATION				
Alternative 1: Proposed Action Ross Project (See SEIS Section 4.3.1.)	During the four phases of the Proposed Action, the unavoidable impacts to transportation, which would be related specifically to traffic volumes, would range from SMALL to LARGE. Because the Ross Project is located in a rural area of Wyoming, where traffic is sparse, the increase in traffic as a result of the Proposed Action could create SMALL to LARGE impacts. However, with the mitigation measures the Applicant has proposed, during all phases of the Ross Project, transportation impacts would be SMALL to MODERATE.	There would be no long-term irreversible or irretrievable environmental impacts from increases in transportation by the Proposed Action. Once the Ross Project has been decommissioned and the site restored, all traffic impacts of the Proposed Action would cease as all related traffic would be zero.	Transportation impacts to the vicinity of the Proposed Action would be SMALL to MODERATE, with mitigation. These impacts would include increased traffic counts and a slightly higher probability of vehicular accidents. All of these short-term impacts would cease after the decommissioning of the Ross Project.	The long-term impacts of the Proposed Action with respect to transportation resources would be enhanced road quality. During the Ross Project, the Applicant would ensure that the roads leading to and from the Project would be well built and maintained. Although the maintenance that would be provided by the Applicant would cease after the Proposed Action is complete, the remaining roads, however, would be of better quality than they are currently.
Alternative 3: North Ross Project (See SEIS Section 4.3.3.)	The construction and operation of the Central Processing Plant (CPP) at the north site would, in general, have the same transportation impacts as the Proposed Action, and these impacts would be SMALL to LARGE. With mitigation measures, the impacts would be SMALL to MODERATE.	There would be no irreversible impacts to land use in the area as a result of Alternative 3. All transportation impacts would cease at the conclusion of Alternative 3. Access roads would be removed, or they would be left as desired by the respective landowner(s).	There would also be short-term impacts to transportation in Alternative 3, and they would be the same as for the Proposed Action.	The long-term impacts of Alternative 3 would be the same as those in the Proposed Action, where roads would be improved by the Applicant and these improvements would remain after Alternative 3 was decommissioned and its site restored.

Table 8.1
Summary of Environmental Consequences of the Proposed Action and Alternatives (Cont.)

	Unavoidable Adverse Environmental Impacts	Irreversible and Irretrievable Commitment of Resources	Short-Term Impacts and Uses of the Environment	Long-Term Impacts and the Maintenance and Enhancement of Productivity
GEOLOGY AND SOILS				
Alternative 1: Proposed Action Ross Project (See SEIS Section 4.4.1.)	There would be SMALL unavoidable adverse potential environmental impacts to the geology and soils at the Proposed Action. Wind and water erosion are possible, but the Applicant would mitigate the potential for erosion with best management practices (BMPs) specifically related to erosion. (Fugitive dust is discussed under "Air Quality" in this Table.) There would be few geology and soils impacts during the Proposed Action.	There would be no irreversible or irretrievable commitment of geology or soil resources during the Proposed Action. No permanent changes would occur to the overall geology and soils of the Ross Project.	There would be some short-term potential impacts to soils under the Proposed Action, such as loss due to erosion. With the mitigation measures proposed by the Applicant, however, these potential impacts are unlikely, even over the short term.	There would be no long-term impacts to geology or soils during the Proposed Action at the Ross Project.
Alternative 3: North Ross Project (See SEIS Section 4.4.3.)	Alternative 3, as in the Proposed Action, would have the potential for wind or water erosion. With mitigation, however, little erosion would be expected. Greater soil disturbance for construction of the surface impoundments for Alternative 3 would be expected because of the site-specific topographic conditions. There would be SMALL impacts to the geology and soils at the Ross Project area.	There would be no irreversible or irretrievable impacts to geology or soils during Alternative 3.	There would be small impacts related to the potential for eroding soils during Alternative 3; these would be mitigated and, consequently, SMALL.	There would be no long-term impacts to geology or soils in Alternative 3.

Table 8.1
Summary of Environmental Consequences of the Proposed Action and Alternatives (Cont.)

	Unavoidable Adverse Environmental Impacts	Irreversible and Irretrievable Commitment of Resources	Short-Term Impacts and Uses of the Environment	Long-Term Impacts and the Maintenance and Enhancement of Productivity
WATER RESOURCES: SURFACE WATER				
Alternative 1: Proposed Action Ross Project (See SEIS Section 4.5.1.)	There would be SMALL unavoidable adverse impacts on surface water over the lifecycle of the Ross Project. Surface water would be potentially impacted by sediment from stormwater run-off and crossing water features with roads and pipelines. The moisture conditions of wetlands situated along the Little Missouri River and adjacent to the Oshoto Reservoir would potentially be impacted. Accidental leaks, spills, and other releases of fluids would potentially impact surface water quality. With mitigation, however, little sedimentation would be expected and accidental releases would be contained.	There would be no irreversible or irretrievable impacts to surface water by the Proposed Action. Small amounts of surface water would be used for construction activities and dust control, but this water would be replaced by normal precipitation.	There would be some short-term potential impacts to surface water under the Proposed Action, such as increased sedimentation. With the mitigation measures proposed by the Applicant, however, these potential impacts are unlikely, even over the short term.	There would be no long-term impacts to surface water during the Proposed Action at the Ross Project area.
Alternative 3: North Ross Project (See SEIS Section 4.5.1.)	The construction and operation of the CPP at the north site in Alternative 3 would, in general, have the same surface water impacts as the Proposed Action; however the mitigation measures required to protect the two ephemeral drainages and the steeper land slopes at the North Ross Project would involve more engineering. With mitigation measures, the impacts would be SMALL.	As with the Proposed Action, there would be no irreversible or irretrievable impacts to surface water during Alternative 3.	The potential short-term impacts to surface water during Alternative 3 would be the same as for the Proposed Action.	There would be no long-term impacts to surface water at the North Ross Project under Alternative 3.

Table 8.1
Summary of Environmental Consequences of the Proposed Action and Alternatives (Cont.)

	Unavoidable Adverse Environmental Impacts	Irreversible and Irretrievable Commitment of Resources	Short-Term Impacts and Uses of the Environment	Long-Term Impacts and the Maintenance and Enhancement of Productivity
WATER RESOURCES: GROUND WATER				
Alternative 1: Proposed Action Ross Project (See SEIS Section 4.5.1.)	There would be SMALL unavoidable adverse impacts to water quality and MODERATE unavoidable adverse impacts to water quantity on ground-water resources over the lifecycle of the Ross Project. Aquifers would be impacted by water withdrawal and injection of lixiviant. Lowering of water levels would be seen within and outside the Project area. The water quality of the aquifers outside and below the exempted aquifer would temporarily be impacted if excursions of lixiviant were to occur. Ground water above the exempted aquifer would potentially by impacted by leaks from wells and releases at the surface. With mitigation, however, little potential for excursions, leaks and accidental releases would be minimized. There would be SMALL to MODERATE impacts to the ground water quantity and SMALL impacts to ground water quality at the Ross Project area.	There would be no irreversible or irretrievable impacts to ground water from the Proposed Action. The lower ground water levels in the ore-zone aquifer and the aquifer overlying the ore zone would be replaced by normal recharge over time. Excursions would be remediated by pumping out contaminated water. The water quality of the exempted aquifer would consequently be restored.	Lowering of water levels would be a short-term impact to ground water from the Proposed Action. Based upon historical experience with uranium-recovery projects, excursions of lixiviant often occur and create short-term impacts to water quality. The mitigation measures proposed by the Applicant and required by permit and license conditions, such as water-management actions to minimize water usage from the aquifers, tests to ensure integrity of the wells, and early detection of excursions short-term impacts would reduce these impacts at the Ross Project.	There would be no long-term impacts to ground water by the Proposed Action. The water levels would rebound through normal aquifer recharge and restoration activities would return the water-quality to aquifer-restoration target values.

Table 8.1
Summary of Environmental Consequences of the Proposed Action and Alternatives (Cont.)

	Unavoidable Adverse Environmental Impacts	Irreversible and Irretrievable Commitment of Resources	Short-Term Impacts and Uses of the Environment	Long-Term Impacts and the Maintenance and Enhancement of Productivity
WATER RESOURCES: GROUND WATER *(Continued)*				
Alternative 3: North Ross Project (See SEIS Section 4.5.3.)	The construction and operation of the CPP at the north site would have the same mitigated impacts to ground water as the Proposed Action except, during Alternative 3, fewer mitigation measures to minimize the potential impacts to the shallow, unconfined aquifer from spills and leaks of byproduct liquid waste from the impoundments would be required. The greater depth to the shallow aquifer below the impoundments at the north site would eliminate the need for the containment barrier wall to prevent ground-water flow in the area below the impoundments as in the Proposed Action. Mitigation of the potential for leaks from the impoundments at the north site would rely upon leak detection and monitoring systems as well as remediation if contaminants reached the ground water.	As descr bed for the Proposed Action, there would be no irreversible or irretrievable impacts to ground water under Alternative 3.	The short-term impacts during Alternative 3 would be the same as described for the Proposed Action.	There would be no long-term impacts to ground water at the North Ross Project under Alternative 3.

Table 8.1
Summary of Environmental Consequences of the Proposed Action and Alternatives (Cont.)

	Unavoidable Adverse Environmental Impacts	Irreversible and Irretrievable Commitment of Resources	Short-Term Impacts and Uses of the Environment	Long-Term Impacts and the Maintenance and Enhancement of Productivity
ECOLOGY				
Alternative 1: Proposed Action Ross Project (See SEIS Section 4.6.1.)	There would be SMALL unavoidable impacts related to the ecology of the Proposed Action. Some wildlife could be displaced during Project activities, especially during construction and decommissioning. The areas where human activities would be conducted could interrupt wildlife, bird, and protected species that presently occur on the site. In addition, some vegetative impacts could occur as the land is used for uranium recovery. However, once facility decommissioning and site restoration have been completed all of these impacts would diminish, or baseline conditions would be re-established, and the ecology of the Project area would be restored over time.	There would be no irreversible or irretrievable commitment of ecological resources during the Proposed Action. There are no Greater-sage-grouse leks on the Ross Project itself. If one or more were to become present, the Applicant would be required to alter its activities, as appropriate, at the Ross Project.	There would be some short-term impacts to the ecology of the Proposed Action which would include the disruption of some species of vegetation as well as the potential for wildlife, including birds, to move elsewhere, away from Ross Project activities and noise. These impacts would cease when the decommissioning and reclamation of the Ross Project area are complete and the local habitat is restored.	There would be no long-term impacts to the area of the Ross Project. At the time of its closure, a decommissioning plan would be required, and this plan would require restoration of the Project area to its former baseline conditions.
Alternative 3: North Ross Project (See SEIS Section 4.6.3.)	Alternative 3 would have the potential for the same impacts to the local ecology as the Proposed Action. These impacts, however, would be SMALL.	There would be no irreversible or irretrievable impacts to the local ecology during Alternative 3.	There would be small, short-term impacts related to the disturbance of native vegetation and nearby wildlife during Alternative 3; these would be SMALL and would be the same as the Proposed Action.	There would be no long-term impacts to ecology under Alternative 3.

Table 8.1
Summary of Environmental Consequences of the Proposed Action and Alternatives (Cont.)

	Unavoidable Adverse Environmental Impacts	Irreversible and Irretrievable Commitment of Resources	Short-Term Impacts and Uses of the Environment	Long-Term Impacts and the Maintenance and Enhancement of Productivity
AIR QUALITY				
Alternative 1: Proposed Action Ross Project (See SEIS Section 4.7.1.)	There would be unavoidable adverse effects to local air quality due to the emission of combustion gases and fugitive dusts during all of the phases of the Proposed Action. These impacts would be SMALL. Fugitive dusts would become airborne during construction activities (e.g., road construction and site clearing and contouring) as well as during vehicular use to, from, and on the Project area. The Applicant would have an Air Quality Permit, however, that would require air emissions to be mitigated so that emissions would be kept to a minimum.	No permanent changes would occur to the overall quality of air at or near the Ross Project.	There would be SMALL short-term potential impacts to air quality under the Proposed Action. Gaseous and fugitive-dust emissions would be generated during all phases of the Ross Project. With the mitigation measures proposed by the Applicant, and those required by its Air Quality Permit, however, these potential impacts would be short term.	There would be no long-term impacts to air quality at the Ross Project. Once the Proposed Action has been decommissioned and the Project area reclaimed and restored, all air-quality impacts would cease.
Alternative 3: North Ross Project (See SEIS Section 4.7.3.)	Under Alternative 3, the same air-quality impacts would occur as during the Proposed Action. Because the CPP would be located such that gravel-road surfaces would be used slightly more often, there could be slightly more fugitive dust generated under Alternative 3.	No permanent changes would occur to the overall quality of air at or near the North Ross Project.	As for the Proposed Action, Alternative 3 would generate gaseous and fugitive-dust emissions throughout its lifecycle. But these impacts would also be SMALL.	There would be no long-term impacts to air quality as a result of Alternative 3. Once facility at the north site has been decommissioned and the Project area reclaimed and restored, all air-quality impacts would cease.

Table 8.1
Summary of Environmental Consequences of the Proposed Action and Alternatives (Cont.)

	Unavoidable Adverse Environmental Impacts	Irreversible and Irretrievable Commitment of Resources	Short-Term Impacts and Uses of the Environment	Long-Term Impacts and the Maintenance and Enhancement of Productivity
NOISE				
Alternative 1: Proposed Action Ross Project (See SEIS Section 4.8.1.)	There would be would be SMALL to MODERATE unavoidable adverse noise impacts during the Proposed Action. Because the Ross Project has several residences located near its boundaries, the noise generated during the Project's lifecycle would be MODERATE to those nearest receptors. In addition, due to supply deliveries, commuter traffic, and shipments from the Project, there would be increased vehicular noise. The four phases of the south Ross Project facility, including the CPP, surface impoundments, and other structures as well as the installation of wells would cause noise impacts that could not entirely be mitigated for nearby residents; however, the noise impacts would quickly diminish at greater distances from the Ross Project area.	There would be no long-term permanent noise impacts from the Proposed Action. Once the Ross Project has been decommissioned and the site restored, all Project-related noise would cease.	Short-term noise impacts in the vicinity of the Proposed Action would be SMALL to MODERATE. These impacts would include increased construction noise as well as increased vehicular noise. All such impacts, however, would cease after facility decommissioning and site restoration activities are complete.	There would be no long-term impacts by the Proposed Action with respect to noise.
Alternative 3: North Ross Project (See SEIS Section 4.8.3.)	During Alternative 3, there would be SMALL to MODERATE unavoidable adverse impacts with respect to noise. These impacts would be the same as in the Proposed Action, except some noise impacts would be diminished because construction and decommissioning activities at the north site would be a greater distance to the nearest residence.	There would be no long-term permanent noise impacts from Alternative 3. Once the north Ross Project has been decommissioned and the site restored, all Project-related noise would cease.	Short-term noise impacts in the vicinity of Alternative 3 would be SMALL to MODERATE. These impacts would include increased construction noise as well as increased vehicular noise. All such impacts, however, would cease after facility decommissioning and site restoration activities are complete.	There would be no long-term impacts by Alternative 3 with respect to noise.

Table 8.1
Summary of Environmental Consequences of the Proposed Action and Alternatives (Cont.)

	Unavoidable Adverse Environmental Impacts	Irreversible and Irretrievable Commitment of Resources	Short-Term Impacts and Uses of the Environment	Long-Term Impacts and the Maintenance and Enhancement of Productivity
HISTORICAL, CULTURAL, AND PALEONTOLOGICAL RESOURCES				
Alternative 1: Proposed Action Ross Project (See SEIS Section 4.9.1.) **Alternative 3: North Ross Project** (See SEIS Section 4.9.3.)	Impacts on historic and cultural resources during the ISR construction phase would be SMALL to LARGE. To mitigate the impact, NRC, BLM, WY SHPO, Tribes, and the Applicant will develop and execute an agreement that would formalize treatment plans for adversely impacted resources during construction. If other NRHP-elig ble sites cannot be avoided then treatment plans would be developed. If other historic and cultural resources are encountered during the ISR lifecycle, the Applicant would notify the appropriate authorities per an unexpected discovery plan.	If archaeological and historic sites cannot be avoided, or the impacts to these sites cannot be mitigated, this could result in an irreversible and irretrievable loss of cultural resources.	There would be a SMALL to LARGE impact on historic and cultural resources during the ISR construction phase. The development of an agreement between NRC, BLM, WY SHPO, Tribes, and the Applicant would address adverse impacts to cultural and historic sites and historic properties of traditional religious and cultural importance to Native American tribes. If any unidentified historic or cultural resources are encountered, work would stop and appropriate authorities would be notified per the unexpected discovery plan.	If potential impacts from implementation of the proposed action are not mitigated, then long-term impacts to cultural and historic resources would result.

Table 8.1
Summary of Environmental Consequences of the Proposed Action and Alternatives (Cont.)

	Unavoidable Adverse Environmental Impacts	Irreversible and Irretrievable Commitment of Resources	Short-Term Impacts and Uses of the Environment	Long-Term Impacts and the Maintenance and Enhancement of Productivity
VISUAL AND SCENIC RESOURCES				
Alternative 1: Proposed Action Ross Project (See SEIS Section 4.10.1.)	There would be SMALL unavoidable impacts related to the visual and scenic resources at the Proposed Action. During the construction and decommissioning phases, in particular, earth-movement activities could be seen by a few of the nearby residents, who could experience MODERATE impacts. The lights of the Project could also be seen by some of the nearest neighbors. There would be many mitigation measures related to the lights of the Ross Project which would be implemented by the Applicant to diminish as much as possible the light emanating from the Proposed Action.	There would be no irreversible or irretrievable commitment of visual or scenic resources caused by the Proposed Action. All visual-resource impacts would be eliminated upon the Ross Project's facility decommissioning and site restoration activities. These activities would include restoring the baseline contours of the Ross Project area.	There would be some short-term impacts to the visual resources at the Proposed Action. These would include changes to the topography of the area; the presence of man-made structures; and light during nighttime hours. These impacts would cease when the decommissioning and reclamation of the Ross Project area are complete, when the baseline topography is restored and all lights are removed.	There would be no long-term impacts to the visual and scenic resources of the Ross Project area. At the time of the Ross Project's decommissioning, a decommissioning plan would be implemented, and this plan would require the restoration of the Project area to its former baseline conditions.
Alternative 3: North Ross Project (See SEIS Section 4.10.3.)	Alternative 3 would have the potential for visual-resource impacts as the Proposed Action. These SMALL impacts would be even less than those of the Ross Project, because the natural topography of the north area would shield construction and operation activities related to the uranium-recovery facility. Thus, the closest residences' views would be less impacted as would their experience of light in the nighttime skies.	There would be no irreversible or irretrievable impacts to the visual and scenic resources caused by Alternative 3.	There would be small, short-term impacts related to the disturbance of native vegetation and nearby wildlife during Alternative 3; these would be SMALL and would be the same as the Proposed Action.	There would be no long-term impacts to ecology in Alternative 3.

Table 8.1
Summary of Environmental Consequences of the Proposed Action and Alternatives (Cont.)

	Unavoidable Adverse Environmental Impacts	Irreversible and Irretrievable Commitment of Resources	Short-Term Impacts and Uses of the Environment	Long-Term Impacts and the Maintenance and Enhancement of Productivity
SOCIOECONOMICS				
Alternative 1: Proposed Action Ross Project (See SEIS Section 4.11.1.)	There would be SMALL unavoidable impacts related to the socioeconomics at the Proposed Action. These impacts, however, would be related to the increased revenues that local jurisdictions would collect related taxes on the Ross Project. Other socioeconomic factors such as impacts to employment, income, demographics, housing, and education as well as demand for social and health services would be SMALL, while the impacts relate to local tax revenues would be MODERATE during the construction and operation phases of the Ross Project.	There would be no irreversible or irretrievable commitment of socioeconomic resources during the Proposed Action. Socioeconomic impacts would be diminished upon the Ross Project's decommissioning. For example, the increased need for housing, although SMALL, would be eliminated after the Ross Project terminates.	There would be some short-term impacts to socioeconomic resources by the Proposed Action. These would include SMALL to MODERATE potential changes to local employment, income, demographics, housing, and education as well as demand for social and health services. All tax payments would also be eliminated at the conclusion of the Ross Project.	There would be no long-term impacts to socioeconomic resources by the Ross Project area. After the Ross Project's decommissioning, any increases that would have occurred in the employment, income, demographics, housing, and education sectors would have integrated and every worker would be able to relocate as s/he wishes. The demand for social and health services would be eliminated as would all tax payments in the finance sector.
Alternative 3: North Ross Project (See SEIS Section 4.11.3.)	There would be the same SMALL to MODERATE socioeconomic impacts to the area surrounding Alternative 3. As the socioeconomic variables evaluated for this SEIS do not depend upon the geography of the Ross Project, the North Ross Project would accrue the same impacts as the Proposed Action.	There would be no irreversible or irretrievable commitment of socioeconomic resources during Alternative 3 as for the Proposed Action.	There would be some short-term impacts to socioeconomic resources caused by Alternative 3. These would include the same SMALL potential changes to local employment, income, demographics, housing, and education as well as demand for social and health services and MODERATE impacts to local jurisdictions.	There would be no long-term impacts to socioeconomic resources as a result of Alternative 3.

Table 8.1
Summary of Environmental Consequences of the Proposed Action and Alternatives (Cont.)

	Unavoidable Adverse Environmental Impacts	Irreversible and Irretrievable Commitment of Resources	Short-Term Impacts and Uses of the Environment	Long-Term Impacts and the Maintenance and Enhancement of Productivity
ENVIRONMENTAL JUSTICE				
Alternative 1: Proposed Action Ross Project (See SEIS Section 4.12.)	There are no minority and low-income populations located within four miles of the Ross Project area. Consequently, an environmental justice-analysis has not been performed for this Proposed Action.			
Alternative 3: North Ross Project (See SEIS Section 4.12.)				
PUBLIC AND OCCUPATIONAL HEALTH AND SAFETY				
Alternative 1: Proposed Action Ross Project (See SEIS Section 4.13.1.)	There would be a SMALL impact on public and occupational health. Construction and decommissioning would generate fugitive dust emissions that would not result in a significant dose to the public or site workers. The emissions from construction equipment would be of short duration and would readily disperse into the atmosphere.	Not applicable	There would be a SMALL impact from radiological exposure. Dose calculations under normal operations showed that the highest potential dose within the proposed project area is 5 percent of the 1 mSv [100 mrem] per year public dose limit specified in NRC regulations. The radiological impacts from accidents would be SMALL for workers if procedures to deal with accident scenarios are followed, and SMALL for the public because of the facility's remote location. The nonradiological public and occupational health impacts from normal operations, accidents, and chemical exposures would be SMALL if handling procedures are followed.	There will be no long-term impact to public and occupational health following license termination.
Alternative 3: North Ross Project (See SEIS Section 4.13.3.)				

Table 8.1
Summary of Environmental Consequences of the Proposed Action and Alternatives (Cont.)

	Unavoidable Adverse Environmental Impacts	Irreversible and Irretrievable Commitment of Resources	Short-Term Impacts and Uses of the Environment	Long-Term Impacts and the Maintenance and Enhancement of Productivity
WASTE MANAGEMENT				
Alternative 1: Proposed Action Ross Project (See SEIS Section 4.14.1.)	There would be SMALL unavoidable adverse impacts as a result of waste management at the Proposed Action. The management of all waste streams (ordinary solid and domestic wastes, solid and liquid hazardous wastes, and all byproduct wastes) would be SMALL during all phases of the Ross Project. Many of the waste streams would be shipped off site to regulated (i.e., permitted or licensed as appropriate) facilities, which have undergone careful scrutiny by the respective regulatory agencies. In addition, discharge of small amounts of excess permeate into the Class I deep-injection wells at the Project would comply with Underground Injection Control (UIC) Permit from the Wyoming Department of Environmental Quality (WDEQ).	There would be no irreversible or irretrievable resources committed to waste management, except for the respective liquid wastes which would be injected into the Deadwood and Flathead Formations approximately 8,000 feet below the surface. However, these aquifers are not potable and have not been identified as a source of oil and gas resources, the injection of waste would not impact the aquifer's future use.	There would be few short-term impacts due to waste management at the Proposed Action. These short-term impacts, such as to transportation as well as public and occupational health and safety, which are described in those respective resource areas, would cease when waste is no longer generated or managed at the Ross Project. In addition, during the operation of the Proposed Action, there would be two double-lined surface impoundments over a 6.5 ha [16 ac] area; the presence of these impoundments could have impacts to wildlife and birds. However, control features, such as an avian deterrent system would be operated throughout the Ross Project. These surface impoundments would be completely removed during facility decommissioning.	All Ross Project wastes would be either shipped offsite by the conclusion of the decommissioning phase, or would be disposed of in the Class I deep-disposal wells. During all phases of the Proposed Action permanent disposal or storage of both radiological and nonradiological wastes would represent a long-term, but SMALL, impact on the productivity of the land allocated for these activities.
Alternative 3: North Ross Project (See SEIS Section 4.14.3.)	There could be SMALL unavoidable waste-management impacts at Alternative 3. These would be the same as those indicated above for the Proposed Action. The volume of demolition waste could be somewhat less than the Proposed Action's, because the containment barrier wall would not have been constructed.	This Alternative would also employ a deep-disposal well, so that the same commitment of the Deadwood and Flathead aquifer would occur.	The short-term impacts of Alternative 3's management of wastes would be the same as those for the Proposed Action.	The long-term impacts of Alternative 3's management of wastes would be the same as those for the Proposed Action.

9 LIST OF PREPARERS

This section documents all individuals who were involved with the preparation of this Draft Supplemental Environmental Impact Statement (DSEIS). Contributors include staff from the NRC and its consultants. Each individual's responsibilities and affiliation are as follows.

9.1 U.S. Nuclear Regulatory Commission Contributors

Johari Moore: SEIS Project Manager
M.S., Nuclear Engineering and Radiological Science, University of Michigan, 2005
B.S., Physics, Florida A & M University, 2003
Years of Experience: 7

Ashley Waldron: SEIS Co-Project Manager
B.S., Biology, Frostburg State University, 2009
Years of Experience: 3

John Saxton: Safety Project Manager
Connecticut Licensed Environmental Professional, 2001
M.S., Geology, University of New Mexico, 1989
B.S., Geological Engineering, Colorado School of Mines, 1983
Years of Experience: 25

9.2 Attenuation Environmental Company Team

Doris Minor: Project Manager and Task Co-Manager
Attenuation Environmental Company
Radiological Public and Occupational Health and Safety/Solid and Radioactive Waste Management
B.A., English Literature, University of Washington, 1978
M.S.E., (Nuclear) Engineering, University of Washington, 1984
Years of Experience: 28

Dr. Kathryn Johnson: Assistant Project Manager and Task Co-Manager
Attenuation Environmental Company
Geochemistry/Geology/Water Resources/Liquid Waste Management/Facility Design and Operation
Ph.D., Geology, South Dakota School of Mines & Technology, 1986
M.S., Chemistry, Iowa State University, 1977
B.S., Chemistry and Mathematics, Black Hills State College, 1975

Jerry Boese
Ross & Associates Environmental Consulting Ltd., dba Ross Strategic
Comment Responses
M.P.A., John F. Kennedy School of Government, Harvard University, 1989
B.S., Biology and Economics, Yale University, 1975
Years of Experience: 32

9.2 Attenuation Environmental Company Team (*Continued*)

Dr. Tony Burgess
Attenuation Environmental Company
Ground-Water Resources
Ph.D., Geology (Engineering Geology and Hydrogeology), University of Durham, United Kingdom, 1970
B.Sc., Geology University of Durham, United Kingdom, 1966
Years of Experience: 43

Dr. Cheryl Chapman, P.E.
RESPEC Consulting and Services Inc.
Tribal Coordination and Consultation
Ph.D. in Biological Sciences, South Dakota State University, Brookings, South Dakota, 2007
B.S., Civil and Environmental Engineering, South Dakota School of Mines & Technology, Rapid City, South Dakota, 1978
B.S., Mathematics, South Dakota School of Mines & Technology, Rapid City, South Dakota, 1978
Years of Experience: 32

Lauren Evans, P.E.
Pinyon Environmental Engineering Resources Inc.
Land Use/Transportation/Ecological Resources/Meteorology
B.S., Geological Engineering, Colorado School of Mines, 1982
Years of Experience: 30

Linn Gould
Erda Environmental Services Inc.
Environmental Justice/Cumulative Impacts
M.P.H., University of Washington, 2003
M.S., Soil Science, University of Wisconsin-Madison, 1988
B.A., Geology, Smith College, 1980
Years of Experience: 25

Dave Heffner, P.E.
Aspect Consulting L.L.C.
Chemical Engineering
M.S., Chemical Engineering, Georgia Institute of Technology, 1981
B.S., Environmental Engineering, Rensselaer Polytechnic Institute, 1979
Years of Experience: 30

Mary Kenner
RESPEC Consulting and Services Inc.
Tribal Coordination and Consultation
B.S., Interdisciplinary Science, South Dakota School of Mines & Technology, 1998
Years of Experience: 28

9.2 Attenuation Environmental Company Team (*Continued*)

Scott Kindred, P.E.
Aspect Consulting L.L.C.
Ground-Water and Surface-Water Resources/Geology/Soils
M.S., Civil Engineering, Massachusetts Institute of Technology, 1987
B.S., Geology, Brown University, 1983
Years of Experience: 20

Charles McClendon
The Delphi Groupe, Inc.
Radiological/Mining/Industrial Hygiene/RSO/NEPA
B.S., Civil (Mining) Engineering, Colorado School of Mines, 1976
M.B.A., Nova University, Dade County, Florida 1985
Years of Experience: 35

Dave McCormack
Aspect Consulting L.L.C.
Geology/Soils
M.S., Geology, Northern Arizona State University, 1989
B.S., Geological Sciences, University of Washington, 1983
Years of Experience: 23

Christian J. Miss
SWCA Environmental Consultants
Historical, Cultural, and Paleontological Resources/Tribal Consultation
B.A., Biology, Case Western Reserve University, 1970
M.A., Anthropology, Idaho State University, 1978
Years of Experience: 40

Jessie Piper
SWCA Environmental Consultants
Historical, Cultural, and Paleontological Resources/Tribal Consultation
B.A., Anthropology, University of Arizona, 1989
M.A., Anthropology, University of Arizona, 1992
Years of Experience: 20

Debby Reber
SWCA Environmental Consultants
Visual Resources
B.A., Natural Resource Management, University of California, Chico, 1984
Years of Experience: 15

9.2 Attenuation Environmental Company Team (*Continued*)

Owen Reese, P.E.
Aspect Consulting L.L.C.
Surface-Water Resources
M.S., Civil Engineering, University of Washington, 1997
B.S., Civil Engineering, University of Washington, 1996
Years of Experience: 15

Barbara Trenary
Attenuation Environmental Company
Air Quality/Noise/Non-Radiological Public and Occupational Health and Safety
B.S., Environmental Sciences, Industrial Hygiene, Colorado State University, 1979
Years of Experience: 33

Jeff Vitucci
Robert D Niehaus Inc.
Socioeconomics/Cost-Benefit Analysis
M.A., Economics, University of California, Santa Barbara, 1978
B.A., Environmental Studies, San Jose State University, 1974
Years of Experience: 34

Elly Weber
Pinyon Environmental Engineering Resources Inc.
Ecological Resources
M.S., Environmental Science, Texas Christian University, 2004
B.S., Biology, Texas Christian University, 2001
Years of Experience: 5

Richard Weinman
Attenuation Environmental Company: Assistant Project Manager
NEPA Analysis/Regulatory Compliance/Land Use Planning
B.A., English, New York University, 1965
M.A., English, Brandeis University, 1966
J.D., Law, University of Puget Sound (Seattle University) School of Law, 1978
Years of Experience: 33

10 DISTRIBUTION LIST

The NRC is providing copies of this Draft Supplemental Environmental Impact Statement (DSEIS) to the organizations and individuals listed as follows. NRC will provide copies to other interested organizations and individuals upon request.

10.1 Federal Agency Officials

James Bashor
Bureau of Land Management
Newcastle Field Office
Newcastle, WY

John Keck
National Park Service
Devils Tower National Monument
Devils Tower, WY

Dr. John T Eddins
Advisory Council on Historic Preservation
Washington, D.C.

10.2 Tribal Government Officials

Donnie Cabniss
Apache Tribe
Tribal Historic Preservation Office
Anadarko, OK

John Murray
Blackfeet Tribe
Tribal Historic Preservation Office
Browning, MT

Lynette Gray
Cheyenne and Arapaho Tribes
Tribal Historic Preservation Office
Concho, OK

Steve Vance
Cheyenne River Sioux Tribe
Tribal Historic Preservation Office
Eagle Butte, SD

Alvin Windy Boy, Sr.
Chippewa Cree Tribe
Tribal Historic Preservation Office
Box Elder, MT

Ira Matt
Confederated Salish & Kootenai Tribe
Tribal Historic Preservation Office
Pablo, MT

Emerson Bull Chief
Crow Tribe
Tribal Historic Preservation Office
Crow Agency, MT

Wanda Wells
Crow Creek Sioux Tribe
Tribal Historic Preservation Office
Ft. Thompson, SD

Wilfred Ferris
Eastern Shoshone Tribe
Tribal Historic Preservation Office
Fort Washakie, WY

James B. Weston
Flandreau-Santee Sioux Tribe
Tribal Historic Preservation Office
Flandreau, SD

Morris E. Belgard
Fort Belknap Tribe
Tribal Historic Preservation Office
Harlem, MT

Darrell "Curley" Youpee
Fort Peck Tribes
Tribal Historic Preservation Office
Poplar, MT

Amie Tah-bone
Kiowa Indian Tribe of Oklahoma
NAGPRA Representative
Carnegie, OK

Clair S. Green
Lower Brule Sioux Tribe
Cultural Resources
Lower Brule, SD

Darlene Conrad
Tribal Historic Preservation Office
Northern Arapaho Tribe
Fort Washakie, WY

Conrad Fisher
Northern Cheyenne Tribe
Tribal Historic Preservation Office
Lame Deer, MT

Richard Iron Cloud
Oglala Sioux Tribe
Tribal Historic Preservation Office
Pine Ridge, SD

Russell Eagle Bear
Rosebud Sioux Tribe
Tribal Historic Preservation Office
Rosebud, SD

Rick Thomas
Santee Sioux Tribe of Nebraska
Tribal Historic Preservation Office
Niobrara, NE

Dianne Desrosiers
Sisseton-Wahpeton Oyate
Tribal Historic Preservation Office
Sisseton, SD

Darrell Smith
Spirit Lake Tribe
Tribal Historic Preservation Office
Fort Totten, ND

Wašté Wiŋ Young
Standing Rock Sioux Tribe
Tribal Historic Preservation Office
Fort Yates, ND

Elgin Crows Breast
Mandan, Hidatsa & Arikara Nation
Three Affiliated Tribes
Tribal Historic Preservation Office
New Town, ND

Bruce F. Nadeau
Turtle Mountain Band of Chippewa
Tribal Historic Preservation Office
Belcourt, ND

Lana Gravatt
Yankton Sioux Tribe
Tribal Historic Preservation Office
Wagner, SD

10.3 State Agency Officials

Mary Hopkins
Wyoming State Historic Preservation Office
Cheyenne, WY

Mark Rogaczewski
Wyoming Department of Environmental Quality
Land Quality Division

Tanner Shatto
Wyoming Department of Environmental Quality
Air Quality Division

John Passehl
Wyoming Department of Environmental Quality
Water Quality Division

Scott Talbott
Wyoming Game and Fish Department
Sheridan, WY

10.4 Local Agency Officials

Crook County Commissioners
Sundance, WY

10.5 Other Organizations and Individuals

Shannon Anderson, Esq.
Powder River Basin Resource Council
Sheridan, WY

Geoffrey H. Fettus, Esq.
Natural Resources Defense Council
Washington, DC

Ralph Knode
Strata Energy, Inc,
Gillette, WY

Anthony Thompson, Esq.
Thompson & Pugsley, PLLC
Washington, DC

Echo Bohl
Crook County Library Hulett Branch
Hulett, WY

Pamela Jespersen
Crook County Library Moorcroft Branch
Moorcroft, WY

APPENDIX A
CONSULTATION CORRESPONDENCE

CONSULTATION CORRESPONDENCE

The Endangered Species Act of 1973, as amended, and the National Historic Preservation Act of 1966, as amended, require that Federal agencies consult with applicable State and Federal agencies and groups prior to taking action that may affect threatened and endangered species, essential fish habitat, or historical and archaeological resources. This appendix lists consultation documentation related to these federal acts.

Table A.1
Chronology of Consultation Correspondence

Author	Recipient	Date of Letter	ADAMS Accession Number
U.S. Nuclear Regulatory Commission (L. Camper)	Fort Peck Tribal Executive Board	November 19, 2010*	ML103160580
U.S. Nuclear Regulatory Commission (L. Camper)	Fort Belknap Community Council	February 9, 2011**	ML110400321
Turtle Mountain Band of Chippewa Indians (K. Ferris)	U.S. Nuclear Regulatory Commission (A. Bjornsen)	April 14, 2011	ML111080059
U.S. Nuclear Regulatory Commission (K. Hsueh)	Sisseton-Wahpeton Lakota THPO (D. Desrosiers)	August 11, 2011***	ML112220386
U.S. Nuclear Regulatory Commission (K. Hsueh)	U.S. Department of the Interior, Fish and Wildlife Service (M. Sattelberg)	August 12, 2011	ML112200151
Apache Tribe of Oklahoma (L. Guy)	U.S. Nuclear Regulatory Commission (A. Bjornsen)	August 19, 2011	ML11336A224
U.S. Nuclear Regulatory Commission (A. Persinko)	Wyoming State Historic Preservation Office (M. Hopkins)	August 19, 2011	ML112150393
U.S. Nuclear Regulatory Commission (A. Persinko)	Advisory Council on Historic Preservation (J. Fowler)	August 19, 2011	ML112150427

Table A.1
Chronology of Consultation Correspondence (Cont.)

Author	Recipient	Date of Letter	ADAMS Accession Number
U.S. Department of the Interior, Fish and Wildlife Service (M. Sattelberg)	U.S. Nuclear Regulatory Commission (K. Hsueh)	September 13, 2011	ML112770035
Advisory Council on Historic Preservation (C. Hall)	U.S. Nuclear Regulatory Commission (A. Persinko)	September 13, 2011	ML112770035
Wyoming Game and Fish Department (J. Emmerich)	U.S. Nuclear Regulatory Commission (A. Bjornsen)	September 22, 2011	ML112660130
U.S. Nuclear Regulatory Commission (L. Camper)	National Park Service, Devils Tower National Monument (D. FireCloud)	December 5, 2011	ML113120356
U.S. Nuclear Regulatory Commission (K. Hsueh)	Strata Energy, Inc. (M. James)	December 6, 2011	ML113200121
Advisory Council on Historic Preservation (C. Vaughn)	U.S. Nuclear Regulatory Commission (A. Persinko)	December 12, 2011	ML113480465
U.S. Nuclear Regulatory Commission (K. Hsueh)	Fort Peck Tribe (D. Youpee)	December 22, 2011***	ML113420504
Strata Energy, Inc. (M. James)	U.S. Nuclear Regulatory Commission (K. Hsueh)	January 12, 2012	ML120720266
U.S. Nuclear Regulatory Commission (K. Hsueh)	Advisory Council on Historic Preservation (C. Vaughn)	January 31, 2012	ML113490371
Rosebud Sioux Tribe (R. Eagle Bear)	U.S. Nuclear Regulatory Commission (A. Bjornsen)	February 1, 2012	ML120390551

Table A.1
Chronology of Consultation Correspondence (Cont.)

Author	Recipient	Date of Letter	ADAMS Accession Number
Strata Energy, Inc. (R. Knode)	U.S. Nuclear Regulatory Commission (K. Hsueh)	August 31, 2012	ML12248A421
U.S. Nuclear Regulatory Commission (K. Hsueh)	Santee Sioux Tribe of Nebraska (R. Thomas)	September 20, 2012***	ML12264A220
WWC Engineering (B. Schiffer)	U.S. Nuclear Regulatory Commission (J. Moore)	October 16, 2012	ML12311A338
U.S. Nuclear Regulatory Commission (K. Hsueh)	Kiowa Indian Tribe (J. Eskew)	November 21, 2012***	ML12325A776

*Similar letters sent to Cheyenne River Sioux Tribe (J. Plenty), Crow Creek Sioux Tribe (L. Thompson, Jr.), Lower Brule Sioux Tribal Council (M. Jandreau), Oglala Sioux Tribal Council (T. Two Bulls), Rosebud Sioux Tribal Council (R. Bordeaux), Santee Sioux Nation (R. Trudell), Standing Rock Sioux Tribe (R. Thunder), Three Affiliated Tribes Business Council (M. Wells), Northern Cheyenne Tribe (L. Spaug), Cheyenne and Arapaho Tribes (D. Flyingman), Arapaho Business Committee (H. Spoonhunter), Crow Tr bal Council (C. Eagle), and Eastern Shoshone Tribe (I. Posey).

**Similar letters sent to Standing Rock Lakota Tribal Council (C. Murphy), Crow Tribal Council (C. Eagle), Apache Tribe of Oklahoma (H. Kostzuta), Sisseton-Wahpeton Lakota (A. Grey, Sr.), Yankton Lakota Tribe (R. Courneyor), Blackfeet Tribal Business Council (W. Sharp), Crow Creek Sioux Tribal Council (B. Sazue), Lower Brule Sioux Tr bal Council (M. Jandreau), Spirit lake Tr bal Council (M. Pearson), Oglala Lakota Tribal Council (T. TwoBulls), Shoshone Business Council (I. Posey), Northern Cheyenne Tribal Council (G. Small), Rosebud Sioux Tribal Council (R. Bordeaux), Fort Peck Tribal Executive Board (A. Stafne), Cheyenne and Arapahoe Tr bes of Oklahoma (J. Boswell),Turtle Mountain Chippewa Tribal Council (R. Marcellias), Santee Sioux Nation (R. Trudell), Arapaho Business Council (H. Spoonhunter), Three Affiliated Tribes Business Council (M. Levings), Kiowa Indian Tr be of Oklahoma (D. Tofpi), Flandreau Santee Lakota Executive Committee (G. Bouland), Confederated Salish & Kootenai (E. Moran), and Cheyenne River Lakota Tribal Council (J. Plenty).

***Similar letters sent to the Tribal Historic Preservation Officers: Fort Peck Tribal Executive Board (D. Youpee), Fort Belknap Community Council (D. Belgard), Standing Rock Lakota Tribal Council (W. Young), Crow Tribal Council (D. Old Horn), Yankton Lakota Tribe (L. Gravatt), Blackfeet Tribal Business Council (J. Murray), Crow Creek Sioux Tribal Council (W. Wells), Lower Brule Sioux Tribal Council (C. Green), Spirit lake Tr bal Council (A. Shaw), Oglala Lakota Tribal Council (W. Mesteth), Shoshone Business Council (W. Ferris), Northern Cheyenne Tr bal Council (C. Fisher), Rosebud Sioux Tr bal Council (R. Eagle Bear), Cheyenne and Arapahoe Tr bes of Oklahoma (D. Hamilton and), Santee Sioux Nation (L. Ickes), Arapaho Business Council (D. Conrad), Three Affiliated Tr bes Business Council (E. Crows Breast), Kiowa Indian Tribe of Oklahoma (J. Eskew), Flandreau Santee Lakota Executive Committee (J. Weston), Confederated Salish & Kootenai (C. Burke), and Cheyenne River Lakota Tribal Council (S. Vance), Sisseton-Wahpeton Lakots (D. Desrosiers).

APPENDIX B
VISUAL IMPACTS ANALYSIS

APPENDIX B: VISUAL IMPACT ANALYSIS

Scenic Quality Inventory Point B-1

Photograph from Scenic Quality Inventory Point C-1 to North

Table B.1 Scenic Quality Inventory and Evaluation		
Key Factor	**Rating Criteria**	**Score**
Landform	Low rolling hills, foothills, or flat valley bottoms or few or no interesting landscape features.	1
Vegetation	Some variety of vegetation, but only one or two major types.	3
Water	Present/Little Missouri River and the Oshoto Reservoir are occasionally visible.	1
Color	Some intensity or variety in colors and contrast of the soil, rock, and vegetation, but not a dominant scenic element.	3
Influence of Adjacent Scenery	Adjacent scenery has little or no influence on overall visual quality.	0
Scarcity	Interesting within its setting, but fairly common within the region.	1
Cultural Modifications	Modifications add variety, but are very discordant and promote strong disharmony.	-2
	TOTAL SCORE	7

Scenic Quality Inventory Point B-2

Photograph from Scenic Quality Inventory Point B-2 to East

Table B.2 Scenic Quality Inventory and Evaluation		
Key Factor	**Rating Criteria**	**Score**
Landform	High vertical relief as expressed in prominent cliffs, spires, or massive rock outcrops, or severe surface variation or highly eroded formations including major badlands or dune systems; or detail features dominant and exceptionally striking and intriguing such as glaciers.	5
Vegetation	A variety of vegetative types as expressed in interesting forms, textures, and patterns.	5
Water	Present, but not noticeable.	0
Color	Some intensity or variety in colors and contrast of the soil, rock, and vegetation, but not a dominant scenic element.	3
Influence of Adjacent Scenery	Adjacent scenery greatly enhances visual quality (Devils Tower).	5
Scarcity	One of a kind, or unusually memorable, or very rare within region. Consistent chance for exceptional wildlife or wildflower viewing.	5
Cultural Modifications	Modifications add little or no visual variety to the area, and introduce no discordant elements.	0
	TOTAL SCORE	**23**

Scenic Quality Inventory Point B-3

Photograph from Scenic Quality Inventory Point B-3 to South

Table B.3 Scenic Quality Inventory and Evaluation		
Key Factor	**Rating Criteria**	**Score**
Landform	Low rolling hills, foothills, or flat valley bottoms; or few or no interesting landscape features.	1
Vegetation	Little or no variety or contrast in vegetation.	1
Water	Present, but not noticeable.	0
Color	Subtle color variations, contrast, or interest; generally mute tones.	1
Influence of Adjacent Scenery	Adjacent scenery has little or no influence on overall visual quality.	0
Scarcity	Interesting within its setting, but fairly common within the region.	1
Cultural Modifications	Modifications add little or no visual variety to the area and introduce no discordant elements.	0
	TOTAL SCORE	**4**

Scenic Quality Inventory Point B-4

Photograph from Scenic Quality Inventory Point B-4 to South

Table B.4
Scenic Quality Inventory and Evaluation

Key Factor	Rating Criteria	Score
Landform	Low rolling hills, foothills, or flat valley bottoms; or few or no interesting landscape features.	1
Vegetation	Some variety of vegetation, but only one or two major types.	3
Water	Present, but not noticeable.	1
Color	Some intensity or variety in colors and contrast of the soil, rock and vegetation, but not a dominant scenic element.	3
Influence of Adjacent Scenery	Adjacent scenery has little or no influence on overall visual quality.	0
Scarcity	Interesting within its setting, but fairly common within the region.	1
Cultural Modifications	Modifications add variety but are discordant and promote disharmony.	-1
	TOTAL SCORE	8

Table B.5
Scenic Quality Inventory and Evaluation
Average of Four Views

Key Factor	Score
Landform	2.00
Vegetation	3.00
Water	0.50
Color	2.50
Influence of Adjacent Scenery	1.25
Scarcity	2.00
Cultural Modifications	-0.75
AVERAGE	**10.50**

NRC FORM 335 (12-2010) NRCMD 3.7

U.S. NUCLEAR REGULATORY COMMISSION

BIBLIOGRAPHIC DATA SHEET

(See instructions on the reverse)

1. REPORT NUMBER (Assigned by NRC, Add Vol., Supp., Rev., and Addendum Numbers, if any.)

NUREG-1910, Supplement 5 Draft

2. TITLE AND SUBTITLE

Environmental Impact Statement for the Ross ISR Project in Crook County, Wyoming: Supplement to the Generic Environmental Impact Statement for In-Situ Leach Uranium Milling Facilities (Draft For Comment)

3. DATE REPORT PUBLISHED

MONTH	YEAR
March	2013

4. FIN OR GRANT NUMBER

5. AUTHOR(S)

6. TYPE OF REPORT

Technical

7. PERIOD COVERED (Inclusive Dates)

8. PERFORMING ORGANIZATION - NAME AND ADDRESS (If NRC, provide Division, Office or Region, U. S. Nuclear Regulatory Commission, and mailing address; if contractor, provide name and mailing address.)

Division of Waste Management and Environmental Protection
Office of Federal and State Materials and Environmental Management Programs
U.S. Nuclear Regulatory Commission
Washington, DC 20555-001

9. SPONSORING ORGANIZATION - NAME AND ADDRESS (If NRC, type "Same as above", if contractor, provide NRC Division, Office or Region, U. S. Nuclear Regulatory Commission, and mailing address.)

Same as above

10. SUPPLEMENTARY NOTES

11. ABSTRACT (200 words or less)

By a letter dated January 4, 2011, Strata Energy, Inc. (Strata or the "Applicant") submitted a license application to the Nuclear Regulatory Commission (NRC) for a new source and byproduct materials license for the proposed Ross Project. The Ross Project would be located in Crook County, Wyoming, which is in the Nebraska-South Dakota-Wyoming Uranium Milling Region identified in the NUREG-1910, Generic Environmental Impact Statement (GEIS) for In-Situ Leach Uranium Milling Facilities. The NRC staff prepared this Supplemental Environmental Impact Statement (SEIS) to evaluate the potential environmental impacts of the Applicant's proposal to construct, operate, conduct aquifer restoration, and decommission an in situ uranium-recovery facility at the Ross Project. This SEIS describes the environment that could be affected by the proposed Ross Project activities, estimates the potential environmental impacts resulting from the Proposed Action and two Alternatives, discusses the corresponding proposed mitigation measures, and describes the Applicant's environmental-monitoring program. The NRC staff evaluated site-specific data and information to determine whether the site characteristics and the Applicant's proposed activities were consistent with those evaluated in the GEIS. The NRC staff will respond to public comments received on the draft SEIS in the final SEIS.

12. KEY WORDS/DESCRIPTORS (List words or phrases that will assist researchers in locating the report.)

Uranium Recovery
In-Situ Recovery Process
Uranium
Environmental Impact Statement
Supplemental Environmental Impact Statement

13. AVAILABILITY STATEMENT

unlimited

14. SECURITY CLASSIFICATION

(This Page) unclassified

(This Report) unclassified

15. NUMBER OF PAGES

16. PRICE

UNITED STATES
NUCLEAR REGULATORY COMMISSION
WASHINGTON, DC 20555-0001

OFFICIAL BUSINESS

NUREG-1910
Supplement 5
Draft

Environmental Impact Statement for the Ross ISR Project in Crook County, Wyoming

March 2013

www.ingramcontent.com/pod-product-compliance
Lightning Source LLC
Chambersburg PA
CBHW080232180526
45167CB00006B/2245